The volume deals with the infrared spectra of complexes of two, three, ...molecules in collisional interaction. More than 800 original papers have been published in the field since the discovery of collision-induced absorption by H. L. Welsh and associates in 1949. This volume is the first attempt to present the theoretical and experimental foundations of this basic science of the interaction of radiation with supermolecular systems.

Following a brief introduction and recapitulation of background information (e.g., dipole radiation, molecular collisions and interactions, etc.), the measurements of supermolecular absorption of electromagnetic radiation are reviewed. The focus is on the non-polar gases whose molecules are infrared inactive but the effects described are universal for nearly all fluids. At high enough gas densities, corresponding roughly to atmospheric pressure or higher, absorption by complexes of two or more interacting molecules takes place. Supermolecular absorption in gases may be described in terms of a virial expansion in powers of the gas density, a fact that permits the separation of the two-, three-, ...body spectral components for detailed studies. One chapter reviews in detail our knowledge of the interaction-induced dipole moments of dissimilar atomic and molecular pairs, and of the dipoles of supermolecular systems consisting of three molecules. In subsequent chapters the theory of interaction-induced absorption by rare-gas mixtures and molecular gases is reviewed. *Ab initio* calculations of the interaction-induced absorption, especially in the various rotovibrational bands of hydrogen, are reviewed and compared with the measurements. A final chapter discusses related phenomena and important applications, especially in the planetary sciences and astrophysics.

Specialists and advanced graduate students concerned with the basic principles or applications of the infrared spectroscopy of dense fluids, or with the fundamental interactions of supermolecular complexes with electromagnetic radiation, will find this a valuable text and general reference.

Cambridge Monographs on
Atomic, Molecular, and Chemical Physics
2

COLLISION-INDUCED ABSORPTION IN GASES

Cambridge Monographs on
Atomic, Molecular, and Chemical Physics

1. R. Schinke: *Photodissociation Dynamics*
2. L. Frommhold: *Collision-induced Absorption in Gases*

COLLISION-INDUCED ABSORPTION IN GASES

LOTHAR FROMMHOLD

Department of Physics, University of Texas at Austin

CAMBRIDGE
UNIVERSITY PRESS

CAMBRIDGE UNIVERSITY PRESS
Cambridge, New York, Melbourne, Madrid, Cape Town, Singapore, São Paulo

Cambridge University Press
The Edinburgh Building, Cambridge CB2 2RU, UK

Published in the United States of America by Cambridge University Press, New York

www.cambridge.org
Information on this title: www.cambridge.org/9780521393454

First published 1993
This digitally printed first paperback version (with new Appendix) 2006

A catalogue record for this publication is available from the British Library

Library of Congress Cataloguing in Publication data

Frommhold, Lothar.
Collision-induced absorption in gases / Lothar Frommhold.
p. cm. – (Cambridge monographs on atomic, molecular, and
chemical physics; 2)
Includes bibliographical references and index.
ISBN 0 521 39345 0
1. Gases – Absorption and desorption. 2. Infrared radiation.
3. Molecular spectra. I. Title. II. Series.
QC162.F76 1993
530.4′3 – dc20 92-30351 CIP

ISBN-13 978-0-521-39345-4 hardback
ISBN-10 0-521-39345-0 hardback

ISBN-13 978-0-521-01967-5 paperback
ISBN-10 0-521-01967-2 paperback

Contents

Preface

The existing bibliographies* on collision-induced absorption (CIA) list more than 800 original papers published in the 45 years of history of the field. Furthermore, a number of review articles focusing on one aspect of CIA or another are listed, along with compilations of lectures given at summer schools, advanced research seminars or scientific conferences. A monograph which attempts to review the experimental and theoretical foundations of CIA, however, cannot be found in these carefully compiled listings.

Yet the field is of great significance and continues to attract numerous specialists from various disciplines. CIA is a basic science dealing with the interaction of supermolecular systems with light. It has important applications, for example in the atmospheric sciences. CIA exists in all molecular fluids and mixtures. It is ubiquitous in dense, neutral matter and is especially striking in matter composed of infrared-inactive molecules. As a science, CIA has long since acquired a state of maturity. Not only do we have a wealth of experimental observations and data for virtually all common gases and liquids, but rigorous theory based on first principles exists and explains nearly all experimental results in considerable detail. *Ab initio* calculations of most aspects of CIA are possible which show a high degree of consistency with observation, especially in the low-density limit.

When in 1989 Dr. Alex Dalgarno asked me if I wanted to write a book on collision-induced absorption of gases, I did not hesitate for long. The present monograph is the first attempt to review measurement and theory of CIA in gases at densities which, on the one hand, are high enough so that collision-induced absorption is observable and, on the other, low

* See [343, 344, 191, 192]. (The numbers in brackets refer to the source listing at the end of the book.)

enough so that binary (sometimes along with ternary) interactions prevail. The emphasis is on neutral, non-reactive gases and mixtures, especially those composed of non-polar molecules, and their rotovibro-translational absorption bands in the infrared. In other words, many-body interactions (as they are encountered, for example, in liquids) and CIA involving electronic transitions are not considered to any significant extent. This limitation of scope results in a more manageable volume. Some of the material covered in this volume may be considered a foundation for the treatment of CIA in liquids, especially with regard to the results concerning pairwise additivity of induced dipole moments – and the limitations of such an approach.

The book was written by an experimentalist with a background in molecular collisions and spectroscopy. The author has long felt a need for a book on supermolecular spectroscopy, maybe one that remotely resembles Herzberg's indispensable volumes on the more conventional spectroscopies. The focus of the book is on the simpler molecular systems, especially those involving a H_2 molecule, but attempts are made to present the material in sufficiently general terms so that arbitrary binary and ternary molecular systems are covered. A 'compleat' coverage of the existing $\approx$ 800 original papers on CIA is, of course, not attempted. Rather, the emphasis is on the principles of CIA, a general understanding, and the interplay of fundamental theory and observation. The book is intended as a source of information for the young specialist, the advanced graduate student or worker in related fields, who is interested in the fundamentals or the applications of CIA. I hope that experimentalists as well as theorists will find the volume useful.

I owe much to the late Dr. R. B. Bernstein who first introduced me to the ideas of collision-induced interactions of gases with light in the early seventies. He suggested line shape calculations, which for the analyses of measurements, for applications, and for the comparison of measurements with the fundamental theory are indispensable. During a sabbatical at the *National Bureau of Standards* in 1981, Dr. G. Birnbaum introduced me to the intricacies of CIA, especially of hydrogen and its mixtures with non-polar gases which are so important for the modeling of planetary and certain stellar atmospheres; he remains to this day a friend and mentor in all matters related to collisional induction. I was blessed with a number of brilliant associates, especially Drs. Aleksandra Borysow, Jacek Borysow and Massimo Moraldi, who have made most significant contributions and remain a source of inspiration. I have learned much from many workers in the field, especially from my friends Drs. W. Meyer, J. D. Poll and R. H. Tipping. The bibliographies by Rich and McKellar, and by Hunt and Poll, mentioned above were indispensable in the many years of my involvement with supermolecular processes. Special thanks are due to my wife Margarete for her continued support and encouragement, without which the book could not have been written. The research described in the volume, as far as it originated in Texas, was at times supported by the National Science Foundation. The R. A. Welch Foundation provided valuable, continuing support over the many years of my involvement with supermolecular spectra. I also want to thank all my colleagues and their publishers for the permissions granted to reproduce material previously published elsewhere.

The book was typeset in LaTeX.

Austin, November 1992
L. F.

1

Introduction

Spectroscopy is concerned with the interaction of light with matter. This monograph deals with *collision-induced* absorption of radiation in gases, especially in the infrared region of the spectrum. Contrary to the more familiar *molecular* spectroscopy which has been treated in a number of well-known volumes, this monograph focuses on the *supermolecular* spectra observable in dense gases; it is the first monograph on the subject.

For the present purpose, it is useful to distinguish molecular from supermolecular spectra. In ordinary spectroscopy, the dipole moments responsible for absorption and emission are those of individual atoms and molecules. Ordinary (or allowed) spectra are caused by *intra*-atomic and *intra* molecular dynamics. Collisions may shift and broaden the observable lines, but in ordinary spectroscopy collisional interactions are generally not thought of as a *source* of spectral intensity. In other words, the integrated intensities of ordinary spectral lines are basically given by the square of the dipole transition matrix elements of individual molecules, regardless of intermolecular interactions that might or might not take place. Supermolecular spectra, on the other hand, arise from interaction-induced dipole moments, that is dipole moments which do not exist in the individual (i.e., non-interacting) molecules. Interaction-induced dipole moments may arise, for example, by polarization of the collisional partner in the electric multipole field surrounding a molecule, or by intermolecular exchange and dispersion forces, which cause a temporary rearrangement of electronic charge for the duration of the interaction. The existence of interaction-induced dipole moments and the associated collision-induced absorption (emission), i.e., radiation processes involving more than just one molecule, is a well established fact.

The concept of supermolecules is a most useful one for any detailed treatment of interaction-induced processes. If gases are considered with densities which, on the one hand, are well below liquid density so that many-body effects are minimized but, on the other hand, high enough so that supermolecular effects are observable, a virial expansion of the dielectric and spectroscopic bulk properties in powers of gas density n may be possible. In that case, one need not consider the complex many-body system consisting of all molecules of the fluid. Instead, one has a sequence of relatively simple two-, three-, ... body Hamiltonians which represent the supermolecules of interest. A few leading Hamiltonians may model the spectroscopic properties of the whole system accurately if densities are not too high.

For the purpose of this work, we will call a complex of two or more interacting atoms/molecules a supermolecule. Supermolecules may exist for a short time only, e.g., the duration of a fly-by encounter ($\approx 10^{-13}$ s). Alternatively, supermolecules may be bound by the weak van der Waals forces and thus exist for times of the order of the mean free time between collisions ($\approx 10^{-10}$ s), or longer. In any case, it is clear that, in general, supermolecules possess a spectrum of their own, in excess of the sum of the spectra of the individual (non-interacting) molecules that make up the supermolecule. These spectra are the collision-induced spectra, the subject matter of this monograph.

Conventional molecular spectra arise typically from transitions between the rotovibrational (i.e., bound) states. Collision-induced spectra, on the other hand, involve transitions principally between free (i.e., translational or unbound) states. The translational state of a pair of atoms is given by the energy, E, and angular momentum, L^2, L_z, of relative motion. These quantities are analogous to the vibrational energy, E_v, and rotational angular momentum, J^2, J_z, of bound diatomics; translational energies are, however, continuous while vibrational energies are discrete. The translational spectra of supermolecules involve diffuse lines resembling continuous spectra which are centered at zero frequency and, if molecules are involved, at a number of other frequencies related to the rotovibrational states of the molecules of the complex.

It is a remarkable fact that the translational transitions of virtually all supermolecules are infrared active – even if the individual molecules are not. The only exceptions are supermolecules that possess a symmetry which is inconsistent with the existence of a dipole moment. Pairs of like atoms, e.g., He–He, have inversion symmetry, implying a zero dipole moment and, hence, infrared inactivity. But dissimilar atomic pairs, e.g., He–Ar, or randomly oriented molecular pairs, e.g., H_2–H_2, generally lack such symmetry. As a consequence, more or less significant collision-induced dipoles exist for the duration of the interaction which generate the well known collision-induced spectra.

It was mentioned above that the individual molecules of a complex will usually be unbound particles in collisional interaction, but we note that spectral components involving induced radiative transitions of bound complexes, e.g., HeAr or $(H_2)_2$, the so-called van der Waals molecules, are really inseparable from those of collisional complexes and will, therefore, be of considerable interest here as well.

Molecular and electronic spectra. Two kinds of molecular spectra are generally distinguished: rotovibrational and electronic spectra. Examples of the *rotovibrational* spectra are the purely rotational bands, and the fundamental and overtone bands, of polar molecules like CO; these appear in the far and near infrared, respectively. On the other hand, examples of an *electronic* spectrum are the Schumann–Runge bands of molecular oxygen which correspond to transitions from the ground state, $^3\Sigma_g^-$, to some excited state, $^3\Sigma_u^-$, the highest electronic state shown in Fig. 7.1, p. 357. These bands converge to a limit at 56 850 cm^{-1} (that is 7.047 eV) and become continuous at higher frequencies. Electronic spectra generally appear at the higher frequencies of the electromagnetic spectrum, typically in the visible and ultraviolet.

Supermolecular spectra may also be of the electronic or rotovibrational type. This book deals with the rotovibrational types, which should perhaps be called rovibro-translational spectra to express the significant involvement of translational transitions of supermolecular systems. Even if the molecules by themselves are infrared inactive, the translational motion will generally be infrared active. Supermolecular electronic spectra exist but are not as universal as the rotovibrational induced spectra. Collision-induced electronic spectra will be briefly considered in Chapter 7.

Binary and ternary spectra. We will be concerned mainly with absorption of electromagnetic radiation by binary complexes of inert atoms and/or simple molecules. For such systems, high-quality measurements of collision-induced spectra exist, which will be reviewed in Chapter 3. Furthermore, a rigorous, theoretical description of binary systems and spectra is possible which lends itself readily to numerical calculations, Chapters 5 and 6. Measurements of binary spectra may be directly compared with the fundamental theory. Interesting experimental and theoretical studies of various aspects of *ternary* spectra are also possible. These are aimed, for example, at a distinction of the fairly well understood pairwise-additive dipole components and the less well understood 'irreducible' three-body induced components. Induced spectra of bigger complexes, and of reactive systems, are also of interest and will be considered to some limited extent below.

A collisional complex interacts with radiation for short times only, roughly 10^{-13} s, the duration of a molecular fly-by encounter. We may say that collisional complexes 'exist' for short times only. Nevertheless, supermolecules are a physical reality. Spectra of other species having short lifetimes are, of course, well known. For example, we have radiative transitions involving pre-dissociating molecules (where the final state exists for only $\approx 10^{-12}$ s), radicals (which are chemically unstable and likely to react in the first suitable collision) and van der Waals molecules (which are so weakly bound that the first collision often breaks them up). Collisionally interacting systems and their spectra are just as real as the examples given.

Supermolecular spectra could perhaps be studied with state-selection using adequate molecular beam techniques. That would not be easy, however, because of the smallness of the dipole moments induced by intermolecular interactions. For the purpose of this book, we will mostly deal with bulk spectra, or interaction-induced absorption of pure and mixed gases. A great variety of excellent measurements of such spectra exists for a broad range of temperatures, while state-selected supermolecular absorption beam data are virtually non-existent at this time. Furthermore, important applications in astrophysics, etc., are concerned precisely with the optical bulk properties of real gases and mixtures.

Spectroscopic techniques have been applied most successfully to the study of individual atoms and molecules in the traditional spectroscopies. The same techniques can also be applied to investigate intermolecular interactions. Obviously, if the individual molecules of the gas are infrared inactive, induced spectra may be studied most readily, without interference from allowed spectra. While conventional spectroscopy generally emphasizes the measurement of frequency and energy levels, collision-induced spectroscopy aims mainly for the measurement of intensity and line shape to provide information on intermolecular interactions (multipole moments, range of exchange forces), intermolecular dynamics (time correlation functions), and optical bulk properties.

Real gases. This book deals with the spectra of *real* (as opposed to *ideal*) gases, that is gases composed of real atoms or molecules which interact not only with electromagnetic radiation, but also with the other atoms and molecules of the gas.

One may argue that the traditional spectroscopy is the spectroscopy of *ideal* gases: Intensities of ordinary spectra are generally not thought of as being significantly enhanced or altered by intermolecular interactions. The concept of an ideal (or 'perfect') gas composed of non-interacting particles ('mass points') is familiar from the kinetic theory of gases. It constitutes, of course, a highly simplified, but most useful model that permits an understanding of certain basic properties of gases. For example, the relationship between pressure, p, density, n, and temperature, T,

$$p = nkT,$$

the *ideal gas law*, is well verified experimentally in the low-pressure limit; k designates Boltzmann's constant.

Real gases, on the other hand, consist of atoms or molecules that interact through intermolecular forces. Atoms/molecules attract at distant range and repel at near range; they may be thought of as having a finite size. The theory of real gases accounts for these facts by means of a *virial* expansion,

$$p = A\,n + B\,n^2 + C\,n^3 + \ldots \tag{1.1}$$

In other words, the equation of state may be written as a series in powers of density, with first, second, third, ..., virial coefficients given by

$$\begin{aligned} A &= kT\,, \\ B &= -2\pi kT \int_0^\infty \left[\exp\left(-\frac{V(R)}{kT}\right) - 1\right] R^2\,\mathrm{d}R\,, \\ C &= -\frac{kT}{3} \int\!\!\int f(R_{12})\, f(R_{13})\, f(R_{23})\, \mathrm{d}^3R_{12}\, \mathrm{d}^3R_{13}\,, \\ \ldots & \quad \ldots \end{aligned}$$

etc., with $f(R_{ij}) = \exp(-V(R_{ij})/kT) - 1$ and $\boldsymbol{R}_{23} = \boldsymbol{R}_{13} - \boldsymbol{R}_{12}$. These coefficients account for the effects of monomers and interacting pairs, triplets, ... of atoms or molecules; $V(R)$ designates the pair interaction potential as a function of internuclear separation, R.

In the broad vicinity of the low-pressure (or ideal gas) limit, induced spectra may also be represented in the form of of a virial expansion,

$$I = \tilde{A}\,n + \tilde{B}\,n^2 + \tilde{C}\,n^3 + \ldots \tag{1.2}$$

with coefficients representing the dipole-allowed monomer contributions ($\tilde{A}$), and the induced binary ($\tilde{B}$), ternary ($\tilde{C}$), ..., spectral components. Intensities of spectra of ordinary atoms or molecules vary linearly with density, $I_{\text{allowed}} = \tilde{A}n$: spectral intensities are proportional to the number of sources in the sample volume. Intensities of collision-induced spectra, on the other hand, typically vary with the second or higher powers of density, $I_{\text{induced}} = \tilde{B}n^2 + \tilde{C}n^3 + \ldots$, i.e., these are proportional to the number of pairs, triplets, etc., in the sample volume.

Like the various terms of the virial expansion of the equation of state, the second and higher spectral virial terms in Eq. 1.2 are insignificant at low densities. They become significant, however, at sufficiently high density. At densities approaching liquid density, many-body interactions may be expected to dominate the optical properties. At such densities, each atom or molecule of the liquid is surrounded by a fairly large number of close neighbors, e.g., for liquid argon roughly twelve in the first shell and many more in the more distant 'shells'. Virial expansions are meaningless under such conditions. Other theories appropriate for liquid densities must then be used which, however, will not be considered in this work. We are concerned here mainly with small supermolecular complexes and densities where virial expansions are valid.

We will occasionally use the term 'dense gas' which is meant to describe a gas of a density high enough for induced spectra to appear. In more practical terms, this means roughly atmospheric densities or higher, but we hasten to add that even at much lower densities certain induced features are often discernible as will be seen below (Chapter 3). One atmosphere is certainly not a threshold below which all induced effects miraculously disappear.

1.1 Historical sketch

Spectroscopists are often concerned with the absorption of light in gases. Let I and I_0 be the transmitted and incident intensity of a light beam. The absorption coefficient $\alpha = \alpha(v; T, n)$ is defined by Lambert's law,

$$I = I_0 \exp(-\alpha \, \Delta x) \,, \tag{1.3}$$

where Δx designates the optical path length and T the temperature. In many cases of practical importance, the absorption coefficient is proportional to the number density of the gas, $\alpha \propto n$ (Beer's law). If the absorption coefficient is known as a function of frequency v, it is usually referred to as the absorption spectrum.

Early work. In 1885, Janssen found that in oxygen at pressures of tens or hundreds of atmospheres new absorption bands occur which are unknown from absorption studies at atmospheric pressures; see pp. 357ff. for details. The associated absorption coefficients increase as the *square* of density, in violation of Beer's law. The observed quadratic dependence suggests an absorption by *pairs* of molecules; Beer's law, by contrast, attempts to describe absorption by individual molecules.

Interaction-induced absorption (as the new features were called early on [353]) has stimulated considerable interest. For a long time, explanations were attempted in terms of weakly bound $(O_2)_2$ 'polarization' molecules (that is, van der Waals molecules), but some of the early investigators argued that unbound *collisional pairs* might be responsible for the observed absorption. More recently, a study of the temperature dependence of the induced intensities has provided evidence for the significance of collisional complexes. The idea of absorption by collisionally interacting, unbound molecular pairs was, however, not widely accepted for decades.

A search and a find. Rotovibrational collision-induced absorption was discovered incidentally in a search for the elusive spectra of van der Waals molecules ('dimers'). In his famous dissertation of 1873, J. D. van der

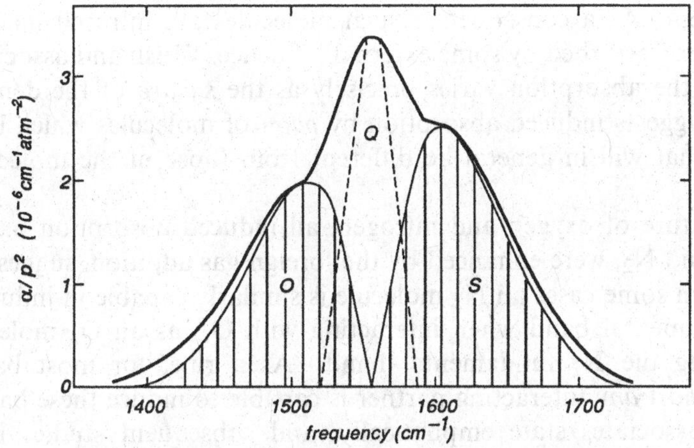

Fig. 1.1. Induced rotovibrational absorption of O_2 pairs. The heavy curve represents the measurement; the light curve is a theoretical envelope ('stick spectrum') of the Raman *O* and *S* branches. The envelope of the *Q* branch is shown as a broken line; after [128].

Waals postulated that intermolecular forces include an attractive component. Ever since then, the existence of weakly bound inert atoms or molecules has been conjectured. The brief remarks made above in connection with induced electronic transitions in oxygen involving hypothetical polarization molecules may serve as an illustration of a situation that has existed for generations. However, until fairly recently, direct evidence for the existence of van der Waals molecules was scant, and a knowledge of energy levels, spectra, etc., of dimers was slow in coming. In 1949, in an attempt to record the infrared vibrational absorption band of $(O_2)_2$ dimers, Welsh and associates [128] discovered instead marked absorption in compressed gases and in liquid oxygen, with a maximum at a frequency roughly equal to the vibrational frequency of the O_2 molecule, 1556 cm^{-1}, see the heavy curve, Fig. 1.1. Similarly, in compressed nitrogen, an absorption band was found with maximum absorption at the N_2 vibrational frequency, 2331 cm^{-1}. (Vibrational bands involving excitation from the ground state to the lowest excited vibrational level are referred to as fundamental bands.) Pressures up to 60 atmospheres were used; the absorption cell was 85 cm long. Subsequent work showed that such induced spectra, and also purely rotational and overtone spectra, exist for virtually all molecular gases. However, these were not the spectra of bound dimers the study had set out to find. The spectral features are diffuse and quite unlike the expected rotovibrational dimer bands.

Homonuclear molecules have a center of inversion symmetry. Molecular charge distributions of such symmetry are inconsistent with a permanent dipole moment. As a consequence, such molecules are infrared inactive unless they are perturbed by some external influence. Welsh and associates showed that the absorption varies precisely as the *square* of the density, a fact that suggests induced absorption by *pairs* of molecules which have symmetries that will in general be different from those of the molecules involved.

In the mixture of oxygen and nitrogen, all induced absorption bands, both of O_2 and N_2, were enhanced by the foreign gas addition, suggesting that at least in some cases an N_2 molecule is similarly capable in inducing the O_2 fundamental band when interacting with O_2, as an O_2 molecule is in inducing the N_2 fundamental band. As a rule, for most bands observed, almost *any* interacting partner is capable to induce these bands. Welsh and associates state emphatically, and subsequent studies have fully confirmed, that the observed absorption must have its origin in dipole transitions induced by intermolecular forces during collisions of *unbound* monomers; bound complexes like O_4 are *not* required for an explanation of the observed absorption bands (but must not be ruled out completely as we will see).

Hence, the process was named collision-induced absorption. Whereas the term 'interaction-induced absorption' used by some early on seems to cautiously leave the question open whether free or bound complexes generate the absorption, Welsh and associates bravely state their conclusion as collisional interactions. Since then, other names have also been used, such as pressure-induced and supermolecular absorption.

Figure 1.2 illustrates the difference between the transitions involved in van der Waals dimer bands which Welsh and associates hoped to find, and the collision-induced absorption spectra that were discovered instead. Intermolecular interaction is known to be repulsive at near range and attractive at more distant range. As a consequence, a potential well exists which for most molecular pairs is substantial enough to support bound states. Such a bound state is indicated in Fig. 1.2 (solid curve *b*). When infrared radiation of a suitable frequency is present, the dimer may undergo various transitions from the initial state (solid curve) to a final state which may have a rather similar interaction potential (dashed curve *b*′) and dimer level spacings. Such transitions (marked *bound–bound*) often involve a change of the rotovibrational state(s) E_{vj} of one or both molecule(s),

$$\Delta E = E_{v_1' j_1'} - E_{v_1 j_1} + E_{v_2' j_2'} - E_{v_2 j_2} , \tag{1.4}$$

as well as of the dimer. The dimer transition frequencies are relatively sharp unless gas pressures are high enough to broaden the dimer lines. Dimer transition frequencies are given by $h\nu = \Delta E + E'_{\text{bound}} - E_{\text{bound}}$. Such processes are known to give rise to absorption spectra of van der

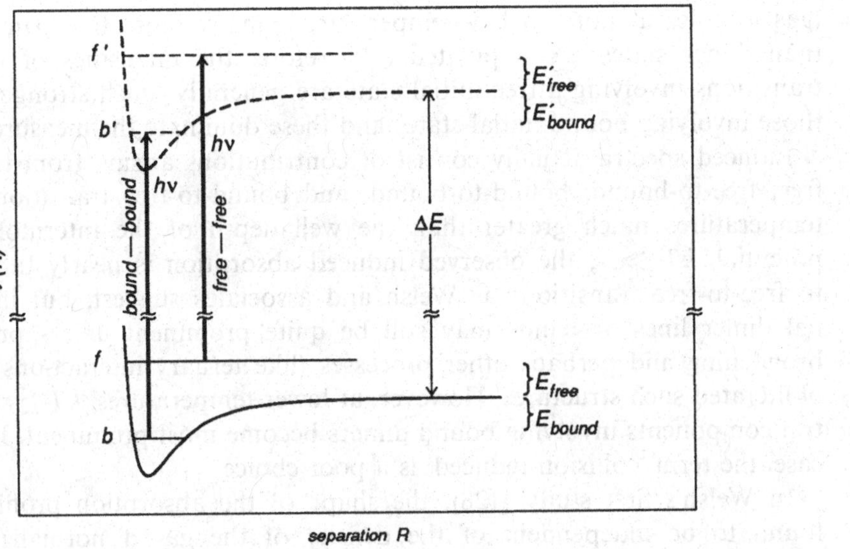

Fig. 1.2. Intermolecular potential curves and radiative transitions of the complex of molecules 1 and 2. The energy spacing $\Delta E = E_f - E_i$ is the difference of the rotovibrational energies of initial and final states of the complex, $E_i = E_{v_1 j_1} + E_{v_2 j_2}$ and $E_f = E'_{v_1 j_1} + E'_{v_2 j_2}$, respectively. The E_{bound} and E_{free} designate bound and free state energies of the complex; a prime indicates final states. Also shown are representative radiative transitions $h\nu$ from bound state to bound state, and from free state to free state, involving rotovibrational transitions in one or both molecules.

Waals molecules in the various rotovibrational bands of the individual molecules, and also at their sums and differences. The actual recording of such dimer bands is a fairly recent accomplishment [49, 268].

Welsh suggested correctly that similar transitions take place even if the molecular pair is not bound. The energy of relative motion of the pair is a continuum. Its width is of the order of the thermal energy, $E_{\text{free}} \approx 3kT/2$. Radiative transitions between free states occur (marked *free–free* in the figure) which are quite diffuse, reflecting the short 'lifetime' of the supermolecule. In dense gases, such diffuse collision-induced transitions are often found at the various rotovibrational transition frequencies, or at sums or differences of these, even if these are dipole forbidden in the individual molecules. The dipole that interacts with the radiation field arises primarily by polarization of the collisional partner in the quadrupole field of one molecule: the free–free and bound–bound transitions originate from the same basic induction mechanism.

We will see below that, in general, van der Waals molecules have relatively few bound states. Under conditions commonly encountered (gas samples at not too low temperatures), many more free pair states than dimer states are populated. Therefore, the intensities of optical transitions involving a free initial state are generally much stronger than those involving bound initial states and these dominate the measurement.

Induced spectra actually consist of contributions arising from free-to-free, free-to-bound, bound-to-bound, and bound-to-free transitions. At temperatures much greater than the well depth of the intermolecular potential, $kT \gg \varepsilon$, the observed induced absorption is nearly fully due to free-to-free transitions as Welsh and associates suggest, but individual dimer lines or bands may still be quite prominent unless pressure broadening and perhaps other processes (like ternary interactions) have obliterated such structures. However, at lower temperatures, $kT \lesssim \varepsilon$, spectral components involving bound dimers become most prominent. In that case, the term 'collision-induced' is a poor choice.

In Welsh's first study [128], the shape of the absorption profile was found to be independent of the density of the gas if normalized by density squared, $\alpha(\nu)/n^2$, at nearly *all* frequencies ν. Pair spectra show no variation with density other than the density squared relationship – up to the point where ternary (or higher-order) spectra affect the measurements (if monomers are infrared inactive in the spectral band considered).

In an early paper, Condon [127] argued that in an external electric field all polarizable molecules will become infrared active, with selection rules known from Raman spectra ('field-induced absorption'). For homonuclear diatomics, this means that rotational transitions occur according to $J \to J + 2$ (S-branch), $J \to J$ (Q-branch). and $J \to J - 2$ (O-branch) if J represents the quantum number of the initial rotational state. Welsh *et al.* argue that collisional induction should have similar selection rules. The external field may be thought of as an intermolecular force field which is electric in nature. For molecules like O_2 and N_2, more or less ordinary Raman O, Q and S branches should be expected. Indeed, the measured spectra show the *envelopes* of these overlapping three Raman branches, Fig. 1.1. The rotational O lines are all to the left of the vibrational transition at 1556 cm^{-1}, and the S lines to the right; the vertical bars in the figure schematically represent the rotational lines observed in Raman studies. The Q branch, on the other hand, consists of a series of densely spaced lines near 1556 cm^{-1} that are often not resolved; only the envelope of the Q branch is indicated in the figure (dashed lines).

Apparently, within each branch, rotational lines are not resolved. This is due to the short interaction time of the collisional pair which renders individual lines rather diffuse, with half-widths that are much greater than the rotational spacings of O_2 and N_2 at temperatures around 300 K.

The spectral line shapes of collision-induced spectra (like in Fig. 1.3) resemble a Lorentzian, but the profile shown in Fig. 1.1 looks quite different. We shall see below that in reality, collision-induced line profiles are all of similar shape. In the case at hand, the real difference is not so much a different shape of the individual collision-induced line. Rather, the spectrum shown in Fig. 1.1 is a composite of many lines, one for each rotovibrational transition of the *O*-, *Q*- and *S*- branches. As a consequence, only the envelope is discernible which shows shoulders normally absent in the Raman spectra. In Fig. 1.1, by contrast, we are looking at a (typical) single induced line.

Welsh *et al.* point out that collision-induced absorption is a universal phenomenon found essentially in *all* compressed molecular gases and condensed matter. Molecules are surrounded by electric fields. Furthermore, molecules are polarizable. To a greater or lesser extent, all molecular pairs show, therefore, collision-induced absorption if only the gas densities are high enough. Accordingly, collision-induced rotational bands occur, as well as rotovibrational fundamental and overtone bands, because both the fields surrounding molecules and the polarizability of molecules are generally modulated by the molecular rotovibrational frequencies. The so-called translational bands occur from the transitions that leave the rotovibrational states of the molecular partners unchanged.

Subsequent work has revealed that collision-induced absorption is observable even in mixtures of monatomic gases, albeit not in unmixed monatomic gases. In rare gas mixtures, the translational absorption profile occurs in the microwave and far infrared regions. In mixtures of molecular gases, such translational absorption profiles are sometimes discernible but they are generally masked by the induced rotational bands mentioned.

In short, a new, universal supermolecular spectroscopy, absorption due to collisionally interacting pairs, was discovered – an unexpected find of great significance for both the basic and applied sciences. In the roughly forty years of its existence, roughly eight hundred papers reporting original research have been published in the field, and hundreds of authors from various disciplines like physics, chemistry, statistical mechanics, astronomy, liquid and solid state, laser science, industrial research, and the space program have made significant contributions [344, 191]. Collision-induced spectroscopy is ubiquitous. It is also a fundamental science of molecular interactions and it manifests itself in many ways of interest in the pure and applied sciences.

Collision-induced microwave spectra. Measurements of the dielectric loss by resonant cavity techniques at 9 and 24 GHz were first reported by Birnbaum and Maryott [33]. The cavity was at room temperature and filled with carbon dioxide gas at densities up to 100 amagat. The loss, which at not too low frequencies increases as the *square* of density, was attributed to transient dipoles induced primarily by the molecular quadrupole fields which polarize collisional partners [255], much as this was concluded above for collision-induced absorption by pairs of linear molecules. Similar microwave loss spectra were thereafter discovered for other non-polar gases, including nitrogen, methane, hydrogen, ethylene, etc., and in binary mixtures of gases, like carbon dioxide and argon, etc. [131].

Collision-induced emission. Emission spectra of ordinary atoms and molecules correspond to downward transitions, from an initial energy level higher than the final one, whereas absorption involves the inverse transition. Both exist in supermolecules as well and have recently been seen in shocktube studies. Emission spectra are generally much richer than absorption spectra and may include 'hot bands,' which involve transitions between excited vibrational (or electronic) states [116].

Beyond the binary systems. Spectroscopic signatures arising from more than just two interacting atoms or molecules were also discovered in the pioneering days of the collision-induced absorption studies. These involve a variation with pressure of the normalized profiles, $\alpha(\omega)/n^2$, which are pressure invariant only in the low-pressure limit. For example, a splitting of induced Q branches was observed that increases with pressure: the intercollisional dip. It was explained by van Kranendonk as a correlation of the dipoles induced in subsequent collisions [404]. An interference effect at very low (microwave) frequencies was similarly explained [318]. At densities near the onset of these interference effects, one may try to model these as a three-body, spectral 'signature', but we will refer to these processes as many-body intercollisional interference effects which they certainly are at low frequencies and also at condensed matter densities.

True induced spectra of *ternary* complexes have also been identified and separated from the binary components in more recent years, by varying gas densities and making use of the virial expansion, Eq. 1.2.

Interesting *line narrowing* has been observed of quadrupole-induced lines of hydrogen–rare gas mixtures. These have been explained by van Kranendonk and associates [428] in terms of the mutual diffusion coefficient of H_2 in a rare-gas environment, as an effective lengthening of the interaction times of H_2–atom complexes.

Many-body effects are evident in the induced spectra of liquids and solids; these spectra may bear a superficial resemblence to the binary spectra, but significant quantitative differences exist which distinguish the various many-body spectra from the binary ones.

1.2 Interaction-induced absorption: the characteristics

We have seen that supermolecular spectra differ from the spectra of single atoms or molecules in a variety of ways which we briefly summarize.

a) Integrated induced intensities do not feature the linear density dependence characteristic of monomeric spectra (often referred to as Beer's law). Instead, at densities well below liquid densities, these vary as a virial series of *density squared, cubed*, ..., terms if binary, ternary, ..., complexes contribute, Eq. 1.2.

b) Collision-induced spectra feature very *diffuse* line profiles. According to Heisenberg's uncertainty principle,

$$\Delta\nu \, \Delta t \gtrsim 1/4\pi \,, \tag{1.5}$$

the product of line width $\Delta\nu$ and lifetime Δt is never smaller than a number of order of unity. Typical line widths are roughly $\Delta\nu \approx 10^{13}$ Hz, which reflect the short lifetime ($\Delta t \approx 10^{-12}$ s) of the supermolecule's brief, radiatively active existence. The duration of an average collision is given by the range of the induced dipole function, divided by the root-mean-square speed of relative motion of the collisional pair.

c) Spectral bands: we may formally describe the absorption of a photon $h\nu$ by a non-reactive molecular pair, A and B, by an expression like

$$A + B + h\nu \rightarrow A^* + B^* + \Delta E_{\text{transl}} \,, \tag{1.6}$$

where the asterisks indicate some possible rotovibrational or electronic excitation in one or more of the molecular partners. The photon energy $h\nu$ matches the sum of changes of the internal energies, E_A, E_B, and the energy of relative motion, E_{transl}. In other words,

$$h\nu = \Delta E_A + \Delta E_B + \Delta E_{\text{transl}} \,. \tag{1.7}$$

The first two terms to the right represent rotovibrational (or electronic) transition frequencies (in energy units) that may be positive, zero, or negative. We will see that for fixed molecular excitations, ΔE_A, ΔE_B, the variations in the change of translational energies determine the *line shape* of induced profiles, see Fig. 1.3. (We note that if the system A+B is a bound van der Waals molecule, AB, insted of a collisionally interacting pair, A–B, the term E_{trans} must be replaced by the rotovibrational energy of the van der Waals molecule.) Specifically, the line width is given by the average $\overline{\Delta E}_{\text{transl}}$ which,

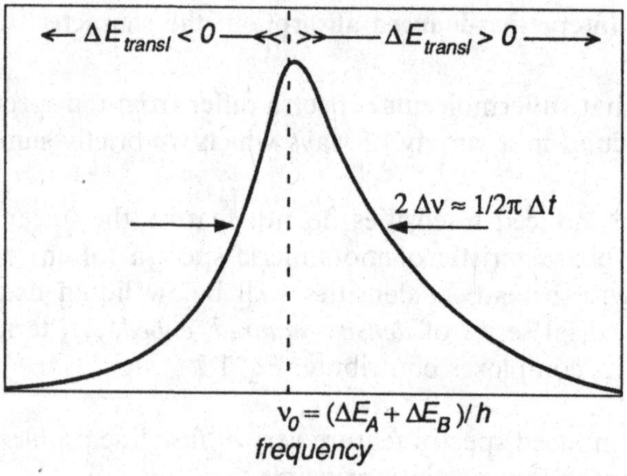

Fig. 1.3. Collision-induced profile – schematic. Shown is the transition frequency ν_0 ('line center'), a sum of molecular transition frequencies, $(\Delta E_A + \Delta E_B)/h$. At higher frequencies, translational energies of relative motion of the collisional pair are increased, at lower frequencies decreased; see Eq. 1.7. The average width is given by Eq. 1.5.

according to the uncertainty relation, is of the order of $h\,\Delta\nu$, roughly 100 cm^{-1} or 0.01 eV for systems of practical interest. The line centers of induced lines may be thought of as being defined by $\Delta E_{\rm transl} = 0$, that is, induced bands occur at sums of molecular frequencies, regardless of whether these transitions are or are not dipole-allowed in the non-interacting molecules A, B.

If both $\Delta E_{\rm A}$ and $\Delta E_{\rm B}$ equal zero, we have a purely *translational line* of the supermolecule. These can be found at frequencies comparable to the width $\Delta\nu$, that is in the microwave and far-infrared regions of the spectrum.

If just one of the $\Delta E_{\rm A}$, $\Delta E_{\rm B}$ is non-zero, collision-induced *rotovibrational* or *electronic bands* occur whose line centers coincide with a transition frequency of one of the molecules.

Induced spectra may appear at frequencies corresponding to molecular transitions that may be allowed or forbidden in the isolated molecules. Induced spectra associated with allowed transitions can sometimes be studied but this is not always easy in the presence of strong, allowed lines. Induced spectra of forbidden transitions ($\tilde{A} = 0$ in Eq. 1.2), on the other hand, are quite common and are generally easy to recognize.

Induced lines have also been seen at frequencies given by sums and differences of transition frequencies of the molecules involved. These arise from *simultaneous transitions* in the interacting molecules, an interesting feature that unmistakeably points at the supermolecular nature of these spectra; in that case, neither ΔE_A nor ΔE_B vanish.

d) Line strengths. The induced dipoles which generate the collision-induced spectra are generally weak. More specifically, induced dipoles amount to typically $\approx 0.01 \ldots 0.1$ Debye* at near range, falling off rapidly to zero with increasing separation. For comparison, we mention that permanent dipoles of typical polar molecules amount to several Debye. Spectral intensities are proportional to the dipole moment squared. It thus appears that an 'allowed' rotational line of a polar molecule may be stronger than an induced line by several orders of magnitude. On the other hand, at not too low gas densities, many more pairs (or triplets, ...) exist than monomers, and for dense systems induced spectra may actually be strong because of large numbers of complexes.

In the presence of dipole-allowed lines or bands, the induced components are rarely discernible without ambiguity, because of the generally great differences of intensity. There is, however, little doubt that dipole-allowed monomer transitions will in general be accompanied by a (relatively weak and very diffuse) induced background if gas densities are high enough. Induced components of allowed lines may be discernible in the far wings. If the allowed lines are very weak, i.e., comparable in intensity to the induced spectra, interesting Fano profiles arising from an interference of allowed and induced components have been observed.

e) Clarification: Supermolecular absorption is the *excess* absorption by complexes of interacting atoms and/or molecules *over* the simple sum of the absorption spectra of the individual (non-interacting) atoms or molecules of the complex.

f) Other remarks: A supermolecule – like any ordinary molecule – is described by its Hamiltonian and the eigenfunctions. A supermolecule of Nth order consists of N molecules and is represented by an N particle Hamiltonian. For the purposes of this monograph, the supermolecules of interest will be of a low order, $N = 2$ or 3, but for dense systems like liquids and solids, one has $N \gg 1$. We note that in theoretical studies of many-body systems the concept of pairwise additive interactions (e.g., of potential and dipole moments) is widely

* 1 Debye = 0.3934267 atomic units of dipole strength = 3.335611×10^{-30} C m.

used. It is good to keep in mind that this assumption is independent of the size N of the supermolecule. In other words: liquid models ($N \gg 1$) may be constructed with the assumption of pairwise additivity, but this does not mean that liquids may be modeled by binary supermolecules. Rarefied gases, on the other hand, may be viewed as an assembly of monomers (described by a one-particle Hamiltonian), pairs (described by a two-particle Hamiltonian), etc., which 'know' nothing of the presence of the other systems present, according to the theory of virial expansions which will be used extensively below.

A typical binary molecular encounter may be assumed to start with widely separated particles approaching each other. In time, a point of closest approach is reached, whereupon the separations increase indefinitely. Collision-induced dipoles are, therefore, vanishingly small initially and finally when the collisional partners are widely separated, $\boldsymbol{\mu}(t) \approx 0$, and go in time through some maximum at the point of closest approach. The spectrum arising from such time behavior is obtained by the Fourier transform of the induced dipole *versus* time function, see Chapter 5 for details. If the collisional partners do not change their rotovibrational or electronic states during the interaction, the resulting spectral line is a diffuse structure at zero frequency. However, one or more collisional encounters will change rotovibrational states with a finite probability. In that case, the dipole function $\boldsymbol{\mu}(t)$ may be thought of as oscillating in time with the molecular transition frequencies, because the molecular multipole moments and the polarizabilities are functions of the vibrational (rotational) coordinates. In that case, the Fourier transform contains diffuse lines or bands at these monomer transition frequencies, and at sums and differences of these if both partners undergo monomer transitions. These bands will appear in the far infrared if purely rotational transitions occur, in the infrared if vibrational transitions occur, and generally in the visible or ultraviolet, if electronic transitions take place.

Supermolecules exist as free and as bound systems, that is as pairs (or bigger complexes) in collisional interaction and as van der Waals molecules, respectively. Almost any combination of atoms or molecules may form bound van der Waals molecules. (Among the few exceptions are H_2–He pairs, H–He pairs and probably He pairs which do not exist as bound van der Waals molecules.) Like ordinary molecules [180] supermolecules may, therefore, be expected to have rotovibrational bands in the infrared and electronic bands in the visible and ultraviolet regions of the electromagnetic spectrum. Besides the familiar bands, induced spectra generally show relatively strong contributions arising from transitions involving the free (i.e., unbound) states, namely free → free, free → bound and bound → free transitions that are associated with continuous spectra. Such continua are by no means unknown from the spectroscopy of ordinary atoms and molecules (e.g., bremsstrahlung, ionization and dissociation continua, etc.). To a large extent, collision-induced spectra consist of these types of continua. In fact, these are so characteristic of supermolecular spectra that in a first approximation one often may neglect the transitions involving bound states altogether and still have a reasonable representation of collision-induced spectra when a low-resolution spectroscopic apparatus is used for the recording. This is so largely because under conditions commonly encountered in such studies, vastly more free monomers exist in a given volume than bound van der Waals systems. The continua may be considered simple generalizations of the bound–bound transitions. Certainly, for a given system all supermolecular spectra arise from the same induced dipole mechanism, regardless whether free or bound states are involved. It should be mentioned, however, that the adjective 'collision-induced' in collision-induced absorption is somewhat of a misnomer because of the generally inseparable spectral contributions of bound states; the terms interaction-induced or supermolecular are correct and, therefore, preferrable.

1.3 Significance

The point was made above that the spectra of real gases possess a virial expansion just like the equation of state: in a given band, spectral intensities can be written in terms of a series in powers of density, Eq. 1.2. The coefficients $\tilde{A}$ of the leading term represent the 'allowed' monomer spectra, $\tilde{B}$ the induced binary components (of both bound and unbound pairs), $\tilde{C}$ the induced ternary components, and so forth. Contrary to the virial expansion of the equation of state, the leading spectral coefficients $\tilde{A}$ may be zero if the monomer spectra are forbidden. In this case, the second virial term can be investigated in great detail with little interference if gas densities are low enough to avoid significant ternary contributions. Virial coefficients always have been of the greatest interest for the study of molecular interaction. The second spectral virial coefficient depends on the induced dipole surface and the interaction potential. This dependence on two functions of intermolecular interaction has at times been considered a disadvantage, but it provides new, additional information on intermolecular interactions.

Ternary virial coefficients consist of two parts. One part is defined in terms of the pairwise additive functions of the interaction, the other by the irreducible ones. Since the latter are not well known, a vast field of study, nearly untouched, is being opened up by collision-induced spectroscopy.

Molecular multipole moments, quantities important for collisional induction, can be determined from the analysis of collision-induced spectra.

The spectra of liquids and solids are known to have strong induced components. Liquids and solids are, however, so dense that many-body terms dominate the spectra; the binary and ternary spectral components which are the main topic of this work (and which are usually measurable in compressed gases at densities much lower than liquid state) will often resemble the spectra of liquids and solids, but a critical comparison will reveal important qualitative and quantitative differences. Nevertheless, a study of binary spectra will help to illuminate important aspects of the theoretical descriptions of liquid spectra and may be considered a basic input into the theory of liquid interactions with radiation.

Of a special astronomical interest is the absorption due to pairs of H_2 molecules which is an important opacity source in the atmospheres of various types of cool stars, such as late stars, low-mass stars, brown dwarfs, certain white dwarfs, population III stars, etc., and in the atmospheres of the outer planets. In short: absorption of infrared or visible radiation by molecular complexes is important in dense, essentially neutral atmospheres composed of non-polar gases such as hydrogen. For a treatment of such atmospheres, the absorption of pairs like H–He, H_2–He, H_2–H_2, etc., must be known. Furthermore, it has been pointed out that for technical applications, for example in gas-core nuclear rockets, a knowledge of induced spectra is required for estimates of heat transfer [307, 308]. The transport properties of gases at high temperatures depend on collisional induction. Collision-induced absorption may be an important loss mechanism in gas lasers. Non-linear interactions of a supermolecular nature become important at high laser powers, especially at high gas densities.

1.4 Organization of the material

This book is concerned mostly with the spectroscopy of *binary* systems, but *ternary* spectra will also be reviewed in some detail. Absorption in fluids at liquid densities, and solids, on the other hand, is controlled by many-body interactions and cannot be considered here in any detail.

We will mainly be concerned with two- and three-body atomic and molecular systems whose components preserve their identity during the radiative encounters. In other words, we will consider *non-reactive* atomic or molecular systems, such as interacting helium and argon atoms, He–Ar, or hydrogen pairs, H_2–H_2, in their electronic ground states.

The subject matter of the book are the rovibro-translational spectra of molecular complexes. Supermolecular spectra involving electronic transitions will be briefly considered in Chapter 7. We are concerned mostly with *neutral* environments where the effects of collisional induction are most clearly seen.

We emphasize the *line shape* problem perhaps a little more than usual in the spectroscopic literature. Collision-induced spectra have little structure. Yet, the diffuse line and band spectra extend over wide frequency bands and must often be subtracted, say from the complex spectra of planetary or stellar atmospheres, for a more detailed analysis of other, less well known components. The subtraction requires accurate knowledge of the profile and its variation with temperature, composition, etc., often over frequency bands of hundreds of cm^{-1}.

Content. After a brief overview of molecular collisions and interactions, dipole radiation, and instrumentation (Chapter 2), we consider examples of measured collision-induced spectra, from the simplest systems (rare gas mixtures at low density) to the more complex molecular systems. Chapter 3 reviews the measurements. It is divided into three parts: translational, rototranslational and rotovibrational induced spectra. Each of these considers the binary and ternary spectra, and van der Waals molecules; we also take a brief look at the spectra of dense systems (liquids and solids). Once the experimental evidence is collected and understood in terms of simple models, a more theoretical approach is chosen for the discussion of induced dipole moments (Chapter 4) and the spectra (Chapters 5 and 6). Chapters 3 through 6 are the 'backbone' of the book. Related topics, such as redistribution of radiation, electronic collision-induced absorption and emission, etc., and applications are considered in Chapter 7.

At the end of Chapters 2 through 7, short bibliographies are given, which list relevant monographs, proceedings of conferences and study institutes, and review articles. Throughout the text, references to original works are given by numbers in brackets; these are listed in the back.

The reader may find that a large part of the book deals with hydrogen, H_2–X, H_2–X–X, where X stands for an atom or simple molecule such as He, Ar, H_2, etc. In fact, many of the basic features of collision-induced spectra are best demonstrated on the basis of data obtainable with hydrogen and its isotopes, or with mixtures containing hydrogen. It is, therefore, no accident that a wealth of information and reliable data exist on systems containing hydrogen. Furthermore, a significant fraction of all matter in the universe is hydrogen in its various states; astrophysicists want to know all that can be known about the spectra of hydrogen; the collision-induced spectra are an essential part. Most facts to be learned from studies of the collision-induced hydrogen spectra may be directly generalized to other systems and are really not specific to hydrogen.

2
Recapitulation

In this chapter we summarize some background information concerning molecular collisions, dipoles and radiation, spectroscopy, and statistical mechanics that will be needed later. This Chapter should be skipped in a first reading. It is hoped that a reader who comes back to this Chapter later with specific questions will find the answers here – or, at least, some useful reference for further study.

2.1 Intermolecular potentials

The ideal gas law, Eq. 1.1 with $B = C = \ldots = 0$, may be derived with the assumption of non-interacting 'point particles'. While in the case of rarefied gases at high temperatures this assumption is successful in that it predicts the relationship between pressure, density and temperature of a gas in close agreement with actual measurement, it was clear that important features of gaseous matter, such as condensation, the incompressibility of liquids and solids, etc., could not be modeled on that basis. As early as in 1857, Clausius argued convincingly that intermolecular forces must be repulsive at short range and attractive at long range. When in 1873 van der Waals developed his famous equation of state, a significant improvement over the ideal gas law, he assumed a repulsion like that of hard spheres at near range, and attraction at a more distant range.

The true nature of the intermolecular forces could be understood only after quantum mechanics was developed. When atoms or molecules with closed shells approach each other closely enough so that their electronic clouds overlap, the energy increases. At close range, the Pauli principle forces electrons into higher states; the effect is a net repulsion. To some degree, the Coulomb forces of the nuclei also contribute to the repulsion at near range. At larger separations, where overlap is negligible, various

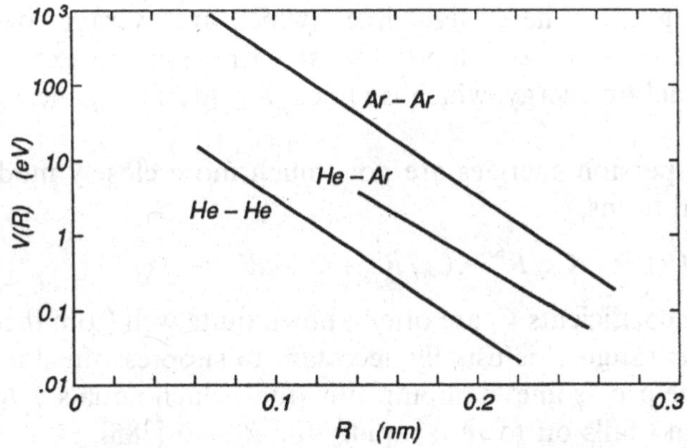

Fig. 2.1. Measured repulsive parts of the interatomic potentials of He–He, He–Ar, and Ar–Ar (schematic; averages of several measurements); after [370].

electric forces may in general be identified even if the interacting partners are electrically neutral. We mention multipole–multipole interactions, multipole-induced interactions, and the dispersion forces. Gravitational forces are roughly 30 orders of magnitude weaker than the electric forces mentioned and can thus be ignored in most cases.

From the inception of quantum mechanics, from about 1930 to the late sixties, most research on intermolecular forces was based on two assumptions, namely 1. that the pair potentials could be represented by simple functions, such as the two-parameter Lennard-Jones model,

$$V(R) = 4\epsilon\left[(\sigma/R)^{12} - (\sigma/R)^6\right],$$

where ϵ and σ represent well depth and the root of the function $V(R) = 0$ for $R = \sigma$; and 2. that the intermolecular energy of a molecular complex is well represented by the sum of their pair interaction energies. It is now abundantly clear that both of these assumptions are wrong. In fact, it has been argued that these assumptions actually have obstructed significant progress in the field (Maitland *et al.*, 1981). However, since about 1970 substantial progress has been possible with the advent of new, accurate measurements of various bulk properties, molecular beam scattering, the spectra of van der Waals molecules, etc. Vastly improved *ab initio* computations have also been possible in recent years. Direct 'inversions' of measured data have emerged that permit the unbiased construction of intermolecular potentials, without any *ad hoc* assumption concerning the analytical form of the potential function.

It is now clear that the repulsive energy branch of rare gas pairs is of an exponential form, unlike the R^{-12} term of the Lennard-Jones model. A few examples of measured repulsive branches of interatomic potentials are shown in Fig. 2.1. The straight lines (which are averages over measurements obtained in several laboratories) correspond to exponentially decreasing interaction energy, which is typical for interacting inert neutrals at near range.

Moreover, dispersion energies are now much more closely modeled by a sum of several terms,

$$V(R) \sim -C_6/R^6 - C_8/R^8 - C_{10}/R^{10} - \cdots$$

for large R. The coefficients C_i are often known quite well from theory. We note that at near range, it is usually necessary to suppress this long-range term with the help of some 'damping function' which equals 1 for large separations R and falls off to zero rapidly for $R \to 0$ [385].

Other potential models have been successfully used, especially the simple three-parameter $n(R) - 6$ potential,

$$V(R) = \epsilon \left(\frac{m}{n-m} (R_{\min}/R)^n - \frac{n}{n-m} (R_{\min}/R)^m \right) \tag{2.1}$$

with $m = 6$ and $n = n(R) = 13.0 + \gamma[(R/R_{\min}) - 1]$. The $n(R) - 6$ model gives remarkably accurate representations of the potential functions of the inert gases. In this expression, γ is a constant and $R_{\min}$ designates the position of the minimum of the potential function. Most atoms and molecules to be considered here have well depths ϵ between roughly 10 and a few hundred K. The $R_{\min}$ are typically between 0.1 and 0.5 nm, and $\gamma = 4$ or 5 (Maitland *et al.*, 1981).

If molecules are involved, isotropic potential functions are in general not adequate and angular dependences reflecting the molecular symmetries may have to be accounted for. In general, up to five angular variables may be needed, but in many cases the anisotropies may be described rigorously by fewer angles. We must refer the reader to the literature for specific answers (Maitland *et al.*, 1981) and mention here merely that much of what will interest us below can be modeled in the framework of the isotropic interaction approximation.

Molecules may vibrate and when they vibrate, their interaction with other molecules is modified. Vibrating molecules often appear bigger and more anisotropic. For selected systems, for example for hydrogen–rare gas pairs, vibrational dependences have been carefully modeled [227]. However, relatively few molecular interaction potentials are well known and for specific cases, one will have to search the recent literature for state of the art models.

Non-pairwise additivity. A significant component of the energy $V(1,2,3)$ of three interacting atoms is given by the sum of the pair potentials, $V(1,2) + V(1,3) + V(2,3)$. However, it is now generally accepted that the so-called Axilrod–Teller term, a long-range, irreducible (classical)

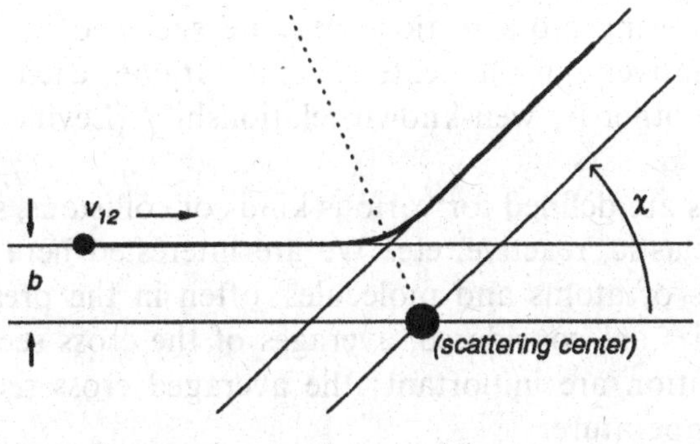

Fig. 2.2. Definition of scattering angle χ and impact parameter b of classical trajectories. The point of closest approach is at the intersection of the trajectory with the dotted line.

component, is also important and should be added to the former for a better representation of $V(1, 2, 3)$. Evidence for a quantal, near-range, irreducible component also exists. Similar non-additive interaction terms of higher order may exist in bigger molecular complexes that are of interest for the study of dense gases, liquids and solids (Maitland *et al.* 1981).

2.2 Molecular collisions

Binary molecular collisions are generally described in terms of various scattering cross sections Q. In the case of a beam scattering experiment, we have a beam of particles, type 1, of velocity v, incident on a target chamber of volume V which is filled with a gas of $N_2 = V n_2$ particles of type 2, with negligible velocites, $|v_2| \ll v$. The total number of beam molecules scattered into all angles per unit time is given by

$$\dot{N}_1 = n_1\, v\, n_2\, V\, Q_{12}\,.$$

The beam intensity is given by $I = n_1\, v$, the incident number of molecules 1 per unit area per unit time. The n_1, n_2, are particle number densities. The 'total' cross section $Q_{12} = Q_{12}(v_{12})$ has units of area; it is a function of the relative speed, v_{12}, which in this case is given by v.

We define a scattering (or deflection) angle, χ, as in Fig. 2.2, an azimuth, φ, and a solid angle, Ω, so that $\mathrm{d}^2\Omega = \sin\chi\ \mathrm{d}\chi\ \mathrm{d}\varphi$. The differential scattering cross section is defined as the number $\Delta\dot{N}$ of particles scattered every second into the solid angle $\Delta\Omega$ around χ, φ,

$$\frac{\Delta\dot{N}}{\Delta\Omega} = n_1\, v\, n_2\, V\, \frac{\partial Q_{12}}{\partial\Omega}\,.$$

Units of the differential scattering cross section are area per steradian.

Differential scattering cross sections may be specified in the laboratory frame or, alternatively, in the center of mass frame; each may be transformed into the other by well-known relationships (Levine and Bernstein 1974).

Cross sections are defined for various kinds of collisions, such as elastic, inelastic, superelastic, reactive, etc. We are interested here mostly in the elastic collisions of atoms and molecules, often in the presence of light. If bulk properties are considered, averages of the cross sections over the velocity distribution are important; the averaged cross sections become functions of temperature.

Cross section and potential. Collision cross sections are related to the intermolecular potential by well-known classical and quantum expressions (Hirschfelder *et al.*, 1965; Maitland *et al.*, 1981). Based on Newton's equation of motion the classical theory derives the expression for the scattering angle,

$$\chi = \chi(b; E_T) = \pi - 2b \int_{R_0}^{\infty} \frac{\mathrm{d}R}{R^2 \left[1 - V(R)/E_T - (b/R)^2\right]^{1/2}} . \tag{2.2}$$

Here, $V(R)$ designates the interaction potential, b the impact parameter (see Fig. 2.2 for a definition), E_T the translational kinetic energy of relative motion, and R_0 is the largest root of the expression under the square root sign. It is intuitively clear that repulsive potentials are associated with *positive* deflection angles χ while attractive potentials show *negative* angles χ, Eq. 2.2. Intermolecular potentials, which are repulsive at near range and attractive at more distant range, may therefore be expected to have positive scattering angles at small values of the impact parameter b ('hard' collisions), and negative scattering angles at larger b ('distant' or 'fly-by' collisions), Figs. 2.3 and 2.4. For a head-on collision ($b = 0$), the projectile bounces back ($\chi = \pi$); in a very distant collision ($b \gg \sigma$), no deflection occurs ($\chi = 0$).

For hard spheres the deflection function $\chi(b)$ is given by

$$\chi = \begin{cases} 2 \arccos(b/\sigma_{12}) & \text{for } 0 \le b \le \sigma_{12} \\ 0 & \text{for } b > \sigma_{12} . \end{cases}$$

This function is similar to the high-energy curve (h) of Fig. 2.3, except that negative scattering angles do not occur. The scattering angle χ is independent of the translational kinetic energy E_T and relative speed v_{12}. High-energy collisions of real atoms or molecules resemble hard sphere collisions, but their 'size' σ is obviously energy-dependent, unlike the size of hard spheres.

In the classical theory orbiting collisions are defined by the radicant in Eq. 2.2 having exactly two roots. Normally, at a fixed, low enough energy E_T and small b, three roots, $R_0 > R_0' > R_0''$, exist and χ is finite. If the

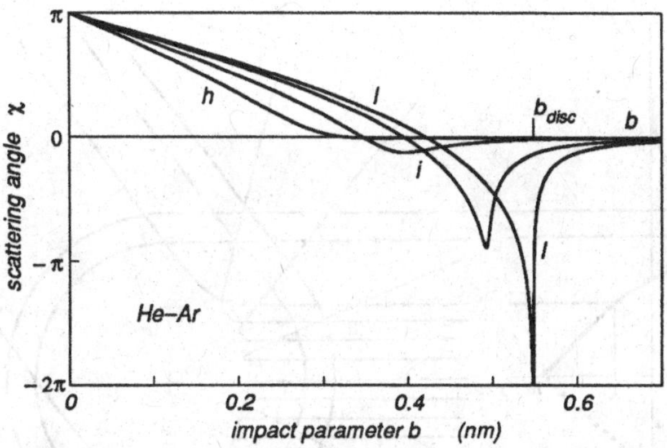

Fig. 2.3. Computed classical deflection function, $\chi(b)$, of He–Ar pairs at a low (l), two intermediate (i) and a high (h) translational speed.

impact parameter b is increased substantially, only one root R_0 persists. Between these extreme cases, a particular $b = b_{\text{disc}}$ value exists for which 'orbiting' occurs; the scattering angle χ goes through $-\infty$. Orbiting are the classical analog of scattering resonances.

For purely repulsive potentials, the scattering cross section is obtained from the inverse of the scattering function, $\chi(b; E_T)$, according to

$$\frac{\partial Q_{12}}{\partial \Omega} = b(\chi) \left| \frac{\partial b}{\partial \chi} \right| \frac{1}{\sin \chi} . \tag{2.3}$$

However, for realistic intermolecular potentials and fixed energy E_T, the function $b(\chi)$ is multivalued for negative χ. The measurement does, of course, not distinguish between negative and positive scattering angles; only the absolute magnitude $|\chi|$ is of practical interest. Furthermore, as the intermediate (i) and low-energy (l) curves of Fig. 2.3 suggest, χ may become smaller than $-\pi$ and may actually have a singularity (orbiting collision at $b = b_{\text{disc}}$). Measurable scattering angles are, however, limited to a range from 0 to π radians, and the cross section must be obtained by summing over all χ, $-\chi$, $\chi + 2\pi$, $-(\chi + 2\pi)$, ..., that are equal to a given angle $\tilde{\chi}$, the observable scattering angle in the range from 0 to π. This fact gives rise to interesting interference patterns when one accounts for the wave nature of the atoms.

Semiclassical scattering theory. A semiclassical theory of scattering has been developed in which the classical atoms are replaced by wave packets which are capable of interference and which follow the classical paths [32]. A number of very striking experimental facts, e.g., glory and rainbow scattering, can thus be modeled by a simple theory. These experimental structures arise from an interference of the various trajectories sketched in Fig. 2.4 that are associated with the same scattering angles (glory

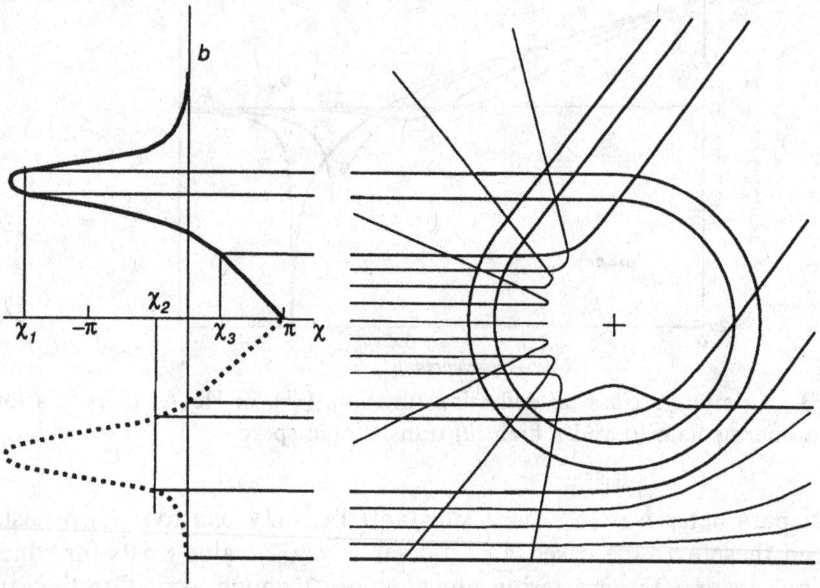

Fig. 2.4. Classical deflection function $b(\chi)$ (to the left) and trajectories (to the right) for atom–atom scattering (schematic); the function $-b(\chi)$ is also plotted (dots). Forward scattering occurs not only at large b values, but also at an intermediate b where $\chi = 0$ (glory scattering; lower part of the figure). Furthermore, the angles $\chi_1 = -2\pi + \chi_0$, $\chi_2 = -\chi_0$ (which occurs twice, for two different b values: lower part of the figure), and $\chi_3 = \chi_0$ all correspond to the same deflection angle, χ_0 (glory scattering). Trajectories that lead to the same deflection angle give rise to striking interference patterns when one accounts for the wave nature of the atoms. Furthermore, the singularity $\mathrm{d}\chi/\mathrm{d}b = 0$ causes a structure in the differential cross section, $\partial Q/\partial\Omega$ at $\chi = \chi_r$ (rainbow scattering), Eq. 2.3.

scattering); the singularity $\mathrm{d}b/\mathrm{d}\chi = \infty$ is also discernible in the cross section, Eq. 2.3 (rainbow scattering).

Quantum theory of scattering. The differential scattering cross section for isotropic potentials is given by the scattering phase shifts η_ℓ as

$$\frac{\partial Q_{12}}{\partial\Omega} = (2\mathrm{i}\kappa)^{-1} \sum_{\ell=0}^{\infty} (2\ell + 1)\, P_\ell(\cos\chi)\, \left[\exp(2\mathrm{i}\eta_\ell) - 1\right] . \qquad (2.4)$$

The $P_\ell(\cos\chi)$ are Legendre polynomials and the magnitude of the wave vector κ is given by $(\hbar\kappa)^2 = 2m_{12}E_T$. This expression is consistent with all experimental facts, including the glory and rainbow interference effects mentioned above. Moreover, scattering or 'shape' resonances occur which involve a temporary complex of two molecules which interact for a time longer than the duration of the fly-by encounter; shape resonances are the quantum analog of the classical orbiting collision. The quantum

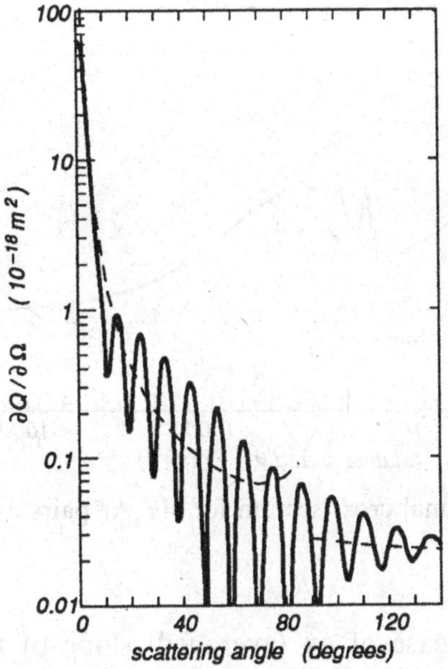

Fig. 2.5. Classical and quantum calculation of the scattering of H_2 by Hg in the idealization where both systems are considered to have zero angular momentum. In actuality, the undulations will be partly washed out by effects associated with the rotation of the molecule, the spin of the protons, and the energy spread in the 'monoenergetic' beam; after [30].

differential scattering cross section differs from the classical result also by a forward scattering component (Hirschfelder *et al.* 1964).

Differential scattering cross section. Figure 2.5 shows a comparison of the computed classical and quantal differential scattering cross sections (H_2–Hg; schematic). There is a strong forward scattering in both the classical and the quantum theory. However, the quantal cross section undulates with increasing scattering angle, which the classical function does not; on the average, quantum and classical results have a clear correspondence. Near 89°, the classical rainbow angle, a singularity is discernible.

Total scattering cross section. The total scattering cross section is given by the integral of the differential scattering cross section,

$$Q_{12} = 2\pi \int_0^{\pi} \frac{\partial Q_{12}}{\partial \Omega} \sin\chi \, d\chi \, . \tag{2.5}$$

It is a function of the translational energy, E_T, or alternatively, of the relative speed v_{12} of the encounter. At the smallest speeds, the attractive intermolecular forces dominate and the cross section is large. In a double logarithmic grid, $\log Q$ *versus* $\log v$, at low speeds we generally find

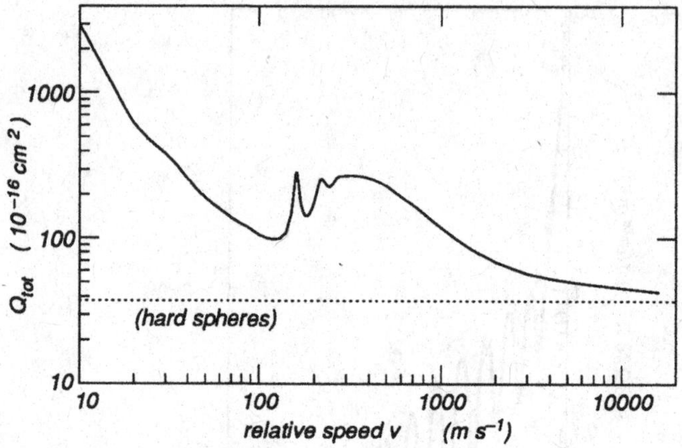

Fig. 2.6. The total collisional cross section for He–Ar pairs as function of speed v_{12} (calculation).

undulations and a decrease of an (averaged) slope of roughly –0.4. At larger speeds, where 'hard' collisions prevail, the cross section decreases rather smoothly, roughly with a slope of –0.2. In any case, the total collisional cross sections as function of speed v_{12} behave unlike classical hard spheres which are independent of speed, $Q_{12} = \pi(\sigma_{12})^2$. Figure 2.6 shows an example of a calculated cross section (He–Ar, SPDF2 potential); the peaks near 160 and 220 m s^{-1} are due to shape resonances.

Mean free path: If a beam of monochromatic particles of type 1 is incident on a rarefied gas sample of molecules, type 2, of the density n_2, the beam intensity falls off exponentially,

$$I_1(x) = I_0 \exp(-x/\lambda_{12}) .$$

In this expression, λ_{12} is the mean free path of the beam particles, which is related to the cross section and gas density according to

$$\lambda_{12} = 1/n_2\, Q_{12} . \tag{2.6}$$

The probability for a collision in dx is given by dx/λ_{12} for d$x \ll \lambda_{12}$.

In a gas or a mixture of two gases 1 and 2, the mean free path must be obtained as an average over the velocity distribution. We quote here a simple formula based on the (crude) hard sphere approximation,

$$\lambda_{12} = \left[n_2(\pi\sigma_{12}^2)\,(1 + m_1/m_2)^{1/2}\right]^{-1} \tag{2.7}$$

with σ_1, σ_2 designating the radii of the colliding atoms and m_1, m_2 their masses. Furthermore, $\sigma_{12} = \sigma_1 + \sigma_2$ is the sum of the hard sphere radii (Chapman and Cowling 1980).

Frequency of collisions. The mean frequency of collisions is similarly expressed in the hard spheres approximation as

$$\nu_{12} = n_2 \left(\pi\sigma_{12}^2\right) \left(8kT/\pi m_{12}\right)^{1/2} , \tag{2.8}$$

where $m_{12} = m_1 m_2/(m_1 + m_2)$ is the reduced mass.

Time scales. For an understanding of spectral line shapes of induced absorption, at not too high gas densities, it is useful to distinguish three different times associated with collisions, namely the average time between collisions, the duration of a molecular fly-by and the duration of the spectroscopic interaction.

A sufficiently rarefied gas, or a mixture of gases, consists of a number of neutral molecules of species 1 and 2 (which may or may not be the same). We may assume a distribution of velocities (measured in the laboratory frame), $f_i(\boldsymbol{v}_i)\ \mathrm{d}^3 v_i$ that may be modeled by a Maxwellian distribution function, with $i = 1$ or 2, as long as the duration of the average collision is short compared to the time between collisions. For binary collisions, one usually transforms from laboratory coordinates, $\boldsymbol{v}_i$, to relative ($\boldsymbol{v}_{12}$) and center-of-mass ($\boldsymbol{v}_{CM}$) velocities,

$$\begin{aligned} \boldsymbol{v}_{12} &= \boldsymbol{v}_1 - \boldsymbol{v}_2 \\ \boldsymbol{v}_{CM} &= (m_1\boldsymbol{v}_1 + m_2\boldsymbol{v}_2)/(m_1 + m_2) , \end{aligned} \tag{2.9}$$

which are the time derivatives of the transforms to center-of-mass and relative position, see p. 207. The m_i are the masses of species i. Energy is conserved,

$$\frac{1}{2}m_1\boldsymbol{v}_1^2 + \frac{1}{2}m_2\boldsymbol{v}_2^2 = \frac{1}{2}m_{12}\boldsymbol{v}_{12}^2 + \frac{1}{2}(m_1 + m_2)\,\boldsymbol{v}_{CM}^2 .$$

If the laboratory speeds $\boldsymbol{v}_i$ are distributed according to Maxwell's law, relative (and also center-of-mass) speeds are likewise,

$$f_{12}(\boldsymbol{v}_{12}) = \left(\frac{m_{12}}{2\pi kT}\right)^{3/2} \exp\left(\frac{-m_{12}\boldsymbol{v}_{12}^2}{2kT}\right) , \tag{2.10}$$

so that $f_{12}(\boldsymbol{v}_{12})\ \mathrm{d}^3 v_{12}$ is the probability of finding a pair of atoms with relative speeds between $\boldsymbol{v}_{12}$ and $\boldsymbol{v}_{12} + \mathrm{d}\boldsymbol{v}_{12}$.* The masses m_i are now replaced by reduced mass (and by total mass in the latter case). The volume elements in velocity space transform as

$$\mathrm{d}^3 v_1\ \mathrm{d}^3 v_2 = \mathrm{d}^3 v_{12}\ \mathrm{d}^3 v_{CM} ,$$

* In our notation, $\mathrm{d}\boldsymbol{v}$ designates the vector with the Cartesian components $\{\mathrm{d}v_x; \mathrm{d}v_y; \mathrm{d}v_z\}$. On the other hand, $\mathrm{d}^3 v = \mathrm{d}v_x\,\mathrm{d}v_y\,\mathrm{d}v_z$ designates a volume element in velocity space; boldface is not used for scalars.

where $d^3v = v^2\, dv\, \sin\vartheta\, d\vartheta\, d\varphi$ is the volume element in velocity space. The polar angle ϑ varies from 0 to π and the azimuthal angle φ from 0 to 2π.

i.) The *mean time between collisions*, τ_{12}, is given by the reciprocal collision frequency, ν_{12}, which in turn is defined in terms of the collisional cross section, $Q_{12}(v_{12})$, the distribution function of relative velocities, Eq. 2.10, and the number density of species 2, n_2, according to

$$\nu_{12} = n_2 \int_0^\infty v_{12}\, Q_{12}(v_{12})\, f_{12}(v_{12})\, d^3v_{12} \Big/ \int_0^\infty f_{12}(v_{12})\, d^3v_{12} \,, \quad (2.11)$$

For rough estimates, the collisional cross section may be assumed to be velocity-independent, $Q_{12}(v_{12}) \equiv Q_0 =$ constant (hard-sphere approximation), so that the mean time between collisions becomes

$$\tau_{12} = \nu_{12}^{-1} = (\bar{v}_{12}\, n_2\, Q_0)^{-1} \quad \text{with} \quad \bar{v}_{12} = [8kT/(\pi m_{12})]^{1/2} \,. \quad (2.12)$$

The average speed is obtained with the assumption of a Maxwellian velocity distribution. The average relative speed of a collisional pair 1–2 is of the order of the root-mean-square speed, which is roughly equal to the speed of sound and varies with temperature as $T^{-1/2}$. For hydrogen, nitrogen and xenon gas the mean time between collisions amounts to roughly 0.07, 0.14 and 0.17×10^{-9} seconds, respectively, at standard temperature (273 K) and pressure (1 atmosphere). The mean time between collisions decreases with temperature T roughly as $T^{-1/2}$ and decreases with density as $1/n$.

ii.) The *duration of a fly-by* encounter is given by

$$\Delta t \approx \sigma/\bar{v}_{12} \,, \quad (2.13)$$

where $\sigma = \sigma_1 + \sigma_2$ designates the sum of the diameters of the interacting atoms (hard sphere approximation). It may be approximated by the root of the intermolecular interaction potential, $V_0(\sigma) = 0$. This duration varies as $T^{-1/2}$ and is density invariant. For hydrogen, nitrogen and xenon gas at standard temperature, this time amounts roughly to $\Delta t = 1.6$, 8, and 22×10^{-13} seconds, respectively.

iii.) A mean *duration of optical interactions* may be defined with the assumption that the square of the induced dipole function, $\mu(R)$, is proportional to the probability of absorbing a photon during an encounter; spectral intensities certainly are proportional to dipole moment squared. With this assumption, a range of optical interactions may be defined, according to

$$R_S = \int_{R_{\min}}^\infty (R - R_{\min})\, (\mu(R))^2\, dR \Big/ \int_{R_{\min}}^\infty (\mu(R))^2\, dR \,. \quad (2.14)$$

In this expression, $R_{\min}$ designates the distance of closest approach which is of the order of the root σ of the interaction potential. The duration of optical interactions is then simply given by

$$\Delta t' = R_S/\bar{v}_{12}\,. \tag{2.15}$$

In Chapter 4 we will see that two types of induced dipole functions are of a special importance: the overlap-induced dipole, Eq. 4.2, an exponential with a range $R_0 \approx 0.1\sigma$, and the multipole-induced dipole, Eq. 4.3, which falls off as R^{-N} $(N = 4, 5, \ldots)$. For these, the optical range becomes

$$R_S^{(ov)} \approx 0.05\sigma \qquad \text{and} \qquad R_S^{(m)} \approx R_{\min}/(2N)$$

with $R_{\min} \approx \sigma$. The range of quadrupolar induction ($N = 4$) is roughly 2.5 times greater than that of overlap induction; but both amount to only a small fraction of the molecular diameter σ which determines the duration of the fly-by encounter. On the other hand, spectroscopically effective molecular encounters are more nearly described by a head-on collision approximation than a straight, distant trajectory approximation; these may typically be of shorter duration than fly-bys. Typical durations vary as $T^{-1/2}$, are density independent and amount to roughly $\Delta t \approx 10^{-12}$ s for the fly-by, and $\Delta t' \approx 10^{-13}$ s for the spectroscopic interactions at room temperature.

Near standard temperature and pressure, the time between collisions is roughly three orders of magnitude greater than the duration of a collision. Under such conditions, a Maxwellian distribution of velocities is a reasonable assumption and the concept of binary interactions appears to be a sound one, if interactions have a range not much greater than the size of the atoms. With increasing density n, the time between collisions decreases as n^{-1}; at densities approaching liquid state (that is: several hundred amagats), the time between collisions and the duration of a collision become comparable. Under such conditions, each particle is simultaneously interacting with several partners. The time scales considered here lose meaning in fluids that are so dense that many-body interactions are significant.

2.3 Dimers and larger clusters

Intermolecular forces are repulsive at close range ($R < R_{\min}$) and attractive at distant range ($R > R_{\min}$; $R_{\min}$ is the position of the minimum of the pair interaction potential). At high enough densities most of the common gases form dimers, also called double molecules or (binary) van der Waals molecules. Higher than binary complexes are expected, too, especially at high densities and temperatures comparable to, or lower than the well depth of the interaction potential. (A few systems which do not form dimers are mentioned briefly on p. 63.) For simplicity we assume for the following a one-component gas composed of isotropic molecules, along with pairwise additive interactions.

The key to a treatment of molecular clusters in situations of thermal equilibrium are the N-particle partition functions. Specifically, the classical two-particle partition function, $Z_2(T)$, is given by [183, 184, 377]

$$Z_2(T) = \frac{v}{2\lambda_1^6} \int_v \exp\left(-V(R)/kT\right) 4\pi R^2 \,\mathrm{d}R \approx \frac{v^2}{2\lambda_1^6} \,; \tag{2.16}$$

$V(R)$ is the pair interaction potential, T the temperature and k the Boltzmann factor. The integral to the right is over the volume v of the container; it is often referred to as the configuration integral, $Q_2(T)$. For large enough volumes, the integral may be replaced by the volume. With the help of Hill's effective potentials, Eq. 2.28, this expression may be written as a sum of free and bound pair state sums, respectively [183, 184],

$$Z_2 = Z_{2f} + Z_{2b} \,, \tag{2.17}$$

with

$$Z_{2b}(T) = \frac{v}{2\lambda_1^6} \int_{v,V(R)\leq 0} \exp\left(-V(R)/kT\right) 4\pi R^2 F(R) \,\mathrm{d}R \tag{2.18}$$

$$Z_{2f}(T) = \frac{v}{2\lambda_1^6} \left\{ \int_{v,V(R)>0} \exp\left(-V(R)/kT\right) 4\pi R^2 \,\mathrm{d}R \right. \tag{2.19}$$

$$\left. + \int_{v,V(R)\leq 0} \exp\left(-V(R)/kT\right) (1 - F(R)) \, 4\pi R^2 \,\mathrm{d}R \right\} .$$

The function $F(R)$ is given by the normalized, incomplete Γ function,

$$F(R) = \frac{2}{\sqrt{\pi}} \int_0^x \exp(-y)\, y^{a-1} \,\mathrm{d}y$$

with $x = -V(R)/kT$ and $a = 3/2$. For the following, we will distinguish three de Broglie wavelengths,

$$\lambda_1^2 = \frac{2\pi\hbar^2}{m_1 kT} \qquad \lambda_0^2 = \frac{2\pi\hbar^2(m_1+m_2)}{m_1 m_2 kT} \qquad \lambda_{\mathrm{CM}}^2 = \frac{2\pi\hbar^2}{(m_1+m_2)kT} \tag{2.20}$$

where m_1, m_2 are monomer masses (assumed here to be equal). In other words, λ_1, λ_0, and λ_{CM} are thermal de Broglie wavelengths of monomers, of a particle of reduced mass, $m = m_1 m_2/(m_1 + m_2)$, and of the center of mass, respectively. We note that $\lambda_1^6 = \lambda_{\mathrm{CM}}^3 \lambda_0^3$.

The corresponding quantum expressions are

$$Z_{2f}(T) = \frac{v}{2\lambda_{\mathrm{CM}}^3 \lambda_0^3} \sum_{\ell=0}^{\infty} g_\ell \, (2\ell+1) \int_v 4\pi R^2 \,\mathrm{d}R \tag{2.21}$$

$$\times \frac{1}{4\pi R^2} \int_0^\infty dE \ |\psi_{E\ell}(R)|^2 \exp(-E/kT)$$

$$Z_{2b}(T) = \frac{v}{2\lambda_{CM}^3} \sum_{E(v,\ell)<0} g_\ell \, (2\ell+1) \exp(-E(v,\ell)/kT) \ . \tag{2.22}$$

The $\psi_{E\ell}(R)$ are the radial free-state wavefunctions (see Chapter 5 for details). The free state energies E are positive and the bound state energies $E(v,\ell)$ are negative; v and ℓ are vibrational and rotational dimer quantum numbers; ℓ is also the angular momentum quantum number of the ℓth partial wave. The g_ℓ are nuclear weights. We will occasionally refer to a third partition sum, that of pre-dissociating (sometimes called 'metastable') dimer states,

$$Z_{2m}(T) = \frac{v}{2\lambda_{CM}^3} \sum_{E(v,\ell)>0} g_\ell \, (2\ell+1) \exp(-E(v,\ell)/kT) \ , \tag{2.23}$$

a sum over the pre-dissociating (quasi-bound) states of positive energy, $E(v,\ell) > 0$. These states are embedded in the continuum and would, therefore, normally have to be included in the expression for Z_{2f}. However, as a practical matter, the longer-lived of these unstable states behave much like bound states and may well be included in the sum, Eq. 2.22. The strongly coupled, high-lying pre-dissociating states, on the other hand, behave more like free states and are included quite naturally in Z_{2f} [149]. Classical expressions for Z_{2m} have been given elsewhere [377].

Examination of Eqs. 2.18, 2.19 or 2.21, 2.22 reveals that the radial integral in the expression for Z_{2f} is roughly proportional to the volume v so that Z_{2f} varies nearly as v^2. The bound states partition function, Z_{2b}, on the other hand, depends roughly linearly on volume.

Dimer concentrations. With the pair partition sums thus defined, we may compute the bound dimer concentrations from the mass action law,

$$\frac{N_{2b}}{N_2} = \frac{Z_{2b}(T)}{Z_2(T)} \approx \frac{Z_{2b}(T)\lambda_1^6}{v^2} \ . \tag{2.24}$$

N_{2b} is the number of bound dimers in the volume v and $N_2 = N(N-1)/2$ is the number of pairs (bound or free) in the volume v. Note that the right-hand side of this expression is roughly proportional to the reciprocal volume, $1/v$. We may, therefore, write the expression to the left in terms of dimer and monomer densities, $n_2 = N_2/v$ and $n_1 = N/v$, respectively [27],

$$\frac{n_2}{n_1^2} = K(T) \qquad \text{with} \qquad K(T) \approx Z_{2b}(T)\,\frac{\lambda_1^6}{2v} \ , \tag{2.25}$$

if $N \gg 1$. Note that $K(T)$ is nearly independent of volume v so that in the integral, Eq. 2.18, the limit $v \to \infty$ may be conveniently substituted.

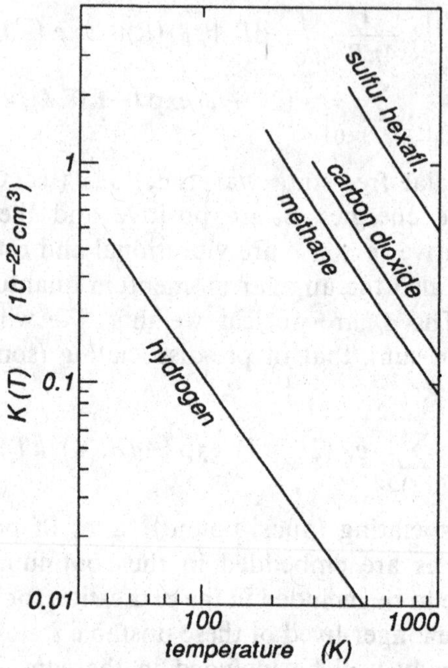

Fig. 2.7. Calculated mass action constant $K(T)$ serves to estimate dimer concentrations in hydrogen, methane, carbon dioxide and sulfur hexafluoride as function of temperature.

Figure 2.7 shows computed mass action constants $K(T)$ for hydrogen, where dimer concentrations are generally small, and a few other gases with generally much higher dimer concentrations, methane, carbon dioxide, and sulfur hexafluoride. The constant $K(T)$ is given as a function of temperature. Dimer concentrations n_2/n_1 at one amagat ($n_1 \approx 2.68 \times 10^{19}$ cm^{-3}) may be as high as 0.5% at the lowest temperatures shown, but if gas densities are increased to 10 or 20 amagats dimer concentrations may amount to 5 or 10% in cases of practical interest.

N-particle systems. The classical partition function $Z_N(T)$ of a canonical ensemble of N molecules is given by [184] (Hirschfelder *et al.* 1956)

$$Z_N(T) = \frac{1}{h^{3N}N!} \int \cdots \int \exp\left(-\mathscr{H}_N/kT\right)\, \mathrm{d}^3p_1 \cdots \mathrm{d}^3p_N\, \mathrm{d}^3R_1 \cdots \mathrm{d}^3R_N\,,$$

with the Hamiltonian

$$\mathscr{H}_N = \sum_{i=1}^{N} \frac{1}{2m}\boldsymbol{p}_i^2 + V\left(\boldsymbol{R}_1 \cdots \boldsymbol{R}_N\right)\,.$$

The $\boldsymbol{p}_i$ and $\boldsymbol{R}_i$ are momentum and position coordinates of particle i. The $\boldsymbol{R}_i$ integrations are taken over the volume v. Z_N is thus a function of N,

v, and T. If the integration over the momenta is carried out, we have

$$Z_N(T) = \frac{1}{N!\lambda_1^{3N}} \times \int \cdots \int \exp\left(-V(\boldsymbol{R}_1 \cdots \boldsymbol{R}_N)/kT\right) \mathrm{d}^3R_1 \cdots \mathrm{d}^3R_N . \tag{2.26}$$

Similar expressions exist for the quantum partition functions.

Free single particle partition sum. The one-particle sum over states of the ideal gas is easily evaluated if the free-particle Hamiltonian (with $V(R) \equiv 0$) is used,

$$Z_1(T) = \frac{1}{h^3}\int_v \exp\left[-\left(p_x^2+p_y^2+p_z^2\right)/2mkT\right] \mathrm{d}^3R\, \mathrm{d}^3p = v/\lambda_1^3 . \tag{2.27}$$

This result is also correct for the center of mass motion of the pair if λ_1 is replaced by λ_{CM}. We will also use this expression for the sum over states $Z_r = Z_2 2\lambda_{CM}^3/v$ of relative motion, where $\lambda_{CM}^3\lambda_0^3$ is substituted for λ_1^6, Eq. 2.16.

Hill's effective potentials. The division of phase space into free and bound pairs is possible by introducing Hill's effective potentials [183, 184],

$$V_b(R) = \begin{cases} V(R) - kT\log_e F(R) & \text{if } V(R) \le 0 \\ \infty & \text{if } V(R) > 0 \end{cases} \tag{2.28}$$

$$V_f(R) = \begin{cases} V(R) & \text{if } V(R) > 0 \\ V(R) - kT\log_e(1-F(R)) & \text{if } V(R) \le 0 , \end{cases}$$

so that

$$\exp(-V(R)/kT) = \exp(-V_b(R)/kT) + \exp(-V_f(R)/kT) .$$

This substitution has permitted us to write Z_2 as a sum of free and bound pair partition functions, Eq. 2.17.

Larger clusters. With the assumptions made, the triplet partition sum is given by [184]

$$Z_3(T) = \frac{1}{3!\lambda_1^9} \times \int_v \exp\left[-\left(V(12)+V(13)+V(23)\right)/kT\right] \mathrm{d}^3R_1\, \mathrm{d}^3R_2\, \mathrm{d}^3R_3 ; \tag{2.29}$$

pairwise additive, isotropic potentials, $V(ik) = V(|\boldsymbol{R}_i - \boldsymbol{R}_k|)$, are here assumed, with $i \neq k$ and $i, k = 1, 2, 3$. We can now use the substitution, Eq. 2.28, to separate the various bound and free state contributions,

$$Z_3 = Z_{3f} + Z_{2b,1f} + Z_{3b} , \tag{2.30}$$

representing the state sums of three free particles, two bound plus one free particles, and three bound particles, respectively. For three free particles we get, for example,

$$Z_{3f}(T) = \frac{1}{3!\lambda_1^9} \times \int_v \exp\left[-\left(V_f(12)+V_f(13)+V_f(23)\right)/kT\right] \, d^3R_1 \, d^3R_2 \, d^3R_3 \; ; \tag{2.31}$$

similar expressions may be obtained for $Z_{2b,1f}$ and Z_{3b}, and also for the partition sums of more than three particles.

Population factors of pairs. The probability of finding a pair of molecules in the state $|E\ell\rangle$ of relative motion is given by a Boltzmann factor like

$$P = g_\ell\,(2\ell+1)\,\exp(-E/kT)\,/Z_r\,. \tag{2.32}$$

The Z_r normalizes the expression; it is chosen so that the sum of P over all free and bound states equals unity. In other words, Z_r equals the pair partition sum of relative motion, that is Eq. 2.16 or the sum of Eqs. 2.21 and 2.22, with the center of mass contribution $v/\lambda_{\mathrm{CM}}^3$ suppressed. (The for like pairs characteristic factor of 1/2 is also commonly suppressed at this point.)

Under conditions of low dimer concentrations, i.e., at low pressures and/or high temperatures, especially for the lighter gases, the free-particle expressions, Eq. 2.27, may be substituted, $Z_r \approx v/\lambda_0^3$.

2.4 Distribution functions

The probability that two atoms 1,2 are separated by the distance R is given by the pair distribution function, $g^{(12)}(R)$, defined as

$$g^{(12)}(R) = V^2\,\langle \delta(\mathbf{R}_2 - \mathbf{q}_1)\,\delta(\mathbf{R}_1 - \mathbf{q}_2)\rangle\ ,$$

with $\mathbf{R} = \mathbf{R}_2 - \mathbf{R}_1$, $R = |\mathbf{R}|$; the $\mathbf{q}_1$, $\mathbf{q}_2$, are the laboratory position coordinates of atoms 1 and 2.

The triplets distribution function, $g^{(112)}(R, R', \cos\vartheta)$, is proportional to the probability that one atom 2 and two atoms 1 form a triangle specified by R, R', $\cos\vartheta$,

$$g^{(112)}(R, R', \cos\vartheta) = V^3\,\langle \delta(\mathbf{R}_3 - \mathbf{q}_{1'})\,\delta(\mathbf{R}_2 - \mathbf{q}_1)\,\delta(\mathbf{R}_1 - \mathbf{q}_2)\rangle\ ,$$

with $\mathbf{R}' = \mathbf{R}_3 - \mathbf{R}_1$; ϑ is the angle subtended by the vectors $\mathbf{R}$ and $\mathbf{R}'$.

The virial expansions of these distribution functions [135] are given by

$$\begin{aligned} g^{(12)}(R) &= g_0^{(12)}(R) + n_1\,g_{(1)}^{(12)}(R) + n_2\,g_{(2)}^{(12)}(R) + \ldots \\ g^{(112)}(R, R', \cos\vartheta) &= g_0^{(112)}(R, R', \cos\vartheta) + \ldots \end{aligned} \tag{2.33}$$

with the numerical density $n_1 = N_L \varrho^{(1)}$. The $g_0^{(12)}$, $g_{(1)}^{(12)}$, and $g_0^{(112)}$ can be written in terms of the auxiliary functions W,

$$\begin{aligned} g_0^{(12)}(R) &= W_{12}(\boldsymbol{R}_1, \boldsymbol{R}_2) \\ g_{(1)}^{(12)}(R) &= \int d^3R' \, \{W_{112}(\boldsymbol{R}', \boldsymbol{R}_2, \boldsymbol{R}_1) - W_{12}(\boldsymbol{R}_2, \boldsymbol{R}_1)\, W_{12}(\boldsymbol{R}', \boldsymbol{R}_1) \\ &\quad -W_{12}(\boldsymbol{R}_2, \boldsymbol{R}_1)\, W_{11}(\boldsymbol{R}', \boldsymbol{R}_2) + W_{12}(\boldsymbol{R}_2, \boldsymbol{R}_1)\} \end{aligned} \tag{2.34}$$

$$g_0^{(112)}(R, R', \cos\vartheta) = W_{112}(\boldsymbol{R}_3, \boldsymbol{R}_2, \boldsymbol{R}_1)\,.$$

The function W_{12} is defined according to

$$W_{12}(\boldsymbol{R}_2, \boldsymbol{R}_1) = \lambda_1^3 \lambda_2^3 \operatorname{Tr}\left\{\exp\left(-\mathscr{H}_{12}/kT\right)\, \delta(\boldsymbol{R}_2 - \boldsymbol{q}_1)\, \delta(\boldsymbol{R}_1 - \boldsymbol{q}_2)\right\}, \tag{2.35}$$

where Tr means the trace over the states of a pair of isolated atoms 1,2. $\mathscr{H}_{12}$ is the Hamiltonian of the pair, λ_1 and λ_2 are the thermal de Broglie wavelengths associated with the two atoms or molecules, 1 and 2.

W_{11} is given by an expression similar to Eq. 2.35, with atom 2 replaced by another atom of species 1. For the three-body function, we get

$$\begin{aligned} &W_{112}(\boldsymbol{R}_3, \boldsymbol{R}_2, \boldsymbol{R}_1) = \\ &\quad \lambda_1^6 \lambda_2^3 \operatorname{Tr}\left\{\exp\left(-\mathscr{H}_{112}/kT\right)\, \delta(\boldsymbol{R}_3 - \boldsymbol{q}_1)\, \delta(\boldsymbol{R}_2 - \boldsymbol{q}_1)\, \delta(\boldsymbol{R}_1 - \boldsymbol{q}_2)\right\} \end{aligned} \tag{2.36}$$

where Tr now means the trace over all the states of a system composed of atom 1 and atom 2.

The W functions should be calculated on the basis of quantum mechanics. However, at sufficiently high temperatures and for massive systems, classical or semi-classical expressions in terms of interatomic potential functions, $V_{12}(R)$, etc., are useful. Specifically, in the classical limit, we may write the pair distribution function as

$$\begin{aligned} W_{12}(\boldsymbol{R}_2, \boldsymbol{R}_1) &\rightarrow \exp\left(-V_{12}(R)/kT\right) \\ W_{112}(\boldsymbol{R}_3, \boldsymbol{R}_2, \boldsymbol{R}_1) &\rightarrow \exp\left(-V_{112}(R, R', \cos\vartheta)/kT\right)\,, \end{aligned}$$

with W_{112} designating the three-body interaction potential; the two members of species 1 are located at $\boldsymbol{R}_3$ and $\boldsymbol{R}_2$, respectively, and 2 is at $\boldsymbol{R}_1$. If we neglect the irreducible three-body contributions to the interaction potential, we can write V_{112} as a sum of pairwise additive components as

$$W_{112}(\boldsymbol{R}_3, \boldsymbol{R}_2, \boldsymbol{R}_1) \rightarrow W_{12}(R)\, W_{12}(R')\, W_{11}(R'')\,,$$

where $R'' = |\boldsymbol{R}_3 - \boldsymbol{R}_2| = (R^2 + R'^2 - 2RR'\cos\vartheta)^{1/2}$.

Quantum effects, if small, may be taken into account by expanding the functions W in powers of $\hbar$. This can be accomplished by using the Wigner–Kirkwood expansion to calculate the traces which appear in Eqs. 2.35 and 2.36. In fact, if $\mathscr{H}^{(n)}$ indicates the Hamiltonian of a system of

n atoms whose positions we collectively indicate by $\boldsymbol{q}^{(n)}$, we have [196, 302]

$$\mathrm{Tr}\left\{\exp\left(-\mathscr{H}^{(n)}/kT\right)\,F\left(\boldsymbol{q}^{(n)}\right)\right\} = \frac{1}{h^n}\int \mathrm{d}^{3n}p\,\mathrm{d}^{3n}q\,F\left(\boldsymbol{q}^{(n)}\right)$$
$$\times \exp\left(-\mathscr{H}(\boldsymbol{p}^{(n)},\,\boldsymbol{q}^{(n)})/kT\right)\left[1+\hbar^2\chi\left(\boldsymbol{p}^{(n)},\,\boldsymbol{q}^{(n)}\right)+\mathcal{O}\left(\hbar^4\right)\right],$$

with χ given by [302]

$$\chi = \frac{1}{24m}\left[\frac{1}{m\,(kT)^3}\left(\boldsymbol{p}\cdot\frac{\partial}{\partial \boldsymbol{R}}\right)V_{12}(\boldsymbol{R}) - \frac{3}{(kT)^2}\frac{\partial^2 V_{12}}{\partial \boldsymbol{R}^2} + \frac{1}{(kT)^3}\left(\frac{\partial V_{12}}{\partial \boldsymbol{R}}\right)\right].$$

The notation $\partial/\partial\boldsymbol{R}$ designates the gradient operator, also called nabla ($\nabla_{\boldsymbol{R}}$). We thus finally obtain from Eqs. 2.35 and 2.36 [296]

$$g_0^{(12)}(R) = \exp\left(-V_{12}(R)/kT\right)\left[1+\hbar^2 A_{12}(R)+\mathcal{O}(\hbar^4)\right] \qquad (2.37)$$

$$\begin{aligned} g_0^{(112)}(R) = {} & \exp\left[-\left(V_{12}(R)+V_{12}(R')+V_{11}(R'')\right)/kT\right] \\ & \times\left\{1+\hbar^2\left[A_{12}(R)+A_{12}(R')+A_{11}(R'')\right.\right. \\ & \left.\left.+B_{112}(R,\,R',\,\cos\vartheta)\right]+\mathcal{O}(\hbar^4)\right\}, \end{aligned} \qquad (2.38)$$

because $g_0^{(12)}(R) = W_{12}(\boldsymbol{R}_2,\,\boldsymbol{R}_1)$ and $g_0^{(112)}(R) = W_{112}(\boldsymbol{R}_3,\,\boldsymbol{R}_2,\,\boldsymbol{R}_1)$. Furthermore, we have set

$$\begin{aligned} A_{12}(R) = {} & \frac{1}{24m_{12}(kT)^2}\times \\ & \left[\frac{1}{kT}\left(\frac{\mathrm{d}V_{12}(R)}{\mathrm{d}R}\right)^2 - 2\frac{\mathrm{d}^2V_{12}(R)}{\mathrm{d}R^2} - \frac{4}{R}\frac{\mathrm{d}V_{12}(R)}{\mathrm{d}R}\right] \\ B_{112}(R,\,R',\,\cos\vartheta) = {} & \frac{1}{12(kT)^3}\left[\frac{\mathrm{d}V_{12}(R)}{\mathrm{d}R}\frac{\mathrm{d}V_{12}(R')}{\mathrm{d}R'}\frac{\cos\vartheta}{m_2}+\right. \\ & +\frac{\mathrm{d}V_{12}(R)}{\mathrm{d}R}\frac{\mathrm{d}V_{11}(R'')}{\mathrm{d}R''}\frac{1}{m_1}\left(\frac{R}{R''}-\frac{R'\cos\vartheta}{R''}\right) \\ & \left.+\frac{\mathrm{d}V_{12}(R')}{\mathrm{d}R'}\frac{\mathrm{d}V_{11}(R'')}{\mathrm{d}R''}\frac{1}{m_1}\left(\frac{R'}{R''}-\frac{R\cos\vartheta}{R''}\right)\right]. \end{aligned}$$

The separation R'' is a function of R, R', $\cos\vartheta$. From these equations it is seen that W_{112} cannot be written as a product of the W functions, as $W_{12}\,W_{12}\,W_{11}$. In fact, even in the case of pairwise-additive potentials, we have, up to order $\hbar^2$,

$$\frac{W_{112}(R,\,R',\cos\vartheta)}{W_{12}(R)\,W_{12}(R')\,W_{11}(R'')} = 1+\hbar^2\,B_{112}(R,\,R',\,\cos\vartheta)\,.$$

Finally, we rewrite Eq. 2.34 to include terms to order $\hbar^2$, as [296]

$$g_{(1)}^{(12)}(R) = 2\pi\,\exp\left[-V_{12}(R)/kT\right]\int_0^\infty (R')^2\,\mathrm{d}R'\int_0^\pi \sin\vartheta\,\mathrm{d}\vartheta \qquad (2.39)$$

$$\times \{\exp(-V_{12}(R')/kT)\, \exp(-V_{11}(R'')/kT)$$
$$\times \left[1 + \hbar^2 \left(A_{12}(R) + A_{12}(R') + A_{11}(R'') + B_{112}(R, R', \cos\vartheta)\right)\right]$$
$$-\exp(-V_{12}(R')/kT) \left[1 + \hbar^2 \left(A_{12}(R) + A_{12}(R')\right)\right]$$
$$-\exp(-V_{11}(R'')/kT) \left[1 + \hbar^2 \left(A_{12}(R) + A_{11}(R'')\right)\right]$$
$$+1 + \hbar^2 A_{12}(R)\Big\}\,.$$

Time-dependent correlation functions. Similar pair and triplet distributions, which describe the time evolution of a system, are also known [318]. These have found interesting uses for the theory of virial expansions of spectral line shapes, pp. 225 ff. below [297, 298].

2.5 Molecular multipoles

A molecule may be viewed as a number N of nuclear charges, $Z_i e$, of a certain arrangement given by their position vectors, $\boldsymbol{R}_i$ for $i = 1 \ldots N$, surrounded by an electronic cloud of charge density $\varrho(\boldsymbol{r})$ of finite dimensions. The potential of the electrostatic field at the point $\boldsymbol{R}$ outside the electronic cloud is given by

$$\Phi(\boldsymbol{R}) = \frac{1}{4\pi\varepsilon_0}\left(\sum_{i=1}^{N} \frac{Z_i\, e}{|\boldsymbol{R} - \boldsymbol{R}_i|} - \int \frac{\varrho(\boldsymbol{r})\, \mathrm{d}^3 r}{|\boldsymbol{R} - \boldsymbol{r}|}\right). \tag{2.40}$$

The integration is over the whole charge cloud. It is common practice to express this in terms of a multipole series (Jackson 1984),

$$\Phi(\boldsymbol{R}) = \frac{1}{4\pi\varepsilon_0} \sum_{\ell,m} \left(\frac{4\pi}{2\ell + 1}\right)^{1/2} Q_{\ell m}\, Y^*_{\ell m}(\widehat{\boldsymbol{R}})\, R^{-(\ell+1)} \tag{2.41}$$

with $\boldsymbol{R} = (R, \widehat{\boldsymbol{R}})$, where $Q_{\ell,m}$ is the mth component of the multipole tensor of order ℓ of the molecular (i.e., the nuclear plus electronic) charge distribution, defined by

$$Q_{\ell m} = \left(\frac{4\pi}{2\ell + 1}\right)^{1/2} \times \left(\sum_i Z_i\, e R_i^{\ell}\, Y_{\ell m}\left(\widehat{\boldsymbol{R}}_i\right) - \int \varrho(\boldsymbol{r})\, r^{\ell}\, Y_{\ell m}(\widehat{\boldsymbol{r}})\, \mathrm{d}^3 r\right). \tag{2.42}$$

In general, the $Q_{\ell m}$ are complex quantities.† The $Y_{\ell m}$ are spherical harmonics; an asterisk designates the conjugate complex. Equation 2.41

† In other works other factors are sometimes used in the definition, Eq. 2.42; caution is advised when comparing expressions from different sources.

simply states that the field (or potential) surrounding a molecule can be written as a superposition of multipole fields (or potentials).

For a determination of the permanent multipole moments, Eq. 2.42 must be averaged over the ground state nuclear vibrational wavefunction; transition elements are also often of interest which are matrix elements between initial and final rotovibrational states. For example, for a diatomic molecule with rotovibrational states $|vJM\rangle$, the transition matrix elements $\langle v'J'M'|Q_{\ell m}|vJM\rangle$ will be of interest; a prime designates final states.

Moments with $\ell = 0$ are monopoles (i.e., the charge q_0 of molecular ions); $\ell = 1$ designates a dipole moment, $\ell = 2$ a quadrupole moment, $\ell = 3$ an octopole moment, $\ell = 4$ a hexadecapole moment, etc. A multipole of order ℓ has in general $2\ell + 1$ spherical tensor components, $Q_{\ell m}$, with $m = -\ell, -\ell + 1, \ldots, \ell$, but this number can be reduced if molecules of high symmetry are considered.

With this notation, the electric charge q_0 of a monopole equals Q_{00}. Cartesian dipole components μ_x, μ_y, μ_z, are related to the spherical tensor components as $Q_{10} = \mu_z$, $Q_{1\pm1} = \mp(\mu_x \pm \mathrm{i}\mu_y)/\sqrt{2}$, with i designating the imaginary unit. Similar relationships between Cartesian and spherical tensor components can be specified for the higher multipole moments (Gray and Gubbins 1984).

Molecular multipole components are best described in the molecular (i.e., body-fixed) frame x, y, z. For example, a dipole when aligned with the z-axis is characterized by a single number, the strength μ of the dipole. The other two independent components can then simply be expressed in terms of Euler angles or, in this case, of azimuthal and polar angle, φ and ϑ, between molecular (x, y, z) and laboratory-fixed frame (X, Y, Z).

For molecules of axial symmetry such as diatomic molecules, if the molecular axis is aligned with the z-axis, we have $Q_{20} = q_2$, the quadrupole strength; all other Q_{2m} vanish.

Symmetry Properties. Under inversion, for $\boldsymbol{R}$ being replaced by $-\boldsymbol{R}$, we have $Q_{\ell m} \to (-1)^{\ell} Q_{\ell m}$. A dipole is odd under inversion and a quadrupole is even. From the properties of spherical harmonics and the definition of the spherical harmonics, it is easy to see that $Q^*_{\ell,m} = (-1)^m Q_{\ell,-m}$. If $\Omega = \alpha, \beta, \gamma$ designates the Euler angles of the rotation carrying the laboratory frame X, Y, Z, into coincidence with the molecular frame, x, y, z, the body-fixed multipole components $Q_{\ell m}$ are related to the laboratory-fixed $Q_{\ell m}$, according to

$$Q_{\ell m} = \sum_n D^{\ell}_{mn}(\Omega)^* \, Q_{\ell n} \,. \qquad (2.43)$$

In this case, Ω denotes the orientation of the molecule in the laboratory-fixed frame. The $D^{\ell}_{mn}(\Omega)$ are rotation matrices. The first two of the Euler angles, α and β, are identical with azimuthal and polar angles, φ and ϑ, respectively.

For ground state atoms and spherical molecules, using Eq. 2.42, we have

$$Q_{\ell m} = \int_0^\infty r^2 \varrho(r)\, r^\ell \,\mathrm{d}r \int Y_{\ell m}(\hat{r})\, \mathrm{d}^2\hat{r} = (4\pi)^{1/2}\, q_0\, \delta_{\ell 0}\, \delta_{m0}\,, \tag{2.44}$$

which vanishes unless $m = \ell = 0$; $Y_{00} = (4\pi)^{-1/2}$. The $Q_{\ell m}$ may be considered a measure of the non-sphericity of the charge distribution.

For linear molecules, we choose the molecular axis to be along the z-axis of the molecular frame. In this case, $Q_{\ell n} = 0$ for all $n \neq 0$ because the $Y_{\ell m}$ depend on φ as exp $im\varphi$. As a consequence, $Q_{\ell n}$ becomes $\mathrm{e}^{-in\alpha}\, Q_{\ell 0}$, which means that only $Q_{\ell 0} = q_\ell$ is invariant under rotation. In other words: there is only one independent multipole moment for every order ℓ. Laboratory-fixed components thus become simply

$$Q_{\ell m} = q_\ell\, Y_{\ell m}(\hat{z})\,, \tag{2.45}$$

where $\hat{z}$ denotes the direction of the molecular axis in the laboratory-fixed frame. For linear molecules that are also symmetric, inversion symmetry relations indicate that only even multipole moments may be non-zero; $q_\ell = 0$ for all odd ℓ.

For some of the more common molecules, the low-order molecular multipole moments are known [166, 378] (Landölt-Bornstein 1974). Collision-induced absorption of molecular gases arises mainly from multipolar induction. Studies of collision-induced absorption in the molecular gases provides, therefore, useful information on multipole moments [38].

2.6 Spectral transforms

In spectroscopy one frequently deals with functions of time, $f(t)$, such as the position $\boldsymbol{r}(t)$ of an electric charge, or a time-varying dipole moment, $\boldsymbol{\mu}(t)$, which lead to emission of electromagnetic radiation if the second time derivative is not vanishing. The frequency spectrum of the associated emission is obtained by Fourier transform of the function of time. If the absolute value of the function, $|f(t)|$, is integrable over all times, $-\infty \leq t \leq \infty$, one defines the Fourier transform according to

$$\mathscr{F}\{f(t)\} = F(\omega) = \int_{-\infty}^{\infty} f(t)\, \mathrm{e}^{-\mathrm{i}\omega t}\, \mathrm{d}t\,. \tag{2.46}$$

The inverse transform is given by

$$\mathscr{F}^{-1}\{F(\omega)\} = f(t) = \int_{-\infty}^{\infty} F(\omega)\, \mathrm{e}^{\mathrm{i}\omega t}\, \frac{\mathrm{d}\omega}{2\pi}\,. \tag{2.47}$$

We note that other authors use definitions of the Fourier and inverse Fourier transforms that differ from Eqs. 2.46 and 2.47, usually by factors or divisors of 2π, or the square root of 2π; the signs of the exponents are often interchanged. Caution is advised if results from different sources are compared.

In physics it is often convenient to represent measurable quantities (which are real) by the real part of a complex function, $m(t)$, as in $f(t) = \mathscr{R}\{m(t)\}$. In such cases the Fourier transform of the real function is not simply the real part of the Fourier transform of the complex function. Rather, if we take $\mathscr{F}\{f(t)\} = F(\omega)$ and $\mathscr{F}\{m(t)\} = M(\omega)$, we have

$$F(\omega) = \frac{1}{2}\,[M(\omega) + M^*(-\omega)]\,. \tag{2.48}$$

The asterisk designates the complex conjugate. Moreover, we note that the above Eqs. 2.46 and 2.47 imply positive as well as negative frequencies. In some physics applications, an appearance of negative frequencies may be confusing; only positive frequencies may have physical meaning. In such cases one may rewrite the above inverse tranform in terms of positive frequencies, using a well-known relationship between the complex exponential function and the sine and cosine functions.

We mention specifically the Fourier transform of the rth time derivative of a function, with $r = 1, 2, \ldots$, which may be expressed as the Fourier transform of the function, $f(t)$, times ω^r,

$$\mathscr{F}\left\{f^{(r)}(t)\right\} = (-i\omega)^r\,\mathscr{F}\{f(t)\}\,, \tag{2.49}$$

provided $f^{(r)}(t)$ exists at all t and all derivatives of lesser order vanish as $t \to \pm\infty$. Borel's convolution theorem states that

$$\mathscr{F}\{f_1(t)\}\;\mathscr{F}\{f_2(t)\} = \mathscr{F}\{f_1(t)\otimes f_2(t)\} \tag{2.50}$$

where the convolution of two functions is defined as

$$f_1(t)\otimes f_2(t) = \int_{-\infty}^{\infty} f_1(\tau)\,f_2(t-\tau)\,\mathrm{d}\tau\,;$$

and

$$\mathscr{F}\{f_1(t)\,f_2(t)\} = \int_{-\infty}^{\infty} F_1(\omega')\,F_2(\omega-\omega')\,\frac{\mathrm{d}\omega'}{2\pi}\,. \tag{2.51}$$

Parseval's theorem is given by

$$\int_{-\infty}^{\infty} \mathscr{F}^*\{f_1(t)\}\;\mathscr{F}\{f_2(t)\}\,\frac{\mathrm{d}\omega}{2\pi} = \int_{-\infty}^{\infty} f_1^*(t)\,f_2(t)\,\mathrm{d}t\,, \tag{2.52}$$

provided $\int |f_1(t)|^2\,\mathrm{d}t$ and $\int |f_2(t)|^2\,\mathrm{d}t$ exist.

The *correlation* of two functions, $f_1(t)$ and $f_2(t)$, is defined as

$$C(t) = \int_{-\infty}^{\infty} f_1^*(\tau)\,f_2(t+\tau)\,\mathrm{d}\tau\,. \tag{2.53}$$

If both $f_1(t)$ and $f_2(t)$ are real, the Fourier transform of the correlation function, $C(t)$, is given by

$$\int_{-\infty}^{\infty} C(t)\, e^{-i\omega t}\, dt = \mathscr{F}^*\{f_1(t)\}\ \mathscr{F}\{f_2(t)\} . \tag{2.54}$$

The correlation of a function with itself is called autocorrelation; in this case we have

$$\mathscr{F}\{C(t)\} = |\mathscr{F}\{f(t)\}|^2 \tag{2.55}$$

(Wiener–Khintchine theorem). The right-hand side of this equation is often called the power spectrum. It is given by the autocorrelation function, Eq. 2.55. The Fourier transform of the autocorrelation function is related to the spectral moments,

$$M_n = \int_{-\infty}^{\infty} \omega^n\, c(\omega)\, d\omega ,$$

according to

$$\begin{aligned} C(t) &= \int_{-\infty}^{\infty} c(\omega)\, e^{i\omega t}\, \frac{d\omega}{2\pi} \\ &= \frac{1}{2\pi} \sum_n \frac{1}{n!}\, (it)^n\, M_n , \end{aligned} \tag{2.56}$$

as the MacLaurin expansion of the exponential exp $i\omega t$ under the integral, Eq. 2.56, shows; the lower expression may be compared to the Taylor series expansion of the correlation function. In this way, we obtain the important relationship of the derivatives of the correlation function at zero time with the spectral moments,

$$\left.\frac{d^n C(t)}{dt^n}\right|_{t=0} = \frac{1}{2\pi}\, i^n\, M_n . \tag{2.57}$$

For this to be valid, the functions $C(t)$ and $c(\omega)$ must fall off to zero sufficiently fast for $t, \omega \to \pm\infty$. Lorentzians, for example, do not satisfy this condition; only the zeroth spectral moment exists in that case.

If we define 'widths' Δt and $\Delta\omega$ of the functions $f(t)$ and $F(\omega)$ by constructing approximate rectangular functions of the same area and height as $f(t)$, $F(\omega)$, it can be shown that

$$\Delta t\ \Delta\omega \geq \frac{1}{2} \tag{2.58}$$

(band width theorem; uncertainty principle in quantum mechanics), provided the functions $|f(t)|^2$ and $|F(\omega)|^2$ fall off to zero fast enough for $t, \omega \to \pm\infty$. Further details concerning Fourier transforms may be found in the literature (e.g., Champeney 1973).

2.7 Dipole radiation

1 Classical theory

An accelerated electric charge emits electromagnetic radiation. If we assume that the displacements of the charge are restricted to a small volume near the origin of a Cartesian frame called the laboratory frame, and if the speed of the charge is much smaller than the speed of light c so that relativistic effects can be neglected, the electric field strength $\boldsymbol{X}(\boldsymbol{R};t)$ at the distant point $\boldsymbol{R} = (x, y, z)$ at time t is given by

$$\boldsymbol{X}(\boldsymbol{R};t) = \frac{1}{4\pi\varepsilon_0}\,\frac{q}{c^2 R}\;\widehat{\boldsymbol{R}} \times \left(\widehat{\boldsymbol{R}} \times \ddot{\boldsymbol{r}}(t')\right) . \tag{2.59}$$

Symbols printed in boldface represent vectors; the hat above a vector signifies a dimensionless unit vector, for example $\widehat{\boldsymbol{R}} = \boldsymbol{R}\,/|\boldsymbol{R}|$; $\boldsymbol{r} = \boldsymbol{r}(t)$ describes the position of the charge q; differentiation with respect to time is indicated by dots so that $\ddot{\boldsymbol{r}}$ signifies acceleration; and $t' = t - R/c$ is the retarded time.‡ Accordingly, the power radiated due to accelerated charge per unit solid angle, Ω, is given by

$$\frac{\mathrm{d}P(t)}{\mathrm{d}\Omega} = \frac{1}{4\pi\varepsilon_0}\,\frac{q^2}{4\pi c^3}\left|\widehat{\boldsymbol{R}} \times \left(\widehat{\boldsymbol{R}} \times \ddot{\boldsymbol{r}}(t')\right)\right|^2 . \tag{2.60}$$

Note that the radiation field is dependent on one angle only, namely the angle ϑ subtended by the acceleration $\ddot{\boldsymbol{r}}$ and radius arm vector $\boldsymbol{R}$; the dependence enters Eq. 2.60 as $(\sin\vartheta)^2$ which is characteristic of the familiar dipole radiation pattern. It is, therefore, straightforward to integrate Eq. 2.60 over a spherical surface $R^2 \int \cdots \mathrm{d}\Omega$ where $\mathrm{d}\Omega = \sin\vartheta\;\mathrm{d}\vartheta\;\mathrm{d}\varphi$, to get the total power emitted,

$$P(t) = \frac{1}{4\pi\varepsilon_0}\,\frac{2q^2}{3c^3}\;|\ddot{\boldsymbol{r}}|^2 \tag{2.61}$$

(Larmor formula). For harmonic oscillations, $\boldsymbol{r}(t) = \boldsymbol{r}_0 \sin 2\pi f t$, the time averages of Eqs. 2.60 and 2.61 may be written§

$$\overline{\frac{\mathrm{d}P}{\mathrm{d}\Omega}} = R^2\,\frac{\mathrm{d}I}{\mathrm{d}\Omega} = \frac{1}{4\pi\varepsilon_0}\,\frac{q^2 r_0^2}{8\pi c^3}\,(2\pi f)^4\,\sin^2\vartheta$$

---

‡ We note that the factor $1/4\pi\varepsilon_0 = 8.987551788 \times 10^9$ V m A^{-1} s^{-1} is characteristic of the International System of Units (*Système International*, SI). If centimeter-gram-second (cgs) units are used instead of the SI units, this factor assumes the value of unity where electric charge squared or dipole moments squared are involved.

§ We use symbols ω, f and ν for frequency. These are related according to $\omega = 2\pi f = 2\pi c\nu$; units are radian s^{-1}, Hz and cm^{-1}, respectively.

and

$$\overline{P} = \frac{1}{4\pi\varepsilon_0} \frac{q^2 r_0^2}{3c^3} (2\pi f)^4 . \tag{2.62}$$

If the charge q is in harmonic or in elliptical motion,

$$\boldsymbol{r}(t) = \boldsymbol{r}_a \sin 2\pi f t + \boldsymbol{r}_b \cos 2\pi f t$$

where $\boldsymbol{r}_a$ and $\boldsymbol{r}_b$ are orthogonal, time-independent vectors, the frequency f of the emitted radiation is the same as that of the motion. Otherwise, *periodic* motion of the charge q may also generate various overtones (Jackson 1984).

We note that the emitted power, Eq. 2.62, can be written as a spectral density, $J(\nu)$, also called line shape, with the help of the δ function, as in

$$J(\nu) = \frac{1}{4\pi\varepsilon_0} \frac{q^2 r_0^2}{3c^3} (2\pi c\nu)^4 \, \delta(\nu - \nu_0) , \tag{2.63}$$

with ν_0 being the (fixed) oscillator frequency; in this case, the integral $\int J(\nu)\, d\nu$ equals the power emitted.

Of a special interest here is a charge in *aperiodic* motion, as in a collisional encounter. In that case, the theory of Fourier transforms is used to describe the continuous spectra that result. Specifically, starting from Eq. 2.60 and making use of Parseval's theorem, Eq. 2.52, the total energy radiated in the aperiodic event per unit solid angle and per unit frequency interval is obtained as

$$\frac{\partial^2 E_{\text{rad}}(\nu)}{\partial\Omega\, \partial\nu} = \frac{1}{4\pi\varepsilon_0} \frac{q^2}{2c^3} \left| \int_{-\infty}^{\infty} e^{-i2\pi c\nu t}\, \widehat{\boldsymbol{R}} \times \left(\widehat{\boldsymbol{R}} \times \ddot{\boldsymbol{r}}(t')\right) dt \right|^2 .$$

Integration over $d\Omega$ yields

$$\frac{dE_{\text{rad}}}{d\nu} = \frac{1}{4\pi\varepsilon_0} \frac{4\pi q^2}{3c^3} \left| \int_{-\infty}^{\infty} e^{-2\pi i c\nu t}\, \ddot{\boldsymbol{r}}(t')\, dt \right|^2 , \tag{2.64}$$

in much the same way as above, in going from Eq. 2.60 to 2.61. For this to be valid, the absolute value of the integrant must fall off to 0 fast enough as the time approaches infinity, $t \to \pm\infty$, see p. 42, ff. For binary collisions, this condition is generally satisfied but under different conditions (e.g., in liquids), this may not be the case and alternative approaches must be chosen for the computations of spectral moments and line shapes (autocorrelation function, p. 46). Equation 2.59 may be rewritten using familiar properties of Fourier transforms of derivatives, Eq. 2.49, as

$$\frac{\partial^2 E_{\text{rad}}(\nu)}{\partial\Omega\, \partial\nu} = \frac{1}{4\pi\varepsilon_0} \frac{q^2 (2\pi\nu)^4}{2c^3} \left| \int_{-\infty}^{\infty} e^{-i2\pi c\nu t}\, \widehat{\boldsymbol{R}} \times \left(\widehat{\boldsymbol{R}} \times \boldsymbol{r}(t')\right) dt \right|^2$$

and

$$\frac{dE_{rad}(\nu)}{d\nu} = \frac{1}{4\pi\varepsilon_0}\frac{4\pi q^2(2\pi c\nu)^4}{3c^3}\left|\int_{-\infty}^{\infty} e^{-2\pi i c\nu t} r(t')\, dt\right|^2 . \tag{2.65}$$

These expressions may be considered the *classical* emission spectrum associatedwith the aperiodic event described by $r(t)$. The prime is a reminder that the retarded time is to be used.

Expressions of the type of the right-hand side of Eq. 2.65 can be expressed in terms of the charge displacement autocorrelation function,

$$C(t) = \frac{q^2}{4\pi\varepsilon_0}\int_{-\infty}^{\infty} r(\tau)\cdot r(\tau+t)\, d\tau , \tag{2.66}$$

using the Wiener–Khintchine theorem (Eq. 2.55),

$$\frac{dE_{rad}}{d\nu} = 4\pi\frac{(2\pi\nu)^4}{3c^2}\int_{-\infty}^{\infty} e^{-i\omega t'}\, C(t')\, dt' .$$

The computation of spectra from the dipole autocorrelation function, Eq. 2.66, does not impose such stringent conditions on the integrand as our derivation based on Fourier transform suggests. Equation 2.66 is, therefore, a favored starting point for the computation of spectral moments and profiles; the relationship is also valid in quantum mechanics as we will see below.

For aperiodic events such as molecular collisions, a characteristic time Δt ('duration') can generally be defined so that the acceleration $\ddot{r}$ is significant only during that time, but is zero for $t \to \pm\infty$. It is well known that in such case the associated emission occurs essentially at low frequencies and features a width, Δf, consistent with Eq. 2.58,

$$\Delta f\ \Delta t \geq \frac{1}{4\pi} .$$

Useful estimates of the width can often be obtained from a knowledge of the spectral band by taking the lower limit of this expression.

Classical dipole radiation. Usually, we will be dealing with radiation emitted (or absorbed) by *dipoles*. In the simplest case, a dipole may be thought of as a point charge q positioned a distance r away from the origin of a Cartesian frame; the dipole moment μ is then given by

$$\mu = q\, r . \tag{2.67}$$

We note that the term 'dipole' implies the existence of another charge, $-q$, that remains fixed at the origin all the time. If the charge q is accelerating, it will emit radiation according to Eq. 2.61, where we replace the product $q\ddot{r}$ by the second derivative of μ with respect to time, $\ddot{\mu}$. Dimensions of the displacement of charge will be of the order of the size of atoms or molecules, usually assumed to be small with regard to both the wavelengths of light involved, and the radius arm vector from the source to the detector, $|\boldsymbol{R}|$.

The classical relationships, Eqs. 2.59 through 2.66, can now be rewritten in terms of the dipole moment, $\boldsymbol{\mu}$ or $\ddot{\boldsymbol{\mu}}$, which we substitute for $q\,\boldsymbol{r}$ and $q\,\ddot{\boldsymbol{r}}$, respectively. Specifically, from Eq. 2.65 we now have the dipole emission spectra

$$\frac{\mathrm{d}E_{\mathrm{rad}}(\nu)}{\mathrm{d}\nu} = \frac{1}{4\pi\varepsilon_0}\,4\pi\,\frac{(2\pi c\nu)^4}{3c^3}\left|\int_{-\infty}^{\infty} \mathrm{e}^{-\mathrm{i}2\pi c\nu t}\,\boldsymbol{\mu}(t')\,\mathrm{d}t\right|^2 , \tag{2.68}$$

and the dipole autocorrelation function, from Eq. 2.66,

$$C(t) = \frac{1}{4\pi\varepsilon_0}\int_{-\infty}^{\infty} \boldsymbol{\mu}(\tau)\cdot\boldsymbol{\mu}(\tau+t)\,\mathrm{d}\tau\,. \tag{2.69}$$

This expression is a useful starting point for a computation of spectral moments and profiles. Equation 2.68 allows the computation of the dipole emission profile if $|\boldsymbol{\mu}(t)|$ falls off to zero sufficiently fast for $t \to \pm\infty$. Equation 2.69, on the other hand, has less stringent conditions on the dipole function itself and is more broadly applicable (dense fluids) when combined with Eq. 2.66.

Absorption. The absorption coefficient, $\alpha(\nu)$, is related to the power due to spontaneous emission by Kirchhoff's law,

$$\alpha(\nu) = J(\nu)\,/cu(\nu)\,. \tag{2.70}$$

where c is the speed of light and u is the density of blackbody radiation. That density is given by Planck's law,

$$u(f)\,\mathrm{d}f = \frac{8\pi h f^3}{c^3}\,\frac{1}{\exp(hf/kT)-1}\,\mathrm{d}f\,. \tag{2.71}$$

However, in a classical theory, at low enough frequencies, one might be able to use the Rayleigh–Jeans law,

$$u(f)\,\mathrm{d}f = \frac{8\pi f^2}{c^3}\,kT\,\mathrm{d}f\,, \tag{2.72}$$

to convert from an emission profile, Eq. 2.68, to the absorption profile; in a purely classical theory Planck's constant h has the value zero.

2 Quantum theory of radiation

Emission and absorption of electromagnetic radiation by molecular systems takes place in *transitions* from an initial quantum state $|i\rangle$ to a final state $|f\rangle$. The dipole transition matrix element associated with such a transition is obtained from the wavefunctions of these states, $\psi_i(\boldsymbol{r})$ and

$\psi_f(\boldsymbol{r})$, according to

$$\langle f\,|\boldsymbol{\mu}|\,i\rangle = \int \psi_f^*(\boldsymbol{r})\,\boldsymbol{\mu}(\boldsymbol{r})\,\psi_i(\boldsymbol{r})\,\mathrm{d}^3r\,. \tag{2.73}$$

In this expression, d^3r stands for $r^2\,\mathrm{d}r\sin\vartheta\,\mathrm{d}\vartheta\,\mathrm{d}\varphi$. The asterisk designates the complex conjugate.

It is well known that classical intensities of dipoles emitting radiation of frequency $\nu_{if} = (E_i - E_f)/hc$ may be converted to quantum expressions by replacing in Eq. 2.62 the classical squared dipole moment $q^2 r_0^2$ by four times the squared dipole transition element,

$$|\boldsymbol{\mu}|^2 \rightsquigarrow 4\,\langle f\,|\boldsymbol{\mu}|\,i\rangle^2$$

(correspondence principle; Born 1970). In this way, we get from Eq. 2.62 the power emitted as

$$P_{if} = \frac{4}{3c^3}\,(2\pi c\nu_{if})^4\,\frac{1}{4\pi\varepsilon_0}\;|\langle f\,|\boldsymbol{\mu}|\,i\rangle|^2\;; \tag{2.74}$$

more precisely, $P_{if} = hc\nu_{if}/\Delta t$ is the energy of the photon, divided by the mean duration of the emission process, an average power. E_i and E_f are the energies of initial and final state; $E_i > E_f$. The expression for the power, P_{if}, may be converted into spectral density by multiplication of Eq. 2.74 by the δ functions,

$$J(\nu) = P_{if}\;\left[\delta(\nu_{if} - \nu) + \delta(\nu_{if} + \nu)\right]\,.$$

In this case, the subscripts i, f of P of frequency may be dropped in the equation.

Einstein's A and B coefficients. Quantized systems, such as atoms and molecules, emit and absorb radiation of frequency $\nu_{ij} = |E_i - E_j|/hc$ in transitions between states $|i\rangle$ and $|j\rangle$ of energy E_i, E_j. Einstein assumed that the probability that a system in state $|i\rangle$ will absorb a photon of energy $hc\nu_{ij}$ is proportional to the density of radiative energy per frequency interval, $u(\nu_{ij})\,\mathrm{d}\nu$. The probability of absorbing a photon in the time interval $\mathrm{d}t$ is given by

$$\mathscr{P}_{i\to j}\,\mathrm{d}t = B_{ij}\,u(\nu_{ij})\,\mathrm{d}t\,, \tag{2.75}$$

where B_{ij} is the Einstein B coefficient of absorption. (We assume at this point that $E_j > E_i$.) On the other hand, emission of a photon of the same frequency by the inverse transition, $j \to i$, is described by a probability that is the sum of a 'spontaneous' and an 'induced' transition probability. Let $\mathscr{P}_{j\to i}$ designate the probability per unit time and per molecule of a transition from the state j to the state i, we may write

$$\mathscr{P}_{j\to i} = A_{ji} + u(\nu_{ij})\,B_{ji}\,, \tag{2.76}$$

where the A_{ji} and B_{ji} are Einstein's A and B coefficients for spontaneous and stimulated emission, respectively. Einstein has shown that the three coefficients are related as

$$B_{ji} = B_{ij} \tag{2.77}$$

$$A_{ji} = \frac{8\pi h(c\nu)^3}{c^3} B_{ji} . \tag{2.78}$$

The A_{ji} may be computed from Eq. 2.74, with $A_{ji} = P_{ij}/\hbar\omega_{ij}$, as

$$A_{ji} = \frac{4}{3\hbar c^3} (2\pi c\nu_{ij})^3 \frac{1}{4\pi\varepsilon_0} |\langle j | \boldsymbol{\mu} | i \rangle|^2 . \tag{2.79}$$

Stimulated emission may be considered the inverse absorption process. In stimulated emission, the photon is emitted in the exact direction of the incident beam; the photon is 'added' to the beam that induced the emission (lasers).

Transition probabilities. The interaction of quantum systems with light may be studied with the help of Schrödinger's time-dependent perturbation theory. A molecular complex may be in an initial state $|i\rangle$, an eigenstate of the unperturbed Hamiltonian, $\mathscr{H}_0 |i\rangle = E_i |i\rangle$. If the system is irradiated by electromagnetic radiation of frequency $\nu = \omega/2\pi c$, transitions to other quantum states $|f\rangle$ of the complex occur if the frequency is sufficiently close to Bohr's frequency condition,

$$\omega_{fi} = (E_f - E_i)/\hbar . \tag{2.80}$$

The probability per unit time for such transition is given by

$$\mathscr{P}_{f\leftarrow i}(\omega) = \frac{\pi}{2\hbar^2} |\langle f | \boldsymbol{X}_0 \cdot \boldsymbol{\mu} | i \rangle|^2 \, \Delta\omega , \tag{2.81}$$

which is known as the golden rule of quantum mechanics. Here $\boldsymbol{X}_0 = X_0\widehat{X}$ specifies amplitude and polarization of the electric field, $\boldsymbol{X}(t) = \boldsymbol{X}_0 \cos\omega t$, and $\boldsymbol{\mu}$ is the electric dipole moment operator. Electric dipole interaction is assumed which is valid if the wavelengths are large compared to the dimensions of the complex.

Absorption coefficient. In every absorption process one photon is lost, so is the energy $\hbar\omega$, and the rate of energy loss, $\dot{E}_{\text{rad}}$, is thus given by

$$\begin{aligned}
\dot{E}_{\text{rad}} &= -\sum_{f,i} \hbar\,\omega_{fi}\,\mathscr{P}_{f\leftarrow i} \\
&= -\frac{\pi}{2\hbar} \sum_{f,i} \omega_{fi} P_i \, |\langle f | \boldsymbol{X}_0 \cdot \boldsymbol{\mu} | i \rangle|^2 \left\{ \delta(\omega_{fi} - \omega) + \delta(\omega_{fi} + \omega) \right\} ,
\end{aligned}$$

where P_i is the probability of finding the molecule (or the molecular complex) in the ith state in the (unperturbed) ensemble. The sums are over all states i and f. In the second δ function we may, therefore, interchange these indices. Since $\omega_{fi} = -\omega_{if}$, we may write

$$\dot{E}_{\text{rad}} = -\frac{\pi}{2\hbar} \sum_{f,i} \omega_{fi} \, (P_i - P_f) \, |\langle f \, | X_0 \cdot \boldsymbol{\mu} | \, i \rangle|^2 \, \delta(\omega_{fi} - \omega) \, .$$

Furthermore, since

$$P_f = P_i \, \exp\left(-\hbar\omega_{fi}/kT\right) \, , \tag{2.82}$$

and since ω_{fi} may be replaced by ω outside of the argument of the δ function, we obtain

$$\dot{E}_{\text{rad}} = -\frac{\pi X_0^2}{2\hbar} \, \omega \, \left[1 - \exp\left(-\hbar\omega/kT\right)\right] \times \sum_{f,i} P_i \left|\left\langle f \left| \widehat{X} \cdot \boldsymbol{\mu} \right| i \right\rangle\right|^2 \delta(\omega_{fi} - \omega) \, . \tag{2.83}$$

In this expression, we have also replaced the electric field vector by the amplitude, X_0, times the polarization vector, $\widehat{X}$. One may often be dealing with random orientations of the dipole moments or polarization so that the substitution

$$\overline{\left|\left\langle f \left| \widehat{X} \cdot \boldsymbol{\mu} \right| i \right\rangle\right|^2} = \frac{1}{3} \, \overline{|\langle f \, |\boldsymbol{\mu}| \, i \rangle|^2} \tag{2.84}$$

may be used.

An absorption *cross section*, $q_{\text{abs}} = -\dot{E}/S$, is obtained from Eq. 2.83 by dividing by the magnitude of the Poynting vector, S, averaged over one period,

$$S = \left(\frac{\varepsilon_0 \varepsilon}{\mu_0}\right)^{1/2} \, \frac{1}{2} \, X_0^2 \, ,$$

where $\mu_0 = 1/\varepsilon_0 \, c^2$ is the permeability of vacuum. The absorption coefficient, which is the reciprocal mean free path for absorption, is given by the product of the absorption cross section and the number density of absorbers, according to

$$\alpha(\omega) = \frac{4\pi^2}{3\hbar c n} \, \frac{1}{V} \, \omega \, \left[1 - \exp\left(-\hbar\omega/kT\right)\right] \, J(\omega) \tag{2.85}$$

where

$$J(\omega) = \sum_{i,f} P_i \, \frac{1}{4\pi\varepsilon_0} \, |\langle f \, |\boldsymbol{\mu}| \, i \rangle|^2 \, \delta(\omega_{fi} - \omega) \, . \tag{2.86}$$

We note that the induced dipole moment $\boldsymbol{\mu}$ is the *total* dipole. In this expression, Eq. 2.85, $n = \sqrt{\varepsilon}$ is the refractive index at the frequency ω. We note that Eq. 2.86 is widely quoted in cgs units, that is without the factor $1/4\pi\varepsilon_0$.

For most treatments, the spectral density, $J(\omega)$, Eq. 2.86, also referred to as the spectral profile or line shape, is considered, since it is more directly related to physical quantities than the absorption coefficient α. The latter contains frequency-dependent factors that account for stimulated emission. For absorption, the transition frequencies ω_{fi} are positive. The spectral density may also be defined for negative frequencies which correspond to emission.

Equations 2.85 and 2.86 may be considered the Schrödinger representation of the absorption of radiation by quantum systems in terms of spectroscopic transitions between states $|i\rangle$ and $|f\rangle$. In the Schrödinger picture, the time evolution of a system is described as a change of the state of the system, as implemented here in the form of the time-dependent perturbation theory. The results hardly resemble the classical relationships outlined above, compare Eqs. 2.68 and 2.86, even if we rewrite Eq. 2.86 in terms of an emission profile. Alternatively, one may choose to describe the time evolution in terms of time-dependent observables, the 'Heisenberg picture'. In that case, expressions result that have great similarity with the classical expressions quoted above as we will see next.

Heisenberg formalism. The Heisenberg view leads to an expression equivalent to the Schrödinger formalism that stresses the time evolution of quantum systems; it has a clear correspondence with classical mechanics; it is most conveniently expressed in terms of the dipole autocorrelation function (Gordon 1968).

With the help of the Fourier expansion of the δ function,

$$\delta(\omega) = \frac{1}{2\pi} \int_{-\infty}^{\infty} \exp(\mathrm{i}\omega t)\; \mathrm{d}t\,,$$

we may write the spectral density, Eq. 2.86, according to

$$J(\omega) = \frac{3}{2\pi} \sum_{i,f} P_i \frac{1}{4\pi\varepsilon_0} \left\langle i \middle| \hat{X}\cdot\boldsymbol{\mu} \middle| f \right\rangle \left\langle f \middle| \hat{X}\cdot\boldsymbol{\mu} \middle| i \right\rangle \qquad (2.87)$$
$$\times \int_{-\infty}^{\infty} \mathrm{d}t\; \exp \mathrm{i}\left\{(E_f - E_i)/\hbar - \omega\right\} t\,.$$

In the Heisenberg representation a time-dependent dipole operator $\boldsymbol{\mu}(t)$ is generated from its value at some previous time t' by a unitary transformation with the time-displacement operator $\exp \mathrm{i}\mathscr{H}_0(t-t')/\hbar$, so that

$$\boldsymbol{\mu}(t) = \mathrm{e}^{\mathrm{i}\mathscr{H}_0(t-t')/\hbar}\; \boldsymbol{\mu}(t')\; \mathrm{e}^{-\mathrm{i}\mathscr{H}_0(t-t')/\hbar}\,, \qquad (2.88)$$

which satisfies the Heisenberg equation of motion,

$$\frac{\mathrm{d}}{\mathrm{d}t}\boldsymbol{\mu}(t) = \frac{\mathrm{i}}{\hbar}\, [\mathscr{H}_0, \boldsymbol{\mu}(t)]\;; \qquad (2.89)$$

[...] designates the commutator. Setting $t' = 0$ and using the completeness relation, $1 = \sum_f |f\rangle\langle f|$, this expression can be rewritten as

$$J(\omega) = \int_{-\infty}^{\infty} e^{-i\omega t}\, C(t)\, dt\,, \tag{2.90}$$

where $C(t)$ is the dipole autocorrelation function,

$$C(t) = \frac{3}{2\pi} \sum_i P_i \frac{1}{4\pi\varepsilon_0} \left\langle i \middle| \widehat{X} \cdot \boldsymbol{\mu}(0)\, \widehat{X} \cdot \boldsymbol{\mu}(t) \middle| i \right\rangle .$$

For randomly oriented radiators, the substitution, Eq. 2.84, will in general be possible so that the correlation function may be written as

$$C(t) = \frac{1}{2\pi} \frac{1}{4\pi\varepsilon_0} \langle \boldsymbol{\mu}(0) \cdot \boldsymbol{\mu}(t) \rangle\,, \tag{2.91}$$

where the angular brackets $\langle \cdots \rangle$ now imply an additional ensemble average. We note that in other work, some or all of the factors shown here will be omitted. The Heisenberg representation of the spectral density function, Eq. 2.90, is the Fourier transform of the autocorrelation function of the dipole moment operator (Gordon 1968). This equation resembles the classical expression for an emitting dipole.

Equations 2.86 and 2.90 are equivalent; these are often taken as the starting point for the theory of spectral moments and line shapes. For the treatment of binary systems, one may start with the Schrödinger expression; when dealing with many-body systems, the correlation function formalism is generally the preferred ansatz.

Note that the above equations for the radiation of a dipole refer to a single dipolar source of radiation. For our spectroscopic interests, we more commonly are dealing with an assembly of many identical such systems. Usually, the total radiation of such an assembly of sources may be assumed to be the incoherent superposition of the individual intensities (see, however, the discussions of the intercollisional effect below, Chapters 3 and 5). We may then simply multiply the contributions of one source by the number N_i of sources in the (initial) state i.

Spectral line shapes. The spectroscopist frequently uses model profiles which may approximate certain spectral line shapes closely when used with understanding. We mention the Lorentzian line shape, Eq. 3.15, the 'natural' profile of spectral lines; the cores of pressure-broadened lines may also be modeled by it. Gaussian line shapes, on the other hand, arise from Doppler broadening. The convolution of Lorentzian and Gaussian profiles is called the Voigt profile. It models spectral line shapes well when pressure broadening shapes the wings and Doppler broadening has affected the core of the lines. A remarkable property of Voigt profiles is the fact that the convolution of two arbitrary Voigt profiles gives another Voigt profile.

In many cases, the profile a spectroscopist sees is just the instrumental profile, but not the profile emitted by the source. In the simplest case (geometric optics, matched slits), this is a triangular slit function, but diffraction effects by beam limiting apertures, lens (or mirror) aberrations, poor alignment of the spectroscopic apparatus, etc., do often significantly modify the triangular function, especially if high resolution is employed.

These line shapes are generally not very useful for collision-induced absorption work, because pressure broadening, Doppler effect and instrumental resolution are here of no great concern. In Chapters 5 and 6 we will consider a number of other *ad hoc* model functions that have acquired a certain significance in collision-induced absorption.

2.8 Instrumentation

The frequencies of interest for studies of collision-induced absorption range from microwave frequencies to the ultraviolet, depending on the systems and specific transitions considered. Light sources, monochromators, detectors and pressure cells are needed for such studies, which are more or less the same as in the conventional spectroscopies.

Microwaves. Among the lowest frequencies of interest in collisional absorption are radio- and microwaves. As will be seen below, the absorption coefficient α is extremely small at low frequencies because absorption falls off to zero frequency as ω^2; see Chapter 5 for details. As a consequence, it has generally been necessary to use sensitive resonator techniques for the measurement of the loss tangent, $\tan\delta = \varepsilon''/\varepsilon'$, where ε' and ε'' are the real and imaginary part of the dielectric constant. The loss tangent is obtained by determination of the 'quality factors' Q_a, Q_0, of the cavity with and without the gas filling, as (Dagg 1985)

$$\tan\delta = \frac{\varepsilon''}{\varepsilon'} = \frac{1}{Q_a} - \frac{1}{Q_0}\,. \tag{2.92}$$

The absorption coefficient is related to the loss tangent by

$$\alpha(\omega) = \frac{\omega}{c}\left(\varepsilon'(\omega)\right)^{1/2}\tan\delta = \frac{\omega}{c}\,\frac{\varepsilon''}{\sqrt{\varepsilon'}}\,. \tag{2.93}$$

For $Q_a \ll Q_0$, the equivalent absorption pathlength is given by $Q_a\lambda/2\pi$ where λ is the wavelength; Q_0 values of 40 000 in temperature-controlled microwave absorption studies at frequencies between ≈ 2 and 5 cm^{-1} have been known (Dagg 1985).

The far infrared region. The standard infrared (IR) light source used to be the glowbar, but in recent years it was realized that synchroton radiation offers higher brightness in the far IR. Laser sources have also successfully been used in recent work on collisional absorption.

One problem of the measurement of weak absorption in the far IR is that short absorption paths must be used. At wavelengths comparable to the beam apertures diffraction effects lead to beam divergence [252]. (The combination of high gas pressures and short absorption paths may not be useful if many-body induction effects must be avoided, and an accurate measurement of α under conditions of weak absorption, $I/I_0 \approx 1$, is difficult [368].)

Multi-path cells. For weakly absorbing gases, and under situations where pressures cannot be raised above a certain point (e.g., vapors; occurence of many-body effects), it is necessary to use very long absorption path lengths to record a high-quality absorption spectrum. White has designed a folded light path absorption cell which permits the use of large angular apertures off the optical axis [423]. A modification of White's design was introduced by Bernstein and Herzberg [29] that permits doubling of the absorption path lengths attainable. This type of multiple-path cell has been shown to be very useful in the infrared. With a cell using mirrors of 0.5 m focal length, based on an $f/10$ aperture design, 75% of the incident radiation was transmitted with 40 transversals (Möller and Rothschild 1971).

Sample cells. Variable temperature. Temperature control has been essential in much of the collision-induced absorption studies. Temperature variation accesses different parts of the intermolecular interaction potential and redistributes the relative importance of overlap and multipolar induction. Furthermore, at low temperatures, collision-induced line shapes are relatively sharp; induced lines may be resolved at low temperatures whose structures may be masked at higher temperatures.

2.9 General references

M. Born, *Atomic Physics.* Hafner 1970.

D. C. Champeney, *Fourier Transforms and their Physical Applications.* Academic Press, New York, 1973.

S. Chapman and T. G. Cowling, *The Mathematical Theory of Non-uniform Gases.* Cambridge University Press 1990.

N. B. Colthup, L. H. Daly, and S. E. Wiberly, *Introduction to Infrared and Raman Spectroscopy.* Academic Press, New York, 1990.

I. R. Dagg, *Collision-induced Absorption in the Microwave Region*, in *Phenomena Induced by Intermolecular Interactions*. G. Birnbaum, ed., Plenum Press, New York, 1985.

R. G. Gordon, Correlation Functions for Molecular Motion. *Adv. Mag. Resonance*, **3**, 1 (1968).

J. O. Hirschfelder, C. F. Curtiss and R. B. Bird, *Molecular Theory of Gases and Liquids*. Wiley, New York, 1964.

J. M. Hollas, *Modern Spectroscopy*. Wiley, New York, 1987.

J. D. Jackson, *Classical Electrodynamics*. Wiley, New York, 1984.

Landölt-Bornstein, *Numerical Data and Functional Relationships in Science and Technology; New Ser., Group II: Atomic and Molecular Physics*, vol. 6. K. H. Hellwege and A. M. Hellwege, eds., Springer, Berlin, 1974.

D. A. McQuarrie, *Statistical Mechanics*. Harper and Row, New York, 1976.

G. C. Maitland, M. Rigby, E. B. Smith, and W. A. Wakeham, *Intermolecular Forces*. Clarendon Press, Oxford, 1981.

K. D. Möller and W. G. Rothschild, *Far-Infrared Spectroscopy*. Wiley-Interscience, New York, 1971.

H. A. Willis, J. H. van der Maas, and R. G. J. Miller, *Laboratory Methods in Vibrational Spectroscopy*. Wiley, New York, 1987.

3
Experimental results

Absorption spectra. The absorption coefficient, α, also called the absorption spectrum if known as a function of frequency ν, is obtained from a measurement of transmitted and incident intensities, $I(\nu)$ and $I_0(\nu)$, of monochromatic light, according to

$$\alpha(\nu) = \frac{1}{L} \log_e \frac{I_0(\nu)}{I(\nu)} \tag{3.1}$$

(Lambert's law), where L is the absorption path length. The absorption coefficient $\alpha(\nu)$ is a function not only of frequency, but also of temperature, density, and, of course, the nature, composition, and state of matter (gaseous, liquid, solid) of the sample as is amply illustrated below. Absolute intensities of absorption spectra may often be determined which are of interest for the comparison of measurements with the fundamental theory and in many applications (atmospheric sciences).

Our main interest is the absorption which arises from complexes of interacting atoms and/or molecules, i.e., the absorption which exceeds the simple sum of the absorption spectra of the individual (non-interacting) atoms and molecules (where such monomer spectra exist). This excess absorption is of a *supermolecular* nature. It will be called interaction-induced, or briefly induced absorption.

Spectroscopists have always known certain phenomena that are caused by collisions. A well-known example of such a process is the pressure broadening of allowed spectral lines. Pressure broadened lines are, however, not normally considered to be collision-induced, certainly not to that extent to which a specific line *intensity* may be understood in terms of an individual atomic or molecular dipole transition moment. The definition of collisional induction as we use it here implies a dipole component that *arises from the interaction* of two or more atoms or molecules, leading at high enough gas density to discernible spectral line intensities in excess of the sum of the absorption of the atoms/molecules of the complex. In other words, we focus on spectral *intensity* that arises from interaction-induced dipole moments.

Collision-induced emission. Induced spectra are usually studied in absorption. However, emission spectra of supermolecular nature will also interest us here, for example, in connection with shock tube measurements and astrophysical studies (stellar and planetary atmospheres).

About units. For our survey of measured spectra and the comparisons with theory that follow in Chapters 5 and 6, it is useful to remember that frequencies are often expressed in units of Hertz, or of cm^{-1}, or in cycles per 2π seconds. In order to avoid confusion we shall distinguish the notations f, $\nu = f/c$, and $\omega = 2\pi f$, respectively, where c designates the speed of light in vacuum. Similarly, gas densities will be expressed as number densities, n, the number of particles per volume, or in units of amagat, $\varrho = n/N_a$, where N_a is the number of particles per cubic centimeter of the gas under consideration; for most gases of interest N_a is about equal to Loschmidt's number, $N_a \approx N_L = 2.686763 \times 10^{19}$ cm^{-3} $amagat^{-1}$, the particle density of an ideal gas at standard temperature and pressure. The values of fundamental constants are taken from [124].

Overview. In this Chapter, we will consider the various types of induced spectra. *Translational* spectra involve transitions between states of relative motion of the collisional pair, without changing the rotovibrational or electronic states of the interacting molecules themselves. Translational spectra occur at zero frequency but, due to their considerable widths, their wings generally extend into the microwave and far infrared regions of the electromagnetic spectrum. In their purest form, translational spectra occur in rare-gas mixtures, but translational components of induced spectra are sometimes discernible in molecular gases and mixtures as well. Translational spectra will be considered in Section 3.1. More typically, molecular systems show *rototranslational* induced spectra in the far infrared, at rotational transition frequencies of the molecules involved even if these are infrared inactive in the isolated molecule. These will be considered in Section 3.2. Furthermore, for molecular systems, *rotovibrational* induced spectra occur which involve vibrational transitions, Section 3.3. If electronic transitions of the interacting atoms or molecules are involved, *electronic* induced spectra at electronic transition frequencies are expected, Chapter 7.

Simultaneous transitions involving two interacting molecules are also known at sums and differences of molecular or electronic transition frequencies. These are considered below, along with the single rotovibrational or electronic transitions where appropriate. *Spectra of van der Waals molecules* are in general closely associated with all of these. Furthermore, we have to distinguish between *binary*, *ternary*, etc., and various *many-body* induced spectra as we will see.

3.1 The translational spectra

The induced spectra of rare gases in the far infrared are particularly simple but show characteristic features common to all collision-induced absorption. These will be considered first.

Collision-induced absorption spectra of pure, monatomic gases like helium or argon have not been recorded. If these exist, higher densities and/or longer pathlengths are needed for a successful recording than were hitherto employed [329]. Of course, in pure gases near the low-density limit where binary interactions prevail, absorption by induced dipoles is not expected because a pair of identical atoms possesses an inversion symmetry which is inconsistent with the existence of a dipole moment. However, at high densities, three- and more-body interaction could conceivably induce dipole moments which in turn would generate translational collision-induced absorption spectra. The fact is that such many-body spectra have not been observed in unmixed rare gases [214], not even in the liquid state [201], which suggests that absorption by three-body and higher-order complexes of identical atoms is very weak under the conditions of the existing measurements.

1 Translational pair spectra

The simplest collision-induced absorption spectra are, therefore, those of mixtures of two monatomic gases, especially those recorded near the low density limit where binary interactions prevail. Examples of collision-induced absorption spectra of interacting pairs of dissimilar atoms have been recorded in the microwave and far infrared regions of the spectrum. As examples, the He–Ar, Ne–Ar and Ar–Kr collision-induced absorption spectra are shown in Fig. 3.1. The He–Ar and Ne–Ar spectra were recorded with a path length of 3 m, at a temperature of 295 K, and with three different total densities from approximately 60 to 90 amagats [75]. The Ar–Kr spectra were taken at total densities up to 200 amagat [104]. Gas densities of the two components of each mixture were roughly equal. Over the range of frequencies shown, absorption varies very nearly linearly with the product of the relevant gas densities, ϱ_1 and ϱ_2. This fact suggests that the observed spectra arise from a complex of two dissimilar atoms. For binary spectra, it is customary to plot the absorption coefficient normalized by the product of gas densities, $\alpha(\nu)/\varrho_1\varrho_2$, a quantity that is invariant under density variation in the binary regime. Data sets taken

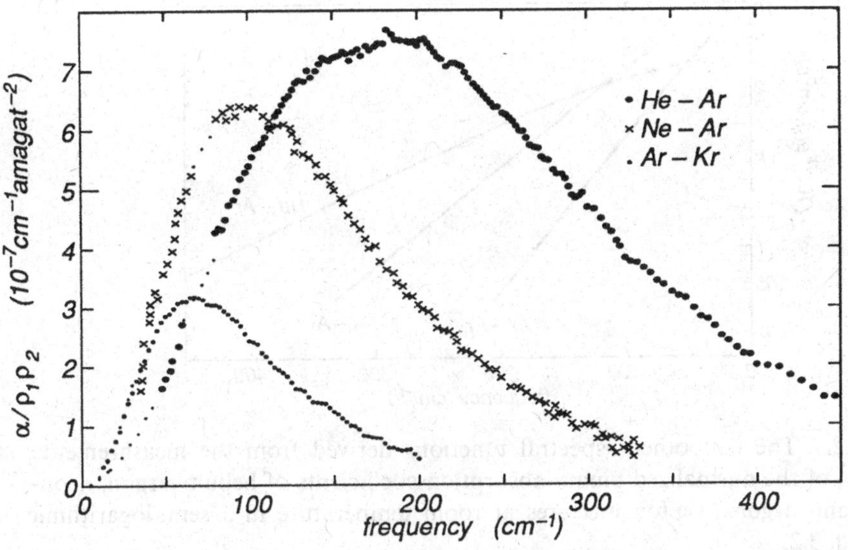

Fig. 3.1. Measurements of the binary absorption coefficient of helium–argon [75], neon–argon [75] and argon–krypton mixtures [104], at room temperature, normalized by the product of relevant densities.

at different total densities can thus be plotted and compared in the same figure.

For the He–Ar and Ne–Ar spectra at low frequencies, for $\nu < 50$ cm^{-1} and 60 cm^{-1} $< \nu <$ 80 cm^{-1}, experimental problems (transmission of windows, etc.,) made a measurement impossible; the small dots are estimates of the absorption obtainable by fitting suitable analytical profiles to the data, such as those mentioned in Section 3.7 below.

The spectra shown in Fig. 3.1 appear as unstructured, broad absorption bands, with a maximum of absorption around 200 cm^{-1} for the lightest system and at lower frequencies for the more massive pairs. Absorption is weak, even at the peaks, and amounts to a mean absorption length of more than $1/\alpha \approx 10^6$ cm (that is 10 km) if both gases are present at partial pressures of just one atmosphere. Absorption of rare gas mixtures increases, however, with increasing densities, with a mean absorption length of centimeters as we approach liquid densities.

For a detailed discussion of such spectra, it is often advantageous to study the shape of the spectral function, $g(\nu)$, which is related to the absorption coefficient, $\alpha(\nu)$, according to

$$\alpha(\nu) = \frac{(2\pi)^3 N_a^2}{3\hbar} \varrho_1 \varrho_2 \, \nu \left[1 - \exp\left(-\frac{hc\nu}{kT}\right)\right] \, Vg(\nu) \,. \tag{3.2}$$

In this expression, $\hbar$ is Planck's constant h divided by 2π, c is the speed

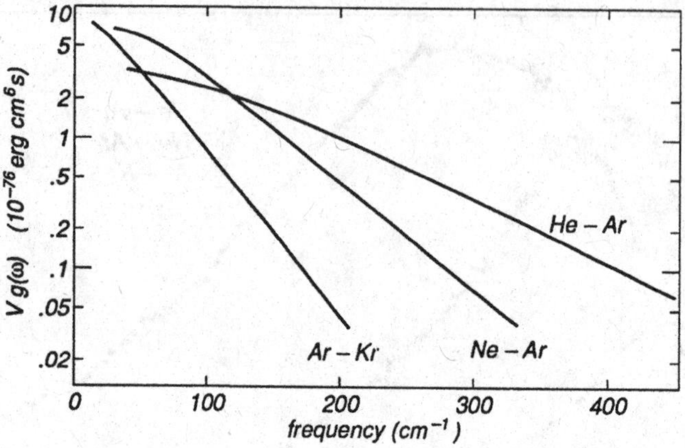

Fig. 3.2. The (smoothed) spectral functions derived from the measurements, Fig. 3.1, of the normalized binary absorption coefficients of helium–argon, neon–argon and argon–krypton mixtures at room temperature in a semi-logarithmic grid, Eq. 3.2.

of light, k is the Boltzmann constant, and V is the sample volume. The spectral function,* $g(\nu)$, is proportional to the reciprocal volume, V^{-1}, so that the product $Vg(\nu)$ is actually not dependent on the volume of the sample. The absorption coefficient, $\alpha(\nu)$, and spectral function, $g(\nu)$, differ essentially by the frequency-dependent factor $\nu\ [1-\exp(-hc\nu/kT)]$ that is typical of absorption spectra in the far infrared. This factor causes absorption to fall off to zero as ν^2 for $\nu \to 0$. The Boltzmann term in this factor arises from stimulated emission; see Chapters 2 and 5 for details.

The spectral functions thus obtained are shown in Fig. 3.2. We will usually display the spectral functions in a semi-logarithmic grid because they are known over a peak-to-wing intensity ratio amounting to several orders of magnitude; in such a grid the far wing intensities are displayed with the same relative accuracy as the line cores, which is impossible in a linear grid. The smooth curves shown in the figure represent averages of hundreds of data points each.

The translational spectral function, $g(\nu)$, may be considered a (very diffuse) spectral line centered at zero frequency which arises from transitions between the states of relative motion of the interacting pair. It is the free-state analog of the familiar vibrational and rotational transitions of bound systems, with the difference that the motion is here aperiodic; the period goes to zero due to the lack of a restoring force. The negative fre-

* We note that in other work, some of the frequency-independent factors appearing in the equation may be included in the definition of the spectral function, $g(\nu)$. In such a case, the spectral function may be scaled relative to our definition and units may differ.

quency wing ($\nu < 0$) of the translational line is, of course, not observable in absorption. Since it corresponds to transitions from a higher to a lower translational energy state, it is related to collision-induced *emission*. The spectral function, $g(\nu)$, is more directly related to the transition matrix elements than the absorption coefficient, α. At low frequencies the latter is strongly affected by stimulated emission.

Near the line centers, the spectral functions have sometimes been approximated by a Lorentzian. The far wings, on the other hand, may be approximated by exponential functions as Fig. 3.2 might suggest. However, better model profiles exist; see Chapters 5 and 6 [421, 102, 320]. Model profiles have been useful for fitting experimental spectra, for an extrapolation of measured profiles to lower or higher frequencies (which is often needed for the determination of spectral moments) and for a prediction of spectra at temperatures for which no measurements exist. We note that van der Waals dimer structures (which appear at low frequencies and low pressures) modify the Lorentzian-like appearance more or less, as we will see.

It is of interest to compare the half-widths at half-intensity of the spectral functions of the three systems shown in Fig. 3.2. These amount to roughly 140, 80 and 50 cm^{-1} for He–Ar, Ne–Ar and Ar–Kr, respectively, which are enormous widths if compared to the widths of common Doppler profiles, etc. The observed widths reflect the short 'lifetimes' of collisional complexes. From the theory of Fourier transforms we know that the product of lifetime, Δt, and bandwidth, Δf, is of the order of unity, Eq. 1.5. The duration of the fly-by interaction is given roughly by the range of the induced dipole function, Eq. 4.30 ($1/a = 0.73$ a.u. for He–Ar), divided by the mean relative speed, Eq. 2.12. We obtain readily†

$$\Delta \nu = \frac{\Delta \omega}{2\pi c} = \frac{2.5}{\pi c \sigma} \overline{v}_{12} , \qquad (3.3)$$

that is $\Delta \nu = 120$, 60 and 37 cm^{-1}, respectively, in reasonable agreement with the observations.

At constant temperature, the observed widths of the spectral functions decrease with increasing mass of the collisional pair. This fact is a simple consequence of the mean translational energy of a pair, $\frac{1}{2} m_{12} \overline{v_{\text{rel}}^2} = \frac{3}{2} kT$, which is the same for all pairs. The interaction time is roughly proportional to the reciprocal root mean square speed, and thus to the square root of the reduced mass.

Temperature variation. The spectra shown in Figs. 3.1 and 3.2 were ob-

† The roots σ of the interaction potentials of He–Ne, Ne–Ar, Ar–Kr are 0.295, 0.305 and 0.345 nm, respectively.

tained at room temperature. From the brief discussion above of half widths of the spectral functions one would expect that temperature variation will affect the line widths. In fact, from Eq. 3.3 one concludes that halfwidths vary roughly as $T^{1/2}$, and a number of measurements of collision-induced absorption spectra in rare gas mixtures at lower and higher temperatures confirm that conclusion [22, 95, 97, 250, 329, 352]. Several examples of the temperature dependences of induced profiles will be shown below of molecular gases and mixtures.

Molecular systems. Translational spectra like the ones shown for rare gas mixtures exist also for molecular gases and mixtures involving molecular gases. However, in that case, the rotational induced band will in general affect the appearance of the translational line since it appears generally at nearly the same frequencies.

For dissimilar rare gas pairs, the dipole is mainly induced by exchange forces; it falls off roughly exponentially with separation R. If, however, systems involving one or more molecules are considered, multipole induction is usually the dominant dipole generating mechanism; multipole-induced dipoles fall off as R^{-n}, where $n = 4$ for quadrupolar induction. We note that exchange and dispersion dipole components are also known to affect molecular induced spectra to some extent; see Chapter 4 for details. The range of multipolar induction is generally greater than that of exchange dipoles which affects the observable widths of the spectra.

Molecules rotate. As a consequence, the induced dipole $\mu(t)$ as function of time is likely to show a modulation by the rotational frequencies which, when Fourier transformed, leads to the appearance of induced rotational lines or bands. These occur at low frequencies in the microwave and far infrared region and are in general superimposed with the translational line, especially at higher temperatures. Only molecules that have a large rotational constant, e.g., H_2 ($B_0 \approx 60$ cm^{-1}), reveal substantial parts of the translational spectra, see Figs. 3.10 and 3.12, pp. 82 and 85, as examples.

Above it was pointed out that in unmixed rare gases binary absorption does not exist because of the inversion symmetry of like pairs. For like molecular pairs inversion symmetry does in general not exist because of the anisotropic structure and vibrational excitations of the individual molecules. In Chapter 5, we will show that in pure hydrogen gas, for example, the translational spectrum arises mainly from orientational ('magnetic') transitions; the translational spectrum of H_2–H_2 is discernible in Fig. 3.10 at low frequencies ($0 < \nu < 250$ cm^{-1}). The translational peak is weak if compared to the strong $S_0(J)$ lines near 354 and 587 cm^{-1}, but its strength is comparable to those of the dissimilar rare gas pairs, Fig. 3.1. The translational H_2–He pair spectra are somewhat stronger, Fig. 3.12, mainly because of the exchange dipole component that exists for H_2–He but not for H_2–H_2.

Translational spectra of most other systems involving molecules are generally hardly discernible, see Figs. 3.17, 3.20 and 3.21, etc., pp. 90, 93 and 93, because of the relatively strong, unresolved induced rotational bands that appear more or less at the same frequencies.

Dimer spectra. Intermolecular forces are attractive at distant range. Consequently, most pairs of atoms/molecules may exist as a bound van der Waals molecule. Exceptions are a few very light systems which have very shallow wells, for example H_2–He, H–He and He–He.

Many common dimers possess just a few rotovibrational levels, but some, like Xe_2, possess thousands of rotovibrational states. Molecular transition frequencies vary from just a few wavenumbers to tens of wavenumbers. Besides these rotovibrational bands, dissociation continua exist, see Chapters 5 and 6 for details. Experimentally, very little is known about these dimer signatures in the translational band, presumably because high-resolution work in the far infrared is difficult. Furthermore, because of the feebleness of induced dipoles, high pressures are commonly necessary for a recording of the absorption spectra. This fact tends to pressure-broaden dimer lines to a point where their observation may be impossible.

Spectral moments. For the analysis of collision-induced spectra and the comparison with theory, certain integrals of the spectra, the spectral moments, are of interest. Specifically, we define the nth moment of the spectral function, $g(v)$, by

$$M_n = \int_{-\infty}^{\infty} v^n \, V g(v) \, dv \, . \tag{3.4}$$

The index n is a small integer, $n = 0, 1, \ldots$ Because of the nearly exponential fall-off of typical spectral functions, these integrals do exist. An evaluation of spectral moments is possible if good measurements over a sufficiently broad frequency band exist. (We note that units of spectral moments specified elsewhere sometimes differ from those implied here, mainly because angular frequency, $\omega^n \, d\omega$, is often substituted for frequency in wavenumbers, $v^n \, dv$.)

The zeroth moment (n=0) gives the total intensity and is related to theory by familiar sum formulae (Chapter 5). For nearly classical systems (i.e., massive pairs at high temperature and not too high frequencies), the first moment (n=1) is very small and actually drops to zero in the classical limit as we will see in Chapter 5. The ratio of second and zeroth moment defines some average frequency squared and may be considered a mean spectral width squared. A complete set of moments ($n = 0, 1, \ldots \infty$) may be considered equivalent to the knowledge of the spectral line shape, but, practically speaking, moments higher than the second have rarely been determined, presumably because of experimental difficulties related to the exponential intensity fall-off of the profiles. Nevertheless, with the knowledge of the lowest three moments and a good choice of a model profile, surprisingly good representations of many spectra of practical significance have been obtained, see p. 275ff. This fact indicates that two or three spectral moments constitute substantial knowledge of such spectra under conditions where the bound states are not of a great significance.

Instead of the moments M_n of the spectral function, integrals of the absorption coefficient $\alpha(\nu)$ are often obtained experimentally‡ that are closely related to the M_n through Eq. 3.2,

$$\gamma_1 = \frac{1}{\varrho_1\varrho_2}\int_0^\infty \alpha(\nu)\,d\nu\,; \tag{3.5}$$

$$\gamma_0 = \frac{1}{\varrho_1\varrho_2}\int_0^\infty \coth\left(hc\nu/2kT\right)\,\alpha(\nu)\,\frac{d\nu}{\nu}\,. \tag{3.6}$$

It is easy to see from Eq. 3.2 that for the rare gas mixtures, these equations are related to the moments M_n, according to

$$\gamma_1 = \frac{4\pi^2 N_L^2}{3\hbar c} M_1 \quad\text{and}\quad \gamma_0 = \frac{4\pi^2 N_L^2}{3\hbar c} M_0\,. \tag{3.7}$$

For classical systems, γ_1 may be expressed in terms of the second moment, M_2, as $\gamma_1 \approx 16\pi^4 c N_L^2 M_2/(3kT)$ if $hc\nu \ll kT/2$.

Table 3.1 lists measured spectral moments of rare gas mixtures at various temperatures. (We note that absorption in helium–neon mixtures has been measured recently [253]. This mixture absorbs very weakly so that pressures of 1500 bar had to be used. Under these conditions, one would expect significant many-body interactions; the measurement almost certainly does not represent binary spectra.) For easy reference below, we note that the precision of the data quoted in the Table is not at all uniform. Accurate values of the moments require good absorption measurements over the whole translational frequency band, from zero to the highest frequencies where radiation is absorbed. Such data are, however, difficult to obtain. Good measurements of the absorption coefficient $\alpha(\nu)$ require ratios of transmitted to incident intensities, $I(\nu)/I_0$, that are significantly smaller than unity and, at the same time, of the order of unity, i.e., not too small. Since in the far infrared the lengths of absorption paths are limited to a few meters and gas densities are limited to obtain purely

‡ Units of spectral moments specified by other workers may differ from those implied here because frequency in wavenumbers, ν, is often replaced by angular frequency, ω. Also, normalization by the product of densities, $\varrho_1\varrho_2$, is sometimes suppressed, especially if many-body interactions are considered. Furthermore, factors of $\hbar/2kT$ or $\hbar c/2kT$ have been used to the right of Eq. 3.6 in the definition of γ_0.

Table 3.1. Binary integrated absorption coefficients of rare gas mixtures.

pair	T	γ_1		γ_0		Ref.
	(K)	(10^{-4} cm^{-2} am^{-2})		(10^{-6} cm^{-1} am^{-2})		
		meas.	calc.	meas.	calc.	
He–Ar	140	1.2	1.50	2.8	3.80	[96]
	165	1.4	1.65	3.5	4.24	[96]
	200	1.6	1.87	3.9	4.87	[96]
	240	1.9	2.11	5.0	5.56	[96]
	295	2.1	2.44	5.1	6.52	[75]
He–Kr	200	4.1		10.6		[96]
	240	5.1		12.5		[96]
	300	5.2		14		[22]
	480	9.1		28		[22]
	620	27				[22]
He–Xe	230	7.3				[352]
	300	9.9		34		[22]
	480	18		73		[22]
	620	32				[22]
Ne–Ar	135	0.51	0.71		5.2	[352]
	165	0.70	0.77	5.0	5.7	[96]
	295	1.06	1.07	7.6	8.3	[75]
	480	4.8	1.5	29	12.2	[22]
Ne–Kr	175	1.6				[352]
	200	1.8		15.1		[96]
	240	1.9		16.4		[96]
	295	1.9		17[a]		[329]
	480	3.9		47		[22]
Ne–Xe	230	3.9				[352]
	295	4.4		35[a]		[329]
	480	7.2		79		[22]
Ar–Kr	170	0.44				[352]
	300	0.28		4.9		[104]
	295	0.50				[329]
Ar–Xe	230	1.33				[352]
	295	1.16				[329]
Kr–Xe	229	0.96				[352]

[a] J. Quazza, thesis (1972), unpublished; data taken from [22].

binary spectra, absorption cannot be measured accurately beyond some upper limit of frequency. As a consequence, estimates based on some analytical or graphical extrapolations must usually be employed (for γ_1, especially at high frequencies). Even for the best existing measurements that limits the precision of the moments.

Similarly, an accurate determination of the moment γ_0 generally requires good absorption data at low frequencies. Since $\coth(hcv/2kT)/v \approx v^{-2}$ for $hcv \ll kT/2$, the integrand, Eq. 3.6, is most significant at the lowest frequencies where absorption is small and difficult to measure. Microwave resonator techniques have been employed to provide good absorption data at low frequencies. These, however, require a very different apparatus and are usually limited in frequency as well. As a consequence, in the translational band, γ_0 may be quite uncertain. Summarizing, one could say that for the spectral moments γ_1 and γ_0 an accuracy of $\pm 10\%$ is admirable; significantly greater uncertainties are probably common. These simple facts must be kept in mind when dealing with measurements of the moments of the translational band.

According to Eq. 3.5, γ_1 may be considered the 'total' absorption in the translational band. We, however, prefer to consider M_0 the total intensity, Eq. 3.4 with $n = 0$, because the spectral function $g(v)$ is more closely related to the emission (absorption) process than $\alpha(v)$. For rare gas mixtures, we have the relationships of Eqs. 3.7. In other words, γ_0 may be considered a total intensity of the spectral function, $g(v)$, and the ratio γ_1/γ_0 is a mean width of the spectral function (in units of cm^{-1}). Both moments increase with temperature as Table 3.1 shows. With increasing temperature closer encounters occur, which leads to increased induced dipole moments and thus greater intensities.

At a given temperature, absolute intensities are greatest for those pairs that differ the most in their electronic structure or nuclear charge, e.g., He–Xe shows stronger absorption than Ne–Xe or He–Kr. This is, of course, consistent with the fact mentioned above that like pairs (He–He) shows no absorption at all, but we point out that the variation of the spectral moments of the various pairs shown is relatively minor if data at the same temperature are considered, Table 3.1.

For any given system, the average width γ_1/γ_0 increases with temperature roughly as $T^{1/2}$ as one may readily conclude from Table 3.1, in agreement with our simple interpretation above related to the duration of collisions, Eq. 3.3.

For He–Ar spectral moments have been computed from first principles, using advanced quantum chemical methods [278]; details may be found in Chapters 4 and 5. We quote the results of the *ab initio* calculations of the moments in Table 3.1, columns 4 and 6. The agreement with measurement is satisfactory in view of the experimental uncertainties. We note that theory and measurement may be compared more closely if spectral *line shapes* are considered instead of moments, see p. 243. In that case, the only errors are those of the measurement of the absorption coefficient, $\alpha(\nu)$; no additional uncertainties due to extrapolation affect the comparison of measured and computed line shapes. Figure 5.5 presents an example of an *ab initio* line shape calculation (solid curve). The comparison with the measurement (dots) shows close agreement where the accuracy of the measurement is greatest, i.e., near the peak of $\alpha(\nu)$. At lower and at higher frequencies, the accuracy of the experimental data suffers somewhat, the more so the smaller the absorption coefficient becomes, and the measurement departs increasingly from the calculated profile. The agreement is, however, within the estimated experimental uncertainties. It would be interesting to see if new, accurate measurements follow the theoretical profile more closely.

Induced dipoles of other pairs have also been obtained by quantum chemical computations [44]. Whereas these computations are not as sophisticated as the ones mentioned above and close agreement with observations is not achieved for some of the systems considered, in the case of Ne–Ar they have resulted in a dipole surface that reproduces the best absorption measurements closely. The Ne–Ar induced dipole may, therefore, be recommended as a reliable, but perhaps semi-empirical surface (because its reliability is judged not solely on theoretical grounds). Spectral moments computed with that surface are also given in Table 3.1.

2 Ternary and many-body translational spectra

Above we have looked at the translational spectra of binary systems. These are obtainable experimentally at sufficiently low densities, especially when absorption of the infrared inactive gases is studied. An induced spectrum may be considered to be of a binary nature if the integrated intensity varies as density squared. In that case, the *shape* of the densities-normalized absorption coefficient, $\alpha/\varrho_1\varrho_2$, is invariant, regardless of the densities employed.[§]

With increasing density, N-body interactions with $N = 3, 4, \dots$ may have a discernible effect on the total intensities as well as on the shape of the absorption profile. One may expect a ternary component, and at higher densities perhaps four-body, etc., spectral components that are superimposed with the binary spectrum. At the highest densities (e.g., liquids and pressurized fluids) every monomer may be assumed in permanent interaction with a substantial number of near neighbors. At 'intermediate' densities, that is well below liquid densities, one may be

[§] See, however, deviations from this behavior over a small band near zero frequency, p. 68ff.

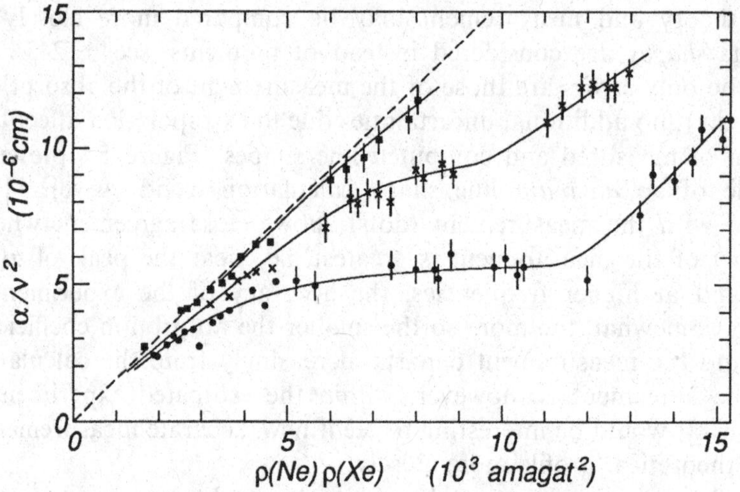

Fig. 3.3. Absorption of neon–xenon gas mixtures at 295 K and the fixed frequency of 4.4 cm^{-1} as function of the product of neon and xenon density [132]. Three ratios of Ne and Xe densities are used: $\kappa = 1.95$ (■); 1.53 (×); and 0.59 (•). Reproduced with permission from the National Research Council of Canada, from [132].

successful in separating binary from higher-order spectral components on the basis of their different variations with density.

We will look next at the variations of measured spectra due to many-body interactions. These manifest themselves in two different ways. One is a relatively sharp 'intercollisional dip' near zero frequency. The other is a diffuse spectral component which leads to line narrowing.

Intercollisional interference. We note that at the lowest frequencies the simple proportionality between absorption coefficient and product of gas densities breaks down. Under such conditions, certain many-body interactions affect the observations and modify the shape or intensities of the binary spectra, often quite strikingly. An example is shown in Fig. 3.3, a measurement of the absorption in a neon–xenon mixture in the microwave region, at the fixed frequency of 4.4 cm^{-1}. Because of the frequency-dependent factor of $g(\nu)$ that falls off to zero frequency as ν^2, absorption is extremely small at such frequencies, Eq. 3.2. As a consequence, it has generally been necessary to use sensitive resonator techniques for a measurement of the absorption at microwave frequencies [131].

Figure 3.3 shows the microwave collision-induced absorption coefficient of a neon–xenon gas mixture as function of the product of the Ne and Xe densities; the ratios of Ne to Xe densities are $\kappa = 1.95$, 1.53 and 0.59

[132]; temperature and frequency were 295 K and 4.4 cm^{-1}. The measured absorption is divided by frequency squared, so that in effect the spectral function, $g(\nu)$, times the product of neon and xenon density is plotted in the figure, see Eq. 3.2 with $\nu[1 - \exp(-hc\nu/kT)] \approx \nu^2$ for $hc\nu \ll kT$. If the observed absorption were strictly due to Ne–Xe pairs, the data should fall on a straight line, perhaps the dashed line shown. At the lower values of $\varrho_1\varrho_2$, the absorption measured with the largest density ratio of $\varrho_1/\varrho_2 = \kappa = 1.95$ nearly follows such a straight line (squares ▪). If, however, the number of xenon atoms per neon atom (or, rather, the Ne–Xe collision frequency) is increased while the product of neon and xenon densities is kept constant, substantially reduced absorption is observed as the lower two curves indicate ($\kappa = 1.53$ (×) and 0.59 (•), respectively). In other words, the linear relationship that is characteristic of binary interactions breaks down for small κ, when the reciprocal average time between Ne–Xe collisions approaches the microwave frequency. At lower microwave frequencies, the breakdown of the linear relationship occurs at an even greater κ (Dagg 1985).

This breakdown of the linear relationship between the absorption coefficient α and the product of densities, $\varrho_1\varrho_2$, indicates that the observed absorption is not a binary process. Specifically, for the case at hand, one can no longer assume that the measured absorption consists of an *incoherent superposition* of the pair contributions. Rather, the correlations of the dipoles that are induced in subsequent binary collisions lead to a partially destructive interference, an absorption *defect* that occurs if the product of the time τ_{12} between Ne–Xe collisions, and microwave frequency, f, approaches unity [404]. We note that for the spectra shown above, Figs. 3.1 and 3.2, the product $f\tau_{12}$ is substantially greater than unity at all frequencies where experimental data are shown and, consequently, incoherent superpositions of the waves arising from different induced dipoles occur. The 'intercollisional' absorption defect is limited to low frequencies (Lewis 1980).

Figure 3.4 is a schematic illustration of the fact that dipoles induced in successive collisions tend to be more or less antiparallel. This anticorrelation of dipoles induced in subsequent collisions leads to the absorption defect and causes the breakdown of the pair behavior illustrated in Fig. 3.3, if the product $f\tau_{12}$ is of the order of unity or less. If, on the other hand, $f\tau_{12} \gg 1$, superposition occurs with widely varying, random phase differences which render an interference effect inefficient.

The time between collisions of dissimilar atoms (Ne and Xe) may be estimated by Eq. 2.12; the subscripts 1 and 2 stand for Ne and Xe, respectively. The elastic scattering cross section amounts to $q_{12} \approx 10^{-15}$ cm^2 for thermal Ne–Xe collisions. The frequency of $\nu = 4.4$ cm^{-1} corresponds to $\nu c = 1.3 \times 10^{11}$ Hz. From this information, we find that the

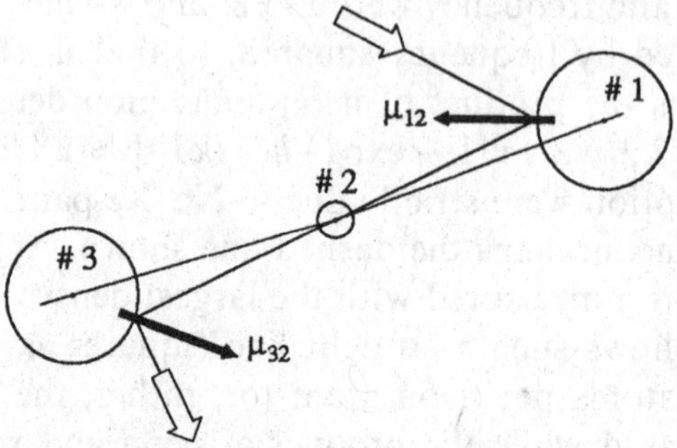

Fig. 3.4. The dipoles induced in successive collisions of a light particle (small circle) with massive ones; the induced dipoles μ_{12} and μ_{32} tend to be more or less antiparallel.

product

$$\nu\tau_{12} = \left(\frac{\pi m_{12}}{8kT}\right)^{1/2} \frac{\nu}{n(\mathrm{Xe})\, Q_{12}} \tag{3.8}$$

equals unity for a Xe density of 2×10^{27} m^{-3} or 75 amagats. For a given ratio κ of Ne and Xe densities, the breakdown of the linear relationship between α and the product $\varrho(\mathrm{Ne})\varrho(\mathrm{Xe})$ is, therefore, to be expected at values of the product $\varrho(\mathrm{Ne})\varrho(\mathrm{Xe}) = \kappa$ (75 amagat)^2, that is 3400, 8700, and 11 000 amagat^2 for the lowermost, middle, and uppermost curve, Fig. 3.4, respectively. The deviation of the lowermost curve from the dashed straight line is about 20% at 3400 am^2, just what the deviation of the middle curve at 8700 am^2 is. The data of the uppermost curve do not extend to 11 000 am^2 but at the highest density shown, it deviates from the dashed line by much less than 20%, which is consistent with the rough estimate. We note that for each of the curves shown, at densities squared higher than these limiting values, the product of frequency and time between collisions, $\nu\tau_{12}$, falls off rapidly to much smaller values than unity which causes even stronger destructive interference.

The spectral profile of the intercollisional dip has been recorded as function of frequency, at constant densities, for a few gases and mixtures [130]. The lower part of Fig. 3.5 shows the low-frequency part of the spectral function, obtained for the 1:1 mixture of helium and argon at a total pressure of 160 atmospheres and room temperature [130]. The sudden drop in absorption below 10 cm^{-1} was seen to vary with density whereas, at the higher frequencies, no such variation of the shape is observed, other than the simple scaling of the entire profile with density squared. The time between collisions of He and Ar atoms is readily obtained as $\tau_{12} = 0.9 \times 10^{-12}$ s. The product $f\tau_{12}$ equals unity for the

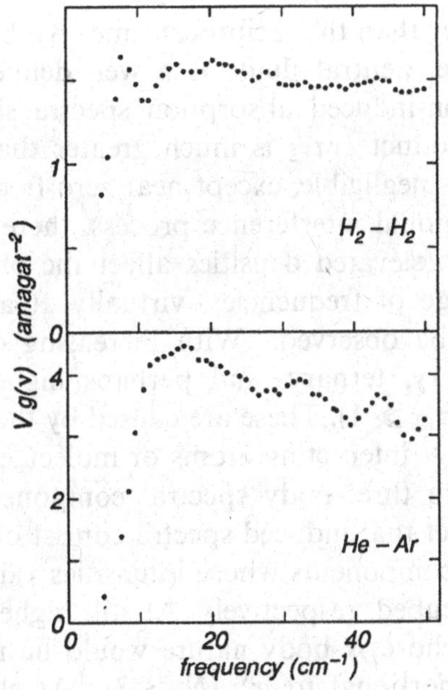

Fig. 3.5. *Upper part:* The low-frequency portion of the spectral function for hydrogen at 140 atmospheres showing the intercollisional dip. *Lower part:* The low-frequency portion of the spectral function for a 1:1 mixture of helium and argon at the total pressure of 160 atmospheres showing the intercollisional dip. Reproduced with permission from the National Research Council of Canada, from [130].

frequency of ≈ 10 cm^{-1}, and for lower frequencies absorption decreases rapidly just as this was seen above. Similar studies exist at much higher pressures where the intercollisional profile is even broader; for helium–argon mixtures in the 1000 atmospheres range, observed dips are ≈ 50 cm^{-1} wide [249] (Lewis 1985). Similar observations were reported for helium–xenon and argon–xenon mixtures [249].

Translational spectra involving molecules show similar dips, see as an example the upper part of Fig. 3.5 which was recorded with hydrogen at room temperature. A more refined treatment of the intercollisional process must take into account not only two arbitrarily selected consecutive collisions. Rather, correlations to all orders are important which render the intercollisional interference a many-body process, not exactly the three-body mechanism our simplified discussion above seems to suggest.

Diffuse component due to three-body interactions. The intercollisional interference process is a many-body effect arising from the correlations of dipoles induced in consecutive collisions. This effect is limited to a certain narrow frequency band defined by $\tau_{12} c\nu \lesssim 1$, that is to frequencies, $c\nu$, comparable to, or smaller than the reciprocal time, τ_{12}, between collisions (which in not too dense, neutral fluids is a well defined quantity). In the examples of collision-induced absorption spectra shown above, for most frequencies the product $cv\tau_{12}$ is much greater than unity and the interference effect is thus negligible, except near zero frequency.

Besides this intercollisional interference process, there are other three-body processes which at elevated densities affect the observable spectra over a much wider range of frequencies, virtually at all frequencies at which absorption may be observed. With increasing density, one will be able to discern binary, ternary, and perhaps higher-order spectral contributions (even if $cv\tau_{12} \gg 1$). These are caused by the dipoles induced in systems consisting of N interacting atoms or molecules, with $N > 2$.

At gas densities where three-body spectral components are just discernible, one might expect that induced spectra consist of a superposition of two- and three-body components whose intensities vary proportionally to density squared and cubed, respectively. At still higher densities, components might appear whose N-body nature would be revealed by their intensity variations proportional to ϱ^N ($N > 3$). At elevated densities, analyses of collision-induced absorption spectra have indeed revealed the presence of spectral components of an intensity that varies according to ϱ^N, with $N = 3$ and, in a few cases using higher densities, with $N > 3$.

3 Virial expansions

The question arises whether collision-induced absorption spectra observed at elevated density can be cast in a form that separates the various many-body components into meaningful two-, three-, etc., body *virial* coefficients.

It is well known that, for example, the *equation of state* may be written as an expansion in terms of powers of density, Eq. 1.1. The resulting virial coefficients A, B, C, ... describe the various contributions arising from one-, two-, three-, ..., body processes to the *pressure* of a gas. In the low-density limit, the pressure (understood to be force per unit area) arises from the change of momentum of the particles bouncing off a wall of the container, which is given by the first virial coefficient, A, of the equation of state. Besides the forces exerted by the walls of the container, each particle of a gas experiences also the intermolecular forces. The effects on pressure of binary interactions are summarized in the second virial coefficient, B, and in general, N-body interactions are described by the Nth virial coefficient, in familiar ways [270]. We note that, while the virial expansion of pressure has a significant one-body (or A) term, the leading term of collision-induced absorption will be a two-body term, $\tilde{B}$, if infrared inactive gases and mixtures are considered.

Returning to our topic of collision-induced absorption, the interesting point to be made here is this: while an expansion of the spectra, $\alpha(\nu)$ or $g(\nu)$, in powers of the densities, ϱ_1 and ϱ_2, at fixed frequencies ν is always possible, the resulting spectral components $g^{(N)}(\nu)$ can be related theoretically to pair-, triplet-, etc., properties *only under certain circumstances.*† If these conditions are not satisfied, the expansion of measured spectra may still be possible, but the coefficients $g^{(N)}$ thus defined are *not* virial coefficients in the usual sense: the coefficient of ϱ^N may not describe strictly the induced N-body component.

For the spectral moments, Eqs. 3.4–3.6, it has long been known that virial expansions exist [400, 402],

$$M_n = M_n^{(11)}\,\varrho_1\varrho_2 + M_n^{(21)}\,\varrho_1^2\varrho_2 + M_n^{(12)}\,\varrho_1\varrho_2^2 + \dots \tag{3.9}$$

The coefficients $M_n^{(ik)}$ describe the $(i + k)$-body contribution involving i atoms of species 1 and k atoms of 2. At not too high densities, the virial expansion of spectral moments provides a sound basis for the study of the spectroscopic three-body (and possibly higher) effects. We note that theoretically terms like $M_n^{(30)}\,\varrho_1^3$ and $M_n^{(03)}\,\varrho_2^3$ should be included in the expansion, Eq. 3.9. These correspond to homonuclear three-body contributions which, however, were experimentally shown to be insignificant in the rare gases and are omitted, see p. 58 for details.

Measurements. Evidence for many-body processes beyond the intercollisional dip is presented in Fig. 3.6 which shows the variation of the moment $\gamma_1 = \int \alpha(\nu)\,d\nu$ with the product of the densities [329], $\varrho_1\varrho_2$. For strictly binary interactions, a straight line is expected and is indeed observed at the lower values of the product of densities. Above a certain threshold that is different for each system shown, superlinear dependences are observed that indicate the emergence of spectral components arising from higher than binary interactions.

Measurements such as these can be conducted to determine the three-body virial coefficients, $M_n^{(12)}$ and $M_n^{(21)}$ of collision-induced absorption. To that end, it is useful to measure the variation of γ_1 (and also of γ_0, Eq. 3.6, where possible) with small amounts of gas 1 mixed with large amounts of the other gas 2, and with small amounts of 2 mixed with 1, to determine the ternary spectral moments $M_n^{(12)}$ and $M_n^{(21)}$ separately, with a minumum of interference from the weaker terms. In a mixture of helium and argon, for example, two different three-body coefficients can be determined, those of the He–Ar–Ar and the He–He–Ar complexes.

The data presented in Fig. 3.6 suggest that a diffuse many-body effect exists which affects the binary spectra over a much broader range of

† Recent work has defined the conditions under which such separation is possible [297, 298]; see p. 229ff.

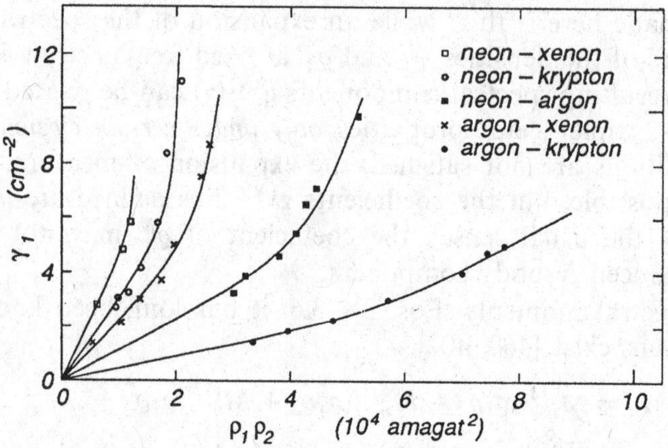

Fig. 3.6. The spectral moment γ_1 as function of the product of densities, for various rare-gas mixtures at room temperature; only one density was varied for each system: the neon densities were fixed at 77, 31 and 46.5 amagats for the neon–argon, neon–krypton and neon–xenon mixtures, respectively; and the krypton and xenon densities were fixed at 152 and 50 amagats, respectively, in their mixtures with argon. The departures from the straight lines seen at intermediate densities squared indicate the presence of many-body interactions. Reprinted with permission by Pergamon Press from [329].

frequencies than the intercollisional dip. The observed upturns seem to suggest that the N-body coefficients of γ_1 with $N > 2$ are positive and thus distinctly different from the intercollisional dips which are associated with 'negative intensities.' We note in passing that for a fit of the data presented in Fig. 3.6, Marteau included a four-body component in the analysis on a trial basis [252]. Indeed, at the highest densities four-body (and perhaps higher) contributions may not be negligible in the measurement. For the determination of accurate three-body contributions, measurements at lower densities are preferred, as in Table 3.2.

Table 3.2 shows the results of measurements of the two significant ternary moments in helium–argon mixtures at 165 K [95]. One series of measurements used a ratio of partial pressures of argon and helium close to 2.3 and argon densities from 156 to 280 amagat; and another one used a density ratio of 0.31. The densities used in these measurements were thus substantially lower than the highest densities of the work displayed in Fig. 3.6. These data should therefore be less affected by higher-order interactions. The two three-body coefficients determined from these series of measurements are positive and of the order of $\gamma_1^{(3)} \approx 10^{-7}$ cm^{-2}amagat^{-3}, Table 3.2, to be compared with the binary coefficient of $\approx 10^{-4}$ cm^{-2}amagat^{-2}, from Table 3.1 (p. 65). In other words, at densities of helium or argon around 100 amagats, a ternary contribution

Table 3.2. Ternary integrated absorption coefficients of the translational band. Measurement from [95]; calculation [296].

system	T (K)	$\gamma_1^{(3)}$ (10^{-8} cm^{-2} am^{-3}) meas.	$\gamma_1^{(3)}$ (10^{-8} cm^{-2} am^{-3}) calc.	$\gamma_0^{(3)}$ (10^{-9} cm^{-1} am^{-3}) calc.
He–He–Ar	165	7	5.6	0.40
He–Ar–Ar	165	10	3.3	−1.2
Ne–Ar–Ar	145		1.10	−2.1
	128		0.70	−2.2
	112		0.65	−2.0
	92		1.88	−0.7

of $\approx 10\%$ to γ_1 is to be expected, a number experimentalists should keep in mind when binary spectra are to be recorded with little interference from three-body processes. We will see below that at lower temperatures relatively greater three-body contributions are expected. Unfortunately, no three-body γ_0 or M_0 measurements have been reported for rare-gas mixtures at low densities.

Theoretical estimates of the three-body moments may be obtained from the well-known pair dipole moments. These do not include the 'irreducible' three-body components which are poorly known. Interestingly, in every case considered to date, the computations of the three-body spectral moments $\gamma_1^{(3)}$ are always *smaller* than the measurements, a fact that suggests significant positive irreducible three-body dipole components for all systems hitherto considered [296, 299]; further details may be found in Chapter 5.

4 Dense fluids

The variations of spectral absorption profiles with density have been measured in mixtures of argon and krypton, and also of neon and argon, at various densities approaching liquid density, around 500 amagats [242, 252]. Under such conditions, each atom is in permanent interaction with a fairly large number of near neighbors, roughly 12 in the 'first shell', and with many more distant ('second shell') neighbors. The resulting spectra may thus be expected to differ from the binary translational spectra shown above, Figs. 3.1 and 3.2.

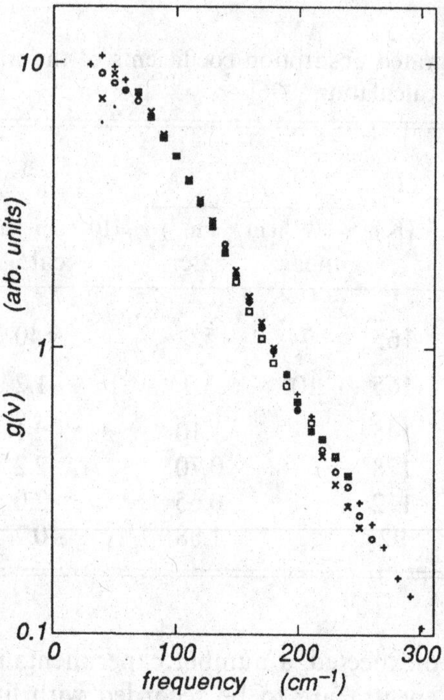

Fig. 3.7. Spectral function of neon–argon mixtures at high densities, obtained with a constant neon density of 77 amagat; the argon densities are 416 (∘), 488 (×), 530 (□), and 553 (·) amagats, respectively. For comparison, the binary profile [75] (+) is also shown. All profiles are normalized so that intensities are equal at peak absorption (near 100 cm^{-1}); after [252].

It is thus perhaps surprising that the profiles obtained at near liquid densities do not dramatically differ from the binary profiles. Figure 3.7 shows the spectral function obtained by Marteau using 77 amagats of neon and varying argon densities, namely 416 (∘); 488 (×); 530 (□); and 553 (·) amagats, respectively. For comparison, Bosomworth and Gush's [75] binary Ne–Ar spectrum is also shown (+). For easy comparison of the spectral profiles, the data have been arbitrarily normalized to unit intensity at the peak of the absorption, at a frequency of about 100 cm^{-1}.

In a semi-logarithmic grid the profiles look much like the binary spectra, Fig. 3.2. The roughly linear fall-off at the higher frequencies > 80 cm^{-1} suggests a near-exponential wing. At the higher frequencies, the high-density data seem to be slightly below the binary profile. However, the noise amplitudes of the measurements are of a magnitude comparable to the observed differences so that the exact trends are not obvious. In any case, in the case of the rare gas mixtures, the variation of line shape with density is small, certainly at the higher frequencies, $\nu > 100$ cm^{-1}.

The most significant differences of the various profiles shown, Fig. 3.7, occur at the lowest frequencies. Albeit the measurements do not extend down to zero frequency, it seems clear that at the lowest frequencies the intercollisional process has affected the profiles. We notice the beginning of a dip similar to the ones seen in Fig. 3.5. At the higher densities the intercollisional wing of the inverted Lorentzian extends to much higher frequencies, the more so the higher the densities are: the intercollisional dip persists to the highest densities.

Similar results were also obtained for argon–krypton mixtures [252]. Apart from the low-frequency region of the intercollisional dip, the variation of the translational line shape is rather subtle; 'reduced' absorption profiles of a number of rare gas mixtures at near-liquid densities (up to 750 amagat) have been proposed which ignore these variations totally [252].

Liquids. The translational absorption profiles of a 2% solution of neon in liquid argon have been measured at various temperatures along the coexistence curve of the gas and liquid phases [107]. Figure 3.8 shows the symmetrized spectral function‡ at four densities. At the lowest density (479 amagat; for $T = 145$ K; curve at top) the profile looks much like the binary spectral function seen in Fig. 3.2, especially the near-exponential wing for frequencies $\nu > 25$ cm^{-1}. With increasing density the intercollisional dip develops at low frequencies, much like the dips seen at much lower densities in Fig. 3.5 – only much broader.

The variation of absorption with density is studied best with the help of Table 3.3, which gives the spectral invariants γ_1, γ_0, normalized by the product of binary densities, $\varrho_1\varrho_2$, obtained in these measurements. These are compared with theoretical two-body moments calculated from first principles [44]; three-body moments are also available, calculated with the assumptions of pairwise-additive three-body induced dipoles [296], Table 3.2 on p. 75. Whereas in the liquid the measured γ_1 are consistently roughly 50% greater than the calculated binary moments, the measured γ_0 fall off to significantly smaller values than the binary values, especially at the highest densities. In other words, the effect of many-body interactions is to increase γ_1 and to reduce γ_0 relative to their binary values (which were seen to be in close agreement with measured moments in the low-density limit). The former observation is consistent with the behavior displayed in Fig. 3.6 which shows that for several rare gas mixtures the integral γ_1 increases with density above the binary values at sufficiently high densities.

‡ Symmetrized spectral functions are sometimes used when dealing with near classical systems. In this case, the function shown in Fig. 3.8 is the same as $Vg(\nu)$ of Eq. 3.2, multiplied by $\tanh(hc\nu/2kT)$; $Vg(\nu)$ and the symmetrized function differ very little at low frequencies.

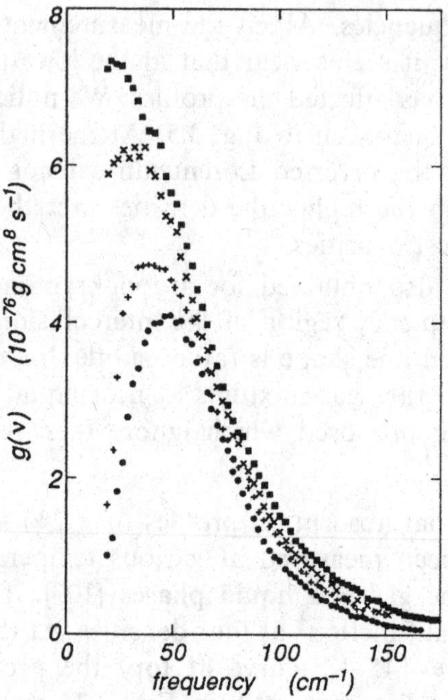

Fig. 3.8. Spectral densities, $g(\nu)$, of a mixture of 2% neon in liquid argon measured at different temperatures along the coexistence curve [107]. Temperatures $T = 145$ K (■); 128 K (×); 112 K (+); and 92 K (•). Reproduced with permission from the National Research Council of Canada, from [107].

Table 3.3. Spectral moments of the neon–argon liquid mixture along the coexistence curve; measurement [107] compared with binary values calculated from first principles. (Calculated ternary moments are given in Table 3.2 above.)

T (K)	ϱ_{Ar} (amagat)	γ_1 $(10^{-4}\text{cm}^{-2}\text{am}^{-2})$ meas.	calc.[a]	γ_0 $(10^{-5}\text{cm}^{-1}\text{am}^{-2})$ meas.	calc.[a]
92	766	0.94	0.65	0.29	0.46
112	687	0.94	0.67	0.35	0.48
128	609	1.00	0.70	0.48	0.51
145	479	1.11	0.73	0.53	0.54

[a] binary component only, for comparison.

The decrease of γ_0 with density, on the other hand, is related to interference. A general explanation must be sought in terms of the symmetries that exist at high densities: a neon atom in a cage of argon atoms interacts very little with radiation, on account of the near inversion symmetry, which is inconsistent with a dipole.

In Table 3.2 the ternary coefficients, $\gamma_1^{(3)}$, are listed which are calculated with the assumption of pairwise-additive induced dipoles. The values are *positive* like the measurement suggests. Also given in the Table are the ternary values, $\gamma_0^{(3)}$, which are *negative*, just like the measurement suggests when compared with the calculated binary moments of the liquid. However, the magnitude of both calculated ternary coefficients is certainly not sufficient to account for the difference between measured and calculated binary moments, Table 3.3; a simple calculation reveals that only a small fraction of the differences between measured and calculated moments may be explained in terms of ternary contributions. In other words, there are either significant four-body (and possibly higher-order) contributions to be accounted for, or else non-additive, 'irreducible' components affect the measurements significantly – or perhaps a combination of these. No definite answers exist at this time.

Measurements similar to those shown here have also been obtained in liquid helium–argon [107] and argon–krypton mixtures [104]. Summarizing the observations of translational spectra in rare gas mixtures near liquid density, one notes that as one approaches liquid density, the absorption profile of the translational spectra of rare gas mixtures tends to acquire wings that are more diffuse than the wings of the binary profiles. At the same time the integrated absorption falls off to values significantly smaller than those of the binary spectra, but does not vanish at the densities commonly considered. The intercollisional effect persists up to the highest densities and amounts to a significant portion of the decreasing of intensities with increasing density.

The translational spectra of pure liquid hydrogen have been recorded with para-H_2 to ortho-H_2 concentration ratios of roughly 25:75, 46:54 and 100:0, Fig. 3.9 [201, 202]. For the cases of non-vanishing ortho-H_2 concentrations, the spectra have at least a superficial similarity with the binary translational spectra; compare with the data shown for low frequencies (< 250 cm^{-1}) of Fig. 3.10 below. A comparison of the spectral moments of the low-density gas and the liquid shows even quantitative agreement within the experimental uncertainties which are, however, substantial.

It is interesting to note the near-absence of absorption in the translational band of pure para-hydrogen. At the temperature of the measurement (20.8 K), all para-H_2 molecules are in the rotational ground state ($J = 0$) which is optically isotropic. The optical properties of these molecules are thus much like those of He atoms. No translational ab-

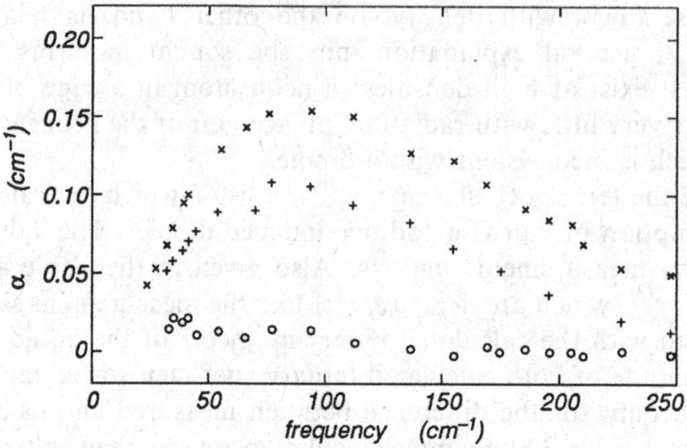

Fig. 3.9. The absorption coefficient of liquid hydrogen at the temperature 20.8 K; after [202]. The para-H_2 concentrations were 25% (×); 44 to 48% (+); and 99.8% (○). Note the absence of absorption in para-H_2.

sorption band was seen in liquid helium (or in any other dense, unmixed monatomic gas) and para hydrogen is no exception. We will see below that absorption in hydrogen proceeds largely via quadrupole induction which is coupled to the rotation of the molecule; see Chapters 4 and 6 for details. Rotational S_0 lines are therefore observed at the appropriate frequencies. The ortho-H_2 molecules, on the other hand, are in one of the orientational sublevels, ($M_J = 0, \pm 1$), of the $J = 1$ rotational state which is optically anisotropic, and induced 'magnetic' transitions ($\Delta M_J = \pm 1$) at zero frequency occur. These generate the observed translational absorption, + and × in Fig. 3.9.

The spectra of a dilute solution of normal and pure para hydrogen in liquid argon have also been recorded in the spectral range from 12 to 700 cm^{-1}, which includes almost the full translational bands. The low-frequency part of the absorption of the solutions of para hydrogen in liquid argon is dominated by the intercollisional interference effects: absorption falls off to very small values as the frequency goes to zero [103]. In this case, the translational band is generated by an isotropic dipole component much like those seen in rare gas mixtures. If, on the other hand, a solution of normal hydrogen in argon is considered, a new line appears at zero frequency, the induced $Q_0(1)$ line of ortho-H_2 that arises from quadrupolar induction, which more or less fills in the intercollisional dip and shows little intercollisional interference. In other words, the near-cancellation of dipoles induced in successive collisions (p. 70) is much reduced for the multipole-induced dipole components. Other spectra of molecular gases at high density will be considered below.

3.2 The rototranslational spectra

If molecular gases are considered, infrared spectra richer than those seen in the rare gases occur. Besides the translational spectra shown above, various rotational and rotovibrational spectral components may be expected even if the molecules are non-polar. Besides overlap, other induction mechanisms become important, most notably multipole-induced dipoles. Dipole components may be thought of as being modulated by the vibration and rotation of the interacting molecules so that induced supermolecular bands appear at the rotovibrational frequencies. In other words, besides the translational induced spectra studied above, we may expect rotational induced bands in the infrared (and rotovibrational and electronic bands at higher frequencies as this was suggested above, Eq. 1.7 and Fig. 1.3). Lines at sums and differences of such frequencies also occur and are common in the fundamental and overtone bands. We will discuss the rotational pair and triplet spectra first.

1 Rototranslational pair spectra

Molecular hydrogen has always played a special role in the collision-induced spectroscopies, for several reasons. First, the hydrogen rotational transition frequencies are widely separated so that translational, rotational and vibrational induced line spectra can all be studied with little interference, certainly at the lower temperatures. Second, as a two-electron molecule, it is amenable to fairly rigorous theoretical treatment. Very elaborate theoretical calculations can thus be compared with observations, with great benefit for the understanding of an often complex situation. Third, the H_2 molecule has a small anisotropy that may often be ignored to first approximation in theoretical studies. Actually, para-hydrogen at sufficiently low temperature is not rotationally excited and is, therefore, an isotropic system much like a helium atom; the anisotropy can, however, be turned on and off at will simply by raising and lowering the temperature. Finally, roughly 90% of the known matter in the universe is hydrogen in the ionized, atomic or molecular states; hydrogen is, therefore, one of the most important species in astrophysics. The hydrogen molecule is non-polar, and some of the most important spectra in the near and far infrared and microwave region are collision-induced.

H_2–H_2. In contrast to the case of unmixed rare gases, pure molecular gases like hydrogen generally absorb radiation in the far infrared (and elsewhere) as is shown, for example, in Fig. 3.10, for three temperatures [37]. The data shown in the figure were obtained at gas densities from 10 to 100 amagats, but when the measured absorption coefficients are

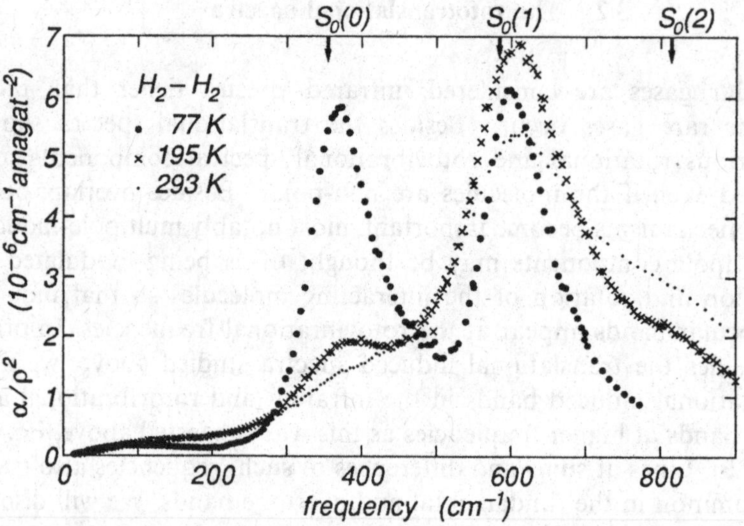

Fig. 3.10. The rototranslational absorption spectrum of H_2–H_2 pairs, recorded in equilibrium hydrogen (i.e., para-H_2 and ortho-H_2 concentrations are in thermal equilibrium proportions) at three temperatures: 77.4 K (•), 195 K (×), and 293 K (·). Various pressures from 10 to 100 atmospheres and a 3 m absorption path length were used; after [37].

normalized by density squared, smooth spectral profiles appear, indicating that the observed absorption arises from *pairs* of molecules.

We have argued above that like rare-gas pairs (Ar–Ar) do not absorb in the infrared because the inversion symmetry which is inconsistent with a dipole moment. Like molecular pairs, on the other hand, absorb in the broad vicinity of the molecular transition frequencies (and at sums or differences of these). At the lower densities where binary interactions dominate, absorption in pure hydrogen is proportional to the density squared (see Fig. 3.15 below) which suggests that H_2–H_2 pairs, just like pairs of virtually all other molecules, do in fact possess an induced dipole moment (except for certain molecular orientations of high symmetry).

An important induced dipole component of pairs involving molecules is multipolar induction. Specifically, the lowest-order multipole consistent with the symmetry of H_2 is the electric quadrupole. Each H_2 molecule may be thought of as being surrounded by an electric field of quadrupolar symmetry that rotates with the molecule. In that field, a collisional partner X is *polarized*, thus giving rise to an induced dipole moment which in turn is capable of emitting and absorbing light. For like pairs, molecule 1 will induce a dipole in molecule 2 and 2 will induce one in 1. In

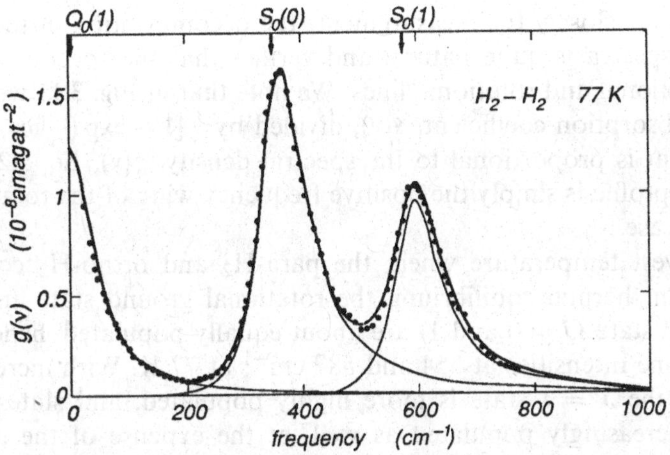

Fig. 3.11. Decomposition of the low-temperature H_2–H_2 absorption spectrum, Fig. 3.10, into translational and rotational lines (after Birnbaum and Cohen, unpublished). The heavy line is the sum of the three line profiles; the measurement is closely reproduced.

general, random orientations of the molecules may be assumed so that a non-vanishing total induced dipole moment results.

A translational 'line' like the one seen above in rare gas mixtures is relatively weak but discernible in pure hydrogen at low frequencies ($\lesssim$230 cm^{-1}), Fig. 3.10. However, if $\alpha(\nu)/[1-\exp(-hc\nu/kT)]$ is plotted instead of $\alpha(\nu)$, the line at zero frequency is prominent, Fig. 3.11: the $Q_0(1)$ line that corresponds to an orientational transition of ortho-H_2. Other absorption lines are prominent, Fig. 3.10. Especially at low temperatures, strong but diffuse $S_0(0)$ and $S_0(1)$ lines appear near the rotational transition frequencies at 354 and 587 cm^{-1}, respectively. These rotational transitions of H_2 are, of course, well known from Raman studies and correspond to $J = 0 \rightarrow 2$ and $J = 1 \rightarrow 3$ transitions; J designates the rotational quantum number. These transitions are infrared inactive in the isolated molecule. At higher temperatures, rotational lines $S_0(J)$ with $J > 1$ are also discernible; these may be seen more clearly in mixtures of hydrogen with the heavier rare gases, see for example Fig. 3.14 below.

Spectra like the ones shown in Fig. 3.10 may be readily decomposed into their line profiles. As an example, we show that the low-temperature measurement may be accurately represented by three identical profiles. Using the so-called BC model profile with three adjustable parameters and centering one at zero frequency (the $Q_0(1)$ line), another one at 354 cm^{-1} (the H_2 $S_0(0)$ line) and the third one at 587 cm^{-1} (the $S_0(1)$ line), one may fit the measurement using least mean squares techniques, Fig. 3.11. The superposition (heavy line type) of the three profiles (thin lines) reproduces closely the measurement; the decomposition of the low-temperature spectra is quite natural and verifies that the spectra consist of the translational and rotational lines. We note that in Fig. 3.11 we have plotted the absorption coefficient, $\alpha(\nu)$, divided by $\nu\,[1 - \exp(-hc\nu/kT)]$, a quantity that is proportional to the spectral density, $g(\nu)$, Eq. 3.2. The translational profile is simply the positive frequency wing of the rotational lines, in this case.

At the lowest temperature where the para-H_2 and ortho-H_2 concentrations are in thermal equilibrium, the rotational ground state and the lowest excited state ($J = 0$ and 1) are about equally populated, hence the comparable line intensities at 354 and 587 cm^{-1} at 77 K. With increasing temperature, the $J = 1$ state is more highly populated, and states with $J > 1$ are increasingly populated as well, at the expense of the $J = 0$ ground state, so that the $S_0(1)$ line shows up much more prominently than $S_0(0)$ at the higher temperatures. Profiles obtained at temperatures $T > 100$ K may similarly be fitted by simple three-parameter model profiles if one accounts for the higher $S_0(J)$ and $Q_0(J)$ lines, $J > 1$, as well. Very satisfactory fits of the laboratory data have resulted [15]. The profiles of the individual lines vary with temperature. Fairly accurate empirical spectra may be constructed, even at temperatures for which no measurements exist, when the empirical temperature dependences of the three BC parameters are known, see Chapter 5 below.

The induced rotational lines are very broad, especially at the higher temperatures, of a width consistent with the short interaction time of H_2–H_2 encounters. In fact, the $S_0(J)$ lines with $J \neq 1$ are just barely resolved at the higher temperatures, Fig. 3.10. At 77 K, the half-widths of the S_0 lines amount to roughly $\Delta\nu = 94$ cm^{-1}. With the assumption of quadrupolar induction, $\boldsymbol{\mu}(R) \sim R^{-4}$, we compute from Eq. 2.14 a range of spectroscopic interaction of $R_S \simeq \sigma/8$, with $\sigma \simeq 0.3$ nm. From this, one computes a mean duration of an inducing collision of $\approx 3\times10^{-14}$ seconds. According to the uncertainty relation, Eq. 2.58, this means a half-width of quadrupole-induced lines in hydrogen at 77 K of not less than 90 cm^{-1}, in agreement with observation. At the higher temperatures, T, this value may be scaled up by the factor $(T/77\,\text{K})^{1/2}$, again in reasonable agreement with the data shown in Fig. 3.10.

A rotation of the H_2 molecule through 180° creates an identical electric field. In other words, for every full rotation of a H_2 molecule, the dipole induced in the collisional partner X oscillates *twice* through the full cycle. Quadrupole induced lines occur, therefore, at twice the (classical) rotation frequencies, or with selection rules $J \to J \pm 2$, like rotational Raman lines of linear molecules. Orientational transitions ($J \to J$; $\Delta M \neq 0$) occur at zero frequency and make up the translational line. Besides multipole induction of the lowest-order multipole moments consistent with

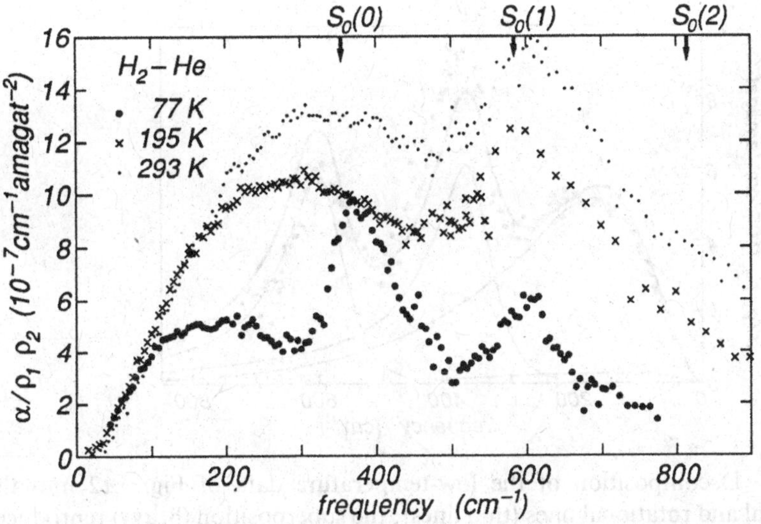

Fig. 3.12. The rototranslational absorption spectrum of H_2–He pairs at three temperatures: 77.4 K (•), 195 K (×), and 293 K (·). The data shown represent the *enhancement* of the absorption due to the addition of helium to hydrogen gas, obtained in 32 mole percent equilibrium hydrogen concentration in helium by subtraction of the H_2–H_2 spectra; after [37].

molecular symmetries, we usually have other induced dipole components. The isotropic overlap-induced dipole familiar from the rare gases may be expected to be of some importance for any dissimilar pair, and other components, like anisotropic overlap and frame distortion, are possible if molecules are involved, see the next Chapter for details. We will return to a full discussion of the hydrogen spectra in Chapter 5.

H_2–X, with X = He, Ne, Ar, Kr, or Xe. At sufficiently low densities, *mixtures* of monatomic and diatomic gases such as helium and hydrogen feature two types of CIA spectra: one arising strictly from H_2–H_2 pairs and the other from H_2–He pairs. (Of course, there are no spectral components arising from He–He interactions.) From exact measurements of the CIA spectra of both, the pure molecular gas and the mixture, the enhancement spectra due to H_2–He pairs are obtained by subtracting the spectrum of pure hydrogen from the total absorption obtained in the mixture of helium and hydrogen [37], Fig. 3.12.

The binary spectra of hydrogen–helium mixtures, Fig. 3.12, differ from the spectra of pure hydrogen, Fig. 3.10, especially by the translational 'line' (familiar from the spectra of rare gas mixtures) whose intensity increases strongly with increasing temperature. Moreover, the rotational line intensities when normalized by the product of helium and hydrogen

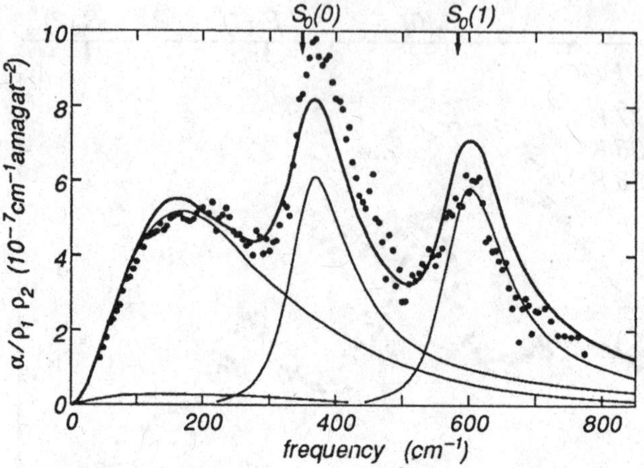

Fig. 3.13. Decomposition of the low-temperature data of Fig. 3.12 into the translational and rotational lines (thin lines); the superposition (heavy) reproduces the measurement; after [123].

densities are weaker by almost one order of magnitude. One also notices a greater scatter of the data points (noise) about the average that seems to be unavoidable in data sets obtained by subtraction of two measurements of comparable intensities.

As mentioned above, for the H_2–He pair two principal induction mechanisms exist: multipole and overlap induction. Multipolar induction occurs mainly in the electric quadrupole field of H_2 which polarizes the collisional partner (He). Qualitatively, the nature of the partner X is of little significance as long as it is polarizable. Quantitatively, one expects highly polarizable species (like Xe) to give rise to much stronger rotational lines than more weakly polarizable atoms (like He). Mainly for that reason, the rotational lines of H_2–He are much weaker than those of H_2–H_2. (Another reason is that for H_2–H_2, molecule 1 induces a dipole in 2 and 2 induces one in 1, thereby doubling the intensities.)

The H_2–He spectra may also be decomposed into translational and rotational line profiles, Fig. 3.13, similarly as this was sketched above for hydrogen, Fig. 3.11. In contrast to the decomposition for H_2–H_2 pairs, for the translational line of H_2–He we must assume different profiles for the translational profile on the one hand, and the rotational profiles on the other [123]. The translational line of H_2–He arises mainly from overlap induction which is of a much shorter range than the quadrupolar induction which shapes the rotational lines. As a consequence, at any given temperature, the translational line is much broader than the rotational lines. In the case of H_2–He, it is also much more intense.

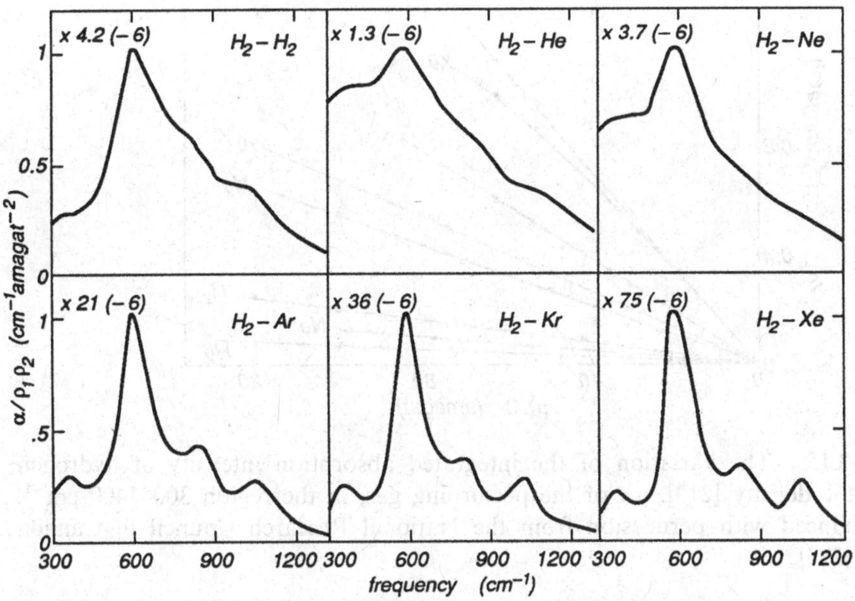

Fig. 3.14. Comparison of the absorption profiles of H_2–X pairs, with X = H_2, He, Ne, Ar, Kr, Xe at room temperature [213]. The spectra are normalized to unit absorption at their peaks, but absolute intensities are indicated in the form of a scaling factor; (–6) stands for $\times 10^{-6}$. Reproduced with permission from the National Research Council of Canada from [213].

Measurements of enhancement spectra exist for several gases and mixtures. Figure 3.14 shows the collision-induced absorption spectra of H_2–X pairs, with X = He, Ne, Ar, Kr, Xe [213]. The translational lines were omitted for technical reasons. Because the spectra are recorded at room temperature, the $S_0(J)$ lines of H_2 are quite diffuse. Most prominent is the $S_0(1)$ line at 587 cm^{-1}, but lines at other rotational transition frequencies of H_2 are also discernible, for example $S_0(0)$ at 354 cm^{-1}, $S_0(2)$ at 815 cm^{-1}, and $S_0(3)$ at 1035 cm^{-1}, especially for the massive pairs.

These spectra arise from binary interactions. Figure 3.15 displays the variation of the integrated intensity, γ_1, Eq. 3.5, normalized by just the hydrogen density, with density of the rare gas X. For all systems H_2–X, straight lines are obtained that go through the origin, except for one system: the hydrogen–xenon mixture shows marked departure from the linear dependence, signaling the presence of three-body spectroscopic processes. A detailed analysis has shown specifically that the H_2–Xe–Xe contributions cause the breakdown of the linear dependence.

In Fig. 3.14, the maxima of the six spectra are normalized to unity and

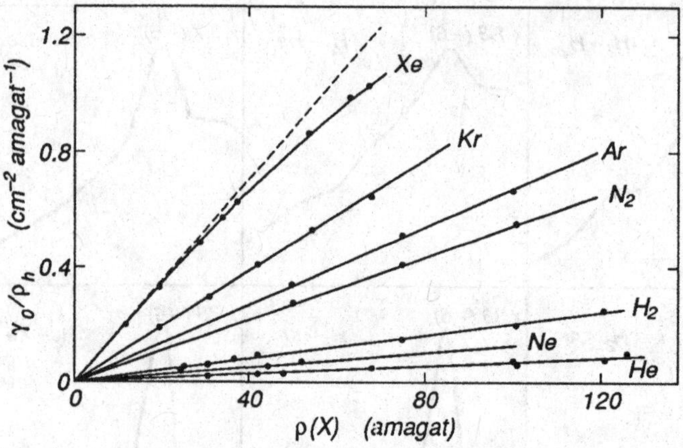

Fig. 3.15. The variation of the integrated absorption intensity of hydrogen with the density [213], ϱ_f, of the perturbing gas, in the region 300–1400 cm^{-1}. Reproduced with permission from the National Research Council of Canada from [213].

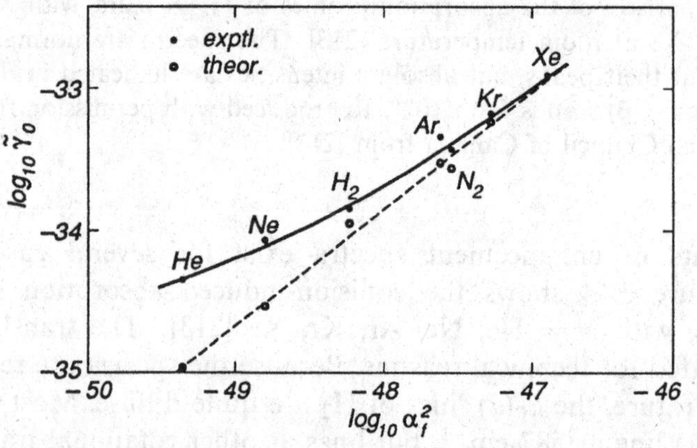

Fig. 3.16. Variation of the experimental values of the spectral moment $\tilde{\alpha}_1$ (solid line), and values calculated with the assumption of quadrupolar induction (dashed line), with the square of the polarizability of the perturbing gas. Reproduced with permission from the National Research Council of Canada from [213].

a scaling factor is given by which the reading should be multiplied to arrive at the correct absorption. The smallest peak absorption is seen for the H_2–He pair (1.27×10^{-6}), followed by that of H_2–Ne, H_2–H_2, etc., up to the maximum for H_2–Xe (74.8×10^{-6}) which is about 60 times greater than the smallest value in the series.

If one assumes quadrupolar induction as the principal induction process, these total intensities of the H_2–X systems should scale as the polarizability of species X squared. A detailed analysis [213] has shown that, indeed, for the more highly polarizable partners of the series, the integrated intensities (i.e., the slopes of the lines shown in Fig. 3.15), are proportional to the polarizabilities squared. Figure 3.16 compares the integrated intensities γ_0 with the sum formulae which were computed with the assumption of pure quadrupolar induction (as in Eq. 4.3). For the more highly polarizable systems (i.e., X = Ar, Kr, Xe), this assumption describes the measurement quite well. However, for the lighter systems (X = He, Ne, H_2), more or less significant deviations of the experimental data are observed which indicates that overlap induction cannot be neglected in those cases. There is little doubt that such overlap component is present in more or less the same magnitude for all systems shown, but it is relatively more important for the less polarizable systems.

We point out that the induced spectral lines of H_2 interacting with the more massive partners, X = Kr or Xe, appear sharper than the lines observed with X = H_2 or He, because at constant temperature the time scales are determined by the reduced mass, $m_{12} = m_1 m_2/(m_1 + m_2)$, of the interacting pair, through the mean *relative* speed, $(8kT/\pi m_{12})^{1/2}$. For the H_2–Xe system, the reduced mass is nearly twice that of the lightest system, H_2–He. Furthermore, the range of the interaction-induced dipole squared is slightly shorter for the lighter systems because of the significant overlap contributions (especially for H_2–He) which shorten the time scale of the interaction even more, Eqs. 2.12 through 2.15.

H_2–X where X is a molecule. If a molecule other than H_2 is chosen as the collisional partner X, new absorption bands appear at the rotovibrational bands of that molecule. As an example, Fig. 3.17 shows the rototranslational enhancement spectra [46] of H_2–CH_4 for the temperature of 195 K. At the higher frequencies ($\nu > 250$ cm^{-1}), these look much like the H_2–Ar spectrum of Fig. 3.10: the H_2 $S_0(J)$ lines at 354, 587, and 815 cm^{-1} are clearly discernible. Besides these H_2 rotational lines, a strong low-frequency spectrum is apparent which corresponds to the (unresolved) induced rotational transitions of the CH_4 molecule; these in turn look like the envelope of the rotational spectra seen in pure methane, Fig. 3.22. This is evident in the decomposition of the spectrum, Fig. 3.17, into its main components [46]: the CH_4 octopole (dashed curve) and hexadecapole (dot-dashed curve) components that resemble the CH_4–CH_4 spectrum of Fig. 3.22, and the H_2 quadrupole-induced component (dotted curve) which resembles the H_2–Ar spectrum, Fig. 3.14. The superposition (heavy curve) models the measurement (big dots) closely. Similar spectra are known for systems like H_2–N_2 [58].

Isotope spectra. Deuterium (D) is an isotope of hydrogen (H) with a mass about twice that of hydrogen; electronic properties of H_2 and D_2,

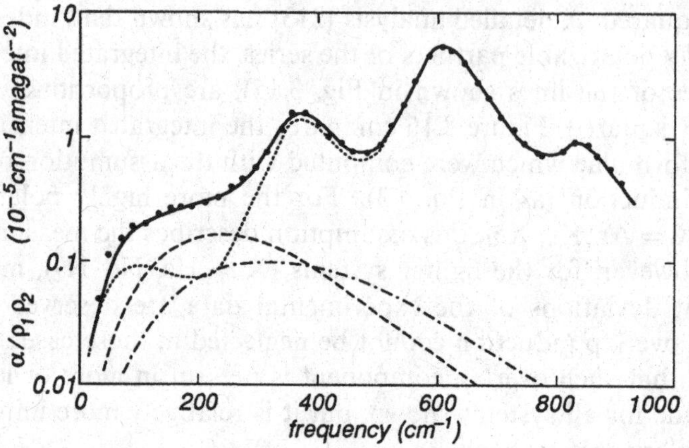

Fig. 3.17. The rototranslational spectrum of H_2–CH_4 at 195 K; experimental points: big dots; H_2 quadrupole-induced component: dotted; CH_4 octopole-induced component: dashed; CH_4 hexadecapole-induced component: dot-dashed; total: heavy curve. Reproduced with permission from the National Research Council of Canada from [46].

such as interaction potential and induced dipole moment, however, are nearly identical. It is interesting to substitute the isotopic molecules D_2 and HD for H_2 in some of the above measurements of collision-induced rototranslational absorption spectra. We give two examples.

Isotope substitution will have four principal effects of interest here but, apart from these, isotopic spectra should have the same total intensities (zeroth-order sum formulae are not mass dependent). Specifically, one would expect that

1) at constant temperature the time scales of the molecular dynamics of the isotopic systems will be greater by the square root of the ratio of reduced isotope mass of the interacting pair to the reduced H_2–X mass, leading to sharper lines;

2) rotational frequencies will be scaled down, by roughly the ratio of the reduced mass;

3) statistical weights of D_2 will be 6 and 3 for even and odd rotational quantum numbers, respectively, whereas for H_2 these nuclear weights are 1 and 3, respectively. Since the rotational energy levels of D_2 are roughly at 1/2 the energy levels of H_2 for the same angular momentum, the relative intensities of the various rotational $S_0(J)$ lines will be strikingly different;

4) moreover, the spectra of van der Waals dimers will be quite different; these will be considered below.

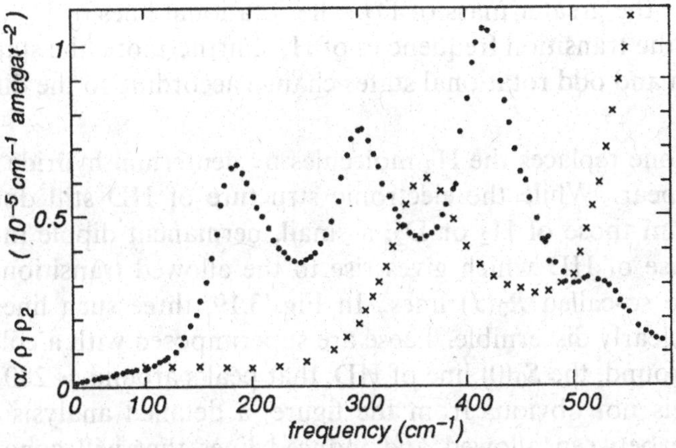

Fig. 3.18. Comparison of the D_2–Ar (•) and H_2–Ar (×) rotational spectra at 165 K, and 142 and 150 amagat argon density for deuterium–argon and hydrogen–argon mixtures, repectively, and a hydrogen concentration of 2 to 10%; after [109].

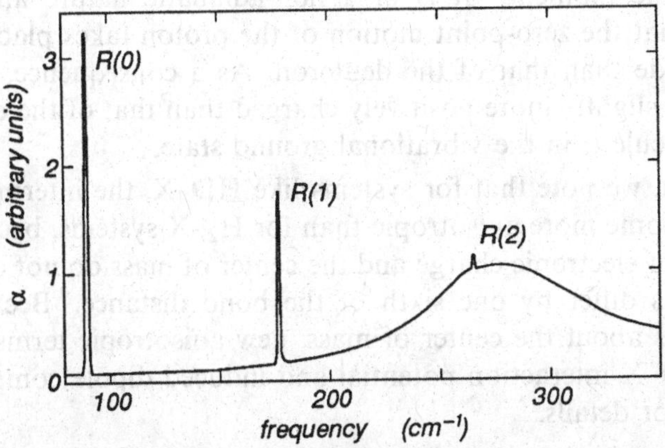

Fig. 3.19. The far infrared absorbance spectrum (in arbitrary units) of pure HD at 43 amagat and 77 K (smoothed). The sharp $R_0(J)$ lines sit upon a broad collision-induced background; after [398].

Figure 3.18 compares the spectrum of D_2–Ar recorded at 165 K at a density of 142 amagat with an H_2–Ar spectrum recorded at the same temperature and 150 amagat argon density [109]. As expected, we see more rotational lines, $S_0(J)$ with $J = 0, \ldots 4$, than for H_2–Ar, and these have different relative intensities. The rotational lines are also sharper, roughly by the factor $1/\sqrt{2}$. The spectral moment M_0 is the same as for H_2–Ar, well within the experimental uncertainties, as it should be.

The major effect of the isotope substitution is a narrowing of all induced lines because of the greater mass of D_2. The rotational lines of D_2 appear roughly at 1/2 the transition frequencies of H_2. Furthermore, the statistical weights of even and odd rotational states change according to the different spin statistics.

If, however, one replaces the H_2 molecules by deuterium hydride (HD), new effects appear. While the electronic structure of HD still does not differ much from those of H_2 or D_2, a small, permanent dipole moment exists in the case of HD which gives rise to the allowed transitions with $J \to J+1$, the so-called $R_0(J)$ lines. In Fig. 3.19, three such lines with $J = 0 \ldots 2$ are clearly discernible. These are superimposed with a collision-induced background, the $S_0(0)$ line of HD, that peaks around ≈ 280 cm^{-1}. Although this is not obvious from the figure, a detailed analysis shows an interference between allowed and induced lines that will concern us below.

The electronic charge distribution of H_2 is inversion symmetric and, therefore, H_2 is necessarily non-polar. The HD molecule possesses a nearly identical electronic cloud. Nevertheless, HD does feature a (weak) permanent dipole moment. It is of a non-adiabatic nature and arises from the fact that the zero-point motion of the proton takes place with a greater amplitude than that of the deuteron. As a consequence, the side of the proton is slightly more positively charged than that of the deuteron if the HD molecule is in the vibrational ground state.

In conclusion, we note that for systems like HD–X, the intermolecular interactions become more anisotropic than for H_2–X systems, because for HD the center of electronic charge and the center of mass do not coincide. The two centers differ by one sixth of the bond distance. Because the molecule rotates about the center of mass, new anisotropic terms appear in both the HD–X interaction potential and induced dipole components; see Chapter 4 for details.

Rototranslational spectra of other systems. The H_2 molecule differs from most other neutral molecules by its large rotational constant, $B_0 = 59$ cm^{-1}. For comparison, rotational constants of N_2, O_2, etc., amount to just a few wavenumbers. As a consequence, the collision-induced spectra of systems containing H_2 generally show well separated H_2 rotational lines, unless temperatures are so high that rotational lines begin to merge. Induced rotational lines of most other gases composed of common molecules are so diffuse and spaced so closely that at temperatures above the liquefaction point individual lines are rarely resolved. Induced rototranslational spectra of most molecular systems containing no H_2 molecules appear as the *envelopes* of the rotational band of the molecules involved. In fact, collision-induced rotational bands may often be approximated by the envelope of the 'stick spectrum', where every rota-

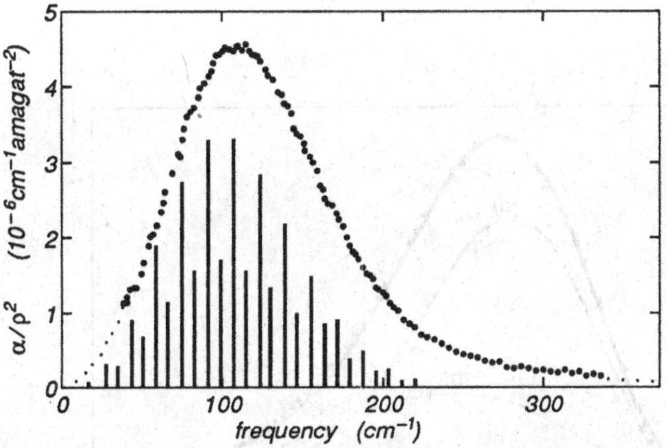

Fig. 3.20. The absorption profile of binary nitrogen pairs, N_2–N_2, at room temperature (dots). The vertical bars represent the rotational $S_0(J)$ lines of the nitrogen molecule; small dots are estimates obtained by extrapolation. Reproduced with permission from the National Research Council of Canada from [75].

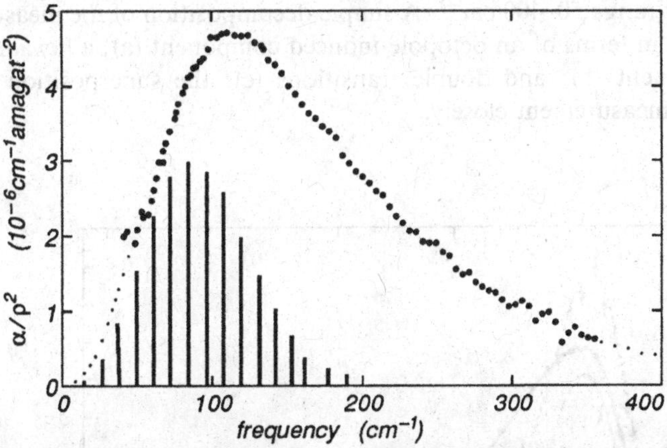

Fig. 3.21. The absorption spectrum of binary oxygen pairs, O_2–O_2, at room temperature (dots). The vertical bars represent the rotational $S_0(J)$ line spectrum of the oxygen molecule; small dots are estimates obtained by extrapolation. Reproduced with permission from the National Research Council of Canada from [75].

tional line is represented by a solid bar of an area given by the integrated intensity and a width equal to the rotational spacing. We mention that for more detailed studies, better model profiles than these rectangular bars are available and may be employed with advantage.

The N_2–N_2 spectra, Fig. 3.20, consist of the unresolved N_2 rotational band at frequencies from 10 to 220 cm^{-1}. Both quadrupole and hexadecapole induction contribute significantly to the observed spectra. Induced

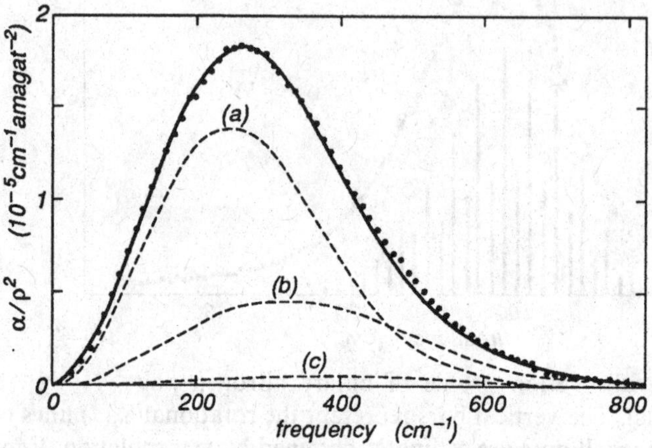

Fig. 3.22. Rototranslational absorption spectra of CH_4–CH_4 pairs [141] at 296 K in the frequency range 50–400 cm^{-1}. A simple decomposition of the measurement (•) is attempted in terms of an octopole-induced component (a); a hexadecapole-induced component (b); and double transitions (c); the superposition (heavy) reproduces the measurement closely.

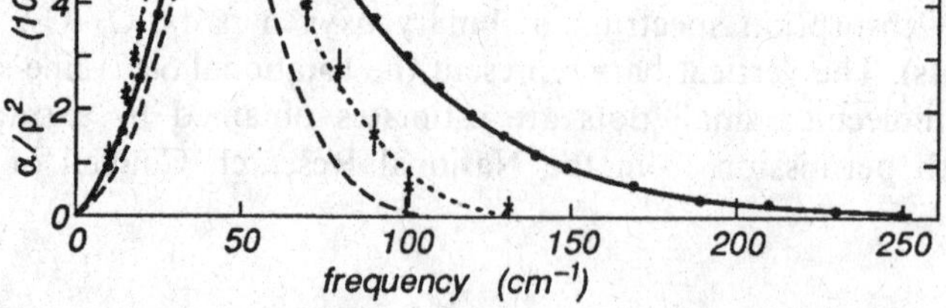

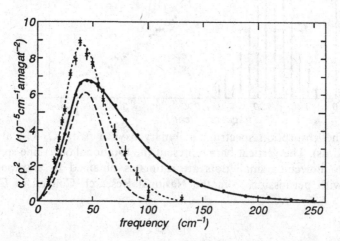

Fig. 3.23. Rototranslational spectrum of CO_2–CO_2 at 296 K [186, 34]. The dashed curve is the profile of the stick spectrum. The dotted curve describes an average through the three-body profile (×) (intensity scale in arbitrary units); peak absorption of that component amounts to 2.2×10^{-7} $cm^{-1}amagat^{-3}$.

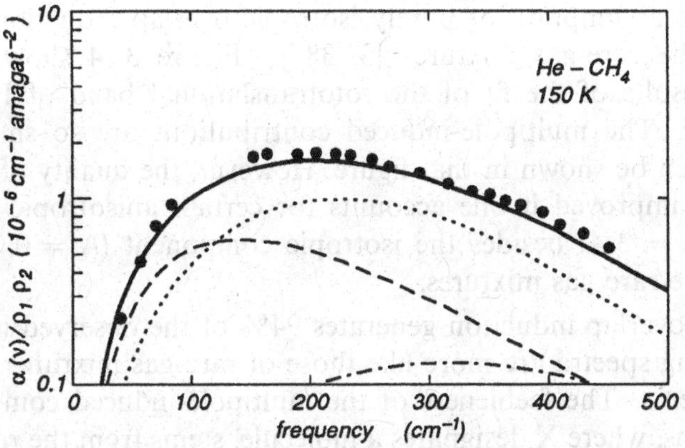

Fig. 3.24. Absorption coefficient, $\alpha(\nu)/\varrho_1\varrho_2$, of He–$CH_4$ at 150 K. The measurements [3] are given as dots (•); the heavy curve represents a fit of the three overlap components: an isotropic dipole component ($\Lambda = 0$, dashed) and overlap components of octopolar and hexadecapolar symmetry ($\Lambda = 3$ and 4; dotted and dash-dotted, respectively); after [387].

spectra consist, therefore, basically of the N_2 $S_0(J)$ lines, with a small admixture of $U_0(J)$ lines (corresponding to $J \to J + 4$ rotational transitions) [320, 53]. In the case of oxygen, Fig. 3.21, an even stronger hexadecapole-induced component exists [375].

Similar rototranslational spectra (which may be roughly approximated by the envelope of their stick spectra) are observed in other gases as well. Figures 3.22 and 3.23 show the binary absorption spectra of pure methane and carbon dioxide. The smooth curves drawn through the data points represent line shape computations based on the multipole-induction model of the induced dipoles involved [75, 56, 141, 186]. A detailed analysis indicates that for the CH_4–X system, CH_4 octopole and hexadecapole induction both contribute roughly in comparable amounts to the observable spectra. Rototranslational spectra of several other systems are known; see, for example, a review [58].

It must be pointed out that for certain systems the stick spectrum is a poor approximation, especially if helium atoms are involved. As an example, we mention the He–CH_4 rototranslational spectra [3, 387]. According to our reasoning above, one might expect relatively strong octopole- and hexadecapole-induced rotational bands like those shown for pure methane, Fig. 3.22. Instead, we get a roughly 16 times stronger absorption than a theoretical estimate based on the multipole moments of CH_4 and the polarizability of He. Furthermore, the absorption of the measurements [3] extends to higher frequencies than the multipole-induced rotational lines. Indeed, the measurements can be modeled roughly from theory with the assumption of purely isotropic overlap induction of the type seen in the rare-gas mixtures [3, 387]. Figure 3.24 shows as an example the results of the fit of the rototranslational band of He–CH_4 pairs at 150 K. The multipole-induced contributions are so small that they cannot even be shown in that figure. However, the quality of the fits is significantly improved if one accounts for certain anisotropic overlap components ($\Lambda = 3, 4$) besides the isotropic component ($\Lambda = 0$, dotted) known from the rare gas mixtures.

In any case, overlap induction generates 94% of the observed intensity and the resulting spectra are more like those of rare-gas mixtures than of molecular systems. The feebleness of the multipole-induced components of He–X systems, where X designates a molecule, stems from the relatively small polarizability of helium atoms. With a polarizability that small, overlap induction becomes the main induction mechanism; above, a similar conclusion was reached, Fig. 3.16. The other systems mentioned above, in contrast, have polarizabilities that are an order of magnitude greater than the helium polarizability so that the resulting induced absorption (which for multipole-induction scales as the square of polarizability) is actually two to three orders of magnitude stronger; overlap-induced dipoles, on the other hand, are more or less comparable for such systems as the data of Table 3.1, p. 65, suggest.

Similar conclusions have been obtained for mixtures of helium and carbon dioxide [222, 224]. Mixtures of carbon monoxide with other rare gases, on the other hand, are more nearly modeled by multipolar induction [7, 224].

Dimers. The spectra considered so far are representative examples of collision-induced absorption of the simpler atomic or molecular systems. They were recorded at fairly high densities (typically $\gg 1$ amagat) and with low resolution ($\approx$ a few wavenumbers), many of them at relatively high temperatures (room temperature). Even if these facts were not much emphasized above, it is important that they be kept in mind because generally, at lower temperatures and densities, interesting structures of van der Waals molecules may be seen if spectra are recorded with sufficiently high resolution ($\lesssim 1$ cm^{-1}) at low density. At the higher temperatures, higher pressures and lower resolution generally used, these spectra are weak, pressure and instrumentally broadened, and rarely discernible.

We remind the reader that induced spectra, especially the low density spectra that are not affected by ternary contributions, are feeble. Spectroscopists use, therefore, high pressures and wide open slits as much as possible for good signal-to-noise ratios. Van der Waals molecules are highly anisotropic systems and their prominent rotational lines show substantial pressure broadening. Moreover, their band spectra typically

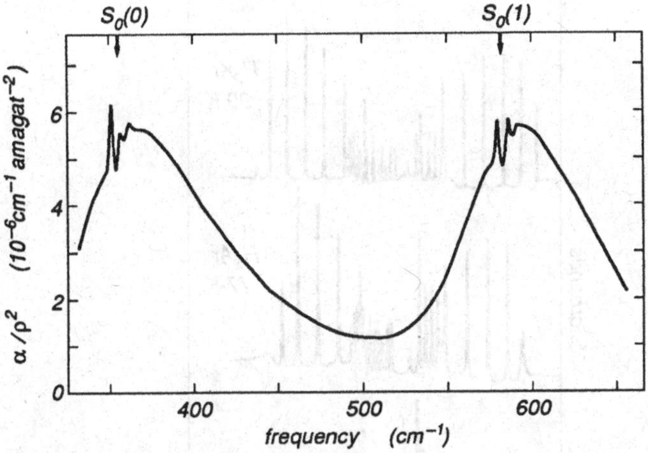

Fig. 3.25. The absorption spectra of equilibrium hydrogen at a temperature of 77 K and the density of 2.63 amagat, measured with a path of 154 m and a spectral resolution of 0.2 cm^{-1}. The arrows indicate the rotational transition frequencies of the free H_2 molecule. The broad features are the collision-induced S_0 lines of hydrogen, and the sharp structure near the peaks is due to the $(H_2)_2$ van der Waals molecule; after [266].

consist of many closely spaced lines that require high resolution. Dimer bands are therefore easily missed unless great care is exercised and advanced instrumentation is employed for their recording.

According to theory, most collision-induced absorption spectra should not only consist of contributions of the free-state to free-state transitions typical of collisional pairs, but also of contributions arising from bound-to-free and bound-to-bound transitions involving van der Waals molecules. In the rotovibrational spectra such dimer bands have been known for some time, but in CIA studies of the rototranslational band, where path lengths have generally been limited to a few meters, dimer features have been seen only recently [268]. The dimer spectra are an integral part of the interaction-induced absorption.

Using 2.63 amagats of hydrogen, a pathlength of 154 m and liquid nitrogen temperature (77 K), McKellar was able to record the $(H_2)_2$ spectral features with a Fourier transform spectrometer. Equilibrium hydrogen (eH_2) at 77 K was made by appropriately mixing para and normal hydrogen. Hydrogen densities from 1.3 to 2.6 amagat were used. The spectra are displayed in Fig. 3.25. We recognize again the collision-induced $S_0(0)$ and $S_0(1)$ lines of H_2 which were seen in Fig. 3.10 above. But besides the diffuse rotational lines of H_2, we notice relatively sharp structures near the peak of each line – something that was not evident in Fig. 3.10, for reasons just given. These arise from transitions involving

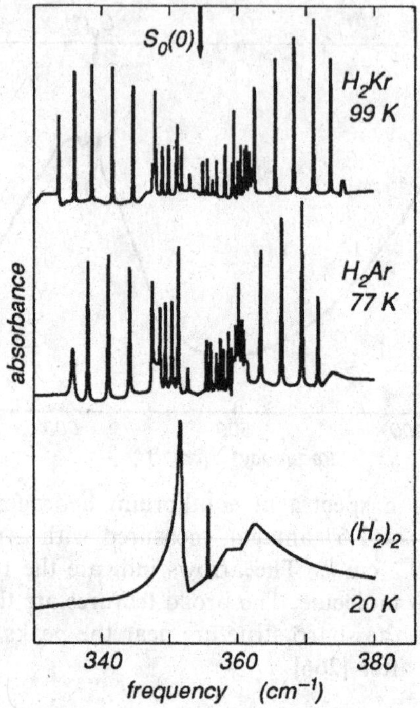

Fig. 3.26. Examples of spectra of some H_2-containing complexes which accompany the $S_0(0)$ pure rotational transition frequency of H_2; after [268].

bound dimer states. These dimer features will be discussed in some detail below from a theoretical standpoint, Chapter 6. Here we simply state that these structures were actually not discovered in the laboratory, but in Voyager's infrared spectra of Jupiter's atmosphere [157, 263, 150], where the conditions for their observation are quite favorable (long pathlengths and low densities at the altitudes where absorption affects the spectra). *Ab initio* calculations of the line profiles [150, 356] generally support the conclusion that the observed structures are due to $(H_2)_2$, but further refinements in the theory are apparently still necessary; these should, for example, account for pressure broadening of the dimer lines; see Chapter 6 for details.

Similar, but generally richer structures are expected for almost any other pair involving a molecule like H_2 that has a sufficiently wide spacing of the (forbidden) rotational lines. Figure 3.26 shows the blown-up regions around the H_2 $S_0(0)$ line of several systems involving an H_2 molecule. The lowermost trace is that of the hydrogen dimer shown above in Fig. 3.25, but at lower temperature (20 K instead of 77 K); the diffuse induced component seen in Fig. 3.25 is suppressed in Fig. 3.26. Just above it we have the signature of the H_2Ar van der Waals molecule; it shows a greater number of resolved lines. The uppermost trace is that of H_2Kr which features even more resolved lines. A more detailed analysis of such spectra will be given in Chapter 6 below. We note that for the relatively simple van der Waals molecules here considered, the number of rotovibrational states increases with increasing well depth of the intermolecular potential, and with increasing reduced mass of the pair. The number of transitions, i.e., of lines of the dimer spectra, increase with increasing number of such levels. This accounts for the richer stuctures seen in the upper parts of the figure.

Theoretically, similar dimer structures appear near all rotational transition frequencies of nearly any molecule, for example if N_2 is substituted for H_2 in such measurements. However, the narrowly spaced N_2 $S_0(J)$ lines are much more numerous and the dimer structures are thus more difficult to resolve. Just a few measurements exist in the purely rotational induced bands of molecules other than H_2. The phenomena described here are, however, nearly universal and should in general be more complex if other molecules than H_2 are involved than the examples shown. Note that in Fig. 3.26 the diffuse induced H_2 $S_0(0)$ line was suppressed. Similar structures have been seen in the other H_2 $S_0(J)$ lines [268].

Spectral moments. The spectral moments of the rototranslational bands are defined according to Eqs. 3.5 and 3.6. For diatom–atom pairs like H_2–Ar, the expressions

$$\gamma_1 \approx \frac{8\pi^3 N_L^2}{3\hbar} \left(M_1^{(i)} + M_1^{(a)} + 6B_0\, M_0^{(a)} \right) \tag{3.10}$$

$$\gamma_0 \approx \frac{8\pi^3 N_L^2}{3\hbar} \left(M_0^{(i)} + M_0^{(a)} \right) \tag{3.11}$$

are often good approximations [123], if quadrupolar induction is the major mechanism that generates the rotational lines. $B_0 = \hbar/4\pi cI$ is the rotational constant of the diatom ($B_0 = 59.34$ cm^{-1} for H_2) and the superscripts (*i*) and (*a*) stand for isotropic (overlap-induced) and anisotropic (quadrupole-induced) component, respectively; see Chapter 5 for details. These expressions differ from spectral moments of rare gas mixtures, Eq. 3.7, to the extent that the contributions of the induced rotational lines (the anisotropic component) are included if molecules are involved. Such moments are determined from measurements of the absorption coefficient in the rototranslational bands, like the ones shown in this Section. This is important for a comparison of the binary components with theory and also if ternary (and perhaps higher) spectral components are to be separated from a dominating binary background. We note that for classical systems the first moment vanishes and is replaced by the second moment, according to $M_2 \approx (2kT/\hbar)M_1$.

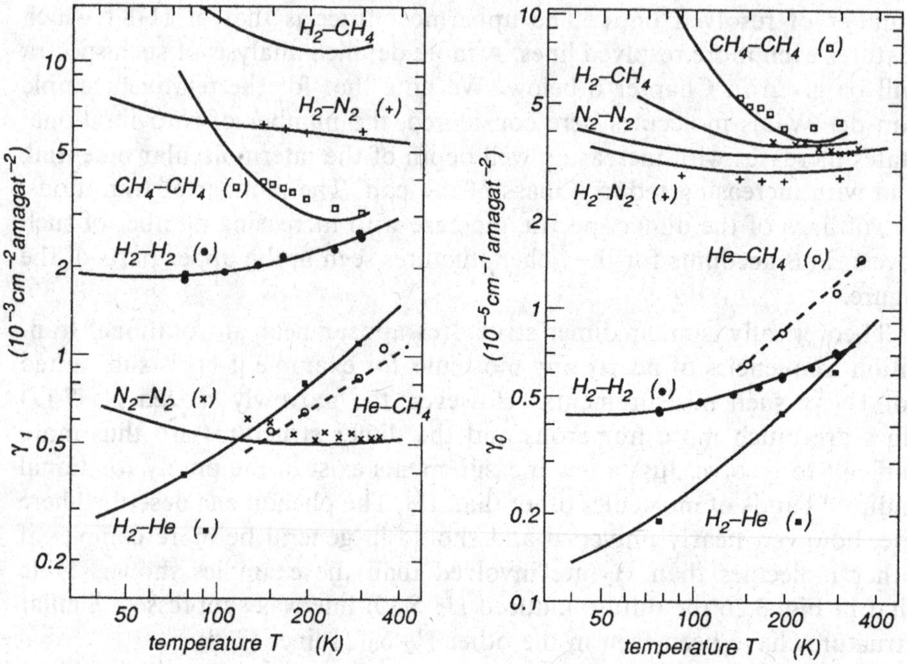

Fig. 3.27. Left: Spectral moments γ_1 of the rototranslational bands of several molecular pairs, as function of temperature. Various measurements (•, ○, etc.) are compared with theoretical data based either on the fundamental theory (H_2–H_2, H_2–He) or on refined multipolar induction models; after [58]. Right: Same as at left, except the spectral moment γ_0 is shown.

Figure 3.27 shows the spectral moments γ_1 and γ_0 of the rototranslational band of various molecular pairs as function of temperature (measurements: dots, circles, etc.). Absorption in pure hydrogen, and the absorption enhancement of H_2 by He, is fairly weak, Fig. 3.27, when compared to the other systems shown. However, especially for H_2–He, absorption increases rapidly with temperature, which is related to the substantial overlap induction component of the dissimilar pair. Overlap induction is short-ranged, increasing rapidly with increasing temperature. The He–CH_4 pair seems to mimic the H_2–He behavior; detailed studies have indicated that for He–CH_4 overlap induction is indeed the dominating influence [2, 21, 387]. The other molecular systems shown absorb more strongly, largely because multipolar induction is strong due to the fact that collisional partners are much more polarizable than He or H_2.

Also shown in the figures are the theoretical temperature variations of the spectral moments (the curves). These are obtained from first principles in the case of H_2–He and H_2–H_2 [279, 282]; measurement and theory are in very close agreement.

For the other gases, no *ab initio* data are available. Fortunately, many of these involve highly polarizable partners so that multipolar induction is the dominating induction mechanism. The multipole moments and polarizabilities of the simpler molecules are in general known. With this information, fairly accurate approximations of the absorption spectra and their moments are possible, a small empirical overlap component is usually included in the analysis [54, 55, 53, 56]. The calculated curves generally agree within a few percent with the measurements. This fact suggests that theory is capable of predicting the correct spectral moments (and line shapes) of the simpler binary systems, even for temperatures for which no measurements exist [58].

Other measurements of spectral moments of the rototranslational bands of binary systems are given in Fig. 3.15. Many more measurements exist for various gases and mixtures, at various temperatures [215, 422]; a complete listing is here not attempted.

2 *Ternary rototranslational spectra*

Above we have stated that over a substantial range of gas densities, essential parts of the profiles of collision-induced absorption spectra are invariant if normalized by density squared, α/ϱ^2, in pure gases, or by the product of densities, $\alpha/\varrho_1\varrho_2$, in mixed gases. Induced spectra that show this density-squared dependence may be considered to be of a binary origin. Above, we have seen examples that at very low frequencies many-body effects may cause deviations from the density-squared behavior at any pressure, over a limited frequency band near zero frequency (intercollisional effect). Furthermore, with increasing densities, a diffuse N-body effect with $N > 2$ more or less affects most parts of the observable spectra. It is interesting to study in some detail how the three-body (and perhaps higher-order) interactions modify the binary profiles.

It has been known since the early days of collision-induced absorption that spectral moments may be represented in the form of a virial expansion, with the coefficients of the Nth power of density, ϱ^N, representing the N-body contributions [402, 400]. The coefficients of ϱ^N for $N = 2$ and 3 have been expressed in terms of the induced dipole and interaction potential surfaces. The measurement of the variation of spectral moments with density is, therefore, of interest for the two-body, three-body, etc., induced dipole components.

The measurement of spectral moments requires the recording of complete spectra, including regions of high and low absorption where accurate measurements are difficult. In ordinary spectroscopy, these difficulties are often alleviated through the use of variable absorption path lengths and pressure variation. In the far infrared where the wavelengths are comparable to the beam apertures, absorption path lengths cannot be increased beyond a certain limit. Furthermore, for a study of the density dependence of absorption, pressure variation affects the various N-body components in different ways and is thus not really an independent option when virial properties are to be investigated. It is, therefore, of interest that an alternative has been developed which uses least mean squares techniques to fit the measured spectra by analytical expressions. For that purpose, two- or three-parameter model profiles are known that reproduce individual line profiles closely – as illustrated in Figs. 3.11 and 3.13, for example. This procedure may give dependable values of the model profile parameters, and thus of the lowest two or three spectral moments, even if the spectra are not known well at regions of very high or very low absorption. Furthermore, spectral moments of rotational or quadrupole-induced lines may thus be available with little interference from other spectral components, e.g., the isotropic overlap component, which would be an inseparable part in any straight measurements of moments.

Large parts of the rototranslational spectra of hydrogen–argon mixtures could be recorded around the induced H_2 $S_0(0)$ and $S_0(1)$ lines, with good signal to noise ratios, at the temperature of 195 K, at six argon densities from 42 to 185 amagats, using a small (3%) hydrogen admixture [139, 140]. At the lowest density the spectra were shown to be almost purely of a binary origin [139, 280]; the spectra look much like the H_2–Ar spectrum shown in the lower left corner of Fig. 3.14, except that the temperature is lower and the rotational lines are sharper. The H_2 S_0 lines arise basically from quadrupolar induction; a small overlap component is almost negligible [280]. The rotational profiles are well represented by an analytical expression based on the three-parameter BC model (p. 271). The three-parameters M_0, τ_1, τ_2, of the model profile, and thus with the help of Eqs. 5.109 the moments, M_n for $n = 1, 2$, of the spectral function, were determined as functions of the argon density by least mean squares techniques that fit the spectra obtained at six argon densities simultaneously. Second-order polynomials in powers of the argon density were employed to model the density dependences of M_0, τ_1, and τ_2, with a total of nine parameters; the quadratic terms were found to be very small and fits of comparable quality were obtained by a six-parameter simultaneous fit that omitted all density squared terms of these polynomial expressions. The set of six adjustable parameters could be reduced further to a total of three by adopting for the density-independent term the values computed from the fundamental theory for the binary system [280].

The results are shown in Fig. 3.28. Spectral moments M_n have been normalized by the product of hydrogen and argon densities, so that all data shown should represent horizontal lines if only binary interactions occurred. Instead, we see straight lines with negative slopes for M_0 and τ_2,

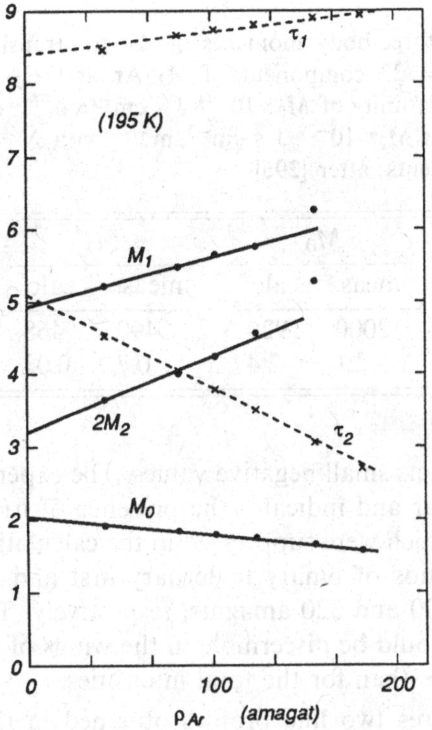

Fig. 3.28. The parameters describing the density dependence of the quadrupole-induced H_2 $S_0(0)$ and $S_0(1)$ lines observed in argon with 3% hydrogen admixture at 195 K [140]; after [152]. Units are 10^{-41} J s cm^{-1} $amagat^{-2}$ for M_0; 10^{-40} J s cm^{-2} $amagat^{-2}$ for M_1; 10^{-37} J s cm^{-3} $amagat^{-2}$ for M_2; 10^{-14} s for τ_1 and τ_2.

and positive slopes for the other quantities. In other words, the presence of a three-body contribution is revealed by the well defined, non-vanishing slopes, Fig. 3.28, which represent the relevant three-body moments. What is more, the method also gives a fairly detailed representation of the variation with density of the *shape* of the S_0 line profiles.

Table 3.4 summarizes the binary and ternary moments obtained from our simultaneous least-mean-squares fit of the measurements at six argon densities. The three-body zeroth moment is negative, in agreement with semi-classical estimates of the coefficient based on pairwise additive three-body dipoles, which are also given in the Table. The ratio of binary and ternary zeroth moment, apart from the sign, amounts to roughly 800 amagats which suggests that at $\approx$ 800 amagats binary and ternary contributions would be of the same magnitude; the ternary component amounts to a 10% contribution of the binary component at only 80 amagats. The ternary first and second moments are positive, a result that is not consistent with theoretical calculations based on pairwise

Table 3.4. Two- and three-body moments of the rototranslational absorption spectrum, of the $\lambda L = 23$ component of H_2–Ar and H_2–Ar–Ar at 195 K. Measurement: Fig. 3.28; units of M_0: 10^{-44} J s cm^{-1}am^{-N}; units of M_1: 10^{-42} J s cm^{-2}am^{-N}; units of M_2: 10^{-39} J s cm^{-3}am^{-N}; with $N = 2$ for binary and $N = 3$ for ternary moments; after [296].

	M_0		M_1		M_2
	meas.	calc.	meas.	calc.	meas.
two-body	2000	1920	490	468	160
three-body	–2.1	–2.4	0.7	–0.03	0.5

additivity which predicts small negative values. The experimental evidence is, however, quite clear and indicates the presence of irreducible induced dipole components which were suppressed in the calculations; see Chapter 6 for details. The ratios of binary to ternary first and second moments amount roughly to 700 and 320 amagats, respectively. This fact suggests that ternary effects should be discernible in the wings of the induced lines at even lower densities than for the total intensities.

Figure 3.29 compares two line profiles obtained in the fit. The solid line represents the rotational line profile in the zero density limit ($\tau_1 = 8.4 \times 10^{-14}$ s, $\tau_2 = 5.1 \times 10^{-14}$ s, from Fig. 3.28). The dashed line shows the profile at 185 amagat of argon ($\tau_1 = 8.94 \times 10^{-14}$ s, $\tau_2 = 2.70 \times 10^{-14}$ s). The amplitudes S are here arbitrarily set to unity, $S = 1$, in both cases so that the areas under the curves are equal. We note a very slight narrowing of the profiles near the line centers; this is caused by τ_1 which increases slightly with density. Moreover, the far wing intensities increase with increasing argon density which is related to the decreasing τ_2 values; see the discussion of the properties of the BC profile, p. 271.

Very similar conclusions were reported for measurements in argon with small admixtures of hydrogen, and with deuterium at 165 K, using much higher argon densities, up to 650 amagats [109], where many-body interactions may have affected the observable profiles. Figure 3.30 shows as an example the density variation observed in the mixture of deuterium and argon; the profiles obtained in hydrogen–argon mixtures are very similar, only $\approx 40\%$ broader because of the smaller mass. In that work an empirical line shape of the rotational lines was directly obtained, without the assumption of a special model profile. The conclusions are virtually the same: a narrowing of the core and a broadening in the wings with increasing argon densities, approximating liquid densities.

Other cases. In nitrogen, a similar line shape analysis was undertaken in

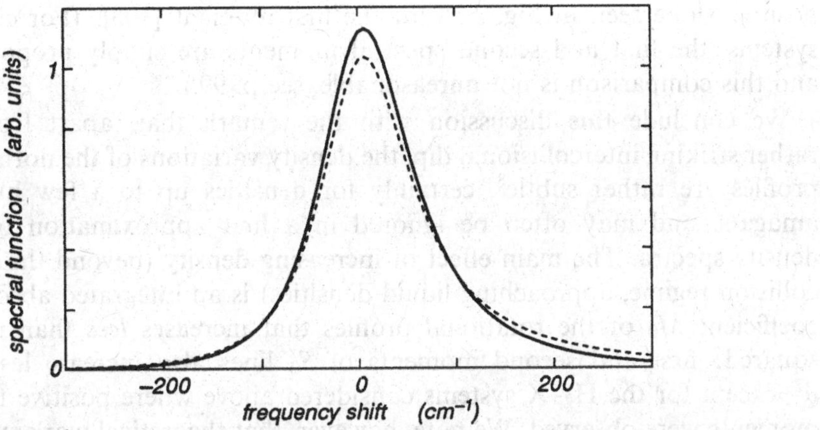

Fig. 3.29. Onset of ternary (H_2–Ar–Ar) interactions: the variation of fitted profiles of the H_2 S_0 lines with density; argon with 3% hydrogen admixture. Solid line: zero density limit (profile of H_2–Ar); dashed line: at 185 amagat of argon where ternary interactions have affected the profile. The functions are scaled so that the areas under the curves are equal; after [152].

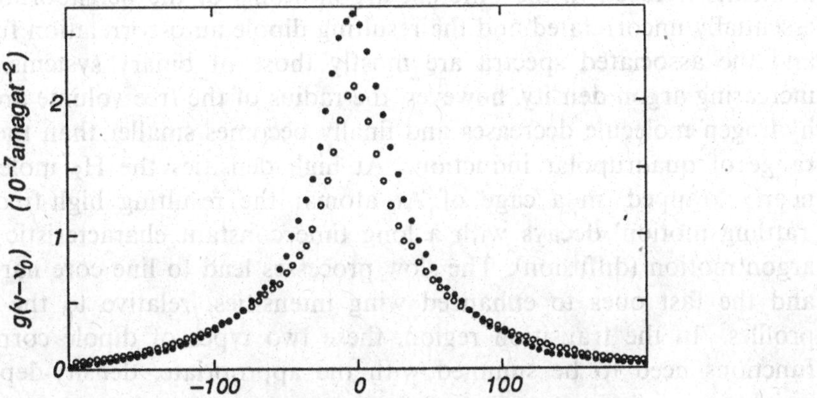

Fig. 3.30. Experimental reduced line shapes of the induced D_2 S_0 lines of deuterium–argon mixtures at 165 K; argon density 142 (•) and 650 amagat (○); after [109].

terms of a single profile representing all $S_0(J)$ lines [110]. With increasing density, narrowing of the rotational lines was observed similar to the hydrogen–argon profiles shown above, Fig. 3.30. The zeroth moment of the quadrupole-induced rotational lines shows a negative three-body component, similar to the one seen in Fig. 3.28. The second moment also shows a *negative* three-body contribution; to be contrasted with the positive slope seen in Fig. 3.28 for the first moment [105]. (For classical systems, the first and second spectral moments are simply proportional and this comparison is not unreasonable, see p. 99.)

We conclude this discussion with the remark that, apart from the rather striking intercollisional dip, the density variations of the normalized profiles are rather subtle[§], certainly for densities up to a few hundred amagats, and may often be ignored in a first approximation of low-density spectra. The main effect of increasing density (beyond the binary collision regime, approaching liquid densities) is an integrated absorption coefficient M_0 of the rotational profiles that increases *less* than density squared; first and second moments of S_0 lines also increase less than ϱ^2, except for the H_2–X systems considered above where positive ternary moments were observed. We note, however, that theoretical work suggests rapid variation of ternary moments with temperature [296]. The variations also include sign changes, and the rules just mentioned may not be true for all systems or all temperatures. The understanding of ternary spectra is in its beginnings.

A simple model has been suggested to explain the observed variation of the profiles with density [428, 109]. In the low-density limit, the dipole moments induced in the different argon atoms of the neighborhood are essentially uncorrelated and the resulting dipole auto-correlation function and the associated spectra are mostly those of binary systems. With increasing argon density, however, the radius of the free volume around a hydrogen molecule decreases and finally becomes smaller than the mean range of quadrupolar induction. At high densities the H_2 molecule is nearly 'trapped' in a cage of Ar atoms; the resulting high-frequency, 'rattling motion' decays with a long time constant characteristic of the argon motion (diffusion). The slow processes lead to line core narrowing and the fast ones to enhanced wing intensities, relative to the binary profiles. In the transition region, these two types of dipole correlation functions need to be summed with the appropriate, density-dependent weights.

Carbon dioxide. Collision-induced absorption in carbon dioxide shows a discernible density dependence beyond density squared, even at densities as low as 20 amagats [34]. Over a range of densities up to 85 amagats the variation of the absorption with density may be closely represented by a (truncated) virial series (as in Eq. 1.2, with $I(\nu)$ replaced by $\alpha(\nu)$) of just two terms, one quadratic and the other cubic in density. The coefficient of ϱ^3 is *negative*. Relative to the leading quadratic coefficient, it is,

---

§ An exception is the intercollisional dip which may be striking, especially in lines generated by isotropic overlap-induced dipoles; rotational lines are less affected.

furthermore, more significant than similar terms known from other gases. The normalized cubic term, the three-body absorption coefficient, $|\alpha^{(3)}|$, divided by density cubed, is plotted as function of frequency in Fig. 3.23. The dotted curve represents a smoothed average of the experimental points (× with error bars). A linear intensity scale was chosen; peak absorption corresponds to -2.2×10^{-7} cm^{-1} amagat^{-3}.

The cubic term, $-\alpha^{(3)}(\nu)$, actually approximates the S_0 stick spectrum (dashed curve) more closely than the quadratic term (solid curve); its band shape is substantially sharper. If one may associate the quadratic and cubic terms with binary and ternary components, respectively, this remarkable fact may be simply related to another fact: long-range induction operators show in general stronger three-body effects than short-range operators. In carbon dioxide quadrupolar induction is strong; quadrupolar induction is long-ranged and generates a spectrum that may reasonably well be approximated by the envelope of the CO_2 $S_0(J)$ stick spectrum. The binary spectrum is, however, shaped also by other induction processes, perhaps hexadecapolar and anisotropic overlap components that are shorter range and associated with different selection rules. As a consequence, the binary spectrum deviates more strongly from the stick spectrum (which is simply the envelope of the rotational S_0 band) than the ternary component. More detailed studies of the three-body components of carbon dioxide are desirable.

3 Dense fluids

Rototranslational spectra have been recorded in many liquids and in compressed gases at or near liquid densities. Liquid hydrogen shows prominently the rotational S_0 lines [202], much like these were seen in compressed hydrogen gas at low temperatures. Liquid hydrogen may or may not contain ortho hydrogen, depending on the preparation. Important differences of the translational spectra between mixtures of para and ortho hydrogen on the one hand, and pure para hydrogen on the other, were discussed above, Fig. 3.9; these are *qualitatively* the same as in dense gases. The *shapes* of these spectra of liquids are similar to the ones observed in the gas, apart from the apparent line narrowing as shown above, Figs. 3.29 and 3.30.

Dilute solutions of para hydrogen in liquid argon, and of normal hydrogen in liquid argon, show similarly the translational spectrum and the H_2 S_0 lines, much like this was seen above in the gas mixtures [103]. In this case, a translational band is always observed that shows a strong intercollisional dip in the para hydrogen–argon mixture. In rarefied gases, the translational band arises from an isotropic overlap-induced dipole moment characteristic for dissimilar pairs; such translational bands

Table 3.5. Spectral moments of liquid nitrogen [230]. For comparison, binary moments $\gamma_1^{(2)}$ times density squared are also given in the last column (from Fig. 3.27).

T (K)	ϱ (amagat)	γ_1 (cm^{-2})	$\gamma_1^{(2)}\varrho^2$ (cm^{-2})
87	743	142	298
91	610	103	199

generally show strong intercollisional interference. In the case of normal hydrogen dissolved in liquid argon, a similar dip is expected but not seen so clearly because of the presence of the $Q_0(1)$ line that appears at zero frequency. (No Q_0 transition is possible for para-H_2 at low temperatures.) This $Q_0(1)$ line has the same shape as the rotational S_0 lines (because both are quadrupole-induced), but it is noticeably narrower than the translational line, and roughly comparable in width with the intercollisional line (which are overlap-induced). Again, in the liquid phase, no new phenomena seem to exist that have not also been seen in compressed gases as far as the *shape* of the observed spectra is concerned.

Detailed studies of the rototranslational absorption spectra of compressed nitrogen show great similarity to the binary spectra obtainable at low densities [110, 252], Fig. 3.20. The rotovibrational band of liquid nitrogen has been determined at several temperatures and densities [102, 201, 230]. Measured profiles look superficially like the binary gas spectra (which may be estimated by temperature extrapolation of low-density spectra recorded at higher temperatures [58]), but absorption of the liquid extends to somewhat higher frequencies. Moreover, the spectral moment γ_1 of the liquid amounts to only about one half of the binary moment of the gas at the same temperature [102], a significant change between the (estimated) binary and the many-body moments, Table 3.5. The small absorption of liquids is related to the strong cancellation of the two-body component by three- and four-body contributions [170]. The rototranslational bands of liquid nitrogen, oxygen, methane, and carbon monoxide have also been recorded [201]. These are similar to the binary gas spectra shown above.

A substantial dependence of the induced absorption on density has been seen in liquid nitrogen when high pressure was applied to the liquid [252]. Furthermore, with increasing density, a shift of the peak absorption to higher frequencies is observed; while the low-frequency profile is not much affected by increasing density, significantly increased absorption is observed at the higher frequencies. This fact has been thought to be related to the N-body cancellation effects ($N > 2$) which are more efficient at low frequencies. In other words: increasing density leads to higher absorption at all frequencies but also to greater cancellation at low frequencies.

3.3 The rotovibrational spectra

Up to this point, we have considered representative examples of induced spectra in the microwave and far infrared regions, below frequencies of roughly 1000 cm^{-1}. Collision-induced absorption is not limited to the far infrared. It is well known that molecules may vibrate. Vibrational transition frequencies are generally much higher than the rotational transition frequencies, namely roughly a few thousand cm^{-1}. One may, therefore, expect induced spectra in the near infrared region if these involve vibrational transitions that may be forbidden in the isolated molecules but are 'activated' (induced) by intermolecular interactions, just as this was seen in the rotational bands above. We will consider induced spectra in the fundamental bands of molecules which appear near frequencies corresponding to transitions from the ground state to the lowest vibrational states, but the so-called overtone and hot bands will also be of considerable interest; these involve transitions from the ground state to higher excited vibrational states, and from vibrational states higher than the ground state, respectively.

Historically, collision-induced absorption was discovered in the fundamental band of oxygen and nitrogen [128], Fig. 1.1. Literally, any molecular complex may be expected to have more or less prominent induced bands in the fundamental band and overtone regions of the molecules involved–besides the rototranslational bands considered above. Induced vibrational spectra are indeed known for many molecular systems and selected examples will be discussed below. Since in virtually all of these spectra rotation and vibration are coupled, we will generally refer to these as rotovibrational induced spectra.

It is instructive to start this survey of measured, induced rotovibrational spectra again with hydrogen because, in this case, well-resolved rotational lines may be expected, certainly at the lower temperatures. The pressure-induced fundamental band of hydrogen has been studied at pressures up to thousands of atmospheres and temperatures from 18 to over 400 K. We will be interested here in the spectra obtained at the lowest gas densities ('pair spectra') but will also consider the modifications of the rotovibrational features observed in dense fluids when ternary interactions become discernible.

Figure 3.31 shows the spectra of pure hydrogen in the fundamental band

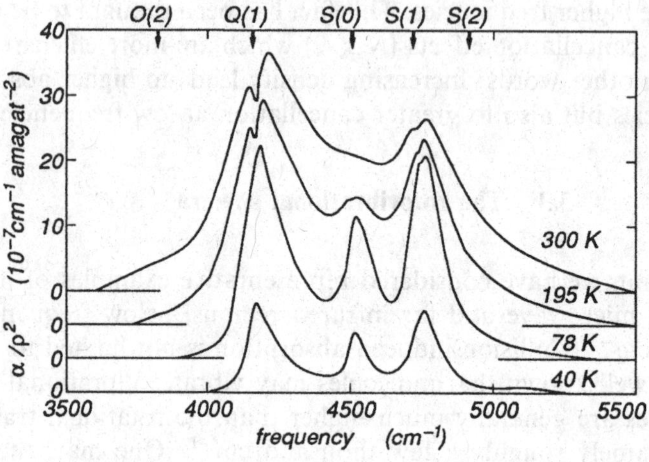

Fig. 3.31. Absorption profiles of H_2–H_2 pairs in the fundamental band of hydrogen at various temperatures and at pressures around 100 amagat; after [187].

of H_2 at four temperatures. The different spectra are shifted vertically for clarity of the display. In the low-temperature spectra shown in the figure one distinguishes clearly three induced lines. The fundamental band consists of a vibrational component, or Q branch, corresponding to the vibrational transitions $v = 0 \rightarrow v' = 1$ and $J = J'$, and several $S_1(J)$ components each of which corresponds to the vibrational transition, accompanied by a rotational transition with $J \rightarrow J' = J + 2$; v and J are vibrational and rotational quantum numbers of initial state of H_2 and a prime indicates the final state. The $Q_1(1)$ line appears at the frequency of 4155 cm^{-1}, the $S_1(J)$ lines with $J = 0$ and 1 at 4498 and 4713 cm^{-1}, respectively; these transitions are infrared inactive in the isolated molecule but are well known from Raman studies.

At the higher temperatures the lines appear more diffuse, a fact related to the mean duration of a collision which decreases with increasing temperature, see p. 61. Besides, O lines appear at elevated temperatures, to the extent that higher rotational states ($J > 1$) are significantly populated with increasing temperature; these O lines correspond to transitions with $J \rightarrow J' = J - 2$.

Interestingly, at the higher temperatures and if low densities are employed, the Q branch has a minumum near the $Q_1(1)$ transition frequency at 4155 cm^{-1}, and there are two maxima (in the early days of collision-induced absorption, these were called the Q_R and Q_P 'lines'); these and the resulting dip between the maxima are clearly visible in the high-temperature spectra, Fig. 3.31. These maxima have a frequency separation that increases nearly linearly with gas density over a large density range. As the temperature is lowered, the intensity of the Q_P component decreases rapidly with respect to the Q_R component until all trace of the splitting disappears. The intensity distribution within the Q branch (really within all lines discernible in the spectrum) becomes highly asymmetric and the low-frequency wing falls off over a narrow frequency band, de-emphasizing more and more the left peak (Q_P) with decreasing temperature. We mention here that at high pressures, where the frequency splitting is large, a third component (Q_Q) appears that is related to orientational transitions of the quadrupole-induced component; its shape is the same as that of the rotational lines, Fig. 3.48. (The Q branch is overlap-induced and is more diffuse than the rotational lines.)

Similar absorption dips, but less striking, seem to occur at the rotovibrational transition frequencies, $S_1(J)$, at the higher temperatures shown in the figure. We will postpone our study of these features temporarily and note here only that all of these features are of a many-body origin, due to the correlations of dipoles induced in subsequent collisions, just as this was seen above, e.g., Figs. 3.3 and 3.5.

The important point to be made here is that at low pressures, a few wavenumbers away from these narrow intercollisional features, the intensities of the spectra vary accurately as density squared. This important fact, which has been carefully verified on many occasions, indicates again the binary nature of the main parts of the spectra observed. The intercollisional dips show a variation with density that differs strikingly from that of the binary spectra. Here, we study the binary parts of the rotovibrational spectra; the parts that show deviations from the density squared dependence will be considered below.

1 *Rotovibrational pair spectra*

At not too high gas densities (say 10 or 50 amagat), a large part of the induced rotovibrational spectra is demonstrably of a binary origin. (We may momentarily ignore small regions around the intercollisional dips where intensities do not follow a density square dependence.) In that sense, the spectra shown in Fig. 3.31 (and Fig. 3.37, etc.) may be considered the binary induced spectra of the rotovibrational bands.

A detailed analysis of the measurements, Fig. 3.33, suggests that the binary rotovibrational spectra are somewhat more complex than the simple discussion above may have suggested. Figure 3.33 shows a measurement like the ones of Fig. 3.31, but at an even lower temperature so that a overlapping of the various lines is minimal. Low temperature favors the detailed analysis of the spectral profiles.

In the vibrational band of pure (unmixed) hydrogen, two major in-

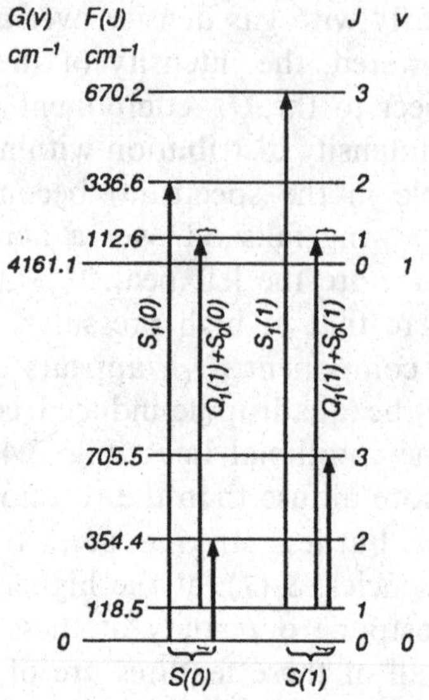

Fig. 3.32. Rotovibrational term scheme of H_2 showing single and double transitions of pairs; after [187].

duction processes occur: overlap- and quadrupole-induction shape the spectra. Accordingly, two types of spectral profiles are expected. The isotropic overlap induction generates most of the Q line intensity, and quadrupole induction shapes the S lines, and also a small component of the Q branch related to orientational ('magnetic') transitions. In other words, the Q branch actually consists of several components; it is a superposition of at least two different profiles. Actually, the theory to be considered in Chapter 6 suggests that *all* lines seen in the figure are composites. A total of eleven components must be expected for the fundamental band of hydrogen, even with the simplifying assumption of having just two induced dipole components to account for. (We will see in Chapter 4 that a few other induced dipole components must also be included for a better reproduction of the measurement from theory.)

Specifically, the eleven profiles include the obvious single transitions, i.e., the rotovibrational transitions in just one of the two colliding H_2 molecules; these are the $S_1(0)$, $S_1(1)$, and $Q_1(1)$ transitions in one of the two interacting molecules. Double transitions in both collisional partners are also taking place, such as the simultaneous transitions $Q_1(1)+S_0(0)$ (which occur near the $S_1(0)$ transition frequency) and $Q_1(1)+S_0(1)$ (near $S_1(1)$), Fig. 3.32. Intensities of all these lines are known from theory (classical multipole approximation, Chapter 6); their superposition reproduces the measurement closely, Fig. 3.33.

The binary enhancement spectra of hydrogen–helium mixtures in the

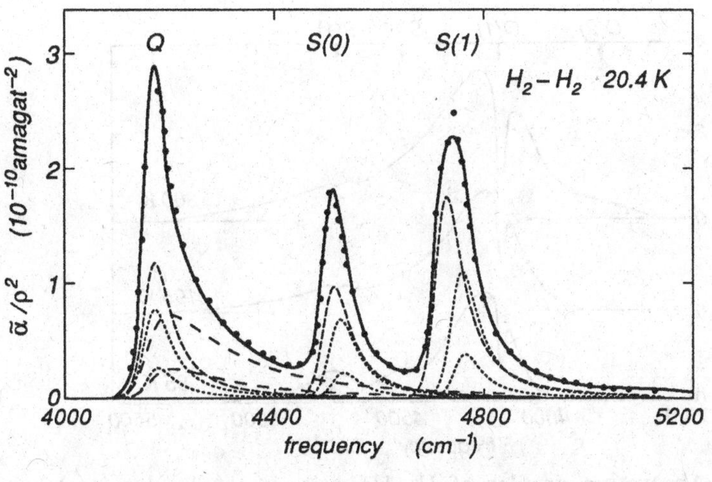

Fig. 3.33. Analysis of the fundamental band of normal hydrogen at 20.4 K into its 11 main components. Overlap-induced components $Q(0)$ and $Q(1)$ (widely spaced dashes) are broader than the quadrupole-induced components (closely spaced dashes, from left to right: $Q_Q(1)$, $S_1(0)$ and $S_1(1)$); and double transitions (dotted: $Q_Q(0)$ and $Q_Q(1)$; $Q_1(1) + S_0(0)$ and $Q_1(0) + S_0(0)$; $Q_1(1) + S_0(1)$ and $Q_1(0) + S_0(1)$. The dots represent the summation of these; the measurement is shown as a heavy line. Reproduced with permission from the National Research Council of Canada from [414].

hydrogen fundamental band are shown for three temperatures in Fig. 3.34. At the lower temperatures, one is able to distinguish again three basic line profiles (like in pure hydrogen, Fig. 3.31) arising from binary interactions, loosely labeled as the Q branch and the $S(0)$ and $S(1)$ lines. The Q line is now relatively more important – something like this was seen in the translational band, Fig. 3.12, which may be considered the Q branch of the rototranslational spectra. It shows, furthermore, a strong intercollisional interference dip that is of a many-body origin. A much weaker but clearly discernible intercollisional dip is seen at the $S_1(1)$ transition frequency. Double transitions are, of course, impossible in a system like H_2–He and the various lines are the $Q_1(J)$ and $S_1(J)$ induced single transitions. The lines appear to be very diffuse, even more so than in pure hydrogen, a feature that may be traced to the fact that in the dissimilar H_2–He system stronger overlap induction occurs than for the H_2–H_2 pairs (Chapter 4); due to the short-range nature overlap-induced lines are generally more diffuse than, for example, quadrupole-induced lines.

The induced H_2 fundamental bands at room temperature of the systems H_2–Ar and H_2–Xe are shown in Fig. 3.35. The $S_1(J)$ and $Q_1(J)$ lines appear more intense, and also sharper, for the more massive system, a fact related to the higher polarizabilities of Ar (11.07 a.u.) and Xe (27.11

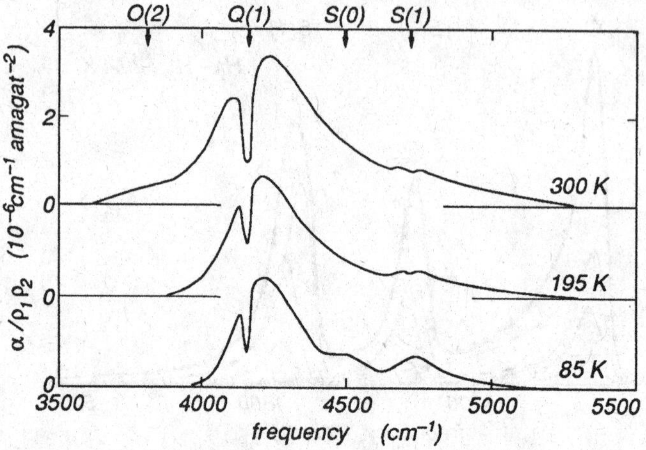

Fig. 3.34. Absorption profiles of H_2–He pairs in the fundamental band of hydrogen at various temperatures and pressures around 100 amagat; after [187].

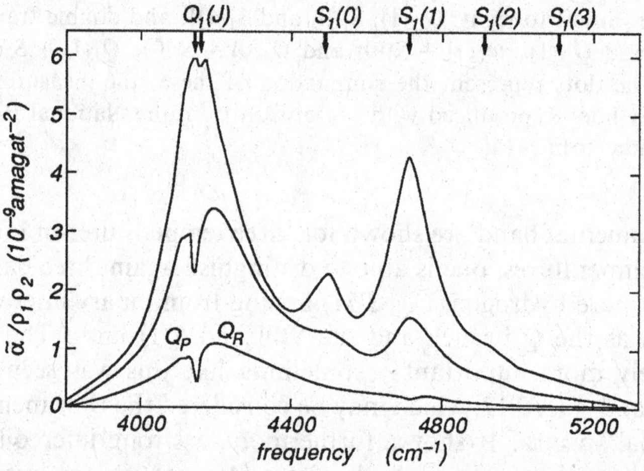

Fig. 3.35. Absorption profiles, $\widetilde{\alpha} = \alpha\nu$, of the enhancement of the H_2 fundamental by helium (bottom), argon (center) and xenon (top) at 298 K. Hydrogen densities were 10.3, 4.5, and 11.1 amagat, and the helium, argon and xenon densities were 105, 113 and 51 amagat, for the lower, middle and upper curve, respectively; after [187].

a.u.) relative to He (1.38 a.u.). High polarizabilities tend to enhance quadrupolar induction over the overlap induction. Since quadrupolar induction is of a long-range nature it produces sharper lines than overlap induction. Besides, the reduced mass of H_2–X pairs increases as X goes from He to Ar to Xe; this fact also slows the root mean square speeds and leads to sharper lines.

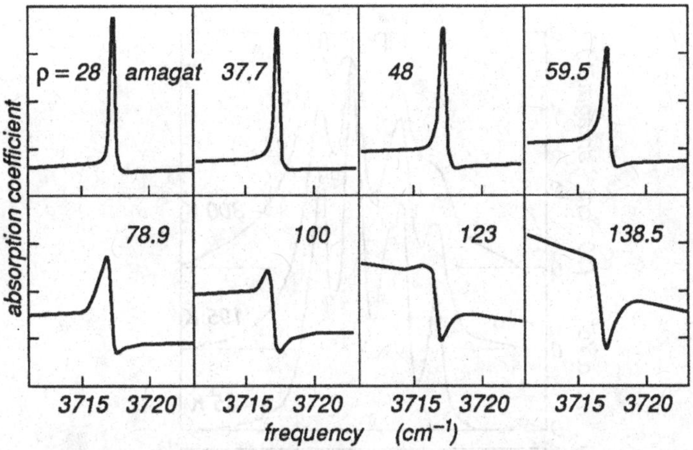

Fig. 3.36. Fano line shapes fitted to the measured absorption profile of the $R_1(0)$ transition of HD at 77 K for densities from 28 to 138.5 amagat. Reproduced with permission by the National Research Council of Canada from [345].

Isotope spectra. Rotovibrational spectra of deuterium, and of deuterium–rare gas mixtures, have also been recorded over a wide range of temperatures and densities [342]. The differences between the H_2–X and D_2–X spectra (with X = H_2 or D_2, respectively, or a rare gas atom) are much like what has been seen above for the rototranslational spectra.

If, however, HD is substituted for H_2, interesting Fano profiles appear in connection with the dipole-allowed $R_1(J)$ lines. Figure 3.36 shows striking density dependences of the $R_1(0)$ line observed in deuterium hydride at 77 K as an example [345]. At low densities (< 28 amagat), the allowed $R_1(0)$ line is more or less of a normal shape. With increasing pressures, a negative 'blue wing' develops and at the highest density (138 amagat) the whole $R_1(0)$ line appears inverted in the collision-induced background. The series of line profiles are shaped by interference of the allowed and induced spectra. The Fano line shapes have been described in terms of phase shift parameters Δ' and Δ'' [178], which depend on the transition considered, and on temperature and gas mixture.

Overtone bands. The induced first overtone band of H_2 is shown in Fig. 3.37 at a variety of temperatures, observed in pure hydrogen gas using long absorption paths. Instead of the three components Q, $S(0)$ and $S(1)$ seen in the fundamental band, we now observe much richer structures, especially at the lower temperatures. This fact suggests that a number of double transitions take place. If one constructs a rotovibrational term scheme of the H_2 molecule, like Fig. 3.32, which includes the lowest rotational levels of the $v = 2$ vibrational state, this is obvious. Various

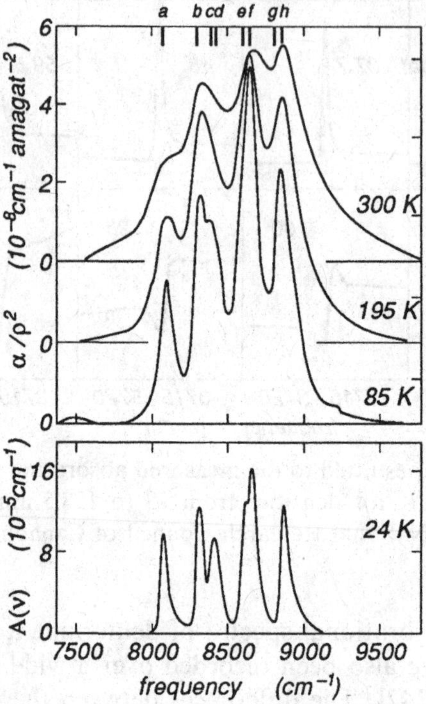

Fig. 3.37. Absorption profiles of normal hydrogen in the first overtone band at gas densities of roughly 100 amagat at various temperatures [187]. Transition frequency labels are *a* for $Q_2(1)$; *b* for $Q_1(1) + Q_1(1)$ (double transition); *c* for $S_2(0)$; *d* for $Q_2(1) + S_0(0)$ (double transition); *e* for $S_2(1)$; *f* for $Q_1(1) + S_1(0)$ (double transition); *g* for $Q_2(1) + S_0(1)$ (double transition); *h* for $Q_1(1) + S_1(1)$; $A(v) = \alpha(v)/\varrho^2$.

rotovibrational single transitions of the type $S_2(J)$ or $Q_2(J)$ in just one of the two interacting H_2 molecules are associated with transition frequencies which are close to but different from certain frequencies of simultaneous transitions of the type $Q_1(J_1) + S_1(J_2)$, $Q_2(J_1) + S_0(J_2)$, $Q_0(J_1) + S_2(J_2)$, etc. There are many more possible double transitions in the overtone band than in the fundamental band and this simple fact explains the rich structures observed; see for example [265].

At the lowest temperature of 20 K, Fig. 3.33, certain lines are nearly pure, for example the lines at the $Q_2(1)$ and $S_2(1)$ transition frequencies which correspond to $v_1 = 0$, $v_1' = 2$, $J_1 = 1$, and $\Delta J_1 = 0$ and 2, respectively, leaving the other molecule unaffected, $v_2 = v_2' = 0$ and $J_2 = J_2'$. Other 'lines' are, however, composite and/or involve simultaneous transitions. We mention the fairly strong $Q_1(1) + S_1(0)$ and $Q_2(1) + S_0(1)$ pair that causes discernible structures next to the induced $S_2(1)$ line. In

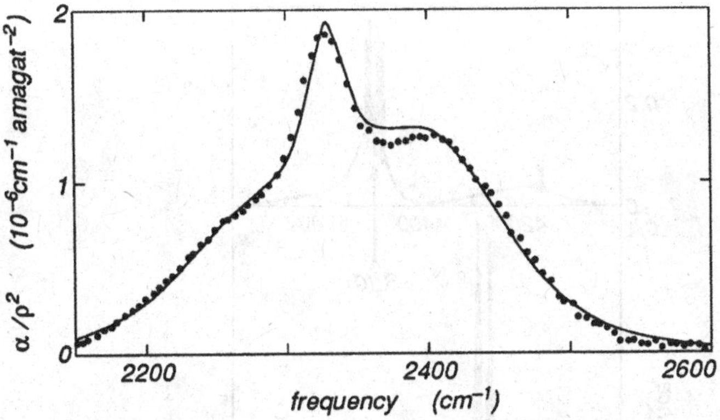

Fig. 3.38. Absorption of N_2–Ar in the fundamental band of N_2 [111]. Nitrogen and argon densities were both 19 amagat; room temperature; measurement (•); the solid line represents a fit based on the stick spectrum and a J-independent quadrupole transition moment. In pure nitrogen, an almost identical spectrum was obtained.

the figure, several of these double transitions are indicated that are more or less resolved at low temperature. Still other lines, for example the prominent one at the combined $Q_1(1) + S_1(1)$ transition frequencies, are purely double transition lines. Interestingly, no overlap-induced Q branch has been seen in the first H_2 overtone band, and no intercollisional dips are observed in the Q branch of these spectra. The examples shown are essentially purely binary spectra.

In hydrogen–helium mixtures, the overtone spectra are so weak that they have not yet been recorded. But in mixtures of the heavier rare gases with hydrogen, it has been possible to record the H_2 overtone spectra. In those cases, the simultaneous transitions are missing, a compelling confirmation of the conclusions concerning simultaneous transitions.

Other systems consisting of molecules other than H_2 have similar rotovibrational spectra. However, the various rotational lines cannot usually be resolved, owing to the smallness of the rotational constants B and the typically very diffuse induced lines. One example, the spectrum of compressed oxygen, was shown above, Fig. 1.1. It consists basically of three branches, the Q, S, and O branch. The latter two are fairly well modeled by the envelope of the rotational stick spectra, similar to that shown in Fig. 3.20, but shifted by the fundamental vibration frequency.

Rotovibrational spectra of many other molecular systems are known, for example those of N_2–N_2 and N_2–Ar [111] which have many features in common. Figure 3.38 shows the latter as an example. Like in the O_2–O_2 rotovibrational band, Fig. 1.1, one recognizes one well developed peak, the Q branch, and asymmetric shoulders to the right (S branch) and left

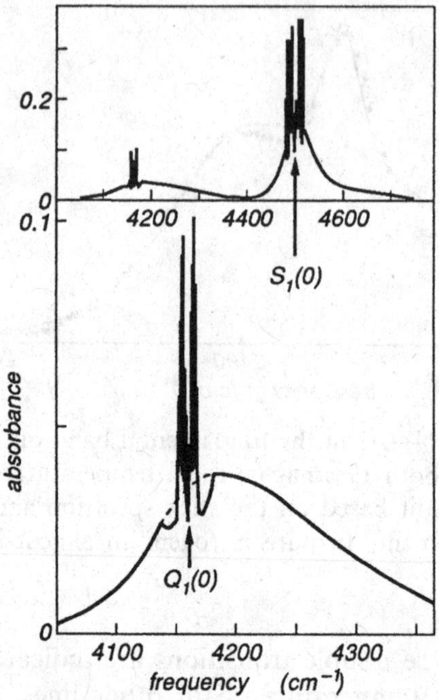

Fig. 3.39. Absorption spectrum of para-H_2 in argon mixture at 77 K and 1.2 amagat; after [267]. The diffuse lines shown in the upper portion of the figure are the collision-induced H_2 $Q_1(0)$ and $S_1(0)$ lines, and the sharp structures near the peaks are due to transitions of bound H_2Ar complexes. The lower trace shows an enlargement of the Q branch region.

(O branch). The measurement is given by the dots. The curve drawn was constructed from the envelopes of the stick spectrum of N_2 (which is well known from Raman studies). The curve approximates the measurement quite closely. In order to fit the experimental line shape one had to take the dependence of the quadrupole transition moments on the rotational quantum number into account.

Dimers. High-resolution spectra obtained by Fourier transform spectroscopy with long path lengths (up to 150 m) and temperatures down to 20 K have shown the bound state-to-bound state bands of a number of van der Waals molecules [267]. Spectra of complexes that contain H_2 (e.g., H_2Ar) can sometimes fully be resolved and rotationally assigned; this provides valuable information concerning molecular interactions. Spectra of heavier complexes (e.g., N_2Ar) may not be fully resolved but can still yield useful information.

The H_2Ar dimer bands of the H_2 fundamental band were first reported in 1965 [221] and have been recorded since with increasing resolution

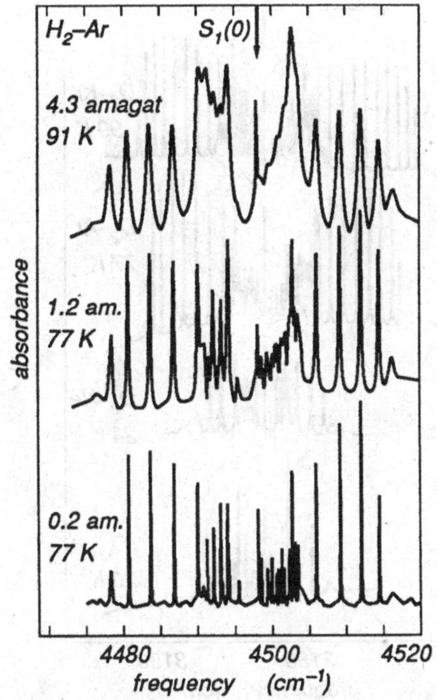

Fig. 3.40. The fine structures centered at the frequency of the H_2 $S_1(0)$ rotovibrational transition due to H_2Ar van der Waals molecules, at 0.2 amagat and 77 K (bottom), 1.2 amagat and 77 K (center) and 4.3 amagat and 91 K (top); after [267].

and detail. The dimer bands occur right and left of the rotovibrational transition frequencies of H_2. They are an inseparable part of the collision-induced absorption spectra. Dimer lines show strong pressure broadening and are easily washed out if high pressures are employed; high temperatures tend to reduce dimer concentrations dramatically which makes a recording of their spectra difficult. Rotation and vibration of the H_2 molecule remain virtually unaffected by the attached Ar atom.

The collision-induced background of the dimer bands seen in the lower part of Fig. 3.39 shows a broad dip of absorption (the difference between the dashed line and the measured trace). Nearly symmetrically with the H_2 $Q_1(0)$ transition frequency (at 4161.1 cm^{-1}) a series of sharp and intense absorption lines is seen that are resolved in the next two figures. The nature of the dip will become apparent below when we consider line shape computations; they are related to the impossibility of bound-to-free and free-to-bound transitions at small frequency shifts relative to the H_2 rotovibrational transition frequencies and are quite general. We simply state that the dimer dip is a two-body spectral feature whereas the

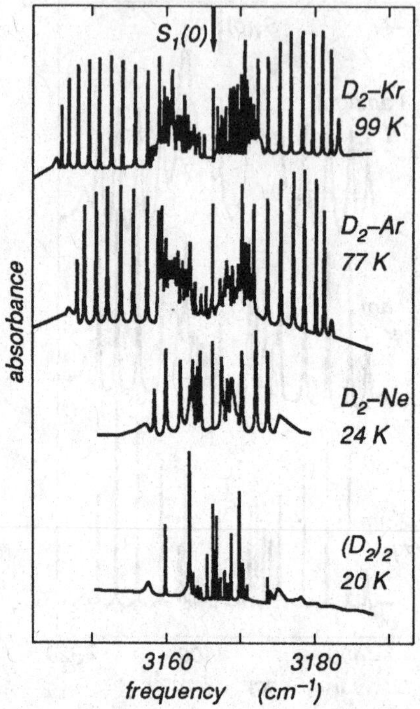

Fig. 3.41. The spectra of van der Waals molecules containing D_2 that are centered at the D_2 $S_1(0)$ rotovibrational transition frequency; after [267]. Bottom: $(D_2)_2$; first trace above bottom trace: D_2Ne; second trace above bottom trace: D_2Ar; top: D_2Kr. A sharp monomer quadrupole line can also be seen in each trace at the position of the D_2 $S_1(0)$ transition frequency.

intercollisional dip arises from many-body interaction; the two kinds of dips occur for totally different reasons.

Figure 3.40 shows the high-resolution spectra of H_2Ar accompanying the H_2 $S_1(0)$ transition for three different experimental conditions; besides the dimer bands, the monomer quadrupole line is also discernible at 4497.8 cm^{-1}, a weak allowed, very sharp line at the H_2 $S_1(0)$ transition frequency (the central line that shows no pressure broadening). The spectrum at the bottom of the figure is obtained at the very low density of 0.2 amagat and shows well resolved, sharp dimer lines. The spectrum in the center was recorded at a higher density (1.2 amagat) and shows clearly pressure broadened dimer lines. The spectrum at the top is recorded at the highest density (4.3 amagat) and shows the strongest broadening. The broadening coefficient is very big, of the order of 0.2 cm^{-1}/amagat, and explains to some extent why even moderate pressures render the bands of van der Waals molecules difficult to resolve.

Figure 3.41 compares the spectra of similar van der Waals dimers, from bottom to top: $(D_2)_2$, D_2Ne, D_2Ar, and D_2Kr of the D_2 $S_1(0)$ line. The lightest system $(D_2)_2$ shows relatively few lines, a reflection of the fact that the bond is relatively weak ($\varepsilon/k \approx 37$ K) and the de Broglie wavelength long (because of the reciprocal reduced mass dependence). (The lighter system $(H_2)_2$ shows even fewer lines and levels; see Fig. 3.25.) It is well known that with increasing mass of the partner X, the attraction between D_2 and X grows stronger and the de Broglie wavelength shorter so that more bound levels exist for those systems. This simple fact leads to much richer van der Waals dimer bands as the figure shows.

Van der Waals molecules of heavier homonuclear diatomics have also been studied, using similar techniques to the ones mentioned above. However, the numbers of bound states generally are much greater for such systems, and the band structures are richer and therefore harder to resolve. Detailed work has shown that for the more massive diatomics molecular rotation is more or less hindered and the level structures are much more complex than the ones seen in the H_2–X systems. Rather uncertain band contour analyses are used in those cases but a few reasonably well resolved band spectra of van der Waals molecules are known [49, 267].

Spectral moments. Spectral moments are defined according to Eqs. 3.5 and 3.6. For the rotovibrational bands of diatom–X pairs, like H_2–Ar or H_2–H_2, the expressions

$$\gamma_1 \approx \frac{8\pi^3 N_L^2}{3\hbar} \left\{ M_1^{(i)} + M_1^{(a)} + (\nu_{vv'} + 6B)\, M_0^{(a)} \right\} \tag{3.12}$$

$$\gamma_0 \approx \frac{8\pi^3 N_L^2}{3\hbar} \left\{ M_0^{(i)} + M_0^{(a)} \right\} \tag{3.13}$$

are often useful approximations; more general expressions must account for anisotropic overlap components, double transitions, higher multipole-induction (beyond quadrupole), for the vibrational state dependence of the interaction potential, and the rotation-dependence of the vibrational matrix elements of molecular polarizability invariants and multipole moments, see Chapter 4 for details. In these expressions, B is the rotational constant and $\nu_{vv'}$ the vibrational frequency associated with transitions between vibrational states $v \to v'$; v and v' are vibrational quantum numbers of the initial and final states of the molecules involved. The superscripts (i) and (a) designate isotropic, overlap-induced and anisotropic (or quadrupole-induced) contributions. These expressions are analogous to those shown for the rototranslational bands, Eqs. 3.10 and 3.11; they formally differ from the latter by the appearance of the vibrational frequency, $\nu_{vv'}$, in Eq. 3.12. The presence of this factor of $M_0^{(a)}$ emphasizes the (anisotropic) zeroth moment of γ_1 much more than in the rototranslational bands so that in the rotovibrational bands the measurement of γ_1

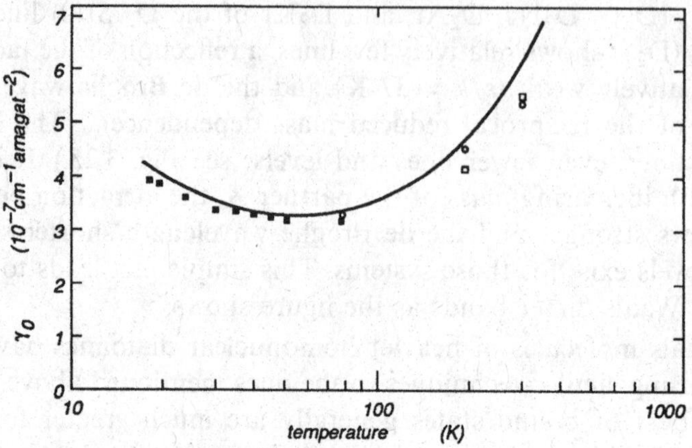

Fig. 3.42. Spectral invariant γ_0 of H_2–H_2 as function of temperature; after [281]. Various measurements are shown (■; ○; □). The solid line is computed from first principles.

Table 3.6. Experimental values of the binary absorption coefficient for the fundamental band of normal hydrogen.

gas	T (K)	γ_0 $(10^{-7}cm^{-1}am^{-2})$	Ref.
H_2–He	300	2.48	[175]
H_2–Ne	300	5.71	[337]
H_2–Ar	300	9.29	[175]
H_2–Kr	300	18.2	[337]
H_2–Xe	300	27.3	[414]
H_2–N_2	300	12.3	[175]
H_2–O_2	300	14.7	[408]
D_2–D_2	300	3.37	[335]

repeats to some extent the information contained in the measurements of γ_0. For classical systems, the first moments M_1 may be replaced by second moments, M_2; see p. 99.

Figure 3.42 shows the measurements (dots, etc.), at various temperatures of the spectral moment γ_0 of H_2–H_2 pairs for the fundamental band, $v = 0 \rightarrow v' = 1$, at $v_{01} = 4161.1$ cm^{-1}. Also shown are computations of that quantity from first principles (curve) [281]. The agreement is well within the experimental uncertainties. Figures 3.43 and 3.44 show similarly the spectral moments of H_2–He pairs. Again, measurement and

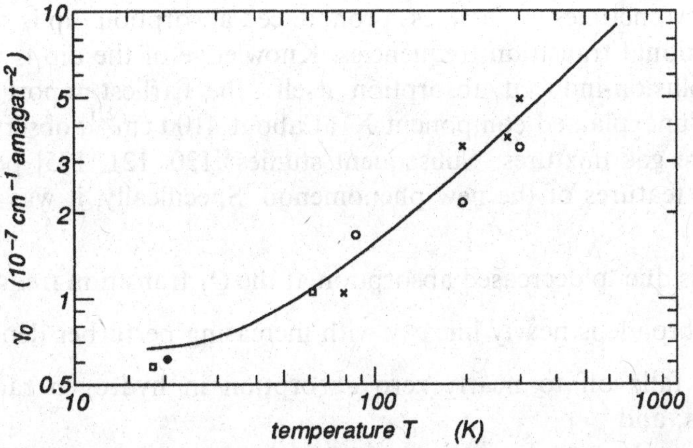

Fig. 3.43. Spectral moment γ_0 of the H_2–He fundamental band as function of temperature; after [151]. Various measurements are shown (•, ○, ×, □). The curve is computed from first principles.

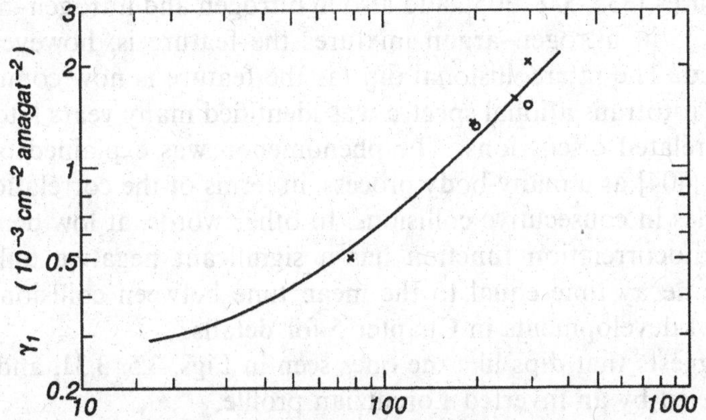

Fig. 3.44. Spectral moment γ_1 of the H_2–He fundamental band as function of temperature; after [151]. Measurements (○, ×); The curve is computed from first principles.

fundamental theory agree within the experimental uncertainties.

As a further illustration, Table 3.6 lists measured spectral moments for a number of other binary systems involving H_2 or D_2 molecules. An even broader listing may be found in a recent review article [342].

2 Ternary and many-body components

Intercollisional dip. In most examples of the rotovibrational spectra shown, one notices a fairly well developed dip at the Q_1 transition frequency.

Furthermore, sometimes a much less pronounced absorption dip is seen at the rotovibrational transition frequencies. Knowledge of the dip is nearly as old as collision-induced absorption itself; the earliest report [129] mentions an 'unexplained component X' at about 4100 cm^{-1}, observed in hydrogen–rare gas mixtures. Subsequent studies [120, 121, 175] pointed out the main features of the new phenomenon. Specifically, it was noted that

- the dip is due to decreased absorption at the Q_1 transition frequency;
- the dip broadens nearly linearly with increasing perturber density;
- the dip falls off to nearly zero absorption in hydrogen–rare gas mixtures; and
- the dip persists and continues to broaden to the highest pressures ($\approx$ 5000 atmospheres).

Later studies showed the same phenomena in deuterium and deuterium–rare gas mixtures [335, 338, 305], and also in nitrogen and nitrogen–helium mixtures [336]; in nitrogen–argon mixtures the feature is, however, not well developed. The intercollisional dip (as the feature is now commonly called) in the rototranslational spectra was identified many years later; see Fig. 3.5 and related discussions. The phenomenon was explained by van Kranendonk [404] as a many-body process, in terms of the correlations of induced dipoles in consecutive collisions. In other words, at low densities, the dipole autocorrelation function has a significant negative tail of a characteristic decay time equal to the mean time between collisions; see the theoretical developments in Chapter 5 for details.

Theory suggests that dips like the ones seen in Figs. 3.5, 3.31, and 3.34, may be modeled by an inverted Lorentzian profile,

$$D(\nu) \cong (1-\gamma) + \frac{\gamma\nu^2}{\nu_{12}^2 + \nu^2}\,, \tag{3.14}$$

with ν_{12} given by the mean time between collisions according to $\nu_{12} = 2\pi/\tau_{12}$ and $\gamma \approx 1$ [407, 237]. The function $D(\nu)$ appears as a factor of the spectral function $g(\nu)$ in Eq. 3.2. This function, Eq. 3.14, has been successfully fitted to a selection of measurements; the value of γ was found to be between 0.9 and 1 in the cases considered [245].

The weak dips near the rotovibrational H_2 $S_1(1)$ line have been studied in some detail [316] and show the same dependence on density variation as those of the Q branch. For these, the parameter γ of Eq. 3.14 is of course much smaller than unity.

An excellent review article was published by Lewis [238] which considers in detail the wealth of experimental facts and data; it also reviews the theoretical developments.

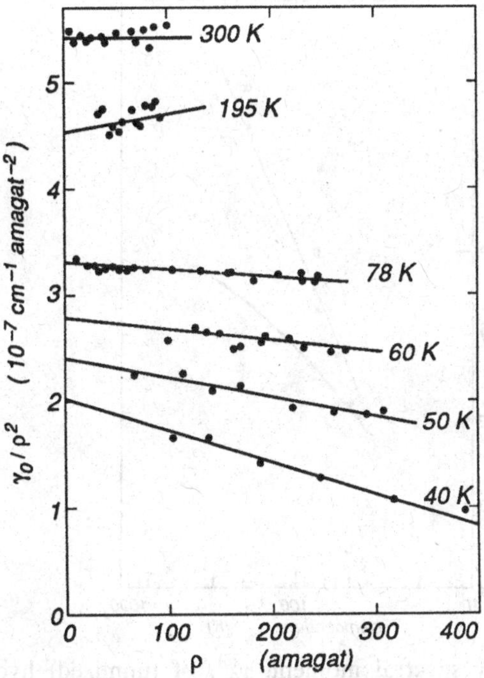

Fig. 3.45. Variation of the normalized spectral moment with density for the H_2 fundamental band in pure hydrogen at various temperatures; after [187].

Diffuse three-body component. When densities are increased to ≈ 100 amagats or more, additional three-body contributions are readily discernible in most gases. Figure 3.45 shows measurements of the spectral integral γ_1, Eq. 3.5, of the H_2 rotovibrational band, divided by density squared, as function of density and temperature in pure hydrogen [187]. These total absorption coefficients, when normalized by density squared, should be independent of density as long as binary processes prevail. However, as may be seen in the figure, one notes a linear dependence, with slopes that are negative at low temperatures, and positive at the higher temperatures. The linear behavior (constant slopes) suggests a ternary spectral contribution. In fact, the in most cases well defined slopes represent the ternary virial spectral coefficient of hydrogen for the fundamental band, the spectral contributions of the ternary system H_2–H_2–H_2. The intercepts with the $\varrho = 0$ axis give the purely binary contributions, Fig. 3.45. Most of the measurements have been made with 'normal' hydrogen, that is with a para- to ortho-H_2 concentration ratio of 1:3. A few data are also given for a 1:1 para- to ortho-H_2 ratio. We note that dramatic variations of such data with the para- to ortho-H_2 ratio are not expected on theoretical grounds [296]. Other works report similar three-body data in reason-

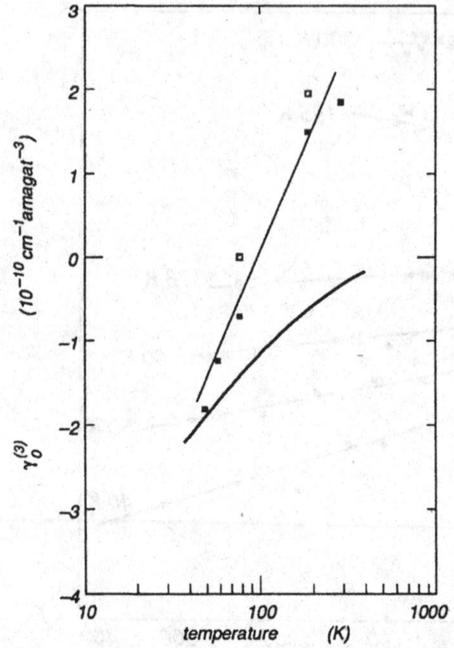

Fig. 3.46. Three-body spectral moment, $\gamma_0^{(3)}$, of (unmixed) hydrogen of the fundamental band at various temperatures [296]. Solid squares represent measurements using a normal para- to ortho-H_2 concentration of 1:3; open squares from measurements with a 1:1 concentration ratio; the thin line is a visual average of Hunt's measurements. The thick line is a calculation of the pairwise-additive contribution of that quantity. The comparison suggests that substantial irreducible contributions have affected the measurements, especially at elevated temperatures.

able agreement [121]. Even the data obtained at much higher densities with pure hydrogen, and for hydrogen–helium, hydrogen–nitrogen and hydrogen–argon mixtures at room temperature are generally consistent [175]. Table 3.7 collects the ternary moments thus obtained.

In the H_2 fundamental band of hydrogen gas of low pressure, both ternary moments, $\gamma_1^{(3)}$ and $\gamma_0^{(3)}$, are *negative* at low temperature, go through zero and turn positive at temperatures greater than ≈ 90 K. Figures 3.46 and 6.2 present these moments as functions of temperature [296]. Moreover, the differences between data taken with differing para- to ortho-H_2 concentrations are reasonably consistent (open and solid squares); the apparent differences at constant temperature are probably just an indication of the actual uncertainties of these difficult measurements. The variations of results obtained in other works [121, 175] with the same gas and at the same temperature, which the reader may discover in Table

Table 3.7. Ternary integrated absorption coefficients.

system	T (K)	$\gamma_1^{(3)}$ (10^{-8} cm^{-2} am^{-3}) meas.	calc.	$\gamma_0^{(3)}$ (10^{-10} cm^{-1} am^{-3}) meas.	calc.	Ref.
H_2–Ar–Ar	298	300				[121]
	298	390				[175]
	300	293		6.0		[187]
	373	600				[121]
H_2–H_2–H_2	40	-134^a	−100	-3.06^a	−2.1	[187]
	50	-77^a	−95	-1.81^a	−1.9	[187]
	60	-57^a	−90	-1.24^a	−1.8	[187]
	78	-23^a	−80	-0.71^a	−1.5	[187]
	78	$\approx 0^b$	−80	$\approx 0^b$	−1.5	[187]
	195	55	−20	1.5	−0.3	[187]
	195	84	−20	1.95	−0.3	[187]
	300	89	−5	1.85	−0.1	[187]
	298	200	−5			[121]
	298	110	−5			[175]
H_2–He–He	200	8				[121]
	298	10	3.70			[121]
	298	55.	3.70			[175]
	300	14	3.70	0.23		[187]
	295	41	3.70			[142]
H_2–N_2–N_2	198	600				[121]
	298	600				[121]
	298	550				[175]
	300	437		9.2		[187]

[a]obtained with an ortho-H_2 to para-H_2 concentration ratio of 1:1.
[b]obtained with an ortho-H_2 to para-H_2 concentration ratio of 3:1.

3.7, do similarly reflect the uncertainties of the measurements. We note, though, that some of these other measurements may give less reliable three-body information, because they were conducted at generally higher densities; these may be affected by (undetected) N-body processes with $N > 3$. Nevertheless, in most cases the consistency of the data from various works seems quite reasonable.

We note that an elaborate set of three-body spectral moments of various rotovibrational bands was compiled elsewhere [342], the values of which are generally much greater than the data listed in Table 3.7. We point out that these more recent data were obtained at low gas densities, typically from 10 to 50 amagats. They are, therefore, barely affected by three-body interactions: pressure variation of the spectral moments M_n/ϱ^2 amounted typically to just one or two percent. While measurements at small densities are desirable to minimize interference by four-body effects, reliable data of this kind must be based on measurable effects that exceed significantly the experimental uncertainties (typically ten percent). On that score, the more recent data are deficient, in our judgement.

For easy reference we also plot theoretical three-body moments of H_2–H_2–H_2 which are computed from first principles based on the assumption of a pairwise-additivity, Figs. 3.46 and 6.2 (heavy curves). The pair dipole moments have been shown to allow a close reproduction of measured binary spectra from first principles in the hydrogen fundamental band, for temperatures from 20 to 300 K; these are believed to be reliable. Interestingly, the pairwise-additive assumption is not sufficient to reproduce the experimental three-body moments from theory, except perhaps at the lowest temperatures. With increasing temperature, rapidly increasing differences between measured and computed moments are observed, a fact which suggests the presence of an irreducible three-body dipole component of the overlap-induced type [296].

Table 3.7 also lists ternary spectral moments for a few systems other than H_2–H_2–H_2. For the H_2–He–He system, the pairwise-additive dipole moments are also known from first principles. The measured spectral moments are substantially greater than the ones calculated with the assumption of pairwise additivity – just as this was seen in pure hydrogen. For the other systems listed in the Table, no *ab initio* dipole surfaces are known and a comparison with theory must therefore be based on the approximate, classical multipole model.

We note that similar conclusions were drawn from the data obtained in the rototranslational bands, and the purely translational bands, pp. 75ff. and 104ff. In all cases considered, the moments calculated with the assumption of pairwise-additivity are smaller than the measurements.

Double transitions. In molecular fluids, simultaneous or 'double' transitions occur at sums and differences of the rotational lines, with the absorption of a single photon. Several of such double transitions have been pointed out above, Figs. 3.32 and 3.37. In general terms, one may say that these occur at sums and differences of the rotovibrational (and/or electronic) transition frequencies of the molecules involved, as was explained in the discussions related to Fig. 1.3.

Van Kranendonk has discussed the role of double and triple collisions in the ternary integrated absorption coefficient [400]. Triple and higher-order collisions in which the rotovibrational states of two or more of the molecules of the complex change during the absorption process can occur only as a result of the non-additive contributions in groups of three or more molecules. On the basis that the induced dipole moments are additive, only single and double transitions can occur [392]. Van Kranendonk found that when the transition in the central molecule is accompanied by a change in the rotational state of one or more of the surrounding molecules, the cancellation or interference effect discussed above becomes ineffective, even if the accompanying transition is only an orientational transition between two degenerate rotational states.

3 *Dense fluids*

In dense fluids and in solids many-body interactions are significant. It is interesting to study how these affect the observable spectra.

Hydrogen. Collision-induced absorption in hydrogen in the H_2 fundamental band has been studied over a wide range of pressures and temperatures down to 20 K. It is evident from the examples shown above that the resolution of the components of the band improves greatly as the temperatures are lowered. The early measurements in the gas phase were followed by studies of the absorption of liquid and solid hydrogen [4], Fig. 3.47; the densities of gas, liquid and solid are nearly the same. The results show that the Q line (between arrows labeled a and b) and the rotational S lines (arrows labeled c and f for $S_1(0)$ and $S_1(1)$, respectively) occur at nearly the same frequencies as with the free H_2 molecule. Rotation of the H_2 molecule in the liquid and solid is more or less unhindered. Other structures can be identified as double transitions $S_1(0)+S_0(0)$, $S_0(0)+S_1(1)$, $S_1(1)+S_0(1)$ (arrows h, i, k). Other (feeble) structures or grouped lines have been thought to be associated with the coupling to lattice vibrations (phonon spectra).

At intermediate densities, the splitting of the Q branch (i.e., the intercollisional dip) of the fundamental band increases linearly with the density. A more rapid increase was observed at the highest densities which may signal significant N-body contributions, with $N > 3$. Interestingly, the components of the overtone and the simultaneous vibrational transitions, like the rotatational S lines of the fundamental band, show no significant splitting or broadening with increasing density [175].

The intercollisional dips of the Q_1 lines were seen at all densities where spectra could be recorded, even if hydrogen is dissolved in liquid argon, etc., but at the highest densities (e.g., pressurized liquids) certain deviations

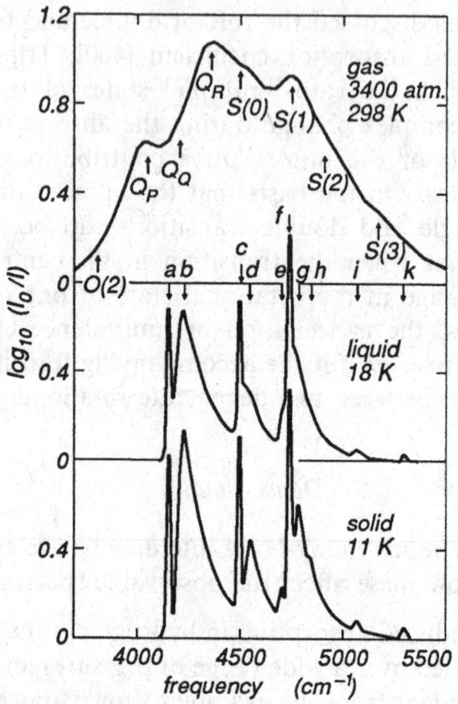

Fig. 3.47. Fundamental absorption band of hydrogen; top: of the compressed gas (at the density of 1014 amagat); center: of the liquid (817 amagat); bottom: of the solid (972 amagat); after [4].

of the profiles and their center frequencies were discovered [245].

The collision-induced fundamental band of H_2 has been investigated in pure hydrogen, and in hydrogen–helium, hydrogen–argon and hydrogen–nitrogen mixtures at pressures up to 1500 atmospheres (corresponding to roughly 400 to 600 amagats, depending on gas and temperature) and temperatures in the range 80 through 376 K [121]. Parallel work used the same gases at room temperature and pressures up to 5000 atmospheres (that is densities up to roughly 1000 or 1500 amagats) [175]. The measurements of the integrated absorption coefficient γ_1 could be represented in both works in the form of a virial series which includes two- and three-body terms; results are listed in Table 3.7, along with other measurements which were conducted at much lower densities. Reasonable consistency is observed in most cases.

Absorption in pure nitrogen has been studied in the N_2 rotovibrational bands in the gaseous, liquid and solid state. As this was observed with other gases and in other bands, the *shapes* of these absorption spectra do not vary much with density [336] but the intensity of the normalized absorption coefficient, $\alpha(\nu)/\varrho^2$, decreases more or less uniformly at

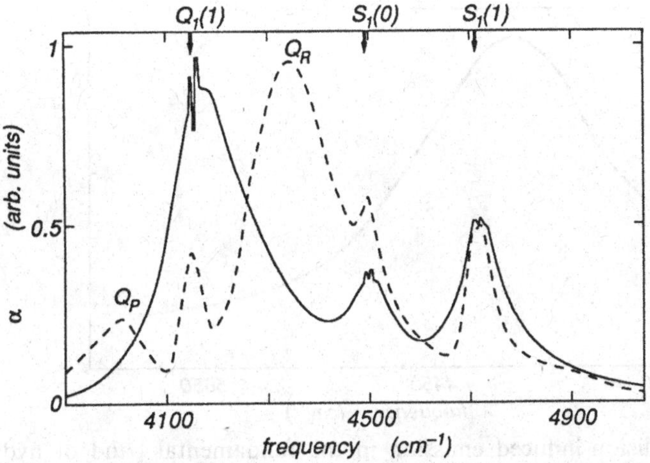

Fig. 3.48. Enhancement of the absorption in the fundamental band in a hydrogen-argon mixture at low and high densities at 152 K; the profiles are normalized to give $S_1(1)$ the same peak intensity. The argon density was 8 amagat (solid line) and 820 amagat (dashed line), respectively. The density splitting of the (overlap-induced) Q branch and the density narrowing of the S lines are apparent. A new (quadrupole-induced) Q line appears inthe wide absorption dip (between Q_P and Q_R) observed at high density. Reproduced with permission from the National Research Council of Canada from [137].

all frequencies with increasing densities, and the moments γ_1 and γ_0 decrease with it. Similar behavior is also observed in nitrogen–argon, nitrogen–hydrogen and nitrogen–helium mixtures at high pressure in the N_2 fundamental band [369]. The absorption spectrum of nitrogen in the N_2 first overtone band has been recorded in highly compressed gases and liquids [409].

The spectra of liquids and solids cannot be studied here in any detail; we refer the reader to other works, such as the two volumes by Gray and Gubbins, and the review articles by Guillot and Birnbaum (1989).

3.4 Collision-induced emission

Dipole moments that absorb will also emit radiation. Especially at elevated temperatures collision-induced emission (CIE) spectra may be expected and have in fact been found in the laboratory using shock tubes. Stellar atmospheres are also known to emit radiation by interatomic (e.g., H–He) and intermolecular (H_2–H_2, etc.) interactions.

Shock tubes consist of a driver and a test section. The driver is pressurized with gas until the diaphragm, which separates driver and

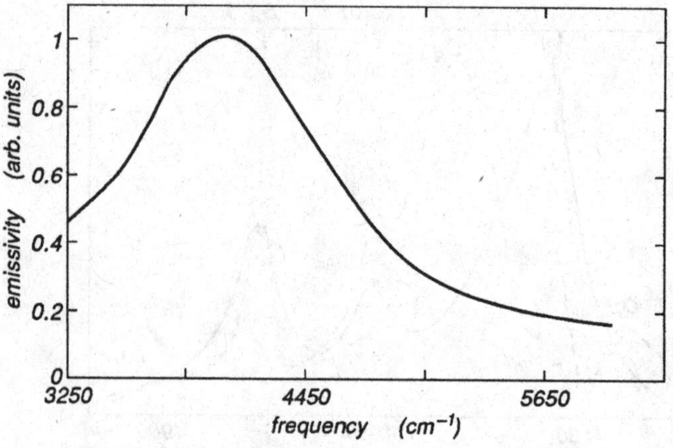

Fig. 3.49. Collision-induced emission in the fundamental band of hydrogen; after [116]. 20% hydrogen in argon at the temperature $T = 2844$ K.

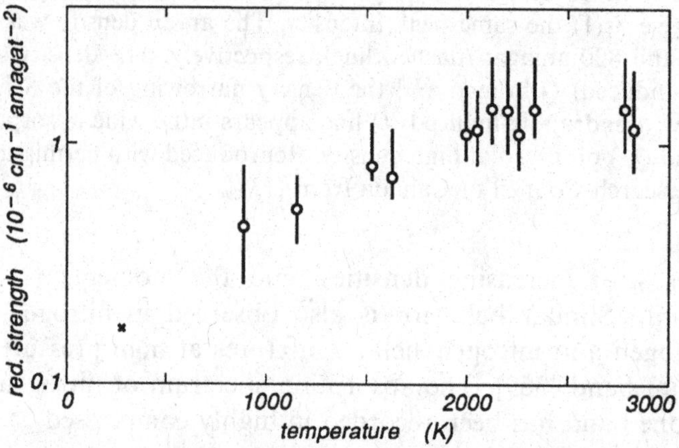

Fig. 3.50. Total band intensities of collision-induced emission in the fundamental band of H_2, after [116]; 20% hydrogen in argon. The symbol × represents emission at 300 K, estimated on the basis of an absorption measurement at that temperature.

test region, ruptures. The test region contains the gas or mixture whose emission is to be studied. When a shock wave propagates down the test region, the gas is compressed and heated. When the shock wave reaches the end of the shock tube it is reflected, thus causing additional compression and heating of the test gas. High densities and temperatures may be maintained for a few milliseconds. At longer times, the cooling effects are significant and the rarefaction wave, which travels down the tube behind the shock will also modify the conditions radically.

For a mixture of 20% hydrogen and 80% argon and pressures between

1 and 10 atmospheres, shock fronts may travel at speeds in excess of 1 km s^{-1}; pressures from 100 to 200 atmospheres and temperatures from 1000 to 3000 K are typical if hydrogen at $\approx$ 100 atmospheres is used as driver. Measurements of collision-induced emission exist in the H_2 fundamental band [219]. Figure 3.49 shows as an example the emission spectrum of a 20% hydrogen in argon mixture at 2844 K in the H_2 fundamental band [116]. The spectrum peaks roughly at the Q line but shows otherwise little discernible structure, a fact that is doubtlessly related to the diffuseness of the lines at this high temperature. Spectral moments γ_1 have been determined, Fig. 3.50. Emission is largely due to H_2–Ar pairs, but a small contribution from H_2 pairs has not been subtracted. Similar data exist for nitrogen [115].

3.5 Dipole autocorrelation function

The dipole autocorrelation function, $C(t)$, is the Fourier transform of the spectral line profile, $g(\nu)$. Knowledge of the correlation function is theoretically equivalent to knowledge of the spectral profile. Correlation functions offer some insight into the molecular dynamics of dense fluids.

Dipole autocorrelation functions have been determined for a number of systems (Buontempo *et al.*, 1980; Barnabei *et al.*, 1985). In the binary regime, correlation functions are roughly exponential functions of time at long times. At short times marked deviations from the exponential form are noted, Fig. 3.51 (top). At long times, theory suggests deviations from the exponential behavior as well; even a change of sign is likely because the induced dipole moment changes sign at long range (dispersion dipole component). With increasing density, the short time behavior deviates even more from the exponential, Fig. 3.51 (lower traces), and a prominent *negative* long-time part develops (intercollisional dip, not shown).

3.6 Induced spectra and allowed transitions

Spectral moments of allowed molecular absorption bands vary in general nearly linearly with the gas density, $\gamma_1/\varrho \approx$ constant. At sufficiently high pressures, a small, linear increase with density of the ratio γ_1/ϱ is, however, discernible, e.g., [8]. This quadratic absorption component is largely due to 'apparent' induced absorption, resulting from the long range interaction of dipoles induced by the incident radiation field [400]. Moreover, a 'true' induced absorption component is believed to exist which arises from collision-induced dipole components (Chapter 4) [146, 210]. It was argued, however, that in most measurements true induced absorption was too weak to be identified positively in this way. Recent experimental and

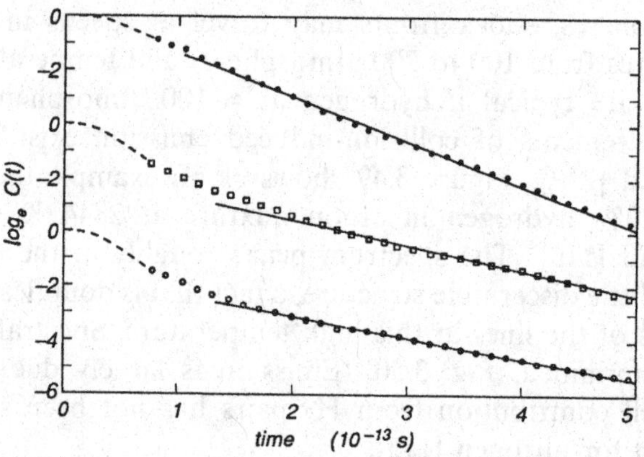

Fig. 3.51. Logarithmic plot of the normalized induced dipole moment correlation function, $C(t)$, for hydrogen–argon mixtures at 165 K. Measurements at 90 amagat (•); 450 amagat (□); and 650 amagat (○). The broken lines at small times represents the portion of $C(t)$ affected by the smoothing of the wings of the spectral profiles. Reproduced with permission by the National Research Council of Canada from [109].

molecular dynamics studies suggest that in dense fluids (liquids), evidence for true induced absorption has emerged [312, 247].

Interference effects between allowed and induced phenomena have received attention [246]. Most of the work reported to date is for liquids and thus outside the scope of this monograph. We mention, though, that the Fano profiles observed in the HD–X spectra, Fig. 3.36, are interesting examples of such interference.

3.7 Empirical line shapes

In conventional spectroscopy, analytical models of line profiles have been of great utility. Specifically, we mention the Lorentzian shape,

$$g_L(\overline{\nu}) = \frac{S}{\pi\Gamma} \frac{\Gamma^2}{\overline{\nu}^2 + \Gamma^2}, \tag{3.15}$$

which describes the 'natural' line width of unperturbed atoms and molecules. The constant S is the line strength (the zeroth moment), $\overline{\nu}$ is the frequency shift relative to the line center, and Γ is the half-width at half-intensity; $\frac{1}{2}\Gamma\,\mathrm{d}t$ is the probability of emission (assumed to be independent of time) in the time interval between t and $t + \mathrm{d}t$. Measurements of the natural line width of spectral lines have been possible by fitting the wings of various spectral lines to a Lorentzian function. Simple theories of pressure broadening of spectral lines also lead to such profiles, provided the Doppler effect is negligible and the mean duration of a collision is small compared to the mean time between collisions. In that case, $\Gamma\, dt$ becomes the probability of a (phase-changing) collision in dt.

The random velocities of atoms and molecules are described by velocity distribution functions which can often be approximated by a Maxwellian distribution (as in Eq. 2.10). If radiating atoms have such a distribution, the resulting line profile is a Gaussian,

$$g_G(\overline{\nu}) = \frac{S}{\Delta\sqrt{\pi}} \exp\left[-(\overline{\nu}/\Delta)^2\right], \tag{3.16}$$

provided natural width and pressure broadening are negligible. In this expression, Δ is related to the thermal Doppler half-width at half-maximum, $2\Delta\sqrt{\ln 2}$. If natural width and/or pressure broadening are *not* negligible, the so-called Voigt profile, the convolution of a Lorentzian and Gaussian function, is useful. Finally, if the mean duration of collisions is not negligible compared to the other parameters, or if the distant wings need to be modeled closely, an exponential function will often describe actual far wings much more closely than the $1/\overline{\nu}^2$ fall-off of the Lorentzian; the logarithmic slope is related to the mean duration of a collision. Such model profiles are standard tools in spectroscopy.

The question arises whether collision-induced profiles may perhaps also be modeled by these or other simple functions, perhaps under circumstances to be defined. Such model functions would be of interest for the analysis of measured spectra, for empirical predictions of spectra at temperatures other than those of the measurements, for frequency extrapolations as may be necessary for accurate determination of spectral moments, etc.

From the beginnings, attempts to model the line shapes of collision-induced absorption spectra were based on the assumption that the various rotational lines of induced spectra, Figs. 3.10 through 3.14, are superpositions of scaled and shifted line profiles, $g_{(c)}(\overline{\nu})$, of a small number of different, e.g., overlap- and quadrupole-induced, types [313, 404],

$$\frac{\alpha(\nu)}{\nu\ [1-\exp(-hc\nu/kT)]} = \sum_{i=0}^{i_{max}} A_i\, g_{(c)}(\nu-\nu_i)\,, \tag{3.17}$$

where the ν_i are rotational transition frequencies and $\overline{\nu}$ designates the frequency shift relative to the line center. The A_i are frequency-independent factors which can be computed from theory with simplifying assumptions. For example, at low temperatures (77 K) the H_2–H_2 spectra, Fig. 3.10, and even at higher temperatures (298 K) the H_2–Xe spectra, feature well separated lines with but little overlap, Fig. 3.14. In those cases, an empirical, individual line profile is fairly well defined, except in the wings where lines overlap.

The conventional Lorentzian, Gaussian, etc., profiles mentioned above are all symmetric, $g(-\overline{\nu}) = g(\overline{\nu})$. In contrast to this symmetry, the individual line profiles in collision-induced absorption have early been recognized as being quite asymmetric, roughly as [120, 215, 188]

$$g(-\overline{\nu}) \approx e^{-hc\overline{\nu}/kT}\, g(\overline{\nu}) , \tag{3.18}$$

a relationship that is sometimes referred to as the principle of detailed balance. According to Eqs. 1.6 through 1.7 above, since hc times the shift $\overline{\nu}$ equals the change of translational energy of the pair, one might assume that for a fixed $|hc\overline{\nu}|$ any upward transition, $\Delta E_{\text{transl}} > 0$, is $\exp(hc\overline{\nu})$ times more likely than the corresponding inverse transition, which reflects the ratio of the Boltzmann population factors of the states of the pair. This simple fact leads to the observed asymmetry of the profiles. We mention, however, that Eq. 3.18 is not necessarily exact as stated and must be used with caution (Chapter 6).

Early attempts have modeled the positive frequency wing ($\overline{\nu} > 0$) with the help of a Lorentzian, Eq. 3.15, because of a perceived similarity of the observed induced lines with the Lorentzian; no theoretical justification was pretended. The negative frequency wing may then be described by $\exp(-hc\overline{\nu}/kT)$ times that Lorentzian so that Eq. 3.18 is satisfied [215, 188, 414, 411]. Systematic deviations from this model were, however, noticed in the wings [75, 244]. Nevertheless, beautiful analyses of various rotational and rotovibrational induced bands were thus possible and significant new knowledge concerning the role of overlap and multipole induction, double transitions, etc., was obtained in this way [422].

In an attempt to model the spectral functions of rare gas mixtures, Fig. 3.2, it was noted that a Gaussian function with exponential tails approximates the measurements reasonably well [75], about as well as the Lorentzian core with exponential tails. Two free parameters were chosen such that at the mending point a continuous function and a continuous derivative resulted; the negative frequency wing was again chosen as that same curve, multiplied by the Boltzmann factor, to satisfy Eq. 3.18. Subsequent work retained the combination of a Lorentzian with an exponential wing and made use of a desymmetrization function [320],

$$g_{LE}(\overline{\nu}) = \begin{cases} \dfrac{2SF}{1+\exp(-hc\overline{\nu}/kT)}\, \dfrac{1}{1+(\tau\overline{\nu})^2} & \text{for } |\overline{\nu}| \le p/\tau \\[2ex] \dfrac{2SC}{1+\exp(-hc\overline{\nu}/kT)}\, \exp(-b|\overline{\nu}|) & \text{otherwise,} \end{cases} \tag{3.19}$$

with $F = \tau p/(1 + 2p \arctan p)$ and $C = (F/(1 + p^2)) \exp(2p^2/(1 + p^2))$. The free parameters are total intensity, S, a time parameter τ, and the dimensionless switch-over parameter p. The desymmetrization function, the first one of the two factors to the right of the equation, specifies just one of many such functions which will give rise to the desired asymmetry, Eq. 3.18. A recent computational study suggests that better desymmetrization procedures exist [70]; these will be discussed in Chapter 5. One- and two-parameter line profiles were given recently by [223].

Theoretical models of correlation functions and line shapes have been proposed which satisfy the principle of detailed balance [35, 36, 41, 232]. These profiles, along with a number of extensions that were later added [69, 295, 47, 48], describe the known profiles well. Especially the BC and K0 functions, Eqs. 5.105 and 5.108, model multipole- and overlap-induced lines of the rototranslational bands closely. These three-parameter functions are simple analytical expressions that are readily computable, even on computers of small capacity (pocket calculators); the parameters can be computed from the lowest three spectral moments, see Chapter 5.

Intercollisional dip. The line shape of the intercollisional dip has been modeled by an inverted Lorentzian as was explained on pp. 124ff.

Principle of corresponding states. The principle of corresponding states, originally introduced by van der Waals and applied since to model intermolecular potentials, transport and equilibrium properties of fluids over a wide range of experimental conditions, was remarkably successful, albeit it is not exact in its original form. An interesting question is whether one could, perhaps, describe the diversity of spectral shapes illustrated above by some 'reduced' profile, in terms of reduced variables. If all known rare-gas spectra are replotted in terms of *reduced* frequencies and absorption strengths,

$$\nu^* = \nu/\nu_{\max} \quad \text{and} \quad \alpha^*(\nu^*) = \alpha(\nu)/\alpha_{\max}\,,$$

where $\nu_{\max}$ is the frequency at maximum absorption, $\alpha_{\max}$, the resulting reduced spectra are similar [22, 417] but not really identical. A reduced empirical dipole moment has been derived from that presentation which we will briefly discuss in Chapter 4.

A better approach actually exists. There is considerable empirical and theoretical evidence that the various spectral functions, $g(\nu)$, of binary systems are indeed closely modeled by a combination of the BC and K0 profiles, Eqs. 5.105 and 5.108. These are functionals of reduced spectral moments, M_0, M_1/M_0, and M_2/M_0, which in the classical limit may be expressed in terms of reduced temperature, to the extent that interaction potentials are describable by reduced potentials.

3.8 Analysis of measured spectra

Measurements of collision-induced spectra reflect certain details about intermolecular interactions. If analyzed with care, such information will enhance knowledge of molecular interactions. Furthermore, for specific applications, laboratory measurements of collision-induced spectra taken at a few fixed temperatures must be interpolated, often even extrapolated to the temperatures of interest. We will, therefore, discuss the tools available for analysis and further use of laboratory measurements, for example, for astrophysical applications.

It is well known that for a given binary system of interacting molecules, the pair interaction potential and induced dipole functions determine spectral shapes and intensities; see Chapters 5 and 6 for details. If for a given pair collision-induced spectra were known over a broad range of temperatures, such knowledge might be considered to be equivalent in principle to the knowledge of the induced dipole moment as function of the intermolecular separation and angular orientations, provided the intermolecular potential function is known from independent studies [248]. The extraction of the induced dipole surface from spectroscopic measurements is, however, not a trivial process; the problems are similar to the 'inverse problem' of extracting the interaction potential from scattering or equilibrium data (Maitland, Rigby, Smith and Wakeham 1981). For reasonably successful work of that kind, certain assumptions are to be made. For example, *ad hoc* functional dependences, such as exponential function for overlap induction, or a dependence on some inverse power of separation, R^{-n}, for multipole induction, or a combination of these as explained in Chapter 4, must be input. At the same time only the one or two most important dipole components are usually included in the analysis, at the expense of other dipole components whose real significance may be difficult to assess empirically. Whereas the parametrized models of the main dipole components may be well chosen to model the contributions for which they stand, the truncated set must somehow simulate the effects of a bigger set of dipole components. As a consequence, the inferred parameters may not necessarily be physically meaningful and should be used with caution. Nevertheless, methods have been developed which are capable of reproducing the original spectral data closely and may serve as a reasonable basis for cautious interpolation, or even extrapolation, of existing laboratory measurements.

If measurements at more than just one temperature exist, a desirable consistency check of inferred dipole models is often possible; induced dipole moments may sometimes be assumed not to depend on temperature. This is certainly true for the dissimilar rare-gas pairs like He–Ar and, perhaps, for several of the simpler molecular systems in the far infrared. However, with increasing temperature, an increasing number of rotational states, and eventually also various bending and vibrational modes are excited so that molecular properties like multipole moments and polarizabilities may change, e.g., under the influence of centrifugal forces. For example, at low temperatures, a molecule like CH_4 possesses tetrahedral symmetry and, therefore, the lowest-order multipole moment is of octopolar symmetry; furthermore, the anisotropy of the polarizability tensor is zero. At temperatures above ≈200 K, however, a rotation-induced rotational spectrum has been seen in the far infrared range of the spectrum [347, 348]. It arises from the rotation-induced dipole moment because of a frame distortion due to centrifugal forces. It increases with increasing rotation of the molecule but drops to zero as the rotational quantum numbers approach zero. At the same time, a molecular Raman spectrum appears [349] that arises from a rotation-induced anisotropy of the polarizability. A CH_4 molecule may possess a rotation-induced quadrupole moment and its octopole moment is likely to be dependent on the rotational state. In other words, little is known of how the induced dipole moment of CH_4–X systems varies with temperature [56]. The assumption of constant multipole moments is of a limited value at elevated temperatures. Moreover, the symmetries of one or both of the interacting molecules may possibly be distorted during a collision, presumably the more so the higher the temperature is. As a consequence, dipoles induced by collisional frame distortion may be temperature-dependent as well. Again, very little is known about such induced dipole components, but it seems clear that temperature-dependent induced dipole components may be significant at high temperatures for certain molecules.

Moment analysis. The first significant connections between the induced dipole moment and the observed spectral intensities have been made with the help of sum rules and spectral moments. The relevant relationships are given in Chapters 5 and 6. We summarize these by saying that certain integrals over the measured spectra, the spectral moments, can be compared with computable thermodynamic averages of the dipole moment squared, and of certain expressions derived from the dipole moment, provided the pair distribution function, g_2, is known, along with the induced dipole function. The computation of the pair distribution function requires knowledge of the interaction potential; very simple classical approximations of that function, along with low-order quantum corrections of the Wigner–Kirkwood type, are known which are often useful for this purpose. In other words, if a suitable analytical model of the induced dipole moment is adopted and the thermodynamic averages are computed, a comparison of computed and measured moments is possible. By suitable adjustment of the one or two parameters of the induced dipole model, the computed and measured data may be made consistent. Empirical dipole models may be thus obtained.

In order to obtain reliable values of the spectral moments, the spectra must be known over a wide enough frequency band such that the spectral integrals can be evaluated accurately; the errors due to extrapolations of the spectral intensities to zero frequency, a range that is often inaccessible with standard spectroscopic equipment, must be made as small as possible. Similarly, especially for the computation of the higher-order moments, the high-frequency wings must be known accurately. These are, however, difficult to measure because of the small absorption observed at zero frequency and in the distant wings. Analytical functions have been used to extrapolate the measurements to zero and infinite frequency but it is clear that not all of the procedures employed in the past have accurately modeled these inaccessible parts of the spectra. A more or less accurate computation of the spectral profile that accounts in detail for the bound dimer contributions at low frequency, and for the appropriate symmetry of the profile at high frequency, is desirable for such extrapolations; see the discussions in Chapter 5. For certain molecular pairs which have only insignificant dimer contributions, analytical models, like the BC and K0 model, Eqs. 5.105 and 5.108, are useful if these reflect the required symmetry of the profile; for other systems, classical or quantum calculations of the profiles are desirable and often possible.

Moment analysis always has been, and is likely to continue to be a most important first device for the development of new knowledge concerning intermolecular interactions and for a connection of measured spectroscopic data with the fundamental theory. If moment analysis is to be viewed critically, one would have to point out that it reduces the measurement of induced spectra to merely one or two numbers, discarding significant information that could be extracted by more sophisticated methods. Spectral moments are certain averages of the spectral distribution function. The lowest-order moments describe total intensity and a characterisitc width of the distribution. One or two low-order moments cannot very well describe the whole distribution – that requires a complete (i.e., an infinite) set of moments. If one could argue that the distribution is Lorentzian, Gaussian, exponential or of some other form that can be modeled by analytical functions of one or two parameters, specifying one or two moments may be sufficient to summarize the information contained in the measurement.

The shapes of collision-induced spectra are, however, not fully characterized by one or two averages of the distribution. The far wings of a collision-induced line show logarithmic slopes that are gently decreasing in magnitude with increasing frequency; this behavior is, however, not modeled by low-order moments. Furthermore, the low-frequency part of a profile is more or less affected by bound state contributions, both structured ones from bound-to-bound transitions (dimer rotovibrational band spectra), and diffuse ones from bound-to-free transitions involving the bound dimer and the free collisional pair. These differ from one system to another and cannot be modeled with sufficient accuracy by any method we know. Experience has shown that for the few systems known that do not form dimers (such as H_2–He), suitable three-parameter profiles permit a fairly close modeling of the spectra. In other words, three moments describe the major information obtainable from such spectra fairly completely. For most other systems, the low-frequency part of the induced profiles is so much affected by the presence of bound-state structures that even after convolution of the data with a fairly broad instrumental function, more than three parameters seem to be required; high-resolution spectra require, of course, many more parameters for a detailed modeling. In short, one or two parameters almost never summarize the information that is contained in the better measurements of collision-induced spectra. Thus the question arises of whether the *shapes* of measured spectral distributions cannot be analyzed directly.

Line shape analyses. From the discussions above one recalls that often spectral moments cannot be determined as accurately as the absorption coefficient as function of frequency, especially where absorption is strong. For example, for the translational and rototranslational spectra near zero frequency, and also for the high-frequency wing, the absorption is difficult, if not impossible, to measure. However, for the determinations of the moments, such data must be included even if these were obtained with lesser precision. For the comparison of measurements with theory, and for the determination of empirical dipole moments, it is therefore desirable to develop methods of analyzing the better known parts of the spectra, without the need for uncertain extrapolations to low and high frequencies. Under favorable conditions, absorption coefficients may be obtained with good precision, perhaps 5 to 10%, while the precision of spectral moments is generally less. These facts suggest that line shape calculations (as opposed to the much simpler moment calculations) might offer significant advantages when theory and measurements are to be compared closely.

Line shapes of a great many collision-induced absorption spectra have been obtained in recent years, using classical and quantum formalisms. These will be discussed in Chapters 5 and 6. Line shape calculations require the same input as the moment calculations, namely the dipole moment and interaction potential. They offer the advantage of generating certain parts of the spectra with remarkable precision, for example the dimer bands, etc., if a quantum formalism is employed. Furthermore, quantum line shape calculations render the symmetrization procedures superfluous which must be employed when classical expressions are used in the analysis.

We note that line shape calculations may be reduced to near trivial proportions if the basic line profile is approximated by one of the better model profiles mentioned in Chapters 5 and 6. This is generally possible, even advantageous, if the van der Waals bound–bound and bound–free transitions do not shape the spectra significantly. In such a case, the spectra are constructed by a simple superposition of the model line profiles, which is done in seconds even if small (desktop type) computers are used. The simplified line shape calculation has been used successfully on many occasions. Early examples are shown in Figs. 3.11, 3.13, and 3.33, but many more are known [58].

3.9 General references

M. Barnabei, U. Buontempo and P. Maselli, Molecular motions in dense fluids from induced rotational spectra. In *Phenomena Induced by Intermolecular Interactions*, G. Birnbaum, ed., p.95, Plenum Press, New York, 1985.

U. Buontempo, S. Cunsolo, P. Dore and P. Maselli, Molecular motions in liquids. In J. van Kranendonk, ed., *Intermolecular Spectroscopy and Dynamical Properties of Dense Systems – Proceedings of the Int. School of Physics 'Enrico Fermi', Course LXXV*, p. 211, 1980.

I. R. Dagg. Collision-induced absorption in the microwave region. In *Phenomena Induced by Intermolecular Interactions*, G. Birnbaum, ed., p.95, Plenum Press, New York, 1985.

J. C. Lewis. Intercollisional Interference – theory and experiment. In *Phenomena Induced by Intermolecular Interactions*, G. Birnbaum, ed., p. 215, Plenum Press, New York, 1985.

Ph. Marteau. Far infrared induced absorption in highly compressed atomic and molecular systems. In G. Birnbaum, ed., *Phenomena Induced by Intermolecular Interactions*, pp. 415–436, Plenum Press, New York, 1985.

H. L. Welsh, *Pressure-induced absorption spectra in hydrogen*, MTP International Review of Science – Physical Chemistry, Series One, vol. 3, A. D. Buckingham and D. A. Ramsay, eds., Butterworths, London, 1972.

J. van Kranendonk. Intercollisional interference effects. In J. van Kranendonk, editor, *Intermolecular Spectroscopy and Dynamical Properties of Dense Systems – Proceedings of the Int. School of Physics 'Enrico Fermi', Course LXXV*, p.77, 1980.

J. van Kranendonk, *Solid Hydrogen*, Plenum, New York and London, 1983.

A. Weber, ed., *Structure and Dynamics of Weakly Bound Molecular Complexes*, NATO ASI Series C: Math. and Phys. Sci., vol. 212, D. Reidel, Dordrecht, 1987.

4
Induced dipoles

In the preceding chapter, we have seen examples of absorption spectra, mostly of non-polar, dense gases. Such absorption arises from dipole moments which atoms or molecules do not possess, unless they are in interaction with other atoms or molecules. For the duration of the interaction, dipoles are induced that absorb and emit radiation. In this Chapter, we will consider the causes and properties of dipole moments induced by intermolecular interactions.

4.1 Overview

Induced dipole moments are generated by the same processes which cause intermolecular repulsion at near range, and attraction at long range. At short range, electron exchange effects generate an 'exchange' or 'overlap' dipole component which falls off roughly exponentially with separation [400, 401, 402],

$$\mu_O \approx \mu_0 \exp\left(-R/R_0\right) , \tag{4.1}$$

much like the repulsive intermolecular forces, Fig. 2.1. At long range, on the other hand, dispersion forces may induce a 'dispersion dipole' component, whose expansion in terms of inverse powers of separation, R, takes the form [225] (Buckingham 1959; Hunt 1985)

$$\mu_D \approx -D_7/R^7 - D_9/R^9 - \dots \tag{4.2}$$

for $R \to \infty$. While these expressions resemble those of van der Waals interaction forces at near and far range, respectively, it must be said that the parameters R_0, $D_7, \dots$ of these expressions differ from the analogous parameters that appear in the analytical models of the intermolecular force: knowledge of one of these functions, the interaction-induced dipole moment or the intermolecular force, does not imply much useful knowledge of the other function.

Dispersion and exchange forces are both quantum mechanical in nature. The former may be described in semi-classical terms of fluctuating dipole-induced dipole interactions and the latter arises from the exchange symmetry of electronic wavefunctions imposed by the Pauli principle. Dispersion forces are attractive and exchange forces are repulsive. The space between the nuclei of interacting atoms shows a slight *excess* of electronic charge if the interacting atoms are widely separated; such enhancement of charge in that region is typical of interatomic attraction. As a consequence, the far side of the atom with the greater number of electrons appears positive, due to the slight defect of negative charge. The net induced dispersion dipole shows, therefore, a polarity according to $He^{-}Ar^{+}$. At near range, on the other hand, a *depletion* of electronic charge occurs in the region between the nuclei as is typical of repulsive interactions. Exchange dipoles are, therefore, of the opposite polarity, $He^{+}Ar^{-}$, because of the opposite redistribution of electronic charge caused by the interaction at near range. We note that both the overlap-induced dipole and the dispersion dipole may be zero if an inversion symmetry exists, as is the case for pairs of identical atoms.

If at least one of the interacting particles is a molecule, further induction mechanisms arise. Molecules are surrounded by an electric field which may be viewed as a superposition of multipole fields. A collisional partner will be polarized in the multipole field and thus give rise to 'induced dipole' components. In the case of symmetric diatoms like H_2 or N_2, the lowest-order multipole is a quadrupole and asymptotically, for $R \to \infty$, the quadrupole-induced dipole may be written as [288, 289]

$$\mu_Q \approx B_{23}(R) \approx \sqrt{3}\, q_2\, \alpha\, / R^4 \,. \tag{4.3}$$

Here α designates the trace of the polarizability tensor of one molecule; $(1/4\pi\varepsilon_0)$ times the factor of α represents the electric fieldstrength of the quadrupole moment q_2. Other non-vanishing multipole moments, for example, octopoles (e.g., of tetrahedral molecules), hexadecapoles (of linear molecules), etc., will similarly interact with the trace or anisotropy of the polarizability of the collisional partner and give rise to further multipole-induced dipole components.

On a scale of the order of atomic size, molecular multipole fields vary strongly with orientation and separation. As a consequence, one will generally find induced dipole components arising from *field gradients* of first and higher order which interact with the so-called dipole–multipole polarizability tensor components, such as the $\boldsymbol{A}$ and $\boldsymbol{E}$ tensors.

Finally, dipoles may arise from collisional 'frame distortion'. Polyatomic molecules like CO_2, CH_4, SF_6, C_2H_6,..., in their ground states are non-polar because of their (inversion-, tetrahedral, hexagonal, ...) symmetry. The symmetry may, however, temporarily be modified if in a collision one of the outer atoms (H, F, O,...) is temporarily displaced from the equilibrium position. In such a case, the molecule acquires momentarily an induced distorted frame dipole component, the interacting molecules 'borrow' a dipole moment from the infrared active bending, stretching, etc., modes they possess. The experimental evidence for distorted frame-induced dipoles appears to be somewhat inconclusive [21, 153, 418]. However, theoretical estimates considering systems like He–CO_2 suggest substantial contributions, comparable to the overlap-induced dipole components [418, 419].

Not all of these induced dipole types may exist in any given system. The components that exist generally couple in different ways to the translational, rotational, vibrational, etc., states of the complex and usually are associated with different selection rules, thus generating different parts of the collision-induced spectra.

Besides the collision-induced dipoles, we will occasionally refer to field-induced dipoles, or to rotation-induced dipoles, that is dipoles induced by an external electric field, or by centrifugal forces distorting certain symmetries of rotating molecules. Moreover, we will be interested in the dipoles induced in binary, ternary, etc., systems as we proceed.

4.2 Theory

The total dipole moment of two interacting, polar molecules is given by

$$\boldsymbol{\mu}(\boldsymbol{r}_1, \boldsymbol{r}_2, \boldsymbol{R}_{12}) = \boldsymbol{\mu}_1(\boldsymbol{r}_1) + \boldsymbol{\mu}_2(\boldsymbol{r}_2) + \boldsymbol{\mu}_{12}(\boldsymbol{r}_1, \boldsymbol{r}_2, \boldsymbol{R}_{12}) + \boldsymbol{\mu}_{21}(\boldsymbol{r}_2, \boldsymbol{r}_1, \boldsymbol{R}_{21})\,. \quad (4.4)$$

Here, the $\boldsymbol{\mu}_i$ are the permanent dipoles of molecules $i = 1$ and 2, and the $\boldsymbol{\mu}_{ij}(\boldsymbol{r}_1, \boldsymbol{r}_2, \boldsymbol{R}_{ij})$ are the dipoles induced by molecule i in molecule j; the $\boldsymbol{R}_{ij}$ are the vectors pointing from the center of molecule i to the center of molecule j and the $\boldsymbol{r}_i$ are the (intramolecular) vibrational coordinates. In general, these dipoles are given in the adiabatic approximation where electronic and nuclear wavefunctions appear as factors of the total wavefunction, $\Phi(\boldsymbol{r}_e^{(N)}, \boldsymbol{r})\,\Psi(\boldsymbol{r})$. Dipole operators $\boldsymbol{\mu}_{\mathrm{op}}$ are defined as usual so that their expectation values shown above can be computed from the wavefunctions. For the induced dipole component, the dipole operator is defined with respect to the center of mass of the pair so that the induced dipole moments $\boldsymbol{\mu}_{ij}$ do not depend on the center of mass coordinates. For bigger systems the total dipole moment may be expressed in the form of a simple generalization of Eq. 4.4. In general, the molecules will be assumed to be in a $^1\Sigma$ electronic ground state which is chemically inert.

For the purposes of computing spectral line shapes and spectral moments, the dipole moment $\boldsymbol{\mu}$ is best expressed in spherical coordinates

[189, 317, 323]. For example, the permanent dipole components of a diatomic system may thus be written

$$\mu_\nu^{(i)}(\boldsymbol{r}_i) = \left(\frac{4\pi}{3}\right)^{1/2} q_1^{(i)}(r_i)\, Y_{1\nu}(\widehat{\boldsymbol{r}}_i)\,, \tag{4.5}$$

with $\nu = 0, \pm 1$ and Y_{lm} designating spherical harmonics as usual; the dipole strength is a function of the vibrational coordinates. Spherical dipole components μ_0, $\mu_{\pm 1}$ are defined in terms of their Cartesian components, μ_x, μ_y, μ_z, according to

$$\mu_0 = \mu_z\,, \qquad \mu_{\pm 1} = \mp\,(\mu_x \pm i\mu_y)\,/\sqrt{2}\,. \tag{4.6}$$

In order for the induced dipole moment, $\boldsymbol{\mu}$, to transform as a vector, the spherical harmonics describing the various orientations have to be coupled in an appropriate way. We write the induced dipole components of a system of two molecules of arbitrary symmetry, according to [141]

$$\mu_\nu = \sum_{(c)} A_{(c)}(r_1^{(N)}, r_2^{(N)}; R)\; Y_{(c)}^{1\nu}(\Omega_1, \Omega_2, \Omega)\,, \tag{4.7}$$

with

$$\begin{aligned} &Y_{(c)}^{1\nu}(\Omega_1, \Omega_2, \Omega) \\ &\quad = \left[\frac{4\pi}{3}(2\lambda_1+1)(2\lambda_2+1)\right]^{1/2} \sum_{M_\Lambda M M_1 M_2} C(\Lambda L 1; M_\Lambda M \nu) \\ &\qquad \times C(\lambda_1\lambda_2\Lambda; M_1 M_2 M_\Lambda)\, D^{\lambda_1}_{M_1 \upsilon_1}(\Omega_1)^*\, D^{\lambda_2}_{M_2 \upsilon_2}(\Omega_2)^*\, Y_{LM}(\Omega)\,. \end{aligned} \tag{4.8}$$

Here, the subscript (c) is short for the set of expansion parameters $(c) = (\lambda_1, \lambda_2, \Lambda, L, \upsilon_1, \upsilon_2)$; r_i is the vibrational coordinate of the molecule i; R is the separation between the centers of mass of the molecules; the Ω_i are the orientations (Euler angles α_i, β_i, γ_i) of molecule i; Ω specifies the direction of the separation $\boldsymbol{R}$; the $C(\lambda_1\lambda_2\Lambda; M_1M_2M_\Lambda)$, etc., are Clebsch–Gordan coefficients; the $D^{\lambda}_{M\upsilon}$ are Wigner rotation matrices. The expansion coefficients $A_{(c)} = A_{\lambda_1\lambda_2\Lambda L;\upsilon_1\upsilon_2}(r_1, r_2; R)$ are independent of the coordinate system; these will be referred to as multipole-induced or overlap-induced dipole components – whichever the case may be.

The functions $Y_{(c)}^{1\nu}$, Eq. 4.8, were obtained by first coupling the spherical harmonics in the variables Ω_1 and Ω_2, and subsequently coupling the result with the spherical harmonics in Ω. This particular coupling scheme we label $(\Omega_1\Omega_2;\Omega)$, it leads to a particular set of coefficients A in Eq. 4.7 which is our standard set. Obviously, two other coupling schemes are possible, namely $(\Omega_2\Omega;\Omega_1)$ and $(\Omega_1\Omega;\Omega_2)$. These lead to sets of coefficients A' and A'' which are related to the set A by certain unitary transforms discussed elsewhere [323, 391].

The expansion parameters that appear in Eqs. 4.7 and 4.8 as arguments of the Clebsch–Gordan coefficients satisfy the triangular relationship,

$$|\lambda_1 - \lambda_2| \le \Lambda \le \lambda_1 + \lambda_2 \quad \text{and} \quad |L-1| \le \Lambda \le L+1 \,. \tag{4.9}$$

Furthermore, the quadruple sum in (4.8) reduces at once to a double sum, because the Clebsch–Gordan coefficients vanish unless

$$M_\Lambda + M = \nu \quad \text{and} \quad M_1 + M_2 = M_\Lambda \,. \tag{4.10}$$

The non-vanishing $A_{(c)}$ coefficients are not all independent. From the Hermiticity of the dipole operator, $\boldsymbol{\mu}$, and the symmetry property,

$$Y^{1\nu\,*}_{\lambda_1\lambda_2\Lambda L;\upsilon_1\upsilon_2} = (-1)^{\nu+\lambda_1+\lambda_2+L+1+\upsilon_1+\upsilon_2}\, Y^{1\nu}_{\lambda_1\lambda_2\Lambda L;-\upsilon_1-\upsilon_2} \,, \tag{4.11}$$

we have

$$A_{\lambda_1\lambda_2\Lambda L;-\upsilon_1-\upsilon_2} = (-1)^{\lambda_1+\lambda_2+L+1+\upsilon_1+\upsilon_2}\, A_{\lambda_1\lambda_2\Lambda L;\upsilon_1\upsilon_2} \,. \tag{4.12}$$

Further symmetries arise when *identical* molecules, or molecules of high symmetry are involved; several examples of practical interest will be considered below. The vibrational matrix elements for transitions between molecular states $(s_1, s_2) \to (s_1', s_2')$

$$B^{s_1 s_1' s_2 s_2'}_{(c)}(R) = \left\langle s_1 s_2 \left| A_{(c)}\left(r_1^N r_2^N ; R\right) \right| s_1' s_2' \right\rangle \tag{4.13}$$

must be computed from the $A_{(c)}$ for each vibrational band, with careful accounting for double transitions.

Diatoms. The simplest systems of interest here are the dissimilar rare-gas pairs. Atoms do not possess rotovibrational states and all dependences on Ω_1, Ω_2, r_1, r_2, must disappear from the Eqs. 4.7 and 4.8. Hence, $\lambda_1 = \lambda_2 = \Lambda = \upsilon_1 = \upsilon_2 = 0$, $L = 1$ and $A_{(c)} \equiv B_{(c)}(R)$. In this case, the sums of Eqs. 4.7 and 4.8 consist of a single term, $\lambda_1\lambda_2\Lambda L = 0001$,

$$\mu_\nu(\boldsymbol{R}) = \left(\frac{4\pi}{3}\right)^{1/2} B(R)\, Y_{1\nu}(\Omega) \,, \tag{4.14}$$

with $\nu = 0, \pm 1$. The dipole is isotropic, mainly of the overlap-induced type, Eq. 4.1, with a usually weak, negative dispersion contribution added, Eq. 4.2. It is zero for like pairs.

Atom–diatom pairs. If molecule i is a linear molecule or diatomic, we may substitute $(4\pi)^{1/2}\, Y_{\lambda_i M_i}(\Omega_i)$ for $(2\lambda_i + 1)^{1/2}\, D^{\lambda_i}_{M_i \upsilon_i}(\Omega_i)$. In the case of an atom, this substitution reduces to 1 if $\lambda_i = 0$, and to 0 for positive λ_i. Equations 4.7, 4.8 and 4.13 thus become [317]

$$\mu_\nu(\boldsymbol{r}, \boldsymbol{R}) = \sum_{\lambda L} A_{\lambda L}(r; R)\, Y^{1\nu}_{\lambda L}(\Omega_1, \Omega) \,, \tag{4.15}$$

$$Y^{1\nu}_{\lambda L}(\Omega_1, \Omega) = \frac{4\pi}{\sqrt{3}} \sum_{mM} C(\lambda L 1; mM\nu)\, Y_{\lambda m}(\Omega_1)\, Y_{LM}(\Omega) \,, \tag{4.16}$$

$$B_{\lambda L}^{vjv'j'}(R) \;=\; \langle vj\,|A_{\lambda L}(r,R)|\,v'j'\rangle\ , \tag{4.17}$$

where we have written λ for Λ. In the case of diatomic molecules, instead of the solid angles Ω_1 and Ω, we will often write the unit vectors $\widehat{\boldsymbol{r}}_1$ and $\widehat{\boldsymbol{R}}$, respectively, which designate the same angular variables.

Diatom–diatom pairs. We are often concerned with dipoles induced by diatomic molecules, or on such molecules. In such cases, the Wigner rotation matrices reduce again to spherical harmonics. Equations 4.7, 4.8, and 4.13 can thus be written [316, 317, 189]

$$\mu_\nu\,(\boldsymbol{r}_1,\boldsymbol{r}_2,\boldsymbol{R}) = \sum_{\lambda_1\lambda_2\Lambda L} A_{\lambda_1\lambda_2\Lambda L}(r_1,r_2;R)\; Y^{1\nu}_{\lambda_1\lambda_2\Lambda L}(\Omega_1,\Omega_2,\Omega)\,, \tag{4.18}$$

$$\begin{aligned} Y^{1\nu}_{\lambda_1\lambda_2\Lambda L}(\Omega_1,\Omega_2,\Omega) = \frac{(4\pi)^{3/2}}{\sqrt{3}} \sum_{MM_1M_2M_\Lambda} & C(\lambda_1\,\lambda_2\,\Lambda;M_1M_2M_\Lambda) \\ \times\, C(\Lambda\,L\,1;M_\Lambda M\nu)\; Y_{\lambda_1M_1}(\Omega_1)\; & Y_{\lambda_2M_2}(\Omega_2)\; Y_{LM}(\Omega)\,, \end{aligned} \tag{4.19}$$

$$\begin{aligned} &B_{\lambda_1\lambda_2\Lambda L}(R;v_1j_1v_2j_2v_1'j_1'v_2'j_2') = \\ &\qquad \langle v_1j_1v_2j_2\,|A_{\lambda_1\lambda_2\Lambda L}(r_1,r_2;R)|\,v_1'j_1'v_2'j_2'\rangle\ . \end{aligned} \tag{4.20}$$

The dependence of these vibrational matrix elements B on the rotational states, j_1, j_1', etc., is sometimes so weak that it can be suppressed.

1 Expansion coefficients

The coefficients A defined in Eq. 4.18 satisfy certain symmetry relationships [323, 391]. From parity considerations, it follows that $\lambda_1+\lambda_2+L$ must be odd. Moreover, because the dipole operator is Hermitian, the expansion coefficients A are all real. The symmetry property

$$\boldsymbol{\mu}(-\boldsymbol{r}_1,-\boldsymbol{r}_2,\boldsymbol{R}) = -\boldsymbol{\mu}(\boldsymbol{r}_1,\boldsymbol{r}_2,\boldsymbol{R})$$

holds for two identical molecules. It implies

$$A_{\lambda_1\lambda_2\Lambda L}(r_1r_2R) = (-1)^{1-\Lambda}\; A_{\lambda_2\lambda_1\Lambda L}(r_2r_1R)\ . \tag{4.21}$$

Homonuclear diatomic molecules are associated with even λ_i, $i=1$ or 2.

Multipolar induction. For the description of the long-range dipole components, we start with the electric potential at the distance $\boldsymbol{R}$ outside the molecule 1 [323, 391]. In a space-fixed coordinate system, the potential is given as

$$\phi\,(\boldsymbol{r}_1,\boldsymbol{R}) = \sum_{\ell m} q_\ell^{(1)}\,(r_1)\; R^{-(\ell+1)}\; \frac{4\pi}{2\ell+1}\; Y_{\ell m}\,(\widehat{\boldsymbol{r}}_1)\; Y^*_{\ell m}\left(\widehat{\boldsymbol{R}}\right) \tag{4.22}$$

where

$$q_\ell^{(1)} = \int \rho^{(1)}(\boldsymbol{r})\, r^\ell\, P_\ell(\cos\vartheta)\, \mathrm{d}^3r$$

is the multipole moment of order ℓ of molecule 1 in a frame with its z axis parallel to $\boldsymbol{r}$; the origin of the frame coincides with the center of mass of the molecule. The function $\rho^{(1)}(\boldsymbol{r})$ is the adiabatic charge density plus the nuclear charge of molecule 1. The q_ℓ depend only on the internuclear separations $\boldsymbol{r}_1$. The spherical components of the electric field thus become

$$X_\nu^{(1)} = 4\pi\left(\frac{\ell+1}{3}\right)^{1/2}(-1)^\ell\, q_\ell^{(1)}\, R^{-(\ell+2)} \tag{4.23}$$
$$\times \sum_{M_1 M} C(\ell,\ell+1,1;M_1 M\nu)\; Y_{\ell M_1}(\widehat{\boldsymbol{r}}_1)\; Y_{\ell+1,M}\left(\widehat{\boldsymbol{R}}\right),$$

and the induced dipole associated with the trace $\alpha^{(2)}$ of the polarizability tensor of molecule 2 is given by $\mu_\nu^{(1)} = \alpha^{(2)}\, X_\nu^{(1)}$. The comparison with Eq. 4.18 shows

$$A_{\lambda_1\lambda_2\Lambda L}(r_1 r_2 R) = \delta_{\lambda_1\ell}\,\delta_{\Lambda\ell}\,\delta_{\lambda_2 0}\,\delta_{L,\ell+1}\,(\ell+1)^{1/2}\,(-1)^\ell\, q_\ell^{(1)}(r_1)\,\alpha^{(2)}(r_2)\,R^{-(\ell+2)}$$

and with $\lambda_1=\ell$, $\lambda_2=0$, $\Lambda=\ell$, and $L=\ell+1$, we have

$$A_{\lambda_1,0,\lambda_1+1,L}(r_1 r_2 R) = (\ell+1)^{1/2}\,(-1)^\ell\, q_\ell^{(1)}(r_1)\,\alpha^{(2)}(r_2)\,R^{-(\ell+2)}\,.$$

A similar expression is obtained for the dipole induced in this way in molecule 1 by molecule 2,

$$A_{0,\lambda_2,\lambda_2+1,L}(r_1 r_2 R) = -\,(\ell+1)^{1/2}\, q_\ell^{(2)}(r_2)\,\alpha^{(1)}(r_1)\,R^{-(\ell+2)}\,,$$

with $\lambda_1=0$, $\lambda_2=\ell$, $\Lambda=\ell$, and $L=\ell+1$. For the case $\ell=0$, the two distinct induction mechanisms just mentioned degenerate and only a single coefficient, A_{0001}, exists; in this case, the two mechanisms give rise to interference [323].

The induced dipole component arising from the anisotropy $\beta^{(i)}(r_i)$ of the polarizability tensor is similarly obtained

$$A_{\lambda_1\lambda_2\Lambda L}(r_1 r_2 R) = \delta_{\lambda_1\ell}\,\delta_{\lambda_2 2}\,\delta_{L,\ell+1}\,(-1)^{\Lambda+1}\left(\frac{2(\ell+1)(2\Lambda+1)}{3}\right)^{1/2}$$
$$\times W(2,1,\ell,\ell+1;1\Lambda)\; q_\ell^{(1)}(r_1)\;\beta^{(2)}(r_2)\;R^{-(\ell+2)}\,.$$

W designates the Racah coefficient and Λ takes on the values $\ell+2$, $\ell+1$, and ℓ. Similarly, the induction of order ℓ of molecule 1 by molecule 2 is given by

$$A_{\lambda_1\lambda_2\Lambda L}(r_1 r_2 R) = \delta_{\lambda_1 2}\,\delta_{\lambda_2\ell}\,\delta_{L,\ell+1}\left(\frac{2(\ell+1)(2\Lambda+1)}{3}\right)^{1/2}$$

$$\times W\left(2,1,\ell,\ell+1;1\Lambda\right)\, q_\ell^{(2)}\left(r_2\right)\, \beta^{(1)}\left(r_1\right)\, R^{-(\ell+2)}\,.$$

It is noteworthy that, again, a degeneracy exists for the case $\ell = 2$. In that case, the two mechanisms just described both contribute to the *same* coefficient A and interference may arise as a consequence [323].

Asymptotic formulae. For a discussion of induced dipoles in highly polarizable species, it is often sufficient to consider the so-called classical multipole induction approximation in its simplest form (i.e., neglecting field gradients and hyperpolarizabilities). In such a case, one needs to know only the vibrational matrix elements of the multipole moments,

$$q_\ell(i) = \left\langle v_i j_i \left| q_\ell^{(i)}(r) \right| v_i' j_i' \right\rangle\,, \tag{4.24}$$

and of the polarizability tensor invariants, trace and anisotropy,

$$\alpha(i) \quad = \quad \left\langle v_i j_i \left| \alpha^{(i)}(r) \right| v_i' j_i' \right\rangle\,, \tag{4.25}$$

$$\beta(i) \quad = \quad \left\langle v_i j_i \left| \beta^{(i)}(r) \right| v_i' j_i' \right\rangle\,, \tag{4.26}$$

respectively, for molecules $i = 1, 2$. Overlap contributions are then typically ignored, or else a few of the main overlap terms are somehow included in a more or less empirical way. A selection of classical (or asymptotic, $R \to \infty$) multipole-induction terms are collected in Section 4.7 at the end of this Chapter.

Other induction mechanisms. The computation of induced dipoles from first principles indicates that, besides the classical long-range terms just discussed, the coefficients $A_{(c)}$ in general possess a short-range component which arises from quantal processes, mainly from overlap induction; a dispersion contribution is also generally expected but may be weak. Also, a number of coefficients $A_{(c)}$ are purely short-ranged, i.e., without a classical multipole term as we will see below. Moreover, other induction mechanisms are known to create dipoles by field gradients and higher derivatives of the field, and through higher-order polarizabilities [89]. These are discussed in the next Section.

Other than linear molecules. If molecules of symmetry other than axial are considered, it is not possible to describe their orientation by an azimuthal and polar angle, φ_i, ϑ_i. Rather, three Euler angles, $\Omega_i = \alpha_i, \beta_i, \gamma_i$, and Wigner rotation matrices are then needed as Eq. 4.8 suggests. In that case, besides the set of parameters λ_1, λ_2, Λ, L that has been used for linear molecules, two new parameters, υ_i with $i = 1, 2$, occur that enter through the rotation matrices. These must be chosen so that the dipole moment is invariant under any rotation belonging to the molecular symmetry group. The rotation matrix $D^{\lambda}_{M\upsilon}(\Omega)$ is expressed as a linear combination of such matrices with different υ (Gray and Gubbins 1984).

2 Classical multipole approximation

While exchange- and dispersion-induced dipole components are of a quantum nature, the multipole-induced dipole components can be modeled by classical relationships, if the quantum effects are small. For many systems of practical interest, multipolar induction generates the dominant dipole components. The classical multipole induction approximation has been very successful, except for the weakly polarizable partners (e.g., He atoms) [193]. It models the dipole induced in the collisional partner by polarization in the molecular multipole fields.

Besides the linear polarization contribution, the hyperpolarizabilities may be accounted for, as well as the field gradients (and higher derivatives of the multipolar fields) taken at the molecular center if the fields are very non-uniform. The total dipole induced by electric fields in a molecule may be written [89]

$$\begin{aligned}\mu_i &= \mu_i^{(0)} + \sum_j \alpha_{ij}\, X_j + \frac{1}{2}\sum_{jk} \beta_{ijk}\, X_j\, X_k + \frac{1}{6}\sum_{jkl} \gamma_{ijkl}\, X_j\, X_k\, X_l \\ &\quad + \frac{1}{3}\sum_{jk} A_{i,jk}\, X_{jk} + \frac{1}{3}\sum_{jkl} B_{ij,kl}\, X_j\, X_{kl} + \cdots . \qquad (4.27)\end{aligned}$$

The subscripts i, j, k, l denote Cartesian vector and tensor components and can be equal to x, y, z. The X_i and X_{ij} are electric field and field gradient at the origin due to external charges; these are computed from expressions like Eq. 4.23. The permanent dipole moment is called $\mu_i^{(0)}$. The α_{ij}, β_{ijk}, γ_{ijkl} and $A_{i,jk}$, $B_{ij,kl}$ and $C_{ij,kl}$ are molecular polarizabilities describing the distortion of the molecule by the external field and field gradient. The tensors $\boldsymbol{\alpha}$, $\boldsymbol{\beta}$, and γ are symmetric in all suffices, the tensor components $A_{i,jk}$ in jk, $B_{ij,kl}$ in ij and kl, and $C_{ij,kl}$ in ij, kl, and in the pairs (ij) and (kl). The second-rank tensor $\boldsymbol{\alpha}$ is the familiar static polarizability, and $\boldsymbol{\beta}$ and γ are hyperpolarizabilities describing the deviations from a linear polarization law. The tensor $\boldsymbol{A}$ determines both the dipole induced by a field gradient and the quadrupole induced by a uniform field. If the molecular origin is a center of symmetry, the dipole moment is zero, along with $\boldsymbol{\beta}$ and $\boldsymbol{A}$. The number of independent constants needed to describe the induced dipole is determined by the symmetry of the molecule. The number of constants required to specify the tensors for various molecular symmetries is given elsewhere [89].

If, for example, the induced dipole model is truncated at the order R^{-6} in the separation R between the molecular centers, account may be made of the dipoles induced by multipoles up to order $\ell = 4$ (hexadecapole). Moreover, dipoles induced by derivatives of the local field at their center through the third, and the lowest order hyperpolarizability contribution may thus be modeled [88, 89].

The merits of the classical multipole approximation for collision-induced absorption were assessed by Hunt (1985). It is clear that at near range, exchange effects cannot be ignored, for example, in energetic ('hard') collisions, or if weakly polarizable species are involved (especially He atoms), or if multipole moments are weak. Another concern is the choice of where to truncate the multipole model: one needs to carry a sufficient number of terms for a realistic modeling [193, 98]. In spite of all these concerns, the classical multipole model has been a most useful starting point for a realistic accounting for the effects of pair and even higher-order induced dipoles.

4.3 Measurement

Collision-induced dipoles manifest themselves mainly in collision-induced spectra, in the spectra and the properties of van der Waals molecules, and in certain virial dielectric properties. Dipole moments of a number of van der Waals complexes have been measured directly by molecular beam deflection and other techniques. Empirical models of induced dipole moments have been obtained from such measurements that are consistent with spectral moments, spectral line shapes, virial coefficients, etc. We will briefly review the methods and results obtained.

Spectroscopic measurement. Specifically, if the induced dipole moment and interaction potential are known as functions of the intermolecular separation, molecular orientations, vibrational excitations, etc., an absorption spectrum can in principle be computed: potential and dipole surface determine the spectra. With some caution, one may also turn this argument around and argue that the knowledge of the spectra and the interaction potential 'defines' an induced dipole function. While direct inversion procedures for the purpose may be possible, none are presently known and the empirical induced dipole models usually assume an analytical function like Eqs. 4.1 and 4.3, or combinations of Eqs. 4.1 through 4.3, with parameters μ_0, R_0, q_2, etc., to be chosen such that certain measured spectral moments or profiles are reproduced computationally.

Such procedures work well if the appropriate functional form of the induced dipole is chosen, if the interaction potential is known, and if high-quality measured spectra are available over a wide range of frequencies, intensities and temperatures – conditions which are rarely met to the degree desirable. Nevertheless, for a few simple systems like rare-gas pairs, highly polarizable molecular gases, etc., more or less satisfactory empirical dipole models were thus obtained.

Method of moments. In rare gas mixtures, the induced dipole consists of just one B component, with $\lambda_1\lambda_2\Lambda L = 0001$, Eq. 4.14. Alternatively, one particular $B_{(c)}$ component may cause the overwhelming part of a measured spectrum, like the quadrupole-induced component in mixtures of small amounts of H_2 in highly polarizable rare gases ($(c) = \lambda_1\lambda_2\Lambda L = 2023$, Eq. 4.59); in a given spectral range, other components (like 0001, 2021, ...) are often relatively insignificant. In such cases, one can write down more or less discriminating relationships between certain spectral moments of low order n that are obtainable from measurements of the collision-induced spectral profile, $g_{\lambda_1\lambda_2\Lambda L}(\omega)$,

$$M_n(\lambda_1\lambda_2\Lambda L;\, T) = \int_{-\infty}^{\infty} g_{\lambda_1\lambda_2\Lambda L}(\omega)\, \omega^n \, d\omega \,, \tag{4.28}$$

and the principal induced dipole components, Eq. 4.30 [177, 122]. For nearly classical systems, i.e., massive systems at high temperatures that require inclusion of many partial waves, $\ell = 0, 1, \ldots, \ell_{\max}$, for an accurate theoretical description,

$$\left(\hbar^2 \ell_{\max}\,(\ell_{\max} + 1)\right)^{1/2} \approx m_r \left(\frac{8kT}{\pi m_r}\right)^{1/2} b_{\rm up} \,, \tag{4.29}$$

with $\ell_{\max} \gg 1$, only even moments ($n = 0, 2, \ldots$) are significant, but in general even and odd moments must both be considered. In this expression, m_r designates the reduced mass of the collisional pair, and $b_{\rm up}$ designates the maximum impact parameter for which deflections still occur, that is roughly an upper bound of the representative interaction range of the order of molecular dimensions, $b_{\rm up} \approx K\sigma$, where K is a number of order of unity, perhaps $K \approx 2$; σ is the root of the interaction potential.

If one now chooses the appropriate induced dipole model, Eqs. 4.1 through 4.3, or a suitable combination of these, with N parameters μ_0, D_7, R_0, ...,* and one has at least N theoretical moment expressions available, an empirical dipole moment may be obtained which satisfies the conditions exactly, or in a least-mean-squares fashion [317, 38]. We note that a formula was given elsewhere that permits the determination of the range parameter, $1/a$, directly from a ratio of first and zeroth moments; it was used to determine a number of range parameters from a wide selection of measured moments [189]. In early work, an empirical relationship between the range parameter and the root, σ, of the potential is assumed, like $1/a \approx 0.11\sigma$. That relationship is, however, generally not consistent with recent data believed to be reliable.

* Alternatively, in the notation to be introduced below, Eq. 4.30, $B^{(0)}$, $B^{(n)}$, a, b, ...

Whereas there is little doubt that the method of moments, as the procedure is called, is basically sound, it is obvious that for reliable results high-quality experimental data over a broad range of frequencies and temperatures are desirable. As importantly, reliable models of the interaction potential must be known. Since these requirements have rarely been met, ambiguous dipole models have sometimes been reported, especially if for the determination of the spectral moments substantial extrapolations to high or to low frequencies were involved. Furthermore, since for most works of the kind only two moments have been determined, refined dipole models that attempt to combine overlap and dispersion contributions cannot be obtained, because more than two parameters need to be determined in such case. As a consequence, empirical dipole models based on moments do not attempt to specify a dispersion component, or test theoretical values of the dispersion coefficient $B^{(7)}$ (Hunt 1985).

Fitting line shapes. In the next Chapter, we will discuss various approaches to computing spectral line shapes. Such computations require as input a reliable model of the interaction potential and of the dipole components. Once a profile is computed on the basis of an imperfect empirical dipole moment, the comparison with spectroscopic measurements may reveal certain inconsistencies which one may more or less successfully correct by small adjustments of the free parameters. After a few iterations, one may thus arrive at an empirical model that is consistent with a spectroscopic measurement [39]. If measurements at various temperatures exist, the dipole model must reproduce all measured spectra equally well.

Multi-parameter dipole models can thus often be determined with remarkable definition. Further advantages of the 'method of fitting line shapes' are that extrapolations of measured spectra to very high or very low frequencies may not be needed (albeit, for a good definition of the empirical dipole function it never hurts to have experimental data over the widest possible frequency band available); measurements at high or low frequencies are sometimes not possible. Furthermore, the methods of moments sometimes gives multiple solutions; if these various solutions are used to compute line shapes, only one of these will usually reproduce the measured profile; the others are often strikingly inconsistent. This fact (which has been noticed on several occasions) illustrates the greater discriminating power of line shape analyses. Furthermore, it has been shown that fitting high-quality experimental profiles, in general, permits a determination of refined, multi-parameter, empirical dipole models: the 'differential' features of spectral profiles, such as slopes, curvatures, and dimer features, apparently carry information about the interactions in general and the dipole surface in particular that tends to be lost if only

Table 4.1. Empirical induced dipole models for the translational band; Eq. 4.30, with $n = 7$ where $B^{(n)} \neq 0$.

system	R_0 (a.u.)	$B^{(0)}$ (10^{-3} a.u.)	a (a.u.)	b (a.u.)	$B^{(n)}$ (a.u.)	Ref.
He–Ar	5.82	10.9	1.418	0	−1546	[39]
He–Ar	5.88	4.1	1.508	0	0	[189]
He–Kr	6.20	5.9	1.504	0	0	[189]
He–Xe	6.61	8.2	1.477	0	0	[189]
Ne–Ar	5.88	7.4	1.657	0.094	−200	[44]
Ne–Ar	5.80	6.4	1.792	0	0	[189]
Ne–Kr	6.10	11.	1.565	0	0	[189]
Ne–Xe	6.46	13.	1.672	0	0	[189]
Ar–Kr	5.80	20.2	1.302	0.115	−637	[39]
Ar–Kr	6.54	6.4	1.616	0	0	[189]
Ar–Xe	7.10	15.	1.416	0	0	[189]

two integrals of the measurement (the lowest two moments) are used in the analysis.

With high-quality measurements, four- and five-parameter dipole models have been successfully determined which permit a significantly better fit of the spectral profiles than models with fewer parameters; the parameters may be $B^{(0)}$, a, b, $B^{(7)}$, defined by Eq. 4.30 below, sometimes supplemented by higher-order dispersion coefficients, $B^{(9)}, \ldots$, or adjustable dispersion damping functions that, at near range, suppress the dispersion contributions (as is common in studies of intermolecular potentials). While there are good reasons to think that over the range of the spectroscopically significant separations such many-parameter models are superior, one must remember that this range is rather limited, especially if the temperature variation is a limited one. Consequently, an empirical coefficient $B^{(7)}$ thus derived may have little to do with the leading dispersion coefficient unless many measurements obtained over a wide range of temperatures have been used for the determination.

It is not always necessary to compute line shapes from quantum theory for a fit of the spectra [189]. Model profiles are known which describe certain measurements well. Furthermore, classical profiles which are easy to compute may be sufficient for massive systems at high enough temperatures.

1 Results

Rare gas pairs. Induced dipole moments of rare-gas pairs thus obtained are listed in Table 4.1; the parameters $B^{(0)}$, a, b, $B^{(n)}$, are defined according to

$$B(R) = B^{(0)} \exp\left[-a(R - R_0) - b(R - R_0)^2\right] + B^{(n)}/R^n\,, \tag{4.30}$$

which is related to the induced dipole moment according to Eq. 4.14. R_0 is a distance of the order of the size of the system; in general it will be chosen to be nearly equal (but not necessarily exactly equal) to the root σ of the potential, $V(\sigma) = 0$. These dipoles give rise to the translational bands in the far infrared; similar data for the $Q(j)$ lines in the fundamental and overtone bands are also known for pairs involving molecules [189, 342].

For a few systems, two empirical models are entered in the table. While the parameters may differ widely, at the spectroscopically significant separations the dipole strengths are consistent. For the isotropic overlap components, significant separations R range from a few percent below σ to roughly $1.2\,\sigma$. Outside of the range of validity, the empirical dipole models may be seriously in error and must not be used.

It is important to remember that often, especially for the He–Ar system, the empirical $B^{(7)}$ coefficients given in Table 4.1 are not necessarily the same as the lowest-order theoretical dispersion coefficients, D_7. An accurate determination of the true dispersion coefficient would require the inclusion of higher-order dispersion terms in the analysis and, moreover, experimental data of a quality that is presently not available. We note that for He–Ar the *ab initio* dipole specified below is probably superior to the empirical model of Table 4.1.

Molecular gases. In the case of molecular gases, or of mixtures involving a molecular gas, one must in general account for several induced dipole components, as the asymptotic expressions of Section 4.7 suggest. Besides these multipole-induced terms, one or more overlap-induced terms are usually necessary, especially if induction in species of low polarizability is considered (He atoms). While for certain systems just one or two dipole components need to be accounted for, for other systems elaborate sets had to be assumed for a satisfactory fit of the spectra. Moreover, for the molecular systems, we have to consider a number of different dipole models for the different bands (rototranslational, rotovibrational and overtones). It is, therefore, impractical to repeat here even the most important measurements. Table 4.2 quotes some sources where such information may be found. Examples of what is expected for a few representative systems are given below. These are based on first principles.

The classical multipolar induction terms, sometimes supplemented by simple overlap corrections, often work surprisingly well, especially for

Table 4.2. Empirical induced dipole moments of systems involving molecules; m and ls stand for moment analysis and line shape analysis, respectively.

pair	rototranslational band		rotovibrational band	
	method	Ref.	method	Ref.
He–CH_4	ls	[387]		
H_2–He	ls	[123]	m	[317]
H_2–He			ls	[189, 339, 362]
HD–He			ls	[189]
D_2–He			ls	[351]
H_2–Ne			ls	[339]
HD–Ne			ls	[189]
D_2–Ne			ls	[351]
H_2–Ar	m	[177, 317]	ls	[189]
HD–Ar			ls	[189]
H_2–Kr	m	[317]	ls	[189]
HD–Kr			ls	[189]
H_2–Xe	m	[317]	ls	[189]
HD–Xe			ls	[189]
H_2–H_2	m	[317]		
H_2–H_2	ls	[138]	ls	[189, 310, 341]
HD–HD			ls	[189, 310, 340]
D_2–D_2			ls	[310]
H_2–N_2			ls	[54]
H_2–CH_4	ls	[55]		
N_2–N_2	ls	[320, 53]		
N_2–N_2	m	[35, 123, 38]		
O_2–O_2	m	[35, 123, 38]		
CO_2–CO_2	m	[186, 38]		
CH_4–CH_4	ls	[42, 56, 57]		
SF_6–SF_6	m	[40]		

the more highly polarizable species and the rototranslational dipole components; for vibrational spectra (overtones) more substantial overlap is generally needed which is generally less dependably known.

For hydrogen and its isotopes (H_2, HD, D_2, ...) in interaction with rare gas atoms or other hydrogen molecules, very accurate *ab initio* calculations exist that were shown to be in agreement with the known measurements of binary spectra; the calculated dipole functions are as good as, or possibly better than, the best empirical models and will be discussed in the next Section.

Other measurements. Induced dipole moments can be measured by most of the familiar methods that are designed to measure permanent dipole moments. We mention in particular the beam deflection method by electric fields, using van der Waals molecules, and molecular beam electric resonance spectroscopy of van der Waals molecules [373, 193, 98].

The dielectric constant ϵ of dense gases may be written in the form of a virial series, [86]

$$\frac{\epsilon - 1}{\epsilon + 2}\widetilde{V} = A_\epsilon + B_\epsilon/\widetilde{V} + C_\epsilon/\widetilde{V}^2 + \cdots \tag{4.31}$$

(Clausius–Mossotti equation). In this expression, $\widetilde{V}$ designates the mole volume and A_ϵ, B_ϵ, C_ϵ,... are the first, second, third,... virial dielectric coefficients. A similar expansion exists for the refractive index, n, which is related to the (frequency dependent) dielectric constant as $n^2 = \epsilon$ (Lorentz–Lorenz equation, [87]). The second virial dielectric coefficient B_ϵ may be considered the sum of an orientational and a polarization term, $B_\epsilon = B_{\text{or}} + B_{\text{po}}$, arising from binary interactions, while the second virial refractive coefficient is given by just the polarization term, $B_n = B_{\text{po}}$; at high enough frequencies, the orientational component falls off to small values and the difference $B_\epsilon - B_n$ may be considered a 'measurement' of the interaction-induced dipole moments [73].

4.4 Computation

Highly developed quantum chemical methods exist to compute with an ever increasing precision molecular and supermolecular properties from first principles. For example, attempts to compute intermolecular interaction potentials and, more recently, induced dipole moments, are well known for the simpler atomic and molecular systems.

The problems of computing induced dipole moments from first principles are similar to those familiar from the calculations of van der Waals potentials. Because induced dipole moments are rather small and arise from minor distortions of the molecular charge distributions, one might think that perturbation techniques are the natural device for their computation. However, the accounting for the exchange contributions at near range by perturbation techniques has not satisfactorily been solved. Standard quantum-chemical methods, on the other hand, i.e., variational self-consistent field (SCF) and configuration interaction (CI) calculations, suffer from the very slow convergence of the dispersion attraction with the size of the basis set, and from basis-set superposition errors which may lead to spurious attractive contributions from unphysical changes of intra-atomic correlation. Due to the inability of the standard methods to account for high-angular-momentum interatomic correlation, van der Waals potentials that closely model the region of the well have been obtained for systems like He–He, H_2–He, etc., while such results are not available for systems like Ne–Ne or He–Ar. The computation of collision-induced dipole moments is, in principle, equally demanding and only a few advanced calculations are known. However, since collision-induced absorption probes the dipole moment mainly in a region where exchange and low-order multipole induction are much stronger than the dispersion contributions, *ab initio* calculations can provide reliable dipole-moment surfaces, certainly for the simple binary systems (Meyer 1985).

Rare gas mixtures – a historical sketch. The exchange dipole was first investigated by Matcha and Nesbet [256] in SCF calculations with double-ζ plus polarization basis sets, optimized at the very short distance of 2 bohr because of numerical problems at larger separations. The results had to be extrapolated exponentially to the separations of interest, roughly to separations R between 3 and 6 bohrs. A comparison with Bosomworth and Gush's measured spectra [74, 75] suggested that the theoretical dipole strengths were not sufficient. The leading dispersion coefficient, D_7, was, therefore, estimated theoretically [233, 113, 114] but the accuracy of the results was difficult to assess. Furthermore, contributions from higher-order terms, D_9, D_{11}, etc., are expected but could not be estimated theoretically, just like the 'damping' of the dispersion terms at small separations as in interatomic potentials.

Dispersion and exchange dipole components were found to be of opposite sign. The inclusion of a dispersion dipole thus made the discrepancy with measurement only worse. Therefore, the exchange dipole was reconsidered by assuming a pure exchange interaction between otherwise unperturbed SCF wavefunctions of the interacting rare-gas pairs. However, for the fairly well established He–Ar spectra, the discrepancy of theory and measurement persisted [225].

Krauss [44] recalculated the overlap dipole from SCF wavefunctions. The Slater-type orbitals (STO) basis set was of a double-ζ quality, augmented by single diffuse s and d functions, optimized for dipole polarizability. The agreement with measurement was significantly improved for Ne–Ar, but for Ar–Kr the theoretical induced dipole appeared to be too large by about 20%. Differences between the pure exchange dipole mentioned [225] and the SCF dipole are surprisingly large, probably on account of some approximations made in the early work. Treatments of exchange and dispersion dipoles based on the electron-gas and Drude model shell displacements [309, 367] cannot be used for quantitative work.

Whereas the agreement of the SCF plus dispersion dipole model with spectroscopic measurements in neon–argon mixtures is impressive [44], it must be pointed out that the agreement is almost certainly fortuitous. In the framework of this model, electron correlation is included only in form of the leading D_7 dispersion coefficient; no estimates of either higher-order terms, $D_9 \ldots$, nor of the effects of damping at near range, are included but are likely to be significant. Furthermore, intra-atomic correlation is known to increase the 'size' of noble gas atoms as the correlated expectation value of the electronic coordinates indicates. As a consequence, the exchange and overlap dipole may be expected to increase, just like intra-atomic correlation enhances the exchange repulsion for He–He and He–H_2 pairs. A corresponding increase in the exchange dipole is quite likely, and it may actually more or less cancel the higher-order terms that were previously neglected. The subtle balance between these effects may well be responsible for the 'SCF plus D_7' dipole model being too small for He–Ar, but too big for Ne–Kr [44].

1 Results

He–Ar dipole. A new investigation of these correlation effects by CI calculations of the He–Ar system was, therefore, undertaken [278] that was designed to avoid the problems mentioned above. Specifically, the Hartree–Fock wavefunction was transformed to localized orbitals in order to study intra-atomic and interatomic (dispersion) correlation terms separately. Furthermore, the Hartree–Fock basis set of Gauß-type orbitals (GTO) (14s, 10p for Ar, 10s for He) was first augmented by 2d,1f sets for Ar and 2p,1d sets for He which are carefully optimized for intra-atomic correlation. Single diffuse d and f sets for Ar, and p and d sets for He, were then added and optimized to account for dipole and quadrupole atomic polarizabilities, as well as the corresponding terms of the dispersion attraction. Various further extensions of the basis set yielded changes of the dipole moment of less than 2% at 5.5 bohr and were judged to be insignificant. At the CI level, superposition errors were avoided by restricting the intra-atomic correlation to different molecular-orbital subspaces which are spanned by the atoms' 'own' basis functions only, after projecting out the occupied Hartree–Fock orbitals. This is readily implemented in the self-consistent electron-pair (SCEP) technique, which allows the use of different orbital sets for different electron pairs, or even of non-orthogonal external orbitals. Finally, the size-consistent coupled electron-pair approximation (CEPA-1) was used to account in an approximate way for higher-order substitutions, since otherwise intra-atomic correlation would reduce the dispersion terms too strongly.

Table 4.3 shows the results of this most recent calculation [278]. At the SCF level, the new results are only a few percent greater than the best previous calculation which, however, is significant for a comparison with

Table 4.3. Induced dipole moment of He–Ar pairs [278]. The polarity is He^+Ar^-

R (a.u.)	SCF (10^{-6}a.u.)	SCF+intra (10^{-6}a.u.)	SCF+inter (10^{-6}a.u.)	total (10^{-6}a.u.)
4.5	30 081	39 510	22 781	30 434
5	15 529	20 918	11 149	15 891
5.5	7 618	10 090	5 132	7 455
6	3 585	4 721	2 217	3 369
6.5	1 633	2 127	892	1 440
7	728	934	327	580
7.5	308	402	100	216

the spectroscopic measurements. Near the collision diameter, interatomic correlation adds a dispersion dipole which amounts to about three times the previous D_7/R^7 term so that the total dipole moment is only slightly smaller than the SCF value, and in near agreement with the empirical dipole model. A counterpoise calculation reveals that the residual basis set superposition error at both the SCF and CI level remains well below 0.5%. Dispersion damping appears to be important at separations below ≈ 5.2 bohr. The overall accuracy of the dipole computation should be better than 5%.

The results given in the last column of Table 4.3 are well represented by the analytical form, Eq. 4.30, with $B^{(0)} = 0.0386$; $a = 1.371$, $R_0 = 4.5$, $b = 0.04832$, and $B^{(7)} = -290$, in atomic units. We note that the $B^{(7)}$ coefficient was determined from a fit of the long-range distributions, $\mu_{tot} - \mu_{intra}$, for separations from 6.5 to 7.5 bohr. It must not be identified with a D_7 dispersion coefficient because, at such separations, contributions from higher-order dispersion terms are not negligible, albeit non-discernible in the data given in the table.

The last column of Table 4.3 or, equivalently, Eq. 4.30 with the parameters as specified, probably represent the best induced dipole model for He–Ar pairs currently available. This model permits a close reproduction of the measured binary spectra of helium–argon mixtures in the far infrared, see Fig. 5.5 on p. 243.

The comparison of spectral line shapes computed on the basis of the *ab initio* dipole surface of He–Ar with absorption measurements has demonstrated the soundness of the data. The agreement indicates that exchange effects due to intra-atomic correlation and higher-order dispersion terms contribute significantly to the induced dipole. However,

Table 4.4. Dipole moment of H–He. The polarity is is H^-He^+ [283].

R (a.u.)	SCF (10^{-6}a.u.)	intra (10^{-6}a.u.)	inter (10^{-6}a.u.)	total (10^{-6}a.u.)
2.75	−160 184.44	−3 600.70	3 212.48	−160 572.66
3.00	−118 999.20	−2 939.48	3 618.84	−118 319.84
3.25	−88 168.25	−2 403.85	3 837.53	−86 734.57
3.50	−65 016.66	−1 954.91	3 804.26	−63 167.31
3.75	−47 646.70	−1 575.15	3 571.83	−45 650.02
4.00	−34 666.38	−1 255.01	3 216.93	−32 704.46
4.50	−17 926.56	−767.47	2 387.38	−16 306.65
5.00	−8 986.38	−446.01	1 634.94	−7 797.45
5.50	−4 377.23	−247.06	1 062.86	−3 561.43
5.75	−3 025.56	−180.96	845.46	−2 361.06
6.00	−2 079.18	−131.20	667.98	−1 542.40
6.25	−1 421.21	−94.09	524.85	−990.45
6.50	−966.68	−66.91	410.82	−622.77
6.75	−654.56	−47.21	320.71	−381.06
7.00	−441.42	−33.08	249.99	−224.51
7.50	−198.91	−15.90	151.80	−63.01
8.00	−89.31	−7.46	92.67	−4.10
9.00	−19.33	−1.51	35.93	15.09
10.00	−5.46	−0.40	14.99	9.13
11.00	−2.33	−0.09	7.04	4.62
12.00	−1.12	0.12	3.59	2.59

these exchange and dispersion contributions may mutually nearly cancel as is the case for the He–Ar system. Their combined contribution may well account for the differences between theory and experiment remaining to this day for other rare gas systems.

H–He dipole. Similar elaborate computations were recently undertaken for the H-He system [283], a system which is of interest for stellar atmospheres [399]. The results are summarized in Table 4.4, in a format similar to that of Table 4.3.

H_2–He dipole. Early work for that system involving a molecule was of the SCF type [426] and gave dipole moments resulting in rototranslational spectra that were consistently weaker by roughly 20% than the spectroscopic measurements [43]. To some extent, this discrepancy is related to the neglect of vibrational averaging. It is well known that the computed

Table 4.5. Cartesian dipole components of the H_2–He system, for three internuclear spacings r; after [279]. The center-of-mass of the H_2 molecule is at the origin and the He atom is at $z = R$. A positive induced dipole moment μ_z corresponds to the polarity $H_2^-He^+$ ($\mu_y = 0$).

r (a.u.)	R (a.u.)	μ_z at 0° (10^{-6} a.u.)	μ_z at 90° (10^{-6} a.u.)	μ_z at 45° (10^{-6} a.u.)	μ_x at 45° (10^{-6} a.u.)
1.449	3	224 785	86 877	146 662	8 407
	4	55 238	21 511	36 401	–469
	5	13 127	3 913	8 101	–1 060
	6	3 347	201	1 674	–694
	7	1 125	–297	380	–404
	8	531	–251	127	–242
	10	198	–107	42	–100
1.111	3	125 332	59 322	89 386	2 369
	4	29 764	13 465	21 038	–973
	5	6 908	2 119	4 388	–828
	6	1 812	–4	875	–478
	7	659	–222	210	–270
	8	333	–169	78	–161
	10	131	–69	30	–66
1.787	3	366 497	112 415	214 701	17 628
	4	92 284	29 816	55 624	667
	5	22 376	6 009	13 065	–1 201
	6	5 590	525	2 815	–920
	7	1 740	–344	627	–554
	8	748	–333	185	–335
	10	268	–147	54	–139

H_2 quadrupole moment, $q_2(r_e)$ at the fixed equilibrium position, and thus the long-range coefficient of the quadrupole-induced dipole component, Eq. 4.3, is about 5% too small relative to the proper vibrational average, $q_2 = \langle v = 0|q_2(r)|v = 0\rangle$ [216, 217, 209]. A 5% difference of the dipole moment amounts to a ≈ 10% difference of the associated spectral intensities. Furthermore, the effects of electron correlation on this long-range coefficient can be estimated. Correlation increases the He polarizability by 5% but decreases the H_2 quadrupole moment by 8% [275], a net change of –3% of the leading induction term $B_{23}^{(4)}(R)$.

Recent work improved earlier results and considered the effects of electron correlation and vibrational averaging [278]. Especially the effects of intra-atomic correlation, which were seen to be significant for rare-gas pairs, have been studied for H_2–He pairs and compared with interatomic electron correlation: the contributions due to intra- and interatomic correlation are of opposite sign. Localized SCF orbitals were used again to reduce the basis set superposition error. Special care was taken to assure that the supermolecular wavefunctions separate correctly for $R \to \infty$ into a product of correlated H_2 wavefunctions, and a correlated as well as polarized He wavefunction. At the CI level, all atomic and molecular properties (polarizability, quadrupole moment) were found to be in agreement with the accurate values to within 1%. Various extensions of the basis set have resulted in variations of the induced dipole moment of less than 1% [279]. Table 4.5 shows the computed dipole components, μ_x, μ_z, as functions of separation, R, orientation (0°, 90°, 45° relative to the internuclear axis), and three vibrational spacings r, in 10^{-6} a.u. of dipole strength [279].

For molecules with inversion symmetry, like H_2, the expansion parameter λ must be even, Eqs. 4.15 through 4.17. (It also must be non-negative.) In order to relate the expansion coefficient $A_{\lambda L}$ to the Cartesian dipole components calculated in a body-fixed frame, we choose the unit separation vector, $\widehat{\boldsymbol{R}}$, to be parallel to the z-axis, hence $M = 0$, $Y_{L0} = [(2L+1)/4\pi]^{1/2}$, and

$$\mu_\nu(\boldsymbol{r};\boldsymbol{R}) = \left(\frac{4\pi(2L+1)}{3}\right)^{1/2} \sum_{\lambda L} A_{\lambda L}(r,z)\; Y_{\lambda\nu}(\widehat{\boldsymbol{r}})\; C(\lambda L1;\nu 0\nu)\,. \tag{4.32}$$

According to Eq. 4.6, we thus have

$$\begin{aligned}\mu_z(\boldsymbol{r},\boldsymbol{R}) &= \sum_\lambda \left\{(\lambda+1)^{1/2}\, A_{\lambda,\lambda+1} - \sqrt{\lambda}\, A_{\lambda,\lambda-1}\right\}\, P_\lambda^0(\widehat{\boldsymbol{r}}\cdot\widehat{\boldsymbol{z}}) \qquad (4.33)\\ \mu_x(\boldsymbol{r},\boldsymbol{R}) &= \sum_\lambda \left\{\sqrt{\lambda}\, A_{\lambda,\lambda+1} + (\lambda-1)^{1/2}\, A_{\lambda,\lambda-1}\right\}\\ &\qquad \left[\frac{(\lambda-1)!}{(\lambda+1)!}\right]^{1/2} P_\lambda^1(\widehat{\boldsymbol{r}}\cdot\widehat{\boldsymbol{z}})\,.\end{aligned}$$

This system of equations has been solved for the $A_{\lambda L}(r,R)$ coefficients from the known Cartesian dipole components, see Table 4.6. For H_2–He, only four $A_{\lambda L}$ are important, $\lambda L = 01$, 21, 23 and 45. The remaining ones amount to less than 1% of μ_x, μ_z, and can safely be neglected.

At long range, we find (for $z = R \to \infty$)

$$A_{\lambda,\lambda+1}(r,z) = \alpha\, q_\lambda(r)\; (\lambda+1)^{1/2}\; R^{-\lambda-2}\,, \tag{4.34}$$

Table 4.6. The $A_{\lambda L}(r, R)$ induced dipole components for H_2–He, for three different vibrational spacings r [279].

R	$A_{\lambda L}(r, R)$			
	$\lambda L = 01$	$\lambda L = 21$	$\lambda L = 23$	$\lambda L = 45$
(a.u.)	(10^{-6} a.u.)			
$r_- =$	1.111 bohr			
3	79 756.8	−13 660.3	13 606.8	1 202.5
4	18 590.5	−2 498.1	4 107.0	235.7
5	3 648.4	−429.2	1 465.2	51.3
6	585.9	−70.8	634.8	11.9
7	67.1	−13.0	326.6	3.5
8	−3.8	−3.4	189.6	1.6
10	−2.9	−0.3	76.5	0.4
$r_0 =$	1.449 bohr			
3	127 956.2	−30 364.7	26 271.5	3 749.0
4	31 700.8	−6 009.3	7 640.9	806.9
5	6 760.9	−1 119.7	2 540.1	171.3
6	1 196.3	−196.3	1 028.6	40.9
7	158.9	−38.1	508.7	13.9
8	2.7	−10.0	290.0	5.3
10	−7.2	−0.8	116.0	1.4
$r_+ =$	1.787 bohr			
3	183 906.3	−56 815.1	45 962.2	10 121.8
4	47 744.8	−11 922.5	13 115.8	2 218.6
5	10 863.3	−2 358.2	4 126.2	461.0
6	2 084.0	−424.2	1 549.8	99.2
7	312.8	−76.5	724.0	29.0
8	15.3	−13.4	400.2	9.2
10	−12.1	−0.1	158.8	2.7

and $A_{\lambda,\lambda-1}(r, z) = 0$; these short-range components ($L = \lambda - 1$) fall off roughly exponentially ('anisotropic overlap').

For a description of the induced transitions between rotovibrational levels $|vj\rangle \rightarrow |v'j'\rangle$ of H_2, radial matrix elements, Eq. 4.17, are needed. Since the H–H interaction potential is well known, the vibrational wavefunctions, $\varphi_{vj}(r)$, which depend on the rotational excitation j, can be computed with the help of digital computers [217]. At fixed intermolecular separations R, the $A_{\lambda L}(r, R)$ are known at three vibrational spacings,

$r_0, r_\pm = r_0 \pm \delta$. One may assume that a low-order polynomial describes the dependence of the $A_{\lambda L}$ on the vibrational coordinate r quite well,

$$A_{\lambda L}(r;R) = \sum_n a_n(R)\,\varrho^n \tag{4.35}$$

$$B_{\lambda L}^{vjv'j'}(R) = \sum_n a_n(R)\,\langle vj\,|\varrho^n|\,v'j'\rangle\ , \tag{4.36}$$

with $\varrho = r - r_0$, $R =$ constant and fixed λL. For the computations shown in Table 4.5, the three points were chosen according to $r_0 = \langle vj|r|vj\rangle = 1.449$ bohr for $v = j = 0$, and $\delta = \langle 00|(r - r_0)^2|00\rangle^{1/2} = 0.338$ bohr, so that $r_+ = 1.787$ and $r_- = 1.111$ bohr. Matrix elements of H_2 of the type $\langle vj|\varrho^n|v'j'\rangle$ are given elsewhere [425]. The last equation can quite generally be written

$$\begin{aligned} B_{\lambda L}^{vjv'j'}(R) &= A_{\lambda L}(r_0,R)\,\langle vj\,|\,v'j'\rangle \\ &+ \langle vj\,|\varrho|\,v'j'\rangle\,\left(A_{\lambda L}(r_+,R) - A_{\lambda L}(r_-,R)\right)/2\delta \\ &+ \left\langle vj\left|\varrho^2\right|v'j'\right\rangle\,\left(A_{\lambda L}(r_+,R) + A_{\lambda L}(r_-,R) - 2A_{\lambda L}(r_0,R)\right)/2\delta^2 \\ &+ a_3\left(\left\langle vj\left|\varrho^3\right|v'j'\right\rangle - \langle vj\,|\varrho|\,v'j'\rangle\,\delta^2\right) + \cdots, \end{aligned} \tag{4.37}$$

with an unknown coefficient a_3. If we were free to choose δ, we could make the factor of a_3 disappear; with the choice of δ made above, the factor of a_3 is small so that the fourth term, and also the higher terms, are quite negligible.

Purely rotational matrix elements. For $v = v' = 0$ and $j = j' = 0$, with the choice of δ given above, we thus have

$$\begin{aligned} B_{\lambda L}(R) = \; & A_{\lambda L}(r_0,R) + \frac{1}{8}\left(A_{\lambda L}(r_+,R) + A_{\lambda L}(r_-,R) - 2A_{\lambda L}(r_0,R)\right) \\ & + 0.0009\,a_3 + 0.0009\,a_4 \ldots \end{aligned} \tag{4.38}$$

The unknown cubic and quartic terms, which are small, enter with a small weight and may safely be neglected in this case. Even the second term in this series amounts to less than 1% of the leading term (sometimes referred to as the r-centroid approximation), except at small separations where R is comparable to the collision diameter ($\approx$ 5.71 bohr) and corrections amount to +2%.

Figure 4.1 shows the four significant induced dipole components for the rototranslational bands (left panel). The isotropic and anisotropic overlap components, B_{01} and $-B_{21}$, dominate at near range (dotted). These fall off roughly exponentially with separation R so that at more distant range, the quadrupole-induction, B_{23}, dominates; it falls off more slowly, like R^{-4}. A weak hexadecapole component, B_{45}, is also present. The dashed lines show the classical (i.e., overlap-free) multipole induction contributions. These differ only at near range from the computed B_{23} and B_{45} components,

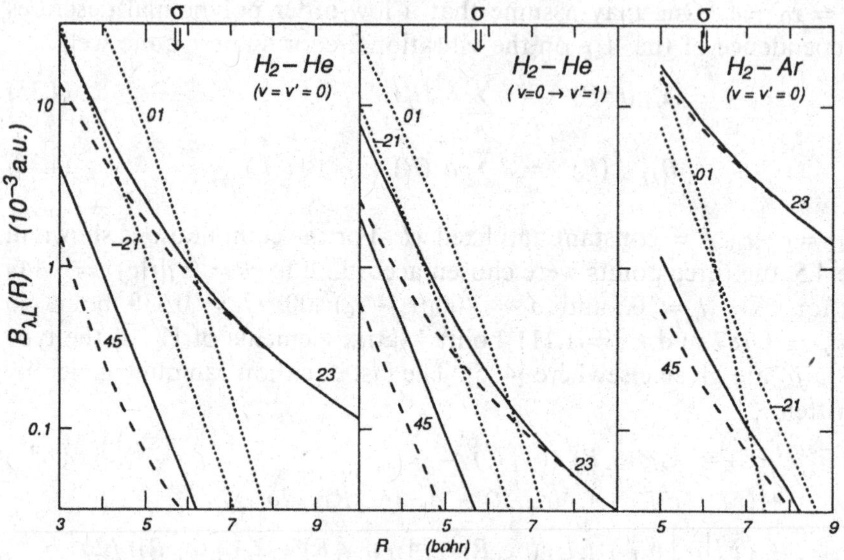

Fig. 4.1. The four significant induced dipole components of H_2–He pairs: ground state vibrational average (left panel) and vibrational transition matrix element $v = 0 \rightarrow v' = 1$ (center panel); and of the ground state vibrational average of H_2–Ar pairs (right panel). Overlap components are dotted; multipole-induced components are represented by solid lines; dashed lines indicate the associated classical multipole component; after [279, 151, 280].

indicating the presence of a small anisotropic overlap induction in these components.

An analytical expression for the $B_{\lambda L}$ coefficients is often desirable. For the purely rotational transitions ($v = v' = 0$) one may write

$$B_{\lambda L}(R) = \frac{B_{\lambda L}^{(n)}}{R^n} + B_{\lambda L}^{(0)}(R) \exp\left[-a_{\lambda L}\left(R - R_0\right) - b_{\lambda L}\left(R - R_0\right)^2\right] \ ; \qquad (4.39)$$

the superscript $vjv'j' = 0000$ is temporarily suppressed but should be attached to all parameters, $B_{\lambda L}^{(n)}$, $B_{\lambda L}^{(0)}$, $a_{\lambda L}$, and $b_{\lambda L}$. Least mean squares techniques permit an easy definition of the values of these parameters [63] and are given in the upper part of Table 4.7.

The long-range coefficient $B_{23}^{(4)}$ describing the classical quadrupole induction is in agreement with the asymptotic value, $(\lambda + 1)^{1/2}\, \alpha q_\lambda$ ($\lambda = 2$) as the comparison of the solid and dashed lines suggests, Fig. 4.1 (left panel). Even the weak hexadecapole term, $B_{45}^{(6)}$, which cannot be computed with great accuracy because of numerical insignificance, is in reasonable agreement with the well-known asymptotic expression.

The dependence of these coefficients on the rotational quantum numbers, j, $j' \neq 0$, was also investigated. For the purely rotational transitions,

Table 4.7. Fit parameters for the $B_{\lambda L}(R)$; with $R_0 = 5.70$ a.u., rotational band ($v = v' = 0$) of H_2–He; to be used with Eqs. 4.39 and 4.40 [63].

	n	$B_{\lambda L}^{(n)}$ (a.u.)	$B_{\lambda L}^{(0)}$ (10^{-6} a.u.)	$a_{\lambda L}$ (a.u.)	$b_{\lambda L}$ (a.u.)
$\langle 00\vert A_{\lambda L}\vert 00\rangle$					
λL=01	7	−82.1	2520	1.671	0.045
23	4	1.17	237	1.644	0.061
21	7	−9.01	−313	1.778	0.042
45	6	1.19	33.3	1.869	0.094
$\langle \bar{0}\vert A_{\lambda L}\vert 0\rangle$					
λL=01	7	−0.0424	2.17	1.569	0.052
23	4	0.000673	0.365	1.607	0.060
21	7	−0.0000353	−0.486	1.700	0.042
45	6	0.00189	0.057	2.010	0.137

this influence introduces variations of less than $\approx 1\%$ [63], for $j, j' = 0 \ldots 3$, and can safely be ignored in most applications. In the lower part of Table 4.7, coefficients permitting the modeling of the dependence on the rotational states are given; as in Eq. 4.40 below.

Matrix elements for the fundamental band. We use again Eq. 4.37 to evaluate the $B_{\lambda L}$ coefficients to account for vibrational transitions of the type $v = 0 \to v' = 1$. In this case, the dipole component displayed in the center of Fig. 4.1 results. The $\lambda L = 23$ long-range component is again closely described by the R^{-4} quadrupolar induction, but at near range relatively strong exchange contributions are noticeable; that is at small separations the difference between the dotted and solid curves becomes bigger than that seen in the rototranslational band.

The fact that matrix elements of the fundamental band are dependent on the rotational quantum numbers j, j' cannot be ignored. As a consequence, many more B coefficients must now be evaluated which, in principle, poses no special problem. The volume of data needed renders the task awkward. Molecular spectroscopists have for generations coped with similar problems which were solved with the so-called Dunham expansion in terms of $j(j+1)$. Specifically, for our purpose, the lowest-order expansion for the fundamental band looks like [63]

$$B_{\lambda L}^{0j1j'}(R) = \langle 00\,|A_{\lambda L}|\,10\rangle + \left\langle \bar{0}\,|A_{\lambda L}|\,1\right\rangle j(j+1) + \left\langle 0\,|A_{\lambda L}|\,\bar{1}\right\rangle j'(j'+1) + \ldots \tag{4.40}$$

Table 4.8. H_2–He fit parameters $B_{\lambda L}^{0j1j'}(R)$, with R_0 = 5.70 a.u., for the fundamental band ($v = 0 \rightarrow v' = 1$); to be used with Eqs. 4.39 and 4.40 [63].

	n	$B_{\lambda L}^{(n)}$ (a.u.)	$B_{\lambda L}^{(0)}$ (10^{-6} a.u.)	$a_{\lambda L}$ (a.u.)	$b_{\lambda L}$ (a.u.)
$\langle 00 \vert A_{\lambda L} \vert 10 \rangle$					
λL=01	7	24.7	−751	1.538	0.051
23	4	−0.205	−119	1.583	0.053
21	7	−2.86	170	1.640	0.018
45	6	−0.511	−21.8	1.873	0.097
$\langle \bar{0} \vert A_{\lambda L} \vert 1 \rangle$					
λL=01	7	0.216	−8.68	1.675	0.056
23	4	−0.00393	−1.01	1.615	0.057
21	7	−0.00494	1.45	1.702	0.022
45	6	−0.00540	−0.132	2.023	0.138
$\langle 0 \vert A_{\lambda L} \vert \bar{1} \rangle$					
λL=01	7	−0.247	7.98	1.683	0.049
23	4	0.00385	0.666	1.729	0.084
21	7	−0.0548	−0.806	1.918	0.089
45	6	0.00502	0.0501	2.309	0.223

The three coefficients in angular brackets are given in Table 4.8. It was previously found that for $j, j' = 0 \ldots 4$, higher-order terms in $j(j+1)$, $j'(j'+1)$ are not needed. The coefficients are functions of separation, R, and can readily be fitted to an analytical expression of the form Eq. 4.39, with $B_{\lambda L} = \langle 00\,|A_{\lambda L}|\,10\rangle$, $\left\langle \bar{0}\,|A_{\lambda L}|\,1\right\rangle$, $\left\langle 0\,|A_{\lambda L}|\,\bar{1}\right\rangle$, respectively. These coefficients are given in the lower part of Table 4.8. The results show that the j, j' corrections amount to 10 or 15% for $j, j' = 1 \ldots 3$ for the main components, $\lambda L = 01$ and 23, and even more for the lesser components. Since the associated spectral intensities vary as the *squares* of dipole strength, these variations are clearly significant for the spectra.

The principal effect of the j, j' corrections amounts more or less to a rescaling of the $B^{0010}(R)$ transition elements. Specifically, the long-range correction ($\lambda L = 23$) looks much like a scaled-down version of $B_{23}^{0010}(R)$, and the overlap correction ($\lambda L = 01$) is more or less a scaled-down version of $B_{01}^{0010}(R)$, so that to a first approximation, scaling factors could be used to model these corrections.

Table 4.9. Fit parameters for the computation of $B_{\lambda L}(R)$; with $R_0 = 6.0$ a.u., for the rotational band ($v = v' = 0$) of H_2–Ar; to be used with Eq. 4.39 [280].

	n	$B^{(n)}_{\lambda L}$ (a.u.)	$B^{(0)}_{\lambda L}$ (10^{-3} a.u.)	$a_{\lambda L}$ (a.u.)	$b_{\lambda L}$ (a.u.)
λL=01	7	−278.846	3.344	1.888	0.094
23	4	9.199	0.799	1.422	0.041
21	7	0	−1.501	1.575	−0.032
45	6	10.126	0.116	1.576	0.057

For the quadrupole-induction matrix element B_{23} and even for the weak hexadecapole-induction term B_{45}, for $R \to \infty$, the coefficients of $R^{-\lambda-2}$ are in close agreement with the asymptotic values $(\lambda + 1)^{1/2}\alpha\langle vj|q_\lambda|v'j'\rangle$. Figure 4.1 (center panel) illustrates this agreement: solid and dashed lines merge for $R \to \infty$.

Overtone band. Similar induced dipole matrix elements for the H_2–He system have been derived from Table 4.5 for the overtone bands [61]. In that case, the numerical accuracy of the resulting B coefficients is limited for large vibrational excitations unless the data are supplemented by larger variations of the vibrational coordinate r.

H_2–Ar dipole. For purely rotational transitions, induced dipole components have also been computed from first principles [280]. The results are shown in Table 4.9 and Fig. 4.1 (right panel). Like for the H_2–He system, we have four significant induced dipole components. The multipole-induced components are now much stronger, on account of the high polarizability of the argon atom relative to that of helium. Near the collision diameter, $\sigma = 6$ bohr, and beyond ($R \geq \sigma$), the quadrupole-induction component clearly dominates the dipole moment – a situation much different from that seen in H_2–He. Again, the dipole components at long range are in agreement with the classical quadrupole-induction expressions, Fig. 4.1.

H_2–H_2 dipole. Early attempts to calculate the induced dipole moments from first principles were described elsewhere [281]. Only in recent times could the substantial problems of such computations be controlled and precise data be generated by SCF and CI calculations, so that that the basis set superposition errors were small enough and the CI excitation level is adequate for the long-range effects. The details of the computations are given elsewhere [282, 281].

A set of 13 relative orientations of the pair has been selected. The ϑ_i

Table 4.10: Part I. Cartesian dipole moment components of H_2–H_2 [281], in 10^{-6} a.u., for three bond distances, r_0=1.449 bohr, r_-= 1.111 bohr, r_+= 1.787 bohr, and 13 relative orientations, $\vartheta_1\varphi_1, \vartheta_2\varphi_2$; number 0 means 0°, 1 = 45°, 2 = 90°, 3 = 135°, respectively.

R	$\vartheta_1\varphi_1, \vartheta_2\varphi_2$												
(a.u.)	00,20	20,00	00,30	10,00	20,30	10,20	10,22	22,30	10,30	00,00	20,20	20,22	10,10
	$\mu_x(r_0r_0)$:												
3.5	0	0	−9602	−9602	−8694	−8694	−1290	−1290	−18419	0	0	0	0
4.0	0	0	−8858	−8858	−7927	−7927	−3276	−3276	−16774	0	0	0	0
4.5	0	0	−7404	−7404	−6552	−6552	−3498	−3498	−13935	0	0	0	0
5.0	0	0	−5869	−5869	−5143	−5143	−3065	−3065	−11000	0	0	0	0
5.5	0	0	−4519	−4519	−3932	−3932	−2474	−2474	−8444	0	0	0	0
6.0	0	0	−3436	−3436	−2972	−2972	−1923	−1923	−6404	0	0	0	0
7.0	0	0	−1983	−1983	−1703	−1703	−1125	−1125	−3684	0	0	0	0
8.0	0	0	−1182	−1182	−1011	−1011	−669	−669	−2192	0	0	0	0
9.0	0	0	−739	−739	−631	−631	−416	−416	−1369	0	0	0	0
	$\mu_z(r_0r_0)$:												
3.5	133881	−133881	73235	−73235	−61181	61181	60809	−60809	0	0	0	0	0
4.0	75776	−75776	41905	−41905	−34438	34438	34326	−34326	0	0	0	0	0
4.5	43444	−43444	23975	−23975	−19822	19822	19791	−19791	0	0	0	0	0
5.0	25834	−25834	14120	−14120	−11897	11897	11891	−11891	0	0	0	0	0
5.5	16167	−16167	8731	−8731	−7529	7529	7530	−7530	0	0	0	0	0
6.0	10652	−10652	5685	−5685	−5015	5015	5017	−5017	0	0	0	0	0
7.0	5272	−5272	2760	−2760	−2527	2527	2529	−2529	0	0	0	0	0
8.0	2973	−2973	1537	−1537	−1441	1441	1442	−1442	0	0	0	0	0
9.0	1824	−1824	935	−935	−890	890	891	−891	0	0	0	0	0

Table 4.10: Part II

R	$\vartheta_1\varphi_1, \vartheta_2\varphi_2$												
(a.u.)	00,20	20,00	00,30	10,00	20,30	10,20	10,22	22,30	10,30	00,00	20,20	20,22	10,10
	$\mu_x(r_0r_-)$:												
3.5	0	0	–8355	–5138	–7854	–4432	–262	–3336	–12941	0	0	0	3300
4.0	0	0	–6632	–5810	–6245	–4999	–2422	–3377	–11846	0	0	0	1036
4.5	0	0	–5116	–5194	–4790	–4447	–2779	–2894	–9773	0	0	0	139
5.0	0	0	–3859	–4226	–3594	–3600	–2479	–2297	–7640	0	0	0	–181
5.5	0	0	–2876	–3289	–2668	–2788	–2009	–1756	–5811	0	0	0	–261
6.0	0	0	–2138	–2510	–1980	–2118	–1562	–1322	–4373	0	0	0	–251
7.0	0	0	–1205	–1451	–1115	–1215	–912	–751	–2493	0	0	0	–171
8.0	0	0	–713	–865	–660	–721	–543	–445	–1480	0	0	0	–106
9.0	0	0	–445	–541	–413	–449	–338	–277	–924	0	0	0	–66
	$\mu_z(r_0r_-)$:												
3.5	133215	–49149	101606	2170	–14932	68640	68565	–14737	36457	66207	15884	15993	37969
4.0	72501	–29437	54027	–1720	–9609	37091	37101	–9529	18319	33015	8174	8239	19348
4.5	39901	–18297	28752	–2901	–6573	20258	20280	–6536	8992	16125	3999	4038	9670
5.0	22627	–11996	15603	–3038	–4789	11333	11352	–4771	4284	7768	1779	1803	4731
5.5	13445	–8279	8812	–2756	–3628	6597	6611	–3618	1974	3726	662	678	2271
6.0	8438	–5928	5252	–2319	–2784	4046	4056	–2779	879	1806	152	162	1081
7.0	3905	–3266	2235	–1482	–1656	1807	1811	–1654	155	472	–123	–119	254
8.0	2142	–1926	1175	–912	–1005	978	980	–1005	23	171	–117	–115	78
9.0	1304	–1200	705	–572	–633	595	596	–633	3	89	–81	–80	36

Table 4.10: Part III

R	$\vartheta_1\varphi_1, \vartheta_2\varphi_2$												
(a.u.)	00,20	20,00	00,30	10,00	20,30	10,20	10,22	22,30	10,30	00,00	20,20	20,22	10,10
	$\mu_x(r_0r_+)$:												
3.5	0	0	−7337	−15532	−6878	−13565	−2419	4040	−21957	0	0	0	−7270
4.0	0	0	−9551	−12586	−8324	−11292	−4154	−1555	−20865	0	0	0	−2935
4.5	0	0	−9067	−9959	−7688	−8978	−4218	−3289	−17802	0	0	0	−1075
5.0	0	0	−7688	−7704	−6400	−6931	−3650	−3429	−14335	0	0	0	−276
5.5	0	0	−6169	−5868	−5068	−5265	−2937	−2995	−11169	0	0	0	44
6.0	0	0	−4816	−4440	−3916	−3973	−2284	−2430	−8564	0	0	0	151
7.0	0	0	−2859	−2553	−2288	−2278	−1337	−1474	−4986	0	0	0	153
8.0	0	0	−1721	−1518	−1364	−1356	−795	−884	−2978	0	0	0	103
9.0	0	0	−1079	−947	−849	−847	−495	−550	−1861	0	0	0	66
	$\mu_z(r_0r_+)$:												
3.5	130553	−258518	27486	−188165	−122173	51943	51035	−121537	−50800	−106898	−16327	−16628	−54381
4.0	76953	−144368	21268	−103802	−67318	30699	30339	−67208	−25591	−54852	−8713	−8894	−28144
4.5	45796	−80445	15130	−56274	−37464	18633	18481	−37487	−12618	−26759	−4531	−4640	−14352
5.0	28364	−45925	10728	−30867	−21412	11967	11899	−21457	−6071	−12774	−2185	−2253	−7225
5.5	18540	−27367	7803	−17536	−12740	8178	8145	−12778	−2820	−6003	−917	−960	−3587
6.0	12704	−17153	5786	−10458	−7953	5841	5823	−7981	−1249	−2787	−290	−319	−1765
7.0	6627	−7837	3294	−4388	−3608	3228	3223	−3622	−186	−589	98	86	−436
8.0	3823	−4239	1950	−2262	−1950	1917	1916	−1957	4	−142	125	119	−131
9.0	2365	−2554	1209	−1335	−1178	1204	1203	−1182	25	−50	96	93	−54

Table 4.10: Part IV

R	$\vartheta_1\varphi_1, \vartheta_2\varphi_2$												
(a.u.)	00,20	20,00	00,30	10,00	20,30	10,20	10,22	22,30	10,30	00,00	20,20	20,22	10,10
	$\mu_x(r_-r_+)$:												
3.5	0	0	−2093	−12265	−1587	−11167	−4308	4639	−13629	0	0	0	−9749
4.0	0	0	−5738	−9053	−4634	−8534	−4098	−836	−13983	0	0	0	−3582
4.5	0	0	−6176	−6755	−4998	−6432	−3455	−2565	−12181	0	0	0	−1015
5.0	0	0	−5463	−5020	−4400	−4795	−2734	−2779	−9844	0	0	0	9
5.5	0	0	−4455	−3717	−3568	−3557	−2092	−2450	−7654	0	0	0	360
6.0	0	0	−3499	−2756	−2786	−2643	−1579	−1992	−5846	0	0	0	432
7.0	0	0	−2082	−1548	−1638	−1494	−899	−1208	−3383	0	0	0	332
8.0	0	0	−1254	−912	−976	−887	−532	−725	−2015	0	0	0	212
9.0	0	0	−786	−567	−607	−555	−331	−451	−1258	0	0	0	134
	$\mu_z(r_-r_+)$:												
3.5	39276	−244824	−48025	−208202	−122686	2119	1674	−122609	−84746	−168238	−31975	−32321	−89138
4.0	25668	−132185	−20372	−109753	−65837	3429	3221	−65928	−42776	−84654	−16794	−17000	−45834
4.5	17309	−71374	−7407	−57273	−35479	3812	3708	−35573	−21159	−41362	−8484	−8608	−23200
5.0	12323	−39341	−1504	−30151	−19398	3843	3787	−19467	−10201	−19917	−3940	−4018	−11550
5.5	9169	−22534	982	−16323	−10905	3596	3564	−10952	−4760	−9513	−1571	−1621	−5657
6.0	6958	−13575	1809	−9236	−6398	3138	3119	−6428	−2138	−4544	−440	−472	−2744
7.0	4098	−5833	1692	−3530	−2607	2112	2105	−2621	−364	−1096	219	205	−662
8.0	2481	−3068	1143	−1735	−1333	1340	1337	−1340	−37	−347	239	232	−199
9.0	1563	−1833	737	−1010	−790	860	858	−793	10	−162	175	172	−85

are the angles subtended by the vectors $\hat{r}_i$ and $\widehat{\boldsymbol{R}}$, and $\Delta\varphi = \varphi_2 - \varphi_1$ is the dihedral angle defined by the planes $\hat{r}_1$, $\widehat{\boldsymbol{R}}$, and $\hat{r}_2$, $\widehat{\boldsymbol{R}}$. These angles assume values of 0°, 45°, 90°, and 135°, which are marked in Table 4.10 by numbers 0, 1, 2, and 3, respectively. This set provides 8 non-redundant Cartesian dipole components if $r_1 = r_2$, and 21 components if $r_1 \neq r_2$. The results are collected in Table 4.10 for 4 non-redundant pairs of bond distances and 9 intermolecular separations.

From these data, it was found that 11 leading A coefficients, Eq. 4.18, could be determined by least mean squares techniques with sufficient numerical significance, namely the $A_{\lambda_1\lambda_2\Lambda L}(r_1 r_2; R)$ with subscripts

$$\lambda_1\lambda_2\Lambda L = 0001,\ 2023,\ 0223, \qquad (4.41)$$
$$2021,\ 0221,\ 2233,\ 2211,\ 4045,\ 0445,\ 2245,\ \text{and}\ 2243.$$

(These results are not shown.) The latter two are already smaller than the leading terms ($\lambda_1\lambda_2\Lambda L$ = 0001, 0223, and 2023) by nearly two orders of magnitude. These and the higher terms can safely be neglected. The A coefficients may be represented by quadratic polynomials in the bond distances, r_1 and r_2. From these, the radial transition matrix elements, Eq. 4.20, are obtained with the help of the H–H matrix elements, $\langle v_1 j_1 v_2 j_2 | \varrho^n | v_1' j_1' v_2' j_2' \rangle$ [425].

As a result, extensive tables of the $B_{(c)}^{(s)}(R)$ coefficients are obtained; the superscript (s) is short for v_1, j_1, v_2, j_2, v_1', j_1', v_2', j_2'. (Results are not shown.) In order to reduce the volume of data, we describe the dependence on the rotational states (j_1, j_1', j_2, j_2') by a Dunham-type expansion,

$$\begin{aligned} B_{(c)}^{(s)}(R) \approx\ & \langle v_1 0 v_2 0 \,|A_{(c)}|\, v_1' 0 v_2' 0\rangle \\ & + \langle \overline{v_1} v_2 \,|A_{(c)}|\, v_1' v_2'\rangle\ j_1(j_1+1) + \langle v_1 \overline{v_2} \,|A_{(c)}|\, v_1' v_2'\rangle\ j_2(j_2+1) \\ & + \left\langle v_1 v_2 \,|A_{(c)}|\, \overline{v_1'} v_2' \right\rangle\ j_1'(j_1'+1) + \left\langle v_1 v_2 \,|A_{(c)}|\, v_1' \overline{v_2'} \right\rangle\ j_2'(j_2'+1) + \ldots \end{aligned} \qquad (4.42)$$

for the two cases $v_1 = v_2 = 0$ and $v_1' = 0$, $v_2' = 1$. The leading coefficients in this expansion are the components $B_{(c)}^{v_1 v_1' v_2 v_2'}(R)$ that correspond to non-rotating molecules ($j_1 = j_1' = j_2 = j_2' = 0$). The remaining terms are corrections that account for the rotational dependences. The case of molecule 1 undergoing a vibrational transition ($v_1' = 1$, $v_2' = 0$) is readily obtained from this by making use of the symmetry relations

$$\begin{aligned} \langle 00 \,|A_{\lambda_1\lambda_2\Lambda L}|\, 01\rangle &= (-1)^{\Lambda-L} \langle 00 \,|A_{\lambda_2\lambda_1\Lambda L}|\, 10\rangle \\ \langle v_1 j_1 v_2 j_2 \,|A_{\lambda_1\lambda_2\Lambda L}|\, v_1' j_1' v_2' j_2'\rangle &= (-1)^{\Lambda-L} \langle v_2 j_2 v_1 j_1 \,|A_{\lambda_2\lambda_1\Lambda L}|\, v_2' j_2' v_1' j_1'\rangle . \end{aligned} \qquad (4.43)$$

For the range of j values of interest here, we found that higher than linear terms can safely be neglected, just as this was seen in related work on the H_2–He system above. The non-redundant coefficients for the RT band

Table 4.11. Fit parameters for the rototranslational band of H_2–H_2 pairs; $R_0 =$ 6 bohr [282].

$\lambda_1\lambda_2\Lambda L$	n	$B^{(n)}$ (a.u.)	$B^{(0)}$ (10^{-6} a.u.)	a (a.u.)	b (a.u.)
2023	4	4.56	209	1.346	0.040
0223	4	–4.56	–209	1.346	0.040
2021	7	–31.2	–348	1.792	0.082
0221	7	31.2	348	1.792	0.082
2233	4	–0.702	33.3	1.566	0.016
2211	7	1.82	–21.8	1.561	0.021
4045	6	4.62	51.4	1.404	0.090
0445	6	–4.62	–51.4	1.404	0.090

are listed in Table 4.11 and those of the fundamental RV band of H_2 in Table 4.12 for various separations R. Coefficients not listed are either negligible, e.g., $\langle \bar{0}0|A|01\rangle$, or related to one of the coefficients given by the relation, Eq. 4.43.

If no vibrational transitions are involved, for the purely rotational band, the rotation dependence is again quite weak and may safely be ignored for most purposes. Figure 4.2 (left-hand plot) shows the most significant induced dipole components for non-rotating molecules.

The most significant induced dipole components are shown in Fig. 4.2 (right-hand plot) as function of separation, R, for the case of non-rotating molecules ($j_1 = j_1' = j_2 = j_2' = 0$). Just as this was seen above for H_2–He, the nearly exponential isotropic overlap component ($\lambda_1\lambda_2\Lambda L = 0001$) dominates at small R, and the quadrupole-induced components, 0223 and 2023, are most significant at large R. The dashed lines correspond to the classical, pure quadrupole-induction term (with the overlap contribution suppressed). These reflect the characteristic R^{-4} dependence. Apart from the fact that some of the pairs, like 0221,2021 that are identical in the rototranslational band, differ distinctly in the fundamental band on account of the vibrational excitation of *one* molecule, the picture looks very similar to the ones shown above. We note that the signs of the B components indicated in the figure (i.e., the sign attached to some labels, as in –0221) correspond to molecule 2 undergoing the vibrational transition. According to Eq. 4.43, another identical set of $|B|$ coefficients exists that corresponds to molecule 1 undergoing the vibrational transition. The comparison of the computed $\lambda_1\lambda_2\Lambda L = 0223$, 2023, and 0445 coefficients

Table 4.12. Fit parameters for the fundamental band of H_2–H_2 pairs, $R_0 = 6$ bohr; molecule 2 undergoes the vibrational transition [281].

$\lambda_1\lambda_2\Lambda L$	n	$B^{(n)}$ (a.u.)	$B^{(0)}$ (10^{-6} a.u.)	a (a.u.)	b (a.u.)
$\langle 00 \vert A \vert 01 \rangle$					
0001	7	38.2	−652	1.521	0.033
2023	4	0.622	−8.11	2.268	0.074
0223	4	−0.853	−118	1.479	0.026
2021	7	−15.5	14.8	1.219	−0.283
0221	7	16.1	153	1.806	0.130
2233	4	−16.3	10.7	1.507	0.028
2211	7	1.05	−8.62	1.431	0.002
4045	6	0.779	−0.316	3.755	0.341
0445	6	−2.20	−29.5	1.430	0.131
$\langle 0\bar{0} \vert A \vert 01 \rangle$					
0001	0	0.0	−0.883	1.517	0.063
2023	4	0.0152	0.645	1.307	0.044
0223	4	−0.0154	−0.883	1.402	0.035
2021	0	0.0	−1.52	1.597	0.014
0221	0	0.0	1.90	1.598	0.018
2233	4	−0.00244	0.122	1.557	0.019
2211	0	0.0	−0.0572	1.665	0.070
4045	6	−0.0176	−0.106	1.884	0.372
0445	6	0.0195	0.200	1.688	0.196
$\langle 00 \vert A \vert 0\bar{1} \rangle$					
0001	7	−0.0161	0.0318	2.657	0.232
2023	4	−0.0148	−0.667	1.339	0.036
0223	4	0.0149	0.616	1.311	0.036
2021	0	0.0	1.50	1.617	0.017
0221	0	0.0	−1.41	1.629	0.017
2233	4	0.00228	−0.105	1.576	0.016
2211	7	−0.00534	0.0679	1.569	0.022
4045	6	−0.0176	−0.0990	1.910	0.332
0445	6	0.0170	0.0602	2.277	0.497

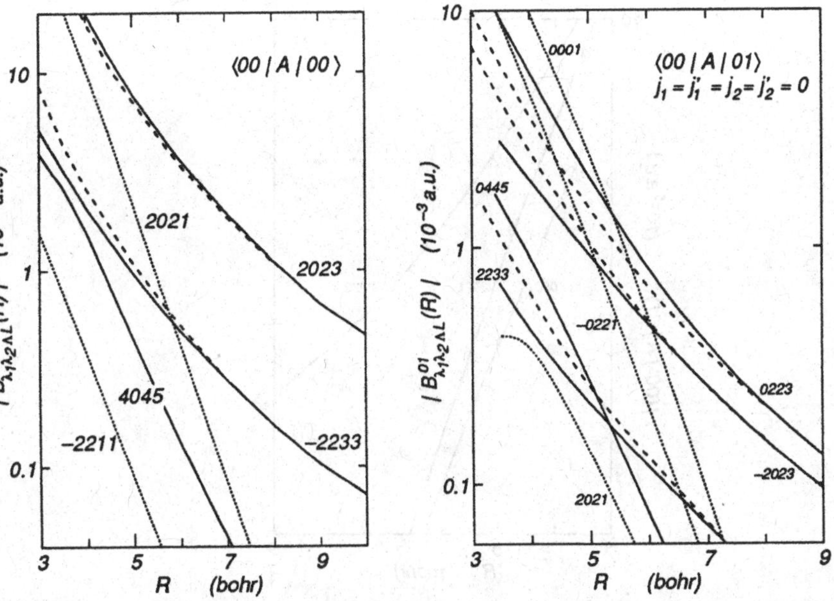

Fig. 4.2. The principal induced dipole components of H_2–H_2: ground state vibrational averages are shown in the left-hand plot. Overlap (dotted), multipole-induced terms (solid lines); classical multipole approximation (dashed). Right-hand plot: ditto, but vibrational transition elements, $v = 0 \to v' = 1$. Molecule 2 undergoes the vibrational transition [281].

B (solid lines) with the classical multipole induction term (dashed lines) shows again agreement at long range as it must; at short range a positive or negative overlap is significant in these terms, especially in Fig. 4.2 (right-hand plot) for the vibrational band.

For the vibrational bands, the j-dependences must again be taken into account. We give $j(j+1)$ expansion coefficients in the lower part of Table 4.12 which, when used as in Eq. 4.42, will modify the data shown in Fig. 4.2 (right-hand plot) and in the upper part of Table 4.12. With these, at least for the smaller j values ($0 \leq j \leq 3$), the actual j-dependence is closely modeled.

Table 4.12 shows that several of these correction coefficients are of nearly the same magnitude but of opposite sign, for example, for the $\lambda_1\lambda_2\Lambda L = 2023$ and 0223 components, $\langle 0\bar{0}|A|01\rangle$ and $\langle 00|A|0\bar{1}\rangle$. Therefore, in Fig. 4.3 we need to give only one representative example of these functions. A similar situation exists for the four correction terms of the 2021,0221 components. The comparison of Figs. 4.2 (right-hand plot) and 4.3 shows that the correction terms belonging to the same $\lambda_1\lambda_2\Lambda L$ set show very much the same dependence as the dipole coefficient for the rotationless case, i.e., the quadrupole induced terms reflect the R^{-4} dependence

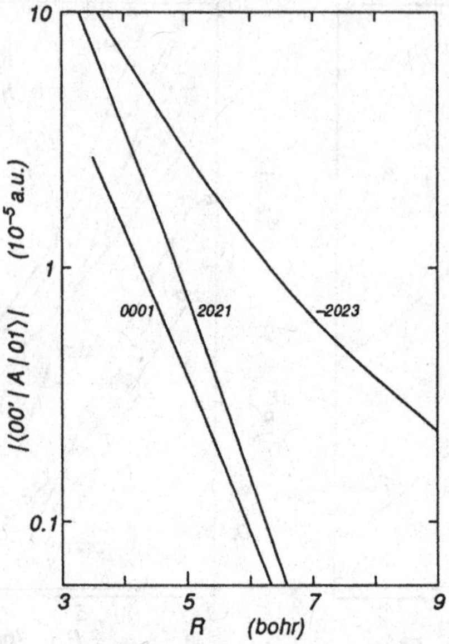

Fig. 4.3. The j_1, j_2... dependences of the principal induced dipole components of H_2–H_2: vibrational transition elements $v = 0 \to v' = 1$. Overlap (dotted) and multipole-induced terms (solid line type) are shown; dashed lines are the classical multipole approximations [281].

and the overlap terms are scaled-down versions of exponential functions similar to the ones shown in Fig. 4.2 (right-hand plot). In other words, the $j_1 j_1' j_2 j_2'$ corrections amount more or less to a *scaling* of the *B*-coefficients. This fact, in turn, has the consequence that the spectral profiles associated with a certain $\lambda_1 \lambda_2 \Lambda L$ are, to a first approximation, simply scaled by a factor that depends on the rotational quantum numbers involved; whereas the *line shape* is not very strongly *j*-dependent, the *line intensities* are.

Overtone band. Transition dipole elements for overtone and hot bands can also be constructed from the data given in Table 4.10, unless high lying vibrational states are involved; in that case, dipole components must be obtained at a wider range of vibrational spacings.

The most striking fact about the dipole matrix elements for the H_2–H_2 overtone band is the large number of components, Tables 4.13 and 4.14. Besides the single vibrational transitions ($v_2 = 0 \to 2$ while $v_1 =$ const, Table 4.13, and *vice versa*), we now have to consider vibrational double transitions ($v_1 = 0 \to 1$ and $v_2 \to 1$, Table 4.14). The associated spectra appear at nearly the same frequencies. If one adds to these the various rotational bands, Eq. 4.40, a very large number of spectral components arises that must be accounted for in the computations of the overtone

Table 4.13. Analytical form (Eqs. 4.39 and 4.40) of the dipole matrix elements for the H_2–H_2 first overtone band, single vibrational transitions [284].

$\lambda_1\lambda_2\Lambda L$	n	$B(n)$ (a.u.)	$B(0)$ (10^{-6} a.u.)	a (a.u.)	b (a.u.)
$\langle 00 \vert B \vert 02\rangle$					
0001	7	−5.40	6.62	−2.818	−0.309
2023	4	−0.0597	−0.629	−2.179	−0.373
0223	4	0.0949	−15.0	−1.415	0.004
2021		0	4.20	−1.800	−0.080
0221		0	15.6	−1.298	−0.011
2233	4	0.0102	−0.0327	−3.122	−0.254
2211		0	0.0203	−3.277	−0.321
4045	6	−0.0562	−0.543	−1.755	−0.175
0445	6	−0.103	−5.23	−1.451	−0.015
2245		0	5.00	−0.898	−0.018
2243	4	−0.0101	1.35	−1.355	0.118
2223	4	0.0020	0.258	−2.088	−0.133
$\langle 0\bar{0} \vert B \vert 02\rangle$					
0001		0	−2.25	−1.615	−0.079
2023	4	0.0000605	−0.113	−1.933	−0.095
0223	4	−0.00107	−0.417	−1.515	−0.022
2021		0	0.0954	−2.106	−0.088
0221		0	0.700	−1.552	−0.031
2233	4	−0.000312	0.0287	−1.448	−0.038
2211	0	0	−0.0131	−1.521	−0.096
4045	6	−0.00243			
0445	6	−0.0140			
$\langle 00 \vert B \vert 0\bar{2}\rangle$					
0001		0	2.30	−1.610	−0.078
2023	4	0.000150	0.125	−1.890	−0.086
0223	4	0.000862	0.424	−1.518	−0.022
2021		0	−0.113	−2.115	−0.105
0221		0	−0.704	−1.548	−0.031
2233	4	0.000284	−0.0277	−1.450	−0.043
2211		0	0.0128	−1.509	−0.101
4045	6	0.00279			
0445	6	0.0140			

Table 4.14. Analytical form (Eqs. 4.39 and 4.40) of the dipole matrix elements for the H_2–H_2 first overtone band, double vibrational transitions [284].

$\lambda_1\lambda_2\Lambda L$	n	$B(n)$ (a.u.)	$B(0)$ (10^{-6} a.u.)	a (a.u.)	b (a.u.)
$\langle 00\|B\|11\rangle$					
2023	4	0.114	6.35	−1.081	−0.399
0223	4	−0.114	−6.35	−1.081	−0.399
2021		0	−18.8	−1.418	−0.084
0221	7	1.00	15.2	−1.498	−0.147
2233	4	−0.0375	4.79	−1.203	0.010
2211		0	−1.43	−1.468	−0.081
4045	6	0.360	−0.0068	−6.462	−0.959
0445	6	−0.360	0.0068	−6.462	−0.959
$\langle 0\bar{0}\|B\|11\rangle$					
0001		0	1.60	−1.628	−0.082
2023	4	0.00282	0.373	−1.486	−0.029
0223	4	−0.00210	0.0324	−2.010	0
2021		0	−0.695	−1.568	−0.026
0221		0	0.143	−1.385	−0.107
2233	4	−0.000554	0.0388	−1.454	−0.015
2211		0	−0.0197	−1.417	−0.031
4045	6	0.0112	0	0	0
0445	6	−0.00269	0.000632	−3.966	−0.318
$\langle 00\|B\|1\bar{1}\rangle$					
0001		0	−1.60	−1.628	−0.082
2023	4	−0.00279	−0.371	−1.488	−0.029
0223	4	0.00206	−0.0335	−1.995	0
2021		0	0.689	−1.570	−0.026
0221		0	−0.137	−1.384	−0.108
2233	4	0.000543	−0.0377	−1.457	−0.014
2211		0	0.0191	−1.426	−0.033
4045	6	−0.0111			
0445	6	0.00262	−0.000789	−3.784	−0.285

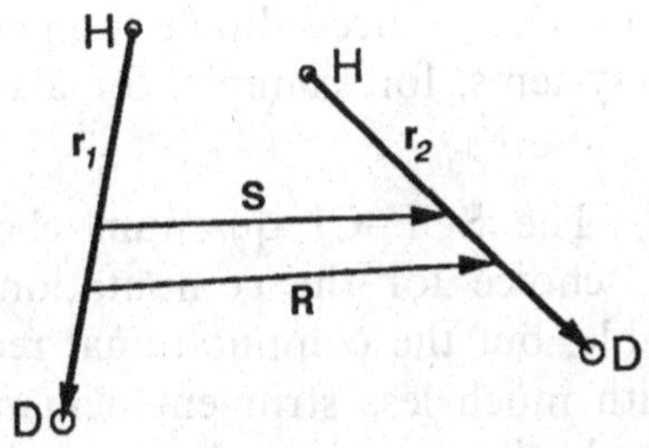

Fig. 4.4. The vector $\boldsymbol{S}$ connects the centers of charge and $\boldsymbol{R}$ the centers of mass of the HD molecules, Eq. 4.44.

spectra. (The situation gets worse for the higher overtones. The rotational state dependence must be accounted for.)

HD–X induced dipoles. The HD molecule differs from H_2 by its greater mass (which is of little concern here), the weak *permanent* dipole moment which arises from a non-adiabatic mechanism, and a center of mass which does not coincide with the center of electronic charge. Dipole moments are defined with respect to an origin that coincides with the center of mass. The presence of the permanent dipole leads to well-known rotational and rotvibrational spectra of $R_v(J)$ lines which show an interesting dispersion shape, arising from the interference of the induced and the allowed dipole spectra. For a theoretical analysis, one needs the induced dipole components of pairs like HD–X, with X = He, Ar, H_2 or HD. These have been obtained previously [59] from the *ab initio* data for the familiar isotopes summarized above, using a simple coordinate transform familiar from potential studies of the isotopes.

We distinguish the vector $\boldsymbol{S}$, which connects the centers of electronic charge of the collision partners, and the vector $\boldsymbol{R}$ which connects their centers of mass, Fig. 4.4. For H_2–H_2, the two vectors are, of course, the same, but for HD–HD these are related by

$$\boldsymbol{S} = \boldsymbol{R} + \frac{1}{6}(\boldsymbol{r}_1 - \boldsymbol{r}_2), \tag{4.44}$$

where the $\boldsymbol{r}_i$ are vibrational coordinates. The induced dipole components of the HD–HD system are then obtained from the known dipole components of the H_2–H_2 systems, according to

$$\mu_{\nu(\text{HD–HD})}(\boldsymbol{r}_1, \boldsymbol{r}_2, \boldsymbol{R}) = \mu_{\nu(\text{H}_2\text{–H}_2)}(\boldsymbol{r}_1, \boldsymbol{r}_2, \boldsymbol{S}) . \tag{4.45}$$

For HD interacting with He, Ar or H_2, a similar transformation can be written down, see [59] for details.

A much used analytical procedure to obtain the HD–X induced dipole component from the knowledge of those of H_2–X has been communicated elsewhere, see for example [323].

Other systems. Calculations of induced dipole moments have been made for a number of other systems, for example, for alkali–inert gas atomic pairs [24, 76, 218, 359].

Long-range treatments. The SCF–CI quantum chemical methods are clearly the methods of choice for the computation of induced dipole surfaces where practicable, but the computational requirements are high and approximations with much less stringent demands are of considerable interest. For molecules interacting at long range, the only quantum correction to the classical multipole model arises from dispersion; in that case perturbation theory gives useful data [193, 50, 51]. In recent years, a 'symmetry adapted perturbation theory' (SAPT) has been developed which starts from correlated monomer wavefunctions. It is believed to account for intra-atomic correlation effects not accounted for in the ordinary perturbation approach [381].

4.5 Miscellaneous remarks

Principle of corresponding states. Intermolecular potentials, like the induced dipole surfaces, are functionals of the intermolecular interactions. The signatures of electron exchange, dispersion and multipole induction are clearly exhibited in both; they have much in common.

For the studies of intermolecular potentials, the principle of corresponding states has been a useful guide. Reasonably accurate models of the equation of state have been proposed that have just two adjustable parameters (e.g., van der Waals equation). It has been argued that the success of such models suggests that all (the isotropic) intermolecular potentials should be of the same form and functions of just two parameters (such as well depth and position of the minimum). While current research does not exactly bear out this conclusion, not even if the scope is limited to rare gas interactions, it is probably fair to say that the idea of the 'principle of corresponding states' is still being tested and tried in many laboratories around the world, for various purposes and motivations.

Turning our attention to the induced dipole function, one wonders if the principle of corresponding states might have any validity for the latter. The question arises whether, or to what extent, the induced dipole function may be written as a universal function with a small number of adjustable scaling parameters. If such a function existed, it would clearly be of interest to workers in the field of collision-induced absorption.

The simplest systems of interest here are the rare gas mixtures. It had been argued that the translational spectra of the rare gas pairs, when expressed in terms of reduced frequencies, $\nu^* = \nu/\nu_{\max}$, and reduced intensities, $\alpha^* = \alpha/\alpha_{\max}$, all look roughly the same [22]: the principle of corresponding states would be satisfied if a well defined, universal spectral shape could thus be obtained – which in our opinion is not the case. With the assumption that indeed a (crude) universal line shape exists, a common reduced dipole function was constructed which was thought to be consistent with the average, normalized spectral moments (and, by implication, with the universal spectral shape) [23]. Moreover, it was noted that the shape of the reduced potential and that of the common dipole function were 'almost identical' [416], at least over the range of the spectroscopically significant separations R, from roughly 0.8 to 1 times σ, with σ being the root of the interaction potential.

In recent years, a dependable dipole function for He–Ar, last column of Table 4.3, has been obtained [278] which we compare with the 'universal' dipole function mentioned [23], Fig. 4.5. The He–Ar interaction potential is one of the better known functions [13] and suggests $R_{\min} = 6.518$ bohr. Both functions were normalized to unity at the separation $R = 5$ bohr in the figure. The comparison shows that at small separations the logarithmic slope of the most dependable dipole function is roughly one half that of the 'universal' μ^*, and μ^* diverges rapidly from $\mu(R)$ for $R \to \sigma$. Similar discrepancies have been noted for other rare gas systems (Ne–Ar, Ne–Kr, and Ar–Kr; [152]). Even if for these other systems the dipole function is not as well known as it is for He–Ar, it seems safe to say that for the rare gas mixtures mentioned the induced dipole function is definitely not identical with the 'universal' function at the distances characteristic of the spectroscopic interactions; the 'universal' dipole function is not consistent with some well established facts and data. We note that the ratio of $\mu^*(R^*)$ and the He–Ar potential is indeed reasonably constant over the range of separations considered (not shown in the figure).

The question raised above was: is the principle of corresponding states applied to induced dipole surfaces (perhaps approximately) valid? If the dipoles were well represented by an exponential function, like Eq. 4.1, the principle would of course be valid. However, it is quite clear that the dispersion part cannot be neglected in the induced dipole models. Moreover, for the best dipole models presently available, we need non-vanishing values b, Eq. 4.30, see p. 162. In other words, the induced dipole surfaces of rare gas pairs seem to require four-parameter functions, like Eq. 4.30, and it is not at all clear how these four parameters may be reduced to but two. The principle, therefore, seems at present to be of little practical value (if it were valid at all).

Dipole and force. In a study of nuclear electric dipole relaxation, Purcell pointed out that even at low gas densities, the force pulses experienced by a molecule cannot be treated as uncorrelated [328]. Instead, the correlation is such that the power spectrum of the net intermolecular force on the

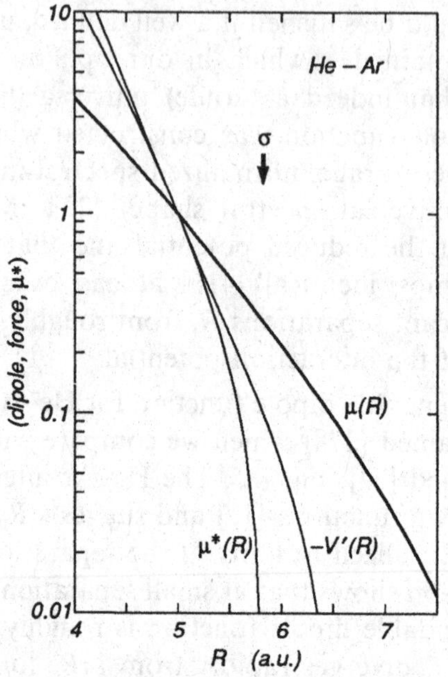

Fig. 4.5. Comparison of the dipole function ($\mu(R)$), 'universal' dipole ($\mu^*(R/R_{\min})$ with $R_{\min} = 6.518$ a.u., and force ($-V'(R)$) for He–Ar. The three functions have been normalized to unity at $R = 5$ bohr; $\sigma = 5.821$ bohr is the root of the potential.

molecule (or a group of molecules) goes to zero at zero frequency. In other words, the autocorrelation function has a negative tail equal in area to the positive peak about zero time. The characteristic time of decay of the negative tail is given by the mean time between collisions. Van Kranendonk, in the successful effort to explain the experimentally observed intercollisional dips, considered the time correlations in the intermolecular force on the molecules [404]. The exponential nature of the interaction potential at near range, and of the induced dipole function, were established at the time. Moreover, force and dipole moment are obviously parallel. With the assumption *that the range parameters of force and dipole function are equal*, van Kranendonk was able to explain the striking intercollisional dips of induced spectra for the first time.

It is clear that intermolecular force and induced dipole function arise from the same physical mechanisms, electron exchange and dispersion. Since at the time neither intermolecular potentials nor the overlap-induced dipole moments were known very dependably, direct tests of the assumptions of a proportionality of force and dipole moment were not possible. However, since the assumption was both plausible and successful, it was widely accepted, even after it was made clear that for an explanation of the intercollisional dip, this assumption is not really necessary [238]. We will here compare dependable models of force and dipole function for He–Ar pairs, Fig. 4.5, to determine if such a proportionality exists in that case. The curves presented in the figure have been normalized to unity at $R = 5$ bohr. The facts shown do not support the assumption of a proportionality. Rather, the logarithmic slopes of the dipole function appear to be between 50 and 70% of those of the force, at the spectroscopically significant separations. We mention that rare gas line shapes computed with the assumption of the dipole being proportional to the force do not at all resemble the observed spectra [152]. The induced dipole function is not proportional to the interaction force in this case, or in any other case where dependable information exists.

4.6 Ternary dipoles

The induced dipole of a gas sample is a function of the internuclear distances, r_i, the molecular orientations, Ω_i, and the positions $\boldsymbol{R}_i$ of the N molecules in the gas,

$$\boldsymbol{\mu} = \boldsymbol{\mu}\left(r_1 \cdots r_N ; \Omega_1 \cdots \Omega_N ; \boldsymbol{R}_1 \cdots \boldsymbol{R}_N\right) = \boldsymbol{\mu}\left(r^{(N)} ; \Omega^{(N)} ; \boldsymbol{R}^{(N)}\right) .$$

The dipole moment induced in a molecule, or in a group of molecules, is a finite range function of the intermolecular separations, R, which falls off faster than R^{-3} for $R \to \infty$. Van Kranendonk has argued that, therefore, it is possible to expand the above equation in a series of cluster functions [400, 402]. If $\boldsymbol{\mu}(1 \cdots n)$ designates the dipole moment induced in the cluster of molecules $1 \cdots n$ when these are present alone in the given volume V, cluster functions $\boldsymbol{U}(1 \cdots n)$ can be defined according to

$$\begin{aligned} \boldsymbol{\mu}(12) &= \boldsymbol{U}(12) \\ \boldsymbol{\mu}(123) &= \boldsymbol{U}(12) + \boldsymbol{U}(13) + \boldsymbol{U}(23) + \boldsymbol{U}(123) \\ \cdots &= \cdots \\ \boldsymbol{\mu}(1 \cdots N) &= \sum_{i<j} \boldsymbol{U}(ij) + \sum_{i<j<k} \boldsymbol{U}(ijk) + \cdots \end{aligned} \tag{4.46}$$

The $\boldsymbol{U}$ functions have the property of being zero unless all of the molecules $1 \cdots n$ are close together, within the range of the induced dipole moments. We may solve these equations for the $\boldsymbol{U}$ functions in terms of the $\boldsymbol{\mu}$ functions. When the solutions $\boldsymbol{U}$ are substituted in the right-hand sides of the above equations, an identity results which represents the cluster development of the $\boldsymbol{\mu}(1 \cdots N)$. Specifically, the $\boldsymbol{\mu}(12)$ represent the dipoles induced in pairs and $\boldsymbol{\mu}(123)$ the dipole induced in three-body complexes.

The latter is understood as the sum of the pairwise additive components,

$$\boldsymbol{\mu}(12) + \boldsymbol{\mu}(13) + \boldsymbol{\mu}(23),$$

and an 'irreducible' three-body term, $\boldsymbol{U}(123)$.

We have seen above that the pairwise-additive parts of the ternary induced dipole component are well known for a number of systems. The irreducible, ternary components, on the other hand, are poorly known, for any system. Even less is known about the N-body irreducible components of the induced dipole for $N > 3$. Attempts exist to model three- and four-body dipole components on the basis of classical relationships, but the evidence indicates that overlap effects are important; the known many-body dipole models do in general not account for quantum effects.

Empirical facts. It has been stated above (p. 58ff.) that unmixed rare gases do not absorb light. Even if the densities are as high as liquid densities, absorption was weak and could not be measured. Pairs of like atoms obviously do not possess a dipole moment, but theory indicates that triplets of like atoms may well possess one [88]. However, the measurements in unmixed gases mentioned above suggest that dipoles induced in higher complexes of like molecules (triplets, etc.) are too weak to cause measurable absorption in the monatomic gases considered.

For a number of molecular pairs (e.g., He–Ar, H_2–He, H_2–H_2, etc.) reliable pair dipole data exist. Some of these are (largely) of the overlap-induced type and others are mainly multipole-induced. The existing data have been used to compute the pairwise additive dipole components for systems like He–He–Ar, He–Ar–Ar, H_2–He–He, etc., for comparison with the few existing measurements of the three-body component of collision-induced absorption [296]. It was found consistently that for all systems considered, the pairwise-additive, ternary dipole suggested *less* absorption than the measurements indicate, especially at the higher temperatures. For unmixed hydrogen, the discrepancy between measured and calculated spectral moments is a function of temperature. At the lowest temperatures ($\approx$ 50 K) measured and calculated values are in near agreement. However, with increasing temperature, the difference between calculated and measured ternary absorption increases rapidly. At $\approx$ 300 K, these differences are significantly greater than the combined uncertainties of measurement and theory, Fig. 3.46.

These facts suggest that the excess absorption is due to irreducible, ternary dipole components. We mention that classical estimates of the long-range, irreducible component in hydrogen (where quadrupolar induction prevails) have suggested a weak *reduction* of the absorption [402]. The observed excess absorption is, therefore, believed to arise from a different, non-classical mechanism. Overlap induction seems to be the most plausible explanation, especially since the rapid increase of absorption with temperature suggests a short-range dipole component. Moreover, the scarce data that exist for He–He–Ar and He–Ar–Ar likewise suggest a *positive* irreducible component of similar magnitude [296].

In recent molecular dynamics studies attempts were made to reproduce the shapes of the intercollisional dip from reliable pair dipole models and pair potentials [301]. The shape and relative amplitude of the intercollisional dip are known to depend sensitively on the details of the intermolecular interactions, and especially on the dipole function. For a number of very dense ($\approx$ 1000 amagat) rare gas mixtures spectral profiles were obtained by molecular dynamics simulation that differed significantly from the observed dips. In particular, the computed amplitudes were never of sufficient magnitude. This fact is considered compelling evidence for the presence of irreducible many-body effects, presumably mainly of the induced dipole function.

In order to estimate the strength of the ternary, irreducible dipole components, Guillot *et al.* used the 'one effective electron model' [171, 173]. For H_2–He–He at liquid state densities, the (admittedly crude) model suggests significant enhancement of the absorption due to overlap-induced, irreducible dipole components in molecular dynamics studies, in qualitative agreement with the above conclusion.

1 Results

Long-range components. The electric dipole moment $\boldsymbol{\mu}$ ($\boldsymbol{R}_{12}$ $\boldsymbol{R}_{13}$ $\boldsymbol{R}_{23}$) of three interacting atoms lies in the plane of the triangle of the atoms and, for like atoms, vanishes for the equilateral configurations. The lowest-order, long-range components of the ternary dipole were calculated, using perturbation methods and, for the atoms bigger than hydrogen, further simplifying assumptions [164, 254, 84, 85]. Such work makes use of the dispersion theory of induced dipoles developed originally for binary dipoles [114], a multipole expansion perturbation theory. The leading terms of the long-range, asymptotic expansions are homogeneous in the tenth power of the inverse internuclear distances for the triatomic systems. The multipole expansion and perturbation treatment is a valid procedure only at sufficiently large internuclear separations. Two types of processes are found to contribute to the longest-range terms in the triatomic dipole:

i.) Two identical spherical atoms induce equal and opposite static dipole moments in each other. Even though the net dipole on the pair vanishes, the effect of the two dipoles is to induce a static dipole on the third atom.

ii.) In the second process, the members of the pair induce static quadrupole moments on each other and the quadrupoles induce a dipole on the third atom.

For dissimilar triplets, basically the same effects are expected.

For atomic pairs, the long-range induced dipole components amount to a small correction (just a few percent) of the total induced dipole; the overlap-induced component generally dominates the total dipole and not much can be learned from the long-range treatment about the overlap induction. It is possible that triatomic systems are similarly given largely by the overlap-induced parts, especially by the irreducible overlap-induced part; in that case, the long-range data are not sufficient to obtain a realistic estimate of triatomic induced dipoles. It is interesting to note that Guillot showed that in liquids, the long-range and near-range triatomic induced dipole moments are of comparable magnitude in the vicinity of the van der Waals minumum of the potential; the superposition leads to a near cancellation of the triatomic dipole component [172].

Near-range components. The dipole moments of three like atoms at near range were estimated using variational methods and a one-electron model. The results are believed to represent the qualitative features of the overlap-induced triatomic induced dipole, which is the dominant dipole at the spectroscopically important near range [1, 173]. The triatomic overlap-induced dipoles were estimated to be just slightly weaker than the binary ones [1]. The observation of very weak ternary absorption would then be a consequence of strong cancellations of long- and short-term induced dipole components [172].

When three dissimilar atoms interact at near range, the triatomic dipole moment originates from three different processes [171, 173]:

i.) A triplet overlap mechanism occurs when the electron clouds of three atoms mutually overlap;

ii.) An exchange quadrupole-induced dipole (EQID) moment occurs when an overlap-induced quadrupole of two colliding atoms induces a dipole on a nearby atom;

iii.) A dipole-induced dipole (DID) mechanism occurs when a colliding pair of dissimilar atoms induces, via its two-body overlap dipole, an induced dipole on a nearby atom.

Theoretical estimates of these were given for some rare gas mixtures elsewhere, based on the one-effective electron model of Jansen [173]. The conclusion was reached that triatomic dispersion dipoles must be added to the near-range induced dipoles just mentioned because of strong cancellations of the near- and distant-range components [172].

4.7 Asymptotic expressions

In this Section, the classical (or asymptotic) multipole expansion terms are summarized for easy reference.† Notation is based on diatom–diatom pairs; the order of the four subscripts is $\lambda_1\lambda_2\Lambda L$, as in $B_{\lambda_1\lambda_2\Lambda L}(R)$.

Monopoles. If the interacting molecules are ionized ($q_0(i) \neq 0$), we have to account for induction by monopoles. In that case, two or three B coefficients are non-vanishing, namely

$$B_{0001}(R) = (q_0(1)\alpha(2) - \alpha(1)q_0(2))\, R^{-2}, \tag{4.47}$$

$$B_{0221}(R) = -\frac{\sqrt{2}}{3}\, q_0(1)\, \beta(2)\, R^{-2}, \tag{4.48}$$

$$B_{2021}(R) = \frac{\sqrt{2}}{3}\, \beta(1)\, q_0(2)\, R^{-2}. \tag{4.49}$$

Dipoles. If the interacting molecules have a permanent dipole moment, $q_1(i) \neq 0$, then besides the permanent dipole terms, $B_{1010} = q_1(1)$ and $B_{0110} = q_1(2)$, we get the induced terms,

$$B_{1012}(R) = -\sqrt{2}\, q_1(1)\, \alpha(2)\, R^{-3} \tag{4.50}$$

$$B_{1212}(R) = \frac{1}{15}\, q_1(1)\, \beta(2)\, R^{-3} \tag{4.51}$$

$$B_{1222}(R) = \frac{1}{\sqrt{15}}\, q_1(1)\, \beta(2)\, R^{-3} \tag{4.52}$$

$$B_{1232}(R) = \frac{2\sqrt{7}}{5\sqrt{3}}\, q_1(1)\, \beta(2)\, R^{-3}. \tag{4.53}$$

For each of these four terms, there is a similar term that has all molecular labels 1,2 interchanged, namely

$$B_{0112}(R) = -\sqrt{2}\, \alpha(1)\, q_1(2)\, R^{-3} \tag{4.54}$$

$$B_{2112}(R) = \frac{1}{15}\, \beta(1)\, q_1(2)\, R^{-3} \tag{4.55}$$

$$B_{2122}(R) = -\frac{1}{\sqrt{15}}\, \beta(1)\, q_1(2)\, R^{-3} \tag{4.56}$$

$$B_{2132}(R) = \frac{2\sqrt{7}}{5\sqrt{3}}\, \beta(1)\, q_1(2)\, R^{-3}. \tag{4.57}$$

Quadrupoles. For quadrupolar induction, $q_2(i) \neq 0$, and

$$B_{2023}(R) = \sqrt{3}\, q_2(1)\, \alpha(2)\, R^{-4} \tag{4.58}$$

† Reproduced from an unpublished manuscript with the kind permission of the authors R. H. Tipping and J. D. Poll; see also [323, 391].

$$B_{0223}(R) = -\sqrt{3}\,\alpha(1)\,q_2(2)\,R^{-4} \tag{4.59}$$

$$B_{2223}(R) = \left(\frac{2}{105}\right)^{1/2} (\beta(1)\,q_2(2) - \beta(2)\,q_2(1))\,R^{-4} \tag{4.60}$$

$$B_{2233}(R) = -\left(\frac{2}{15}\right)^{1/2} (\beta(1)\,q_2(2) + \beta(2)\,q_2(1))\,R^{-4} \tag{4.61}$$

$$B_{2243}(R) = \left(\frac{18}{35}\right)^{1/2} (\beta(1)\,q_2(2) - \beta(2)\,q_2(1))\,R^{-4}. \tag{4.62}$$

Octopoles. The octopole ($q_3(1) \neq 0$) of molecule 1 induces in molecule 2 the dipole components

$$B_{3034}(R) = -2\,q_3(1)\,\alpha(2)\,R^{-5} \tag{4.63}$$

$$B_{3254}(R) = \left(\frac{88}{135}\right)^{1/2} q_3(1)\,\beta(2)\,R^{-5} \tag{4.64}$$

$$B_{3244}(R) = -\frac{1}{\sqrt{5}}\,q_3(1)\,\beta(2)\,R^{-5} \tag{4.65}$$

$$B_{3234}(R) = \frac{1}{\sqrt{27}}\,q_3(1)\,\beta(2)\,R^{-5}. \tag{4.66}$$

Similarly, if $q_3(2) \neq 0$, the octopole field of molecule 2 polarizes molecule 1 so that

$$B_{0334}(R) = -2\,\alpha(1)\,q_3(2)\,R^{-5} \tag{4.67}$$

$$B_{2354}(R) = \left(\frac{88}{135}\right)^{1/2} \beta(1)\,q_3(2)\,R^{-5} \tag{4.68}$$

$$B_{2344}(R) = \frac{1}{\sqrt{5}}\,\beta(1)\,q_3(2)\,R^{-5} \tag{4.69}$$

$$B_{2334}(R) = \frac{1}{\sqrt{27}}\,\beta(1)\,q_3(2)\,R^{-5}. \tag{4.70}$$

Hexadecapole. The hexadecapole moment ($q_4(1) \neq 0$) of molecule 1 induces the dipole components in 2, according to

$$B_{4045}(R) = \sqrt{5}\,q_4(1)\,\alpha(2)\,R^{-6} \tag{4.71}$$

$$B_{4265}(R) = -\left(\frac{26}{33}\right)^{1/2} q_4(1)\,\beta(2)\,R^{-6} \tag{4.72}$$

$$B_{4255}(R) = -\frac{2}{\sqrt{15}}\,q_4(1)\,\beta(2)\,R^{-6} \tag{4.73}$$

$$B_{4245}(R) = -\left(\frac{28}{495}\right)^{1/2} q_4(1)\,\beta(2)\,R^{-6}. \tag{4.74}$$

Similarly, if $q_4(2) \neq 0$, molecule 2 induces dipole components in 1, according to

$$B_{0445}(R) = -\sqrt{5}\,\alpha(1)\,q_4(2)\,R^{-6} \tag{4.75}$$

$$B_{2465}(R) = \left(\frac{26}{33}\right)^{1/2} \beta(1)\,q_4(2)\,R^{-6} \tag{4.76}$$

$$B_{2455}(R) = -\frac{2}{\sqrt{15}}\,\beta(1)\,q_4(2)\,R^{-6} \tag{4.77}$$

$$B_{2445}(R) = \left(\frac{28}{495}\right)^{1/2} \beta(1)\,q_4(2)\,R^{-6} \tag{4.78}$$

32-poles. If $q_5(i) \neq 0$, we get similarly

$$B_{5056}(R) = -\sqrt{6}\,q_5(1)\,\alpha(2)\,R^{-7} \tag{4.79}$$

$$B_{5256}(R) = \frac{1}{\sqrt{13}}\,q_5(1)\,\beta(2)\,R^{-7} \tag{4.80}$$

$$B_{5266}(R) = \frac{1}{\sqrt{3}}\,q_5(1)\,\beta(2)\,R^{-7} \tag{4.81}$$

$$B_{5276}(R) = \left(\frac{12}{13}\right)^{1/2} q_5(1)\,\beta(2)\,R^{-7} \tag{4.82}$$

$$B_{0556}(R) = -\sqrt{6}\,\alpha(1)\,q_5(2)\,R^{-7} \tag{4.83}$$

$$B_{2556}(R) = \frac{1}{\sqrt{13}}\,\beta(1)\,q_5(2)\,R^{-7} \tag{4.84}$$

$$B_{2566}(R) = -\frac{1}{\sqrt{3}}\,\beta(1)\,q_5(2)\,R^{-7} \tag{4.85}$$

$$B_{2576}(R) = \left(\frac{12}{13}\right)^{1/2} \beta(1)\,q_5(2)\,R^{-7} \tag{4.86}$$

Higher multipole moments have rarely been used but expressions of the various dipole components can be obtained in the same way.

As noted above, the asymptotic formulae given here, Eqs. 4.47 through 4.86, are valid for two interacting diatomic molecules, $i = 1$ and 2. For symmetric molecules like H_2, only even λ_i occur. In that case, for example, no octopoles exist. If one of the interacting partners is an atom, the associated λ_i can assume the value 0 only; this reduces the amount of computations needed significantly. Empirical dipole model components have been proposed in the past that were consistent with the asymptotic expressions above, sometimes with exponential overlap terms of the form of Eq. 4.1 added.

Besides the asymptotic terms above, a number of near-range (exponential) induced dipole components do in general exist. For some empirical models, such near-range terms were added for better fits of measured spectra.

4.8 General references

R. D. Amos. Molecular property derivatives. *Adv. Chem. Phys.*, **67**:99, 1987.

G. Birnbaum. Determination of molecular constants from collision-induced far-infrared spectra and related methods. In J. van Kranendonk, ed., *Intermolecular Spectroscopy and Dynamical Properties of Dense Systems – Proceedings of the International School of Physics 'Enrico Fermi', Course LXXV*, p. 111, 1980.

A. D. Buckingham. L'absorption des ondes micrometriques induite par la pression dans des gaz non polaires. *Colloq. Int. C.N.R.S.*, 77:57, 1959.

A. D. Buckingham. Permanent and induced molecular moments and long-range interaction. *Adv. Chem. Phys.*, **12**:107, 1967.

C. G. Gray and K. E. Gubbins, *Theory of Molecular Fluids*, vol. 1, Clarendon Press, 1984.

K. L. C. Hunt. Classical multipole models: Comparison with *ab initio* and experimental results. In G. Birnbaum, ed., *Phenomena Induced by Intermolec. Interactions*, p. 1, Plenum Press, New York, 1985.

W. Meyer. Ab initio calculations of collision-induced dipole moments. In G. Birnbaum, ed., *Phenomena Induced by Intermolec. Interactions*, p. 29, Plenum Press, New York, 1985.

D. E. Stogryn and A. P. Stogryn, Molecular Multipole Moments, *Molec. Phys.*, **11**:371, 1966.

R. H. Tipping and J. D. Poll. Multipole moments of hydrogen and its isotopes. In K. N. Rao, ed., *Molecular Spectroscopy: Modern Research*, vol. 3, ch. 7, p. 421, Academic Press, New York, 1985.

5
Theory: monatomic gas mixtures

The theory of collision-induced absorption developed by van Kranendonk and coworkers [405] and other authors [288, 289, 81, 126, 125] has emphasized spectral moments (sum formulae) of low order. These are given in closed form by relatively simple expressions which are readily evaluated. Moments can also be obtained from spectroscopic measurements by integrations over the profile so that theory and measurement may be compared. A high degree of understanding of the observations could thus be achieved at a fundamental level. Moments characterize spectral profiles in important ways. The zeroth and first moments, for example, represent in essence total intensity and mean width, the most striking parameters of a spectral profile.

While spectral moments permit significant comparisons between measurements and theory, it is clear that some information is lost if a spectroscopic measurement is reduced to just one or two numbers. Furthermore, for the determination of experimental moments, substantial extrapolations of the measured spectra to low and high frequencies are usually necessary which introduce some uncertainty, even if large parts of the spectra are known accurately. For these reasons, *line shape* computations are indispensible for detailed analyses of measured spectra, especially where the complete absorption spectra cannot be measured. Moreover, one might expect that the line shape of the induced spectra, with its 'differential' features like logarithmic slopes and curvatures and the dimer structures, depend to a greater degree on the details of the intermolecular interactions than the spectral moments. Moments are integrals, i.e., averages, of the spectral function and are thus generally less discriminating as to the subtle differences of dipole model and interaction potential, if only a small number of moments is known.

Attempts have been made to construct translational band shapes from more or less sophisticated dynamical models. When such computations are based on realistic potentials and dipole functions, substantial amounts of computer time are usually required, but in this age of computers this is not a serious limitation. Calculated spectral profiles are now routinely used for analyses of measurements and predictions of temperature variation of the absorption under conditions where reliable measurements are difficult or, perhaps, impossible.

In this Chapter, we consider the theory of collision-induced absorption by *rare gas mixtures.* We look at various theoretical efforts and compare theoretical predictions and computations with measured spectra and other experimental facts. The theory of induced absorption is based on quantum mechanics, but in certain cases, the use of classical physics may be justified, or indeed be the only viable choice. The emphasis will be on the computation of induced absorption by non-reactive, small atomic systems in the infrared. Diatomic and triatomic systems show most of the features of collisional absorption without requiring complex theory for their treatment. The theory of induced absorption of small clusters involving molecules will be considered in Chapter 6.

We start with the basic relationships ('Ansatz') of collision-induced spectra (Section 5.1). Next we consider spectral moments and their virial expansions (Section 5.2); two- and three-body moments of low order will be discussed in some detail. An analogous virial expansion of the line shape follows (Section 5.3). Quantum and classical computations of binary line shapes are presented in Sections 5.4 and 5.5, which are followed by a discussion of the symmetry of the spectral profiles (Section 5.6). Many-body effects on line shape are discussed in Sections 5.7 and 5.8, particularly the intercollisional dip. We conclude this Chapter with a brief discussion of model line shapes (Section 5.10).

5.1 Ansatz

The theory of spectral moments and line shape is based on time-dependent perturbation theory, Eqs. 2.85 and 2.86, applied to ensembles of atoms, or equivalently on the Heisenberg formalism involving dipole autocorrelation functions, Eq. 2.90.

We assume that the absorbing gas is of a uniform composition and in thermal equilibrium. The absorption coefficient, which is defined by Lambert's law, Eq. 3.1, is expressed in terms of the probabilities of transitions between the stationary states of the supermolecular system, in response to the incident radiation. Assuming the interaction of radiation and matter may be approximated by electric dipole interaction, i.e., assuming the wavelengths of the radiation are large compared with the dimensions of molecular complexes, the transition probability between the initial and final states, $|i\rangle$ and $|f\rangle$ of each molecular pair, per unit time and per unit angular frequency, is given by (see p. 49 ff.; golden rule)

$$\mathscr{P}_{f\leftarrow i} = \frac{4\pi^2}{3\hbar^2 nc}\, I(\omega)\,\Delta\omega\, \frac{1}{4\pi\varepsilon_0}\, |\langle f\,|\boldsymbol{\mu}|\,i\rangle|^2 .$$

The probability is a function of the incident energy per unit time, per unit area, $I(\omega)\,\Delta\omega$ of the incident radiation in the frequency interval between ω and $\omega+\Delta\omega$. We will also refer to $I(\omega)$ as the spectral intensity of the incident radiation. The matrix element represents the expectation value of the dipole moment operator between initial and final state, $\hbar\omega_{fi} = E_f - E_i$ is Bohr's frequency condition; it is related to the energies of the initial and final states, $|i\rangle$, $|f\rangle$, and n designates the refractive index.

The power absorbed in the frequency band $\Delta\omega$ is given by

$$-\frac{\partial^2 E_{\text{rad}}(\omega)}{\partial t\,\partial\omega}\,\Delta\omega = \sum_{i<f}{}'\,(P_i - P_f)\;\hbar\omega_{fi}\,\mathscr{P}_{f\leftarrow i}\,,$$

where the P_i, P_f are Boltzmann population factors of the states $|i\rangle$, $|f\rangle$ related by Bohr's frequency condition. The term proportional to P_i describes absorption and the one proportional to P_f stimulated emission; the prime attached to the summation sign is a reminder that the summation is restricted to those initial and final states for which the transition frequency ω_{fi} is between ω and $\omega+\Delta\omega$. The introduction of a δ function permits the formal removal of this restriction,

$$-\frac{\partial^2 E_{\text{rad}}(\omega)}{\partial t\,\partial\omega} = \sum_{i<f}(P_i - P_f)\;\hbar\omega_{fi}\,\mathscr{P}_{f\leftarrow i}\,\delta(\omega_{fi}-\omega)\,.$$

Furthermore, we may drop the subscripts of the factor ω_{fi} because of the energy conserving δ function.

The absorption coefficient $\alpha(\omega)$ is given by the ratio of $\partial^2 E_{\text{rad}}/\partial t\,\partial\omega$ and the product of incident intensity, $I(\omega)$, and the volume V of the sample (see also Eq. 2.85), that is

$$\alpha(\omega) = \frac{4\pi^2}{3\hbar^2 ncV}\,\frac{1}{4\pi\varepsilon_0}\sum_{i,f}(P_i - P_f)\;|\langle f\,|\boldsymbol{\mu}|\,i\rangle|^2\;\hbar\omega\,\delta(\omega_{fi}-\omega) \qquad (5.1)$$

for $\omega > 0$. Equation 5.1 is the starting point of most theoretical considerations. We note that the refractive index $n = n(\omega)$ which appears in the denominator is a function of frequency, but may often be considered a constant in a given frequency band. It is usually included with the Polo–Wilson factor, $L(\omega)$, which in other works concerned with liquids appears in the numerator of Eq. 5.1; that factor corrects for the local field and/or the field fluctuations important in dense fluids. $1/n$ is sometimes replaced by $(n^2+2)^2/9n$ [325]. In most theoretical studies [400] the factor $1/n$

(which will be close to unity in most cases of interest here) is suppressed in Eq. 5.1, so that theoretical results generally refer to the *internal* electric field [400]. Clarke (1978) gives a more detailed discussion of the field corrections in liquids.

We note, furthermore, that Eq. 5.1 and most of the following relationships are here expressed in the International System of Units (SI), also called the mks system; conversion to cgs units is readily accomplished by replacing the factor $1/(4\pi\varepsilon_0)$ by unity. For comparison with measurements, it is customary to use frequencies ν in wavenumber units (cm^{-1}). Conversion to these units is accomplished by replacing frequency ω (in units of radians/second) by $2\pi c\nu$, where c designates the speed of light in vacuum; furthermore, we replace $\delta(\omega)$ by the identical $\delta(\nu)/2\pi c$.

It is useful to introduce at this point the spectral density, $J(\omega)$, by rewriting Eq. 5.1 according to

$$\alpha(\omega) = \frac{4\pi^2}{3\hbar c V}\, \omega \left[1 - \exp\left(-\hbar\omega/kT\right)\right] J(\omega)\,, \tag{5.2}$$

with Eq. 2.86

$$J(\omega) = \frac{1}{4\pi\varepsilon_0} \sum_{i,f} P_i\, |\langle f\,|\boldsymbol{\mu}|\, i\rangle|^2\; \delta\left(\omega_{fi} - \omega\right)\,, \tag{5.3}$$

where we have used the relationship $P_i - P_f = P_i[1 - \exp(-\hbar\omega_{fi}/kT)]$.

We note that the collision-induced *emission* profile may be obtained from the spectral density, $J(\omega)$, according to

$$\frac{\mathrm{d}\mathscr{P}(\omega)}{\mathrm{d}\omega}\,\mathrm{d}\omega = \frac{4\omega^3}{3\hbar c^3}\, J(-\omega)\,\mathrm{d}\omega\,, \tag{5.4}$$

see p. 48 for details; the minus sign of the argument of J is a reminder that the final state energy is now lower than the initial state energy. $(\mathrm{d}\mathscr{P}(\omega)/\mathrm{d}\omega)\,\mathrm{d}\omega\,\mathrm{d}t$ is the probability for emission of a photon at times between t and $t + \mathrm{d}t$, in the frequency interval $\mathrm{d}\omega$.

Emission and absorption spectra are thus given by the same basic profile, $J(\omega)$, commonly referred to as the *spectral density*, times some factors that depend on frequency. The exponential in Eq. 5.2 accounts for stimulated emission (see pp. 48ff.). The factor ω of Eq. 5.2 is typical for absorption, just as the factor ω^3 is typical for the emission probability, Eq. 5.4, see also pp. 49ff.

Gordon (1968) pointed out that the spectral profile, Eq. 5.3, presents the Bohr–Schrödinger form of spectroscopy, as transitions between stationary Bohr states, represented by time-dependent Schrödinger states, $|i\rangle$, $|f\rangle$. The Heisenberg form of quantum mechanics gives an equivalent expression that emphasizes the time evolution of the observables rather than that of the states. This formalism leads quite naturally to time-dependent autocorrelation functions and has a clear correspondence with classical mechanics, in contrast with the Schrödinger formalism which does not at all resemble any classical picture. The Heisenberg time-dependent dipole operator is given by

$$\boldsymbol{\mu}(t) = \exp\left(\mathrm{i}\mathscr{H}_S t/\hbar\right)\, \boldsymbol{\mu}\, \exp\left(-\mathrm{i}\mathscr{H}_S t/\hbar\right) ,$$

where $\mathscr{H}_S$ is the Hamiltonian of the system. Gordon shows the equivalence of Eq. 5.3 with the expression

$$J(\omega) = \frac{1}{4\pi\varepsilon_0}\int_{-\infty}^{\infty} \mathrm{d}t\, \exp\left(-\mathrm{i}\omega t\right) \sum_i P_i \,\langle i\,|\boldsymbol{\mu}(0)\cdot\boldsymbol{\mu}(t)|\, i\rangle ,$$

which may be considered the Fourier transform (Eq. 2.90 on p. 52) of the dipole autocorrelation function,

$$C(t) = \frac{1}{4\pi\varepsilon_0}\,\langle\boldsymbol{\mu}(0)\cdot\boldsymbol{\mu}(t)\rangle . \tag{5.5}$$

The dipole moment is the total dipole of the sample, $\boldsymbol{\mu} = \sum_i \boldsymbol{\mu}_i$. The correlation function describes the response of the system to the weakly coupled radiation field. The effects of the field are modeled by the response of the individual atoms or molecules *unaffected* by the weak coupling. The Hamiltonian describes the interaction of the field and matter (first-order perturbatiuon theory). The correlation function describes how the perturbed system approaches equilibrium.

5.2 Spectral moments

Spectral moments are quantities that, on the one hand, may be obtained by integration of the measured absorption,

$$\gamma_n = \begin{cases} \int_0^\infty \alpha(\nu)\,\coth\left(hc\nu/2kT\right)\,\nu^{n-1}\,\mathrm{d}\nu & \text{for even } n \\ \int_0^\infty \alpha(\nu)\,\nu^{n-1}\,\mathrm{d}\nu & \text{for odd } n \end{cases} \tag{5.6}$$

and, on the other, may be computed for small $n = 0, 1, \ldots$ if dipole surface and interaction potential are known (Poll 1980). Spectral moments of low order have always been an important device for the comparison of measurement and theory. We mention that often, instead of the moments γ_n of the absorption coefficient, $\alpha(\nu)$, moments M_n of the spectral density, $J(\nu)$, are specified,

$$M_n = \int_{-\infty}^{\infty} J(\nu)\,\nu^n\,\mathrm{d}\nu , \tag{5.7}$$

which for rare gas mixtures are related to the former by

$$\gamma_n = \frac{4\pi^2}{3\hbar c} M_n \, . \tag{5.8}$$

Moments are averages of the spectral profile, $\alpha(\nu)$ or the spectral density, $J(\nu)$. The zeroth moment ($n = 0$) in essence gives the total intensity of the profile. The first moment ($n = 1$) may be understood as the product of mean width and total intensity; the second moment is the product of the mean squared frequency times total intensity, etc.

We note that in the definition of moments, we have used frequency ν in units of cm^{-1}. Other units of frequency are sometimes chosen; instead of the experimentalist's standard use of frequency in wavenumber units (cm^{-1}), angular frequencies $\omega = 2\pi c\nu$ are the most likely choice of theorists, which leads to different dimensions of the moments, and to the appearence of factors like powers of $2\pi c$. We note, moreover, that in the early days of collisional induction studies the zeroth moment γ_0 was defined without the hyperbolic cotangens function, Eqs. 5.6. For the vibrational bands ($\hbar\omega \gg kT$), the old and new definitions are practically identical. However, for the rototranslational band substantial differences exist. The old definition of γ_0 was never intended to be used for the far infrared [314]; only Eq. 5.6 gives total intensity in that case.

For binary systems, the spectral density $J(\nu)$ is generally replaced by the gas density normalized spectral density, $VG(\nu)$, defined by Eq. 5.60 below. Units thus differ by amagat^2 or cm^6.

1 General relationships

For a canonical ensemble, the density operator is defined as

$$P = \exp\left(-\mathscr{H}_S/kT\right) \Big/ \mathrm{Tr}\left\{\exp\left(-\mathscr{H}_S/kT\right)\right\} \, .$$

$\mathscr{H}_S$ is the Hamiltonian of the system and Tr designates the trace. For the computation of the zeroth moment, we start from Eq. 5.1 and write for the difference of Boltzmann factors

$$P_i - P_f = P_i \, \left[1 - \exp\left(-hc\nu_{fi}/kT\right)\right] \, ,$$

as is appropriate for systems in thermal equilibrium.

The expression, Eq. 5.1, is multiplied by the factor of $\alpha(\nu)$ appearing in Eq. 5.6 for $n = 0$,

$$\frac{1}{\nu} \coth \frac{hc\nu}{2kT} = \frac{1}{\nu} \frac{1 + \exp\left(-hc\nu/kT\right)}{1 - \exp\left(-hc\nu/kT\right)} \, ,$$

and integrated over frequency dν, for $0 < \nu < \infty$. We thus get

$$\begin{aligned}\gamma_0 &= \frac{4\pi^2}{3\hbar^2 cV}\frac{1}{4\pi\varepsilon_0}\int_0^\infty \mathrm{d}\nu \sum_{i,f}(P_i - P_f)\ |\langle i|\boldsymbol{\mu}|f\rangle|^2 \\ &\qquad \times \frac{hc\nu}{2\pi c}\,\delta(\nu_{fi}-\nu)\,\frac{1}{\nu}\,\frac{1+\exp(-hc\nu/kT)}{1-\exp(-hc\nu/kT)} \\ &= \frac{4\pi^2}{3\hbar cV}\frac{1}{4\pi\varepsilon_0}\int_0^\infty \mathrm{d}\nu \sum_{i,f}(P_i + P_f)\ |\langle i|\boldsymbol{\mu}|f\rangle|^2\,\delta(\nu_{fi}-\nu)\ .\end{aligned}$$

For the rare gas mixtures, $|i\rangle$ and $|f\rangle$ represent the initial and final translational states. The integration is over positive frequencies so that only the terms corresponding to $E_i < E_f$ survive,

$$\gamma_0 = \frac{4\pi^2}{3\hbar cV}\frac{1}{4\pi\varepsilon_0}\left\{\sum_{i<f} P_i\,|\langle i|\boldsymbol{\mu}|f\rangle|^2 + \sum_{i<f} P_f\,|\langle i|\boldsymbol{\mu}|f\rangle|^2\right\}\ .$$

This we rewrite in the form of an unrestricted sum,

$$\gamma_0 = \frac{4\pi^2}{3\hbar cV}\frac{1}{4\pi\varepsilon_0}\sum_{i,f} P_i\,|\langle i|\boldsymbol{\mu}|f\rangle|^2\ ,$$

by interchanging i and f in the second term.

Rewriting the square of the matrix element as $\langle i|\boldsymbol{\mu}|f\rangle\langle f|\boldsymbol{\mu}|i\rangle$ and using closure, $\sum_f |f\rangle\langle f| = 1$, we obtain

$$\gamma_0 = \frac{4\pi^2}{3\hbar cV}\frac{1}{4\pi\varepsilon_0}\sum_i P_i\ |\langle i|\boldsymbol{\mu}\cdot\boldsymbol{\mu}|i\rangle|\ , \tag{5.9}$$

a result given in the literature in various equivalent forms, involving the trace of the density matrix, or the dipole autocorrelation function, e.g.,

$$\gamma_0 = \frac{4\pi^2}{3\hbar cV}\frac{1}{4\pi\varepsilon_0}\,\mathrm{Tr}\left\{P\mu^2\right\} \tag{5.10}$$

$$\gamma_0 = \frac{4\pi^2}{3\hbar cV}\frac{1}{4\pi\varepsilon_0}\,\langle\boldsymbol{\mu}(0)\cdot\boldsymbol{\mu}(t)\rangle\Big|_{t=0}\ . \tag{5.11}$$

Equation 5.9 is in essence the ensemble average of the total dipole moment squared. It is given in a form suitable for numerical computation [315].

The computation of the spectral moment γ_1, on the other hand, begins with the integration of Eq. 5.1 over all frequencies,

$$\gamma_1 = \frac{4\pi^2}{3\hbar^2 cV}\frac{1}{4\pi\varepsilon_0}\int_0^\infty \mathrm{d}\nu \sum_{i,f}(P_i - P_f)\ |\langle f|\boldsymbol{\mu}|i\rangle|^2\ \frac{E_f - E_i}{2\pi c}\,\delta(\nu_{fi}-\nu)\ .$$

For rare gas mixtures, $hc\nu_{fi} = E_f - E_i$ is the energy difference between final and initial translational states. The limitation of the integration to positive frequencies implies $E_i < E_f$. Separating the terms P_i and P_f, we get for the integrand

$$\sum_{i<f} P_i \, |\langle i\,|\boldsymbol{\mu}|\, f\rangle|^2 \; (E_f - E_i) - \sum_{i<f} P_f \, |\langle i\,|\boldsymbol{\mu}|\, f\rangle|^2 \; (E_f - E_i) \; .$$

Making use of the symmetry in i and f of the summand and noticing that $hc\nu_{fi} = -hc\nu_{if} = E_f - E_i$, we combine the two terms into one, which assumes the form of an unrestricted sum,

$$\sum_{i,f} P_i \; |\langle i\,|\; \boldsymbol{\mu}|\, f\rangle|^2 \; (E_f - E_i) \; .$$

We thus get

$$\gamma_1 = \frac{2\pi}{3\hbar^2 c^2 V} \, \frac{1}{4\pi\varepsilon_0} \sum_{i,f} P_i \; |\langle f\,|\boldsymbol{\mu}|\, i\rangle|^2 \; (E_f - E_i) \; .$$

Replacing again the square of the dipole matrix elements by the product $\langle i|\boldsymbol{\mu}|f\rangle\langle f|\boldsymbol{\mu}|i\rangle$, the summand may be viewed as a product of the two factors $P_i\langle i|\boldsymbol{\mu}|f\rangle$ and $\langle f|\boldsymbol{\mu}|i\rangle(E_f - E_i)$. The former may be written in terms of the density operator P as $\langle Pi|\boldsymbol{\mu}|f\rangle$ and the latter as $\langle f|\,[\mathscr{H}_S, \boldsymbol{\mu}]\,|i\rangle$, where $[\mathscr{H}_S, \boldsymbol{\mu}] = \mathscr{H}_S\boldsymbol{\mu} - \boldsymbol{\mu}\mathscr{H}_S$ is the commutator of Hamiltonian $\mathscr{H}_S$ and dipole operator, $\boldsymbol{\mu}$, so that

$$\gamma_1 = \frac{2\pi}{3\hbar^2 c^2 V} \, \frac{1}{4\pi\varepsilon_0} \sum_{i,f} \langle i\,|P\boldsymbol{\mu}|\, f\rangle \, \langle f\,|[\mathscr{H}_S, \boldsymbol{\mu}]|\, i\rangle \; .$$

Making use of closure, we get

$$\gamma_1 = \frac{2\pi}{3\hbar^2 c^2 V} \, \frac{1}{4\pi\varepsilon_0} \sum_{i} \langle i\,|P\boldsymbol{\mu} \cdot [\mathscr{H}_S, \boldsymbol{\mu}]|\, i\rangle \; . \tag{5.12}$$

This expression may again be written in a variety of forms, e.g., in terms of the invariant trace, or in terms of the Heisenberg notation,

$$\gamma_1 = \frac{2\pi}{3\hbar^2 c^2 V} \, \frac{1}{4\pi\varepsilon_0} \, \mathrm{Tr}\left\{P\boldsymbol{\mu} \cdot [\mathscr{H}_S, \boldsymbol{\mu}]\right\} \tag{5.13}$$

$$\gamma_1 = \frac{2\pi}{3\hbar c^2 V} \, \frac{1}{4\pi\varepsilon_0} \, \frac{1}{i} \left\langle \boldsymbol{\mu}(0) \cdot \frac{\mathrm{d}\boldsymbol{\mu}(t)}{\mathrm{d}t} \right\rangle\Bigg|_{t=0} , \tag{5.14}$$

with $\mathrm{d}\boldsymbol{\mu}(t)/\mathrm{d}t = (\mathrm{i}/\hbar)\,\left[\mathscr{H}_S, \boldsymbol{\mu}\right]$. The angular brackets of the correlation function, Eq. 5.14, imply an ensemble average; their meaning thus differs from that of the brackets used for the transition matrix elements (as in Eq. 5.12).

We note that moments γ_n with $n > 1$ may be similarly expressed,

$$\gamma_n = \frac{8\pi^3}{3V(2\pi\hbar c)^{n+1}} \, \frac{1}{4\pi\varepsilon_0} \sum_{i,f} P_i \; |\langle f\,|\boldsymbol{\mu}|\, i\rangle|^2 \; (E_f - E_i)^n \; .$$

Using the commutator $[\mathscr{H}_S, \boldsymbol{\mu}]$ n times and repeating the other steps outlined above, one gets at once

$$\gamma_n = \frac{8\pi^3}{3V(2\pi\hbar c)^{n+1}} \frac{1}{4\pi\varepsilon_0} \operatorname{Tr}\Big\{P\boldsymbol{\mu} \cdot \underbrace{[\mathscr{H}_S, \cdots [\mathscr{H}_S, \boldsymbol{\mu}] \cdots]}_{n \text{ times}}\Big\}, \tag{5.15}$$

or

$$\gamma_n = \frac{8\pi^3}{3\hbar V} \mathrm{i}^{-n} (2\pi c)^{-n-1} \frac{1}{4\pi\varepsilon_0} \left\langle \boldsymbol{\mu}(0) \cdot \frac{\mathrm{d}^n \boldsymbol{\mu}(t)}{\mathrm{d}t^n} \right\rangle\Big|_{t=0}, \tag{5.16}$$

in the Schrödinger and Heisenberg notations, respectively.

The expressions, Eqs. 5.9 through 5.16, are quite general, wave mechanical formulae of the spectral moments, which for small order n are suitable for numerical computations; classical approximations may be derived readily from these.

2 *Virial expansion of spectral moments*

Collision-induced absorption takes place by k-body complexes of atoms, with $k = 2, 3, \ldots$ Each of the resulting spectral components may perhaps be expected to show a characteristic variation ($\sim \varrho^k$) with gas density ϱ. It is, therefore, of interest to consider virial expansions of spectral moments of binary mixtures of monatomic gases, i.e., an expansion of the observed absorption in terms of powers of gas density [314]. Van Kranendonk and associates [401, 403, 314] have argued that the virial expansion of the spectral moments is possible, because the induced dipole moments are short-ranged functions of the intermolecular separations, R, which decrease faster than R^{-3}. We label the two components of a monatomic mixture a and b, and the atoms of species a and b are labeled $1, 2, \cdots N$ and $1', 2', \cdots N'$, respectively. A set of k-body, irreducible dipole functions $\boldsymbol{U}_2, \boldsymbol{U}_3, \ldots, \boldsymbol{U}_k$, is introduced (as in Eqs. 4.46), according to

$$\begin{aligned} \boldsymbol{\mu}(1,1') &= \boldsymbol{U}_2(1,1') \\ \boldsymbol{\mu}(12,1') &= \boldsymbol{U}_2(1,1') + \boldsymbol{U}_2(2,1') + \boldsymbol{U}_3(12,1') \\ \cdots &\quad \cdots \\ \boldsymbol{\mu}(12\cdots N, 1'2'\cdots N') &= \sum_{mm'} \boldsymbol{U}_2(m,m') + \cdots \end{aligned} \tag{5.17}$$

The letter m represents the position coordinate $\boldsymbol{R}_m$ of atom m. Here, $\boldsymbol{\mu}(1,1')$ is the dipole moment induced in the dissimilar pair $1,1'$ in total isolation from other atoms, etc. The $\boldsymbol{U}$ functions have the cluster property of being zero unless all the molecules appearing in the argument are interacting at close enough range so that their spectral contributions are significant. The function $\boldsymbol{U}_2$ represents the dipole moment induced in the pair of atoms, and the function $\boldsymbol{U}_3$ represents the non-additive part of the dipole moment in a cluster of exactly three interacting atoms, etc.

The Hamiltonian $\mathscr{H}_S$ of the gas is given by

$$\mathscr{H}(1,2\cdots,1'2'\cdots) = \sum_m K(m) + \sum_{m'} K(m') + V(12\cdots,1'2'\cdots),$$

where the $K(m) = -\hbar^2\nabla_m^2/2m_m$ are the kinetic energy operators of the mth atom, and V is the potential energy of the intermolecular force. The density expansion is obtained by substituting 5.17, ff., into Eqs. 5.9 and 5.12. We define partial distribution functions by the relation [135],

$$P_\kappa(1\cdots\kappa) = V^\kappa \int\cdots\int P(1\cdots N)\,\mathrm{d}(\kappa+1)\cdots\mathrm{d}N,$$

which may be expanded as usual in terms of powers of the number densities, $n_a = N/V$ and $n_b = N'/V$, as

$$\begin{aligned} P_2(1,1') &= P_2^0(1,1') + P_2^{1a}(1,1')\,n_a + P_2^{1b}(1,1')\,n_b + \cdots \\ P_3(12,1') &= P_3^0(12,1') + \cdots \\ \cdots & \quad \cdots \end{aligned}$$

$P_2^0(1,1')$ is the low-density limit of the pair density operator $P_2(1,1')$ [135], etc. Using these definitions and Eq. 5.15, we obtain the virial expansion of the moments γ_n as

$$\gamma_n = \gamma_n^{(ab)}\; n_a n_b + \gamma_n^{(aab)}\; n_a^2 n_b + \gamma_n^{(abb)}\; n_a n_b^2 + \gamma_n^{(aaa)}\; n_a^3 + \gamma_n^{(bbb)}\; n_b^3 + \cdots \tag{5.18}$$

We note that the units of $\gamma_n^{(ab)}$ and γ_n differ by volume squared, V^2, and units of $\gamma_n^{(aab)}$, $\gamma_n^{(abb)}$, etc., differ by V^3 from those of γ_n, because the gas densities n have dimensions of reciprocal volume.

The binary moments, $\gamma_n^{(ab)}$, and the various ternary ones, namely $\gamma_n^{(aab)}$, $\gamma_n^{(abb)}$, $\gamma_n^{(aaa)}$, $\gamma_n^{(bbb)}$, may thus be computed. Poll and van Kranendonk give the first moments, γ_1, according to

$$\gamma_1^{(ab)} = \frac{2\pi V}{3\hbar^2 c^2}\,\frac{1}{4\pi\varepsilon_0}\,\mathrm{Tr}\left\{P_2^0(1,1')\boldsymbol{\mu}(1,1')\cdot[\mathscr{H}(1,1'),\boldsymbol{\mu}(1,1')]\right\} \tag{5.19}$$

$$\begin{aligned} \gamma_1^{(aab)} = {} & \frac{2\pi V}{3\hbar^2 c^2}\,\frac{1}{4\pi\varepsilon_0}\,\mathrm{Tr}\left\{P_2^{1a}(1,1')\boldsymbol{\mu}(1,1')\cdot[\mathscr{H}(1,1'),\boldsymbol{\mu}(1,1')]\right\} \\ & + \frac{2\pi V}{3\hbar^2 c^2}\,\frac{1}{4\pi\varepsilon_0}\,\mathrm{Tr}\Big\{P_3^0(12,1')\Big(\boldsymbol{\mu}(1,1')\cdot[\mathscr{H}(2,1'),\boldsymbol{\mu}(2,1')] \\ & + \boldsymbol{\mu}(1,1')\cdot[\mathscr{H}(12,1'),\boldsymbol{U}_3(12,1')] \\ & + \boldsymbol{U}_3(12,1')\cdot[\mathscr{H}(1,1'),\boldsymbol{\mu}(1,1')] \\ & + \frac{1}{2}\,\boldsymbol{U}_3(12,1')\cdot[\mathscr{H}(12,1'),\boldsymbol{U}_3(12,1')]\Big)\Big\}, \end{aligned} \tag{5.20}$$

along with similar expressions for $\gamma_1^{(abb)}$, $\gamma_1^{(aaa)}$, $\gamma_1^{(bbb)}$. For the kinetic energy of the k-body complex we write $K(1\cdots k) = \sum_{m=1\cdots k} K(m)$; $[\mathscr{H}, \boldsymbol{\mu}]$ may be replaced by $[K, \boldsymbol{\mu}]$, because potential and dipole moment commute.

The ternary moment may be thought of as being a sum of three terms,

$$\gamma_n^{(aab)} = \gamma_n^{(aab)\prime} + \gamma_n^{(aab)\prime\prime} + \gamma_n^{(aab)\prime\prime\prime} , \tag{5.21}$$

and similar for $\gamma_n^{(abb)}$. The singly primed term

$$\gamma_1^{(aab)\prime} = \frac{2\pi V^2}{3\hbar^2 c^2} \frac{1}{4\pi\varepsilon_0} \mathrm{Tr}\left\{P_2^{1a}(1,1')\boldsymbol{\mu}(1,1') \cdot [\mathscr{H}(1,1'), \boldsymbol{\mu}(1,1')]\right\}$$

arises from the density dependence of the pair distribution function (P_2); the doubly primed term

$$\gamma_1^{(aab)\prime\prime} = \frac{2\pi V^2}{3\hbar^2 c^2} \frac{1}{4\pi\varepsilon_0} \mathrm{Tr}\left\{P_3^0(12,1')\boldsymbol{\mu}(1,1') \cdot [\mathscr{H}(2,1'), \boldsymbol{\mu}(2,1')]\right\}$$

represents the interference effect between the induced dipoles $\boldsymbol{\mu}(1,1')$ and $\boldsymbol{\mu}(2,1')$; and the triply primed term

$$\begin{aligned} \gamma_1^{(aab)\prime\prime\prime} = & \frac{2\pi V^2}{3\hbar^2 c^2} \frac{1}{4\pi\varepsilon_0} \mathrm{Tr}\Big\{P_3^0(12,1')\Big(\boldsymbol{\mu}(1,1') \cdot [\mathscr{H}(12,1'), \boldsymbol{U}_3(12,1')] \\ & + \boldsymbol{U}_3(12,1') \cdot [\mathscr{H}(1,1'), \boldsymbol{\mu}(1,1')] \\ & + \frac{1}{2} \boldsymbol{U}_3(12,1') \cdot [\mathscr{H}(12,1'), \boldsymbol{U}_3(12,1')]\Big)\Big\} \end{aligned}$$

gives the spectral contribution arising from the irreducible three-body induced dipole component.

We note that the low-density limit of the density operator of the pair 1,1′ of atoms is given by

$$P_2^0(1,1') = \frac{\lambda_a^3}{V} \frac{\lambda_b^3}{V} \exp\left(-\mathscr{H}(1,1')/kT\right) ; \tag{5.22}$$

the λ are the thermal de Broglie wavelengths of atoms of the type a, b,

$$\lambda_a^2 = 2\pi\hbar^2 / (m_a kT) \quad \text{and} \quad \lambda_b^2 = 2\pi\hbar^2 / (m_b kT) . \tag{5.23}$$

The other terms, P_2^{1a}, P_2^{1b}, $P_3^{(0)}$, ... , may be similarly expressed, Eqs. 2.37 through 2.39.

For an unmixed monatomic gas, one sets $n_b = 0$ to obtain the first term of the density expansion, $\gamma_n = \gamma_n^{(3)} n^3 + \cdots$, with

$$\gamma_1^{(3)} = \frac{\pi V^2}{9\hbar^2 c^2} \frac{1}{4\pi\varepsilon_0} \mathrm{Tr}\left\{P_3^0(123)\ \boldsymbol{U}_3(123) \cdot [K(123), \boldsymbol{U}_3(123)]\right\} . \tag{5.24}$$

In this expression, $\boldsymbol{U}_3(123)$ is the dipole moment induced in a cluster of three like atoms and $\boldsymbol{\mu}(12) \equiv 0$.

The zeroth moments are given by

$$\gamma_0^{(ab)} = \frac{4\pi^2 V}{3\hbar c} \frac{1}{4\pi\varepsilon_0} \mathrm{Tr}\left\{P_2^0(1,1')\mu(1,1')^2\right\} \tag{5.25}$$

$$\gamma_0^{(aab)} = \frac{4\pi^2 V}{3\hbar c}\frac{1}{4\pi\varepsilon_0}\operatorname{Tr}\left\{P_2^{1a}(1,1')\mu(1,1')^2\right\}$$
$$+\frac{4\pi^2 V}{3\hbar c}\frac{1}{4\pi\varepsilon_0}\operatorname{Tr}\left\{P_3^0(12,1')\Big(\boldsymbol{\mu}(1,1')\cdot\boldsymbol{\mu}(2,1')\right.$$
$$+\boldsymbol{\mu}(1,1')\cdot\boldsymbol{U}_3(12,1')+\boldsymbol{U}_3(12,1')\cdot\boldsymbol{\mu}(1,1')$$
$$\left.+\frac{1}{2}\boldsymbol{U}_3(12,1')\cdot\boldsymbol{U}_3(12,1')\Big)\right\}, \quad (5.26)$$

with similar expressions for $\gamma_0^{(abb)}$, $\gamma_0^{(aaa)}$, $\gamma_0^{(bbb)}$. The ternary moments $\gamma_0^{(aab)}$, $\gamma_0^{(abb)}$, Eq. 5.26, may be readily expressed again as a sum of three terms, as in Eq. 5.21, the singly, doubly and triply primed terms arise from the density dependence of the pair distribution function, the interference of $\boldsymbol{\mu}(1,1')$ and $\boldsymbol{\mu}(2,1')$ (or $\boldsymbol{\mu}(1,2')$) and the irreducible three-body induced dipole component, respectively.

For an unmixed rare gas, we obtain the zeroth ternary moment as

$$\gamma_0^{(aaa)} = \frac{1}{6}\frac{4\pi^2 V^2}{3\hbar c}\frac{1}{4\pi\varepsilon_0}\operatorname{Tr}\left\{P_3^0(123)\ \boldsymbol{U}_3(123)^2\right\}, \quad (5.27)$$

because $\boldsymbol{\mu}(12) \equiv 0$.

3 Binary moments

Spectral moments may be computed from expressions such as Eqs. 5.15 or 5.16. Furthermore, the theory of virial expansions of the spectral moments has shown that we may consider two- and three-body systems, without regard to the actual number of atoms contained in a sample if gas densities are not too high. Near the low-density limit, if mixtures of non-polar gases well above the liquefaction point are considered, a nearly pure binary spectrum may be expected (except near zero frequencies, where the intercollisional process generates a relatively sharp absorption dip due to many-body interactions.) In this subsection, we will sketch the computations necessary for the actual evaluation of the binary moments of low order, especially Eqs. 5.19 and 5.25, along with some higher moments.

In the laboratory frame the two-atom Hamiltonian is given by

$$\mathscr{H}(1,1') = -\frac{\hbar^2}{2m_a}\nabla_1^2 - \frac{\hbar^2}{2m_b}\nabla_{1'}^2 + V\left(|\boldsymbol{R}_1 - \boldsymbol{R}_{1'}|\right). \quad (5.28)$$

The m_i designate the atomic mass of species i, and the subscript i of the ∇ symbol indicates differentiation with respect to the coordinates $\boldsymbol{R}_i$. Transformation of $\mathscr{H}(1,1')$ to center of mass and intermolecular separation coordinates,

$$\boldsymbol{R}_{CM} = \frac{1}{m_a+m_b}\left(m_a\boldsymbol{R}_1 + m_b\boldsymbol{R}_{1'}\right)$$

$$R = R_{1'} - R_1 , \tag{5.29}$$

allows the separation of these variables, $\mathscr{H}(1,1') = \mathscr{H}_{CM} + \mathscr{H}$. The resulting center of mass Hamiltonian describes the force-free motion of an unstructured point of mass $(m_a + m_b)$,

$$\mathscr{H}_{\mathrm{CM}} = -\frac{\hbar^2}{2(m_a + m_b)} \nabla^2_{\mathrm{CM}} \cdot$$

It represents the kinetic energy of the center of mass; the resulting uniform motion is of little interest here.

The Hamiltonian of relative motion, on the other hand, is given by

$$\mathscr{H} = -\frac{\hbar^2}{2m} \nabla^2 + V(R) , \tag{5.30}$$

where $m = m_a m_b/(m_a + m_b)$ is the reduced mass of the pair, the ∇ operator implies differentiation with regard to relative position, $\boldsymbol{R}$, and $V(R)$ designates the isotropic interaction potential of the dissimilar pair. The states $|i\rangle$, $|f\rangle$, of the Hamiltonian, $\mathscr{H}$, of relative motion which correspond to the initial and final states of a given spectroscopic transition, are associated with eigenenergies E_i, E_f. These eigenenergies will in general form a continuous spectrum like those of collisional systems. However, bound van der Waals molecules exist for most atomic pairs of interest and discrete eigenenergies must then be considered as well.

Wavefunctions of relative motion are obtained by introducing spherical coordinates R, ϑ, φ and the partial wave expansion, according to*

$$|i\rangle = |E_i, \ell_i, m_{\ell_i}\rangle = \frac{1}{R}\, \psi\,(R; E_i, \ell_i)\ \ Y_{\ell_i m_{\ell_i}}(\widehat{\boldsymbol{R}}) , \tag{5.31}$$

where the $Y_{\ell m_\ell}(\widehat{\boldsymbol{R}})$ designate spherical harmonics of order ℓ as usual. The radial wavefunctions ψ may be chosen to be real; these are solutions of the radial wave equation,

$$\left\{ -\frac{\hbar^2}{2m} \frac{d^2}{dR^2} + V_{\mathrm{eff}}(\ell; R) - E \right\} \psi(R; E, \ell) = 0 . \tag{5.32}$$

The sum of the potential, $V(R)$, and the centrifugal barrier $\hbar^2\ell(\ell+1)/2mR^2$ is called the effective potential,

$$V_{\mathrm{eff}}(\ell; R) = V(R) + \frac{\hbar^2 \ell(\ell + 1)}{2mR^2} , \tag{5.33}$$

see also Fig. 5.4, p. 236. Energy normalization of the wavefunctions is most convenient, Eq. 5.63.

* We have used the letter m both for reduced mass and as magnetic quantum number; a confusion seems unlikely.

Starting with Eq. 5.22 and summing over the center of mass coordinates, $\Sigma_{CM}(\lambda_{CM}^3/V)\exp(-\mathscr{H}_{CM}/kT) = 1$, we get the density operator of the states of relative motion, according to

$$P = \frac{\lambda_0^3}{V}\exp(-\mathscr{H}/kT)\,,$$

because $\lambda_a\lambda_b = \lambda_{\mathrm{CM}}\lambda_0$, where $\lambda_{CM}^2 = 2\pi\hbar^2/(m_a+m_b)kT$ and $\lambda_0^2 = 2\pi\hbar^2/(mkT)$. Since, furthermore, $\boldsymbol{\mu}(1,1') = \boldsymbol{\mu}(\boldsymbol{R})$ and $[\mathscr{H}(1,1'),\boldsymbol{\mu}(1,1')] = [\mathscr{H},\boldsymbol{\mu}(\boldsymbol{R})]$, Eq. 5.19 may be written as

$$\gamma_1^{(ab)} = \frac{2\pi N_{a_a}N_{a_b}}{3\hbar^2c^2}\lambda_0^3\frac{1}{4\pi\varepsilon_0}\,\mathrm{Tr}\left\{\exp(-\mathscr{H}/kT)\;\boldsymbol{\mu}(\boldsymbol{R})\cdot[\mathscr{H},\boldsymbol{\mu}(\boldsymbol{R})]\right\}.$$

A factor of $N_{a_a}N_{a_b}$ has been added here relative to Eq. 5.19 so that gas densities ϱ are now expressed in amagat units; instead of number densities, n_a, n_b, (as in Eq. 5.18), we now use the common expansion

$$\gamma_n = \gamma_n^{(ab)}\,\varrho_a\varrho_b + \gamma_n^{(aab)}\,\varrho_a^2\varrho_b + \gamma_n^{(abb)}\,\varrho_a\varrho_b^2 + \cdots, \tag{5.34}$$

with $\varrho = n/N_a$. N_a is the number density of a gas that corresponds to a density of 1 amagat; for most gases of practical interest, N_a is nearly the same as Loschmidt's number, $N_L = 2.686763\times10^{19}$ cm^{-3}, the density of an ideal gas at standard temperature and pressure. Subscripts a, b refer to the gases a, b, respectively. As a consequence, we now have instead of Eq. 5.8 a relationship of the moments γ_n and M_n

$$\gamma_n^{(ab)} = \frac{4\pi^2}{3\hbar c}N_{a_a}N_{a_b}M_n^{(ab)} \quad\text{and}\quad \gamma_n^{(aab)} = \frac{4\pi^2}{3\hbar c}N_{a_a}^2N_{a_b}M_n^{(aab)}\,, \tag{5.35}$$

and so forth. Writing out the trace appearing in the expression for $\gamma_1^{(ab)}$ in terms of the eigenfunctions of relative motion, Eq. 5.31, we obtain the low-density limit of the pair distribution function [397],

$$g^{(0)}(R) = \lambda_0^3\int_0^\infty \mathrm{d}E\sum_\ell \frac{2\ell+1}{4\pi}\,\mathrm{e}^{-E/kT}\,\frac{1}{R^2}\,|\psi(R;E,\ell)|^2\,. \tag{5.36}$$

The integral over energy must also include a sum over the discrete states if such states exist. The dipole moment may be written as $\boldsymbol{\mu}(\boldsymbol{R}) = \mu(R)\,\widehat{\boldsymbol{R}}$, or, in spherical components,

$$\mu_\nu(\boldsymbol{R}) = \mu(R)\,(4\pi/3)^{1/2}\;Y_{1\nu}\left(\widehat{\boldsymbol{R}}\right),$$

with $\nu = 0, \pm1$. Instead of $\mu(R)$ we may write $B(R)$, according to Eq. 4.14, which for dissimilar rare gas pairs is an identity. Next we use the commutator relations,

$$[\nabla^2, f(R)] \;=\; f'' + \frac{2}{R}f' + 2f'\frac{\partial}{\partial R}$$

$$[\nabla^2, \boldsymbol{R}] = 2\frac{\partial}{\partial \boldsymbol{R}} \cdot$$

Since potential and dipole operators commute, we may write

$$\begin{aligned}\boldsymbol{\mu} \cdot [\mathcal{H}, \boldsymbol{\mu}] &= \boldsymbol{\mu} \cdot \left[-\frac{\hbar^2}{2m} \nabla^2, \boldsymbol{\mu} \right] \\ &= -\frac{\hbar^2}{2m} \left(\boldsymbol{\mu} \cdot \boldsymbol{\mu}'' + \frac{2}{R} \boldsymbol{\mu} \cdot \boldsymbol{\mu}' - \frac{2}{R^2} \mu^2 + 2\boldsymbol{\mu} \cdot \boldsymbol{\mu}' \left(\frac{\partial}{\partial R} \right) \right) .\end{aligned}$$

The contribution arising from the last term (which is proportional to $\partial/\partial R$) may be integrated by parts so that we get finally [314]

$$\gamma_1^{(ab)} = \frac{4\pi^2 N_{a_a} N_{a_b}}{3mc^2} \frac{1}{4\pi\varepsilon_0} \int_0^\infty \left(\left(\frac{\mathrm{d}\mu}{\mathrm{d}R} \right)^2 + \frac{2}{R^2} (\mu(R))^2 \right) g^{(0)}(R)\, R^2\, \mathrm{d}R . \tag{5.37}$$

This expression may be viewed as a sum of two terms. The first one, the integral over the squared derivative of the dipole function, $\mu(R)$, models those contributions which arise from the variation of the dipole strength with the separation R. The second models the contributions due to the variation of direction, $\widehat{\boldsymbol{\mu}}$, as the rather natural separation of the kinetic energy operator into a radial and angular part suggests [314].

We note that a similar derivation for the first moment can be given in terms of Cartesian coordinates [314],

$$\gamma_1 = \frac{\pi N_{a_a} N_{a_b}}{3mc^2} \frac{1}{4\pi\varepsilon_0} \int |\nabla \boldsymbol{\mu}|^2 \; g^{(0)}(R)\, \mathrm{d}x_1\, \mathrm{d}x_2\, \mathrm{d}x_3 ,$$

where $|\nabla \boldsymbol{\mu}|^2 = \sum_{ik} (\nabla_i \mu_k)^* \; (\nabla_i \mu_k)$, $\nabla_i = \partial/\partial x_i$ and $R^2 = x_1^2 + x_2^2 + x_3^2$.

Similar manipulations are possible with the zeroth spectral moment,

$$\begin{aligned}\gamma_0^{(ab)} &= \frac{4\pi^2 \lambda_0^3}{3\hbar c} N_{a_a}\, N_{a_b} \\ &\quad \times \frac{1}{4\pi\varepsilon_0} \sum_{\ell,m} \iint \mathrm{d}E\, \mathrm{d}R\, \mathrm{e}^{-E/kT}\, (\mu(R))^2\, |\psi(R; E, \ell)|^2 \\ &= \frac{4\pi^2 N_{a_a} N_{a_b}}{3\hbar c} \frac{1}{4\pi\varepsilon_0} 4\pi \int_0^\infty (\mu(R))^2\, g^{(0)}(R)\, R^2\, \mathrm{d}R .\end{aligned} \tag{5.38}$$

Equations 5.37 and 5.38 are the desired expressions for the zeroth and first binary spectral moments.

The computation of the pair distribution function, $g^{(0)}(R)$, requires the knowledge of the interaction potential. The expressions for the zeroth and first moments require, furthermore, knowledge of the dipole function, $\boldsymbol{\mu}(R)$. For any given system (i.e., for a given reduced mass, potential and dipole function), the moments γ_n are functions of temperature, T.

Second binary moment. Expressions for the second binary moment are also known [319, 292]. We simply quote here the results [296],

$$\begin{aligned}
\gamma_2^{(ab)} &= \frac{4\pi^2}{3\hbar c m^2} \\
&\times \frac{1}{4\pi\varepsilon_0}\left[\hbar^2 \int_0^\infty g^{(0)}(R)\left\{-\frac{1}{4}\mu\mu^{IV} - \frac{1}{2}\mu^I\mu^{III} + \frac{L_1}{2R^2}\left(\mu^I\right)^2 \right.\right. \\
&\qquad \left. - \frac{L_1}{R^3}\mu\mu^I + \frac{L_1^2}{4R^2}\mu^2\right\} R^2\,\mathrm{d}R \\
&+ m\int_0^\infty g^{(0)}(R)\left\{\mu\mu^I V^I + 2\mu\mu^{II} V\right\} R^2\,\mathrm{d}R \\
&- 2m\int_0^\infty g_E(R)\,\mu\mu^{II}\,R^2\,\mathrm{d}R \\
&\left. + \int_0^\infty g_M(R)\left\{\frac{1}{R^2}\mu\mu^{II} - \frac{1}{R^3}\mu\mu^I + \frac{L_1}{2R^4}\mu^2\right\} R^2\,\mathrm{d}R\right].
\end{aligned} \tag{5.39}$$

In this expression, $L_1 = 2$ for dissimilar rare gas pairs; the superscipts $I \cdots IV$ designate the first through fourth derivatives of potential and dipole component with respect to R. The $g_E(R)$ and $g_M(R)$ are functions similar to the pair distribution function, $g^{(0)}(R)$ of Eq. 5.36, namely

$$\begin{aligned}
g_E(R) &= \lambda_0^3 \int_0^\infty \mathrm{d}E \sum_\ell \frac{2\ell+1}{4\pi}\,\mathrm{e}^{-E/kT}\,E\,\frac{1}{R^2}\,|\psi(R;E,\ell)|^2 \\
g_M(R) &= \lambda_0^3 \int_0^\infty \mathrm{d}E \sum_\ell \frac{2\ell+1}{4\pi}\,\mathrm{e}^{-E/kT}\,\frac{\hbar^2\ell(\ell+1)}{R^2}\,|\psi(R;E,\ell)|^2 .
\end{aligned}$$

We note that Eq. 5.39 has a broader validity; it may be used for molecular systems as long as isotropic interaction may be assumed. In that case, we replace L_1 by $L(L+1)$ and the $\mu(R)$ by the spherical dipole components, $B_{\lambda L}(R)$, etc., as is discussed in Chapter 6.

Third binary moment. Poll [319] has given a formula for the second and third binary moments. The third moment may be expressed as

$$\begin{aligned}
\gamma_3^{(ab)} &= \frac{6\pi\hbar kT}{m^2}\frac{N_{a_a}N_{a_b}}{(2\pi c)^3}\frac{1}{4\pi\varepsilon_0}4\pi\int_0^\infty g^{(0)}(R)\,R^2\,\mathrm{d}R \\
&\quad \times\left\{2\left(\mu^{II}\right)^2 - \mu^I\mu^{III} + \frac{L_1}{R^2}\left(2\left(\mu^I\right)^2 - \mu\mu^{II} - \frac{2}{R}\mu\mu^I\right)\right\} \\
&+ \frac{4\pi\hbar kT}{m^2}\frac{N_{a_a}N_{a_b}}{(2\pi c)^3}\frac{1}{4\pi\varepsilon_0}4\pi\int_0^\infty g_\omega(R)\,R^2\,\mathrm{d}R \\
&\quad \times\left\{\mu^I\mu^{III} - 2\left(\mu^{II}\right)^2 - \frac{1}{R}\mu^I\mu^{II}\right. \\
&\quad \left. + \frac{1}{R^2}\left[L_1\mu\mu^{II} + \left(3 - \frac{1}{2}L_1\right)\left(\mu^I\right)^2\right]\right.
\end{aligned} \tag{5.40}$$

$$
\begin{aligned}
&\quad - \frac{5L_1}{R^3}\mu\mu^I + \frac{\mu^2}{R^4}\left(\frac{3}{2}L_1^2 + L_1\right)\Bigg\} \\
&+ \frac{2\pi\hbar}{m^2}\frac{N_{a_a}N_{a_b}}{(2\pi c)^3}\frac{1}{4\pi\varepsilon_0}4\pi\int_0^\infty g^{(0)}(R)\,R^2\,\mathrm{d}R \\
&\quad \times \left\{V^{II}\left(\mu^I\right)^2 + V^I\left(\mu^I\mu^{II} + \frac{L_1}{R^2}\mu\mu^I\right)\right\} \\
&- \frac{4\pi\hbar}{m^2}\frac{N_{a_a}N_{a_b}}{(2\pi c)^3}\frac{1}{4\pi\varepsilon_0}4\pi\int_0^\infty \left(\frac{\partial g^{(0)}}{\partial\beta} + V(R)\,g^{(0)}(R)\right)R^2\,\mathrm{d}R \\
&\quad \times \left\{2\left(\mu^{II}\right)^2 - \mu^I\mu^{III} + \frac{L_1}{R^2}\left(2\left(\mu^I\right)^2 - \mu\mu^{II} - \frac{2}{R}\mu\mu^I\right)\right\} \\
&+ \frac{\pi\hbar^3}{2m^3}\frac{N_{a_a}N_{a_b}}{(2\pi c)^3}\frac{1}{4\pi\varepsilon_0}4\pi\int_0^\infty g^{(0)}(R)\,R^2\,\mathrm{d}R \\
&\quad \times \Bigg\{\left(4\left(\mu^{III}\right)^2 + 2\mu^{II}\mu^{IV} - \mu^I\mu^V\right) \\
&\quad + \frac{L_1}{R^2}\left(2\mu^I\mu^{III} + 6\left(\mu^{II}\right)^2 - \mu\mu^{IV}\right) \\
&\quad + \frac{L_1}{R^3}\left[2\mu\mu^{III} - 30\mu^I\mu^{II} + \left(\frac{4L_1}{R} + \frac{28}{R}\right)\left(\mu^I\right)^2 + \frac{18}{R}\mu\mu^{II}\right] \\
&\quad + \frac{L_1}{R^5}\left[(-12L_1 - 24)\,\mu\mu^I + \left(L_1^2 + 4L_1\right)\frac{\mu^2}{R}\right]\Bigg\} .
\end{aligned}
$$

For rare gas mixtures, L_1 may again be replaced by 2 throughout; $L_1 = L(L+1)$ in general, where L is the expansion parameter that appears in the dipole representation, Eqs. 4.15, 4.18, etc. In that case, the function $\mu(R)$ is replaced by the spherical dipole component $B_{\lambda L}(R)$, Eq. 4.17. The superscripts $I, II, \ldots V$ indicate one- through five-fold differentiation with respect to R. We have also used $\beta = 1/kT$ above for brevity. The pair distribution function, $g^{(0)}(R) = \sum_\ell g_\ell(R)$ is given by Eq. 5.36 and the function $g_\omega(R)$ is defined as

$$
g_\omega(R) = \frac{\lambda_0^2}{4\pi R^2}\sum_\ell \ell(\ell+1)\,g_\ell(R) .
$$

In the classical limit, $g_\omega(R) = g(R) = \exp(-V(R)/kT)$.

Computation. Quantum pair distribution functions have been computed for various purposes (Hirschfelder, Curtiss and Bird 1964), including for the evaluations of spectral moments [287, 315, 292]. The computer codes consist of the same subroutines required for a computation of line shape. A brief discussion of special considerations in numerical computations of the kind will be found on pp. 234 ff.

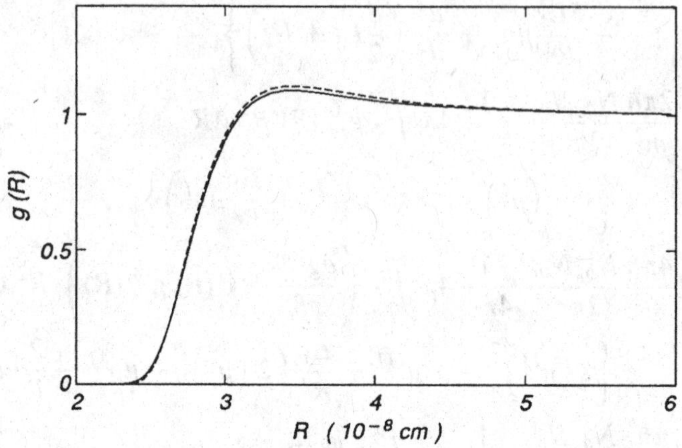

Fig. 5.1. Radial distribution function of He–Ar pairs at 295 K (low-density limit). Solid line based on Eq. 5.36; dashed line: classical approximation.

We mention the obvious fact that for a given system and temperature the number of partial waves needed is a function of the separation R for which $g(R)$ is to be evaluated. Large R require many partial waves; at large R the pair distribution function $g(R)$ is, therefore, largely classical (correspondence principle). The classical and quantal pair distribution functions $g(R)$ differ most significantly at separation less than R_m, the position of the potential minimum. Quantum computations of the pair distribution function may, therefore, be restricted to relatively small separations. For larger separations classical or semi-classical expressions may be employed which will be sketched below. In this way, the number of partial waves required for pair distribution functions need not be much greater than for line shape computations.

Figure 5.1 shows as an example the low-density limit of the pair distribution function for He–Ar pairs at 295 K. The solid line is based on Eq. 5.36 and the dashed line is the classical approximation, $g(R) = \exp(-V(R)/kT)$. The two agree closely in the example shown, but at lower temperatures or for less massive systems the classical and quantal pair distribution functions differ strikingly.

A few results of quantum moment calculations for dissimilar rare gas pairs are shown in Tables 5.1, 5.2 and 3.1, and will be discussed below. The classical fourth binary moment was also reported [79]. Table 5.1 compares quantum, semi-classical and classical calculations based on sum formulae (Eqs. 5.37, 5.38, 5.39) with moments obtained by integration of computed line shapes for the He–Ar system at 295 K.

Classical approximations. While the computation of binary, low-order spectral moments, Eqs. 5.37 and 5.38, poses no special problems, we note that nearly classical systems, i.e., massive diatomic systems at elevated temperatures, require many partial waves for an adequate theoretical description; substantial amounts of computer time are thus necessary in this case if realistic potentials are employed and a precision of the results at the 1% level is desired. For example, on mainframe computers, a computation of the spectral moments for the He–Ar system at 300 K takes a few minutes. For Kr–Xe, on the other hand, straight quantum computations take several times longer. It is, therefore, of interest to develop classical or semi-classical expressions that are adequate for massive systems at elevated temperatures. Such semi-classical formulae are also useful in supplementing quantum calculations at large separations R.

Equations 5.37 and 5.38 may be made classical by substituting the low-density limit of the *classical* pair distribution function for Eq. 5.36,

$$g(R) = \exp\left(-V(R)/kT\right) .$$

This leads to expressions of the zeroth and first spectral moments that may be computed in seconds, even on computers of small capacity. However, few systems of practical interest are actually 'classical' and better approximations are often needed.

We note that in a classical formula Planck's constant does not appear. Indeed, the zeroth moment M_0 of the spectral density, $J(\omega)$, does not depend on $\hbar$, as the combination of Eqs. 5.35 and 5.38 shows. On the other hand, the 'classical' moment γ_1 of the absorption profile, $\alpha(\omega)$, is proportional to $1/\hbar$ because the absorption coefficient α depends on Planck's constant; see the discussions of the classical line shape below, p. 246. In a discussion of classical moments it is best to focus on the moments M_n of the spectral density, $J(\omega)$, instead of the moments, γ_n, of the spectral profile.

The first moment M_1 of $J(\omega)$ vanishes in the classical limit but γ_1 is finite. It is related to M_2 in the classical limit as we will see below.

Let us address briefly the question: when does a system behave like a classical system? For an answer we consider three aspects: symmetry, correspondence principle, and detailed balance.

Symmetry. For the dissimilar rare gas pairs, the subject matter of this Section, exchange symmetry of the atoms does not exist. We will return to a discussion of symmetry in the next Chapter, when discussing binary induced spectra of like molecular pairs.

Correspondence principle. The average number of partial waves required for an adequate treatment of a given system at a given temperature is estimated by the mean angular momentum ℓ of the binary encounter, expressed in units of $\hbar$. One may say

$$\hbar \ell_{\rm av} \approx m \left(\frac{8kT}{\pi m} \right)^{1/2} b_{\rm av} ,$$

where $b_{\rm av}$ is the mean *range* of the interaction. The correspondence principle, applied to the case at hand, states that for $\ell_{\rm av} \gg 1$ the laws of classical physics are applicable. For He–Ar pairs at 300 K, we have roughly $\ell_{\rm av} \approx 25$; i.e., quantum effects due to particle dynamics are small but hardly negligible if the theoretical predictions are to be at least as accurate as good measurements. Lighter systems, e.g., H–He, and systems at lower temperatures are more quantal. We note that the spectra of van der Waals dimers are usually associated with small quantum numbers, $\ell = 0, 1, \ldots$; these are never 'classical'.

Detailed balance. Classical line shapes are symmetric so that all classical, odd spectral moments M_n of the spectral function vanish. The odd moments of actual measurements are, however, non-vanishing because measured spectral density profiles satisfy the principle of detailed balance, Eq. 5.73. This problem of classical relationships may be largely alleviated by symmetrizing the measured profile prior to determining the moments, using the inverse Egelstaff procedure (P-4) discussed on p. 254; this generates a close approximation to the classical profile from the measurement and use of classical formulae is then justified.

A fourth point which concerns the static pair distribution function was discussed above, Fig. 5.1.

Classical moment expressions. Spectral moments expressed in the Heisenberg notation can be immediately interpreted in terms of classical physics. For a discussion of classical moments, we consider the moments M_n of the spectral density, $J(\omega)$, which are related to the moments, γ_n, of the absorption coefficient, $\alpha(\omega)$, according to Eq. 5.8. By combining that equation with Eq. 5.16, we get at once

$$M_n = V\,(2\pi c)^{-n}\, \mathrm{i}^{-n}\, \frac{1}{4\pi\varepsilon_0} \left\langle \boldsymbol{\mu}(0) \cdot \frac{\mathrm{d}^n}{\mathrm{d}t^n} \boldsymbol{\mu}(t) \right\rangle \bigg|_{t=0} ,$$

where the angular brackets denote an equilibrium average. Binary moments are here considered. Instead of Eq. 5.8, Eq. 5.35 is often used which gives rise to additional factors $N_{a_a} N_{a_b}$ so that density units of amagat may be used.

We note that the equilibrium average is independent of time. Hence,

$$\left\langle \boldsymbol{\mu}(0) \cdot \frac{\mathrm{d}}{\mathrm{d}t} \boldsymbol{\mu}(t) \right\rangle \bigg|_{t=0} = \frac{\mathrm{d}}{\mathrm{d}t} \langle \boldsymbol{\mu}(0) \cdot \boldsymbol{\mu}(t) \rangle \bigg|_{t=0}$$

$$= \frac{\mathrm{d}}{\mathrm{d}t} \langle \boldsymbol{\mu}(-t) \cdot \boldsymbol{\mu}(0) \rangle \bigg|_{t=0}$$

$$= -\left\langle \frac{\mathrm{d}}{\mathrm{d}t}\boldsymbol{\mu}(t)\cdot\boldsymbol{\mu}(0)\right\rangle\Big|_{t=0},$$

because classical trajectories are symmetric in time. Repeated application yields

$$M_{2n} = (2\pi c)^{-2n}\, V\, \frac{1}{4\pi\varepsilon_0} \left\langle \left|\frac{\mathrm{d}^n}{\mathrm{d}t^n}\boldsymbol{\mu}(t)\right|^2\right\rangle\Big|_{t=0}$$

and $M_{2n+1} = 0$.

In spherical coordinates, $\boldsymbol{p} = (p_R, p_\vartheta, p_\varphi)$ and $\boldsymbol{q} = (R, \vartheta, \varphi)$, the classical Hamiltonian of relative motion has the form

$$H = \frac{p_R^2}{2m} + \frac{p_\vartheta^2}{2mR^2} + \frac{p_\varphi^2}{2mR^2 \sin\vartheta} + V(R)\,.$$

Since for rare gas pairs $\boldsymbol{\mu}(\boldsymbol{R}) = \mu(R)\widehat{\boldsymbol{R}}$, we may calculate the time derivative of $\boldsymbol{\mu}(\boldsymbol{R})$ from the Poisson bracket

$$\frac{\mathrm{d}\boldsymbol{\mu}}{\mathrm{d}t} = [\boldsymbol{\mu}, H] = \sum_j \left\{\frac{\partial \boldsymbol{\mu}}{\partial q_j}\cdot\frac{\partial H}{\partial p_j} - \frac{\partial \boldsymbol{\mu}}{\partial p_j}\cdot\frac{\partial H}{\partial q_j}\right\}$$

$$= \frac{p_R}{m}\frac{\mathrm{d}\mu}{\mathrm{d}R}\,\widehat{\boldsymbol{a}}_R + \mu(R)\frac{p_\vartheta}{mR^2}\,\widehat{\boldsymbol{a}}_\vartheta + \mu(R)\frac{p_\varphi}{mR^2\sin\vartheta}\,\widehat{\boldsymbol{a}}_\varphi\,,$$

where the $\widehat{\boldsymbol{a}}$ are the unit basis vectors of the spherical coordinate system.

The zeroth binary moment is obtained by the canonical average [79]

$$M_0 = \frac{1}{4\pi\varepsilon_0}\, 4\pi \int_0^\infty \exp\left(-\frac{V(R)}{kT}\right) (\mu(R))^2\, R^2\, \mathrm{d}R\,. \tag{5.41}$$

By expressing the time derivatives in terms of the Poisson bracket and canonically averaging, we get for the second binary moment

$$M_2 = \frac{4\pi kT}{(2\pi c)^2 m}\,\frac{1}{4\pi\varepsilon_0}\int_0^\infty \exp\left(-\frac{V(R)}{kT}\right)\left[\left(\frac{\mathrm{d}\mu}{\mathrm{d}R}\right)^2 + 2\frac{\mu^2}{R^2}\right] R^2\,\mathrm{d}R\,. \tag{5.42}$$

The second derivative of the dipole moment, which is needed for the computation of the fourth moment, is obtained by computing the Poisson bracket once more,

$$\begin{aligned}\frac{\mathrm{d}^2\boldsymbol{\mu}}{\mathrm{d}t^2} = &\left\{\left(\frac{p_R}{m}\right)^2\frac{\mathrm{d}^2\mu}{\mathrm{d}R^2} + \frac{p_\vartheta^2}{m^2R^3}\left(\frac{\mathrm{d}\mu}{\mathrm{d}R} - \frac{\mu}{R}\right)\right.\\ &\left. + \frac{p_\varphi^2}{m^2R^3\sin\vartheta}\left(\frac{\mathrm{d}\mu}{\mathrm{d}R} - \frac{\mu}{R}\right) - \frac{1}{m}\frac{\mathrm{d}V}{\mathrm{d}R}\frac{\mathrm{d}\mu}{\mathrm{d}R}\right\}\widehat{\boldsymbol{a}}_R\\ &+\left\{\frac{2p_Rp_\vartheta}{m^2R^2}\left(\frac{\mathrm{d}\mu}{\mathrm{d}R} - \frac{\mu}{R}\right)\right\}\widehat{\boldsymbol{a}}_\vartheta + \left\{\frac{2p_Rp_\varphi}{m^2R^2\sin\vartheta}\left(\frac{\mathrm{d}\mu}{\mathrm{d}R} - \frac{\mu}{R}\right)\right\}\widehat{\boldsymbol{a}}_\varphi\,.\end{aligned}$$

This expression is squared and canonically averaged, to get the fourth binary moment [79],

$$\begin{aligned} M_4 &= 4\pi \left(\frac{kT}{m}\right)^2 \left(\frac{1}{2\pi c}\right)^4 \frac{1}{4\pi\varepsilon_0} \int_0^\infty R^2\, \mathrm{d}R \exp\left(-\frac{V(R)}{kT}\right) \\ &\times \left\{ 3\left(\frac{\mathrm{d}^2\mu}{\mathrm{d}R^2}\right)^2 + \frac{16}{R^2}\left(\frac{\mathrm{d}\mu}{\mathrm{d}R} - \frac{\mu}{R}\right)^2 \right. \\ &+ \frac{4}{R}\left(\frac{\mathrm{d}^2\mu}{\mathrm{d}R^2} - \frac{1}{kT}\frac{\mathrm{d}V}{\mathrm{d}R}\frac{\mathrm{d}\mu}{\mathrm{d}R}\right)\left(\frac{\mathrm{d}\mu}{\mathrm{d}R} - \frac{\mu}{R}\right) \\ &\left. + \frac{1}{kT}\frac{\mathrm{d}V}{\mathrm{d}R}\frac{\mathrm{d}\mu}{\mathrm{d}R}\left(\frac{1}{kT}\frac{\mathrm{d}V}{\mathrm{d}R}\frac{\mathrm{d}\mu}{\mathrm{d}R} - 2\frac{\mathrm{d}^2\mu}{\mathrm{d}R^2}\right)\right\} . \end{aligned} \tag{5.43}$$

All odd moments vanish, $M_1 = M_3 = \cdots = 0$ as was mentioned above. The sixth and eighth classical moments have also been given [158].

Semi-classical approximations. In classical formulae, quantum effects may be accounted for to low order. For example, the the Wigner–Kirkwood expansion of the pair distribution function may be used [136, 302],

$$\begin{aligned} g(R) &= \exp\left(-\frac{V(R)}{kT}\right) \times \\ &\left\{1 - \frac{\hbar^2}{12m(kT)^2}\left(\nabla^2 V(R) - \frac{1}{2kT}\left(\nabla V(R)\right)^2\right) + \mathcal{O}\left(\hbar^4\right)\right\} . \end{aligned} \tag{5.44}$$

For isotropic potentials, the term in brackets may be written [177]

$$\{\cdots\} = 1 - \frac{\hbar^2}{12m(kT)^2}\left(V'' + \frac{2}{R}V' - \frac{1}{2kT}\left(V'\right)^2\right) + \mathcal{O}\left(\hbar^4\right) .$$

For a computation of the zeroth and first moments, this lowest-order Wigner–Kirkwood correction is often sufficient, but the second and higher spectral moments require additional dynamical corrections [177].

Expressions for the odd moments in terms of even moments, and for the even moments in terms of the odd ones, have been reported [177]. We expand the hyperbolic cotangent function according to

$$\coth x = \frac{1}{x} + \frac{1}{3}x - \frac{1}{45}x^3 + \frac{2}{945}x^5 \mp \cdots ,$$

with $x = hc\nu/2kT$. This we use in Eq. 5.6 to obtain the even moments

$$\gamma_{2n} = \frac{2kT}{hc}\gamma_{2n-1} + \frac{1}{3}\frac{hc}{2kT}\gamma_{2n+1} - \frac{1}{45}\left(\frac{hc}{2kT}\right)^3 \gamma_{2n+3} \pm \cdots \tag{5.45}$$

Similarly, by rewriting Eq. 5.6 for the odd moments,

$$\gamma_{2n+1} = \int_0^\infty \coth\left(\frac{hc\nu}{2kT}\right) \alpha(\nu)\, \nu^{2n-1} f(x)\, \mathrm{d}\nu$$

with $f(x) = (2kT/hc)x \tanh x$ and expanding the hyperbolic tangent in powers of x,

$$x \tanh x = x^2 - \frac{1}{3}x^4 + \frac{2}{15}x^6 - \frac{17}{315}x^8 \pm \cdots ,$$

we express the odd moments in terms of higher-order, even moments, as

$$\gamma_{2n+1} = \frac{hc}{2kT}\gamma_{2n+2} - \frac{1}{3}\left(\frac{hc}{2kT}\right)^3 \gamma_{2n+4} + \frac{2}{15}\left(\frac{hc}{2kT}\right)^5 \gamma_{2n+6} \mp \cdots \quad (5.46)$$

Presently, we are interested in the low-density limit only and superscripts (ab) may be attached to all γ in these two equations. Equations 5.45 and 5.46 may be considered expansions of the moment expressions in powers of $\hbar$; higher-order terms fall off to zero more rapidly as $\hbar \to 0$ than the leading terms. We may consider the higher-order terms to be more quantum mechanical than the low-order terms.

Indeed, by comparing Eqs. 5.42 with 5.37 and accounting for Eq. 5.35, we find at once that the classical second moment, γ_{c2}, is proportional to the first moment, γ_1,

$$\gamma_{c2}^{(ab)} = \frac{2kT}{hc}\gamma_1^{(ab)} ,$$

if the classical pair distribution is used in Eq. 5.37. Compare this result with the leading term of Eq. 5.45 for $n = 1$: $\gamma_2^{(ab)} \to \gamma_{c2}^{(ab)}$ for $\hbar \to 0$, as it must be. Similarly, we may argue that the classical fourth moment, γ_{c4} of Eq. 5.43, is in the classical limit related to the third moment,

$$\gamma_{c4} = \frac{2kT}{hc}\gamma_3 .$$

The right-hand side of this equation is finite for $\hbar \to 0$; it is also computable from Eqs. 5.43 with the help of Eq. 5.35.

Using these relationships in Eq. 5.45, we find that the second moment, corrected to lowest order for quantum effects, is given by [177],

$$\gamma_2 \approx \frac{2kT}{hc}\gamma_1 + \left(\frac{hc}{2kT}\right)^2 \gamma_{c4} + \mathcal{O}(\hbar^4) .$$

In this expression, γ_1 from Eq. 5.37 should be used with the semi-classical pair distribution function, from Eq. 5.44. We thus get [177]

$$\begin{aligned}\gamma_2 \quad \approx \quad & \frac{4\pi kT N_{a_a} N_{a_b}}{(2\pi c)^2 m} \frac{1}{4\pi\varepsilon_0} \int_0^\infty \left[\left(\frac{\mathrm{d}\mu}{\mathrm{d}R}\right)^2 + \frac{2\mu^2}{R^2}\right] \exp\left(-\frac{V(R)}{kT}\right) R^2 \,\mathrm{d}R \\ & + \frac{\pi\hbar^2 N_{a_a} N_{a_b}}{3(2\pi c)^2 m^2} \frac{1}{4\pi\varepsilon_0} \int_0^\infty \exp\left(-\frac{V(R)}{kT}\right) \left\{3\left(\frac{\mathrm{d}^2\mu}{\mathrm{d}R^2}\right)^2 + \frac{20}{R^2}\left(\frac{\mathrm{d}\mu}{\mathrm{d}R} - \frac{\mu}{R}\right)^2\right.\end{aligned}$$

$$-\frac{4}{R}\frac{\mathrm{d}^2\mu}{\mathrm{d}R^2}\left(\frac{\mathrm{d}\mu}{\mathrm{d}R}-\frac{\mu}{R}\right)+\left(\frac{1}{kT}\frac{\mathrm{d}V}{\mathrm{d}R}\right)^2\left[\frac{1}{2}\left(\frac{\mathrm{d}\mu}{\mathrm{d}R}\right)^2-\frac{\mu^2}{R^2}\right]\Bigg\} R^2\,\mathrm{d}R$$
$$+\mathcal{O}(\hbar^4)\cdots \tag{5.47}$$

This is the semi-classical, second, binary moment. While the zeroth and first moments require only static quantum corrections of the Wigner–Kirkwood type, the second and all higher moments require also dynamical corrections involving γ_{c4} and higher moments.

Other Wigner–Kirkwood corrections. The functions $g_E(R)$ and $g_M(R)$ introduced above for the computation of the second moment, Eq. 5.39, have also been given in semi-classical form [292]. Since these are very useful to supplement quantum moment calculations at large separations R, we will simply quote here the results,

$$g_E(R) \approx \exp\left(-\frac{V(R)}{kT}\right)\Bigg\{\frac{3}{2}kT+V(R) \tag{5.48}$$
$$+\frac{\hbar^2}{24m(kT)^2}\Bigg[kT\frac{\mathrm{d}^2V}{\mathrm{d}R^2}+\frac{2kT}{R}\frac{\mathrm{d}V}{\mathrm{d}R}-\frac{3}{2}\left(\frac{\mathrm{d}V}{\mathrm{d}R}\right)^2$$
$$-2V(R)\frac{\mathrm{d}^2V}{\mathrm{d}R^2}-\frac{4V}{R}\frac{\mathrm{d}V}{\mathrm{d}R}+\frac{V}{kT}\left(\frac{\mathrm{d}V}{\mathrm{d}R}\right)^2\Bigg]\Bigg\}$$
$$g_M(R) \approx \exp\left(-\frac{V(R)}{kT}\right) 2mR^2kT$$
$$\times\left\{1+\frac{\hbar^2}{24m(kT)^2}\left[-2\frac{\mathrm{d}^2V}{\mathrm{d}R^2}-\frac{8}{R}\frac{\mathrm{d}V}{\mathrm{d}R}+\frac{1}{kT}\left(\frac{\mathrm{d}V}{\mathrm{d}R}\right)^2+\frac{18kT}{R^2}\right]\right\}.$$

Computations. The semi-classical binary moment expressions are most useful. Their computation requires use of computers but is straightforward. The numerical computations of the derivatives of the functions $\mu(R)$ and $V(R)$ might be a slight problem; derivatives are required to various orders. Especially the more highly refined, semi-empirical potential models are often not differentiable at one or more R values, or they may suffer from a small discontinuity (due to rounding-off of numerical constants) at separations where the different analytical representations are pieced together. These may be of no significance in most of the standard applications, but numerical differentiation at such mending points requires special caution. In order to avoid all problems related to numerical differentiation and still be able to use the most highly developed interaction potentials available without modifications, for each point R for which derivatives are needed, we have resorted to representing the functions $V(R)$, $\mu(R)$, by least mean squares fitted polynomials $\sum_{j=0}^{m} a_j(x-R)^j$ of mth order in the close vicinity

Table 5.1. Comparison of binary spectral moments calculated from classical (C.), semi-classical (S.) and quantum (Q.) calculations, based on line shapes (.LS) and sum formulae (.SF), for He–Ar at 295 K. Moments computed from the classical line shape after desymmetrization procedures P-2 and P-4 (scaled) had been applied are also shown. Computations are based on the *ab initio* dipole, Table 4.3, and an advanced potential [12].

moment:	0th	1st	2nd	3rd	4th	5th	6th
	$(10^{-64+14n}$ erg cm^6 s$^{-n})$						
Q.LS	220.4	15.55	14.03	5.14	5.93	5.58	9.06
Q.SF	217.5	15.25	13.93				
S.SF	216.8	15.28	13.90				
P-2	220.0	15.16	13.54	4.60	5.06	4.16	6.11
P-4s	223.3	15.55	14.40				
C.LS	213.7	0	11.71	0	3.55	0	3.21
C.SF	215.3	0	11.77	0	3.55	0	

of R, and compute the derivatives from the coefficients, $a_j = f^j/j!$. With this numerical smoothing procedure, no problems were encountered even for the highest derivatives that appear above in the moment expressions and Wigner–Kirkwood corrections.

Results. In Table 5.1 we compare a few results of classical, semi-classical and quantum moment calculations. An accurate *ab initio* dipole surface of He–Ar is employed (from Table 4.3 [278]), along with a refined model of the interaction potential [12]. A temperature of 295 K is assumed. The second line, Table 5.1, gives the lowest three quantum moments, computed from Eqs. 5.37, 5.38, 5.39; the numerical precision is believed to be at the 1% level. For comparison, the third line shows the same three moments, obtained from semi-classical formulae, Eqs. 5.47 along with 5.37 with the semi-classical pair distribution function inserted. We find satisfactory agreement. We note that at much lower temperatures, and also for less massive systems, the semi-classical and quantal results have often been found to differ significantly. The agreement seen in Table 5.1 is good because He–Ar at 295 K is a near-classical system.

The last line, Table 5.1, reports the purely classical moments. The zeroth classical moment is a little smaller than the zeroth quantum moment, because of the wave mechanical 'tunneling' of the collisional pair into the classically forbidden region which enhances the intensities. All odd moments of classical profiles are, of course, zero. The second and fourth moments are significantly smaller than the quantum moments, because of dynamical quantum corrections which are significant even for this 'near-classical' system, see p. 218.

Also given in the Table are moments obtained by integration of computed line shapes; results based on a quantum line shape are given in line 1 and a purely classical line shape was used for the computation of data presented in the last line of the Table. We will discuss the line shape calculations below and mention here only that moment calculations are indispensible tests of line shape calculations. Moments, and especially the higher-order moments, are readily obtainable from line shape calculations. Not surprisingly, higher-order moments require strong dynamic quantum corrections, even for near-classical systems like He–Ar at 295 K. For example, the sixth quantum moment, Table 5.1, is nearly three times larger than the corresponding classical moment. Other moments obtained from semi-classical line shape computations will be considered below.

Computed moments have been compared with measurements in Table 3.1, p. 65. The agreement is generally satisfactory.

4 Ternary moments

The coefficients $\gamma_n^{(ab)}$, $\gamma_n^{(aab)}$, $\gamma_n^{(abb)}$ of the virial expansion, Eq. 5.34, are independent of density. Since we neglect the irreducible three-body contributions to the dipole moment here, terms proportional to the third power of ϱ_a and ϱ_b are missing.

Comparing Eq. 5.34 with the general results above, Eqs. 5.21, 5.26 and 5.38 [318], for the zeroth moment we see at once [314, 165]

$$\begin{aligned} \gamma_0^{(aab)\prime} &= \frac{4\pi^2}{3\hbar c} \frac{N_{a_a}^2 N_{a_b}}{4\pi\varepsilon_0} 4\pi \int_0^\infty R^2 \, \mathrm{d}R \, \mu(R)^2 \, g^{(a)}(R) \,, \qquad (5.49) \\ \gamma_0^{(aab)\prime\prime} &= \frac{4\pi^2}{3\hbar c} \frac{N_{a_a}^2 N_{a_b}}{4\pi\varepsilon_0} \\ &\quad \times \int \mathrm{d}^3R \, \mathrm{d}^3R' \, \boldsymbol{\mu}(\boldsymbol{R}) \cdot \boldsymbol{\mu}(\boldsymbol{R}') \, g_0^{(aab)}(\boldsymbol{R}, \boldsymbol{R}') \,, \end{aligned}$$

and similar for $\gamma_0^{(abb)}$. The irreducible components $\gamma_0^{(aab)\prime\prime\prime}$ and $\gamma_0^{(abb)\prime\prime\prime}$ are poorly known and can at present not reliably be obtained from theory; the irreducible components are here suppressed. The $\boldsymbol{R}$ and $\boldsymbol{R}'$ are distances between dissimilar atoms, $\boldsymbol{\mu}$ is the pair dipole moment, $g^{(a)}$ is the coefficient of the density ϱ_a of the virial expansion of the pair distribution function, $g^{(0)}$. $g_0^{(aab)}$ is the zero density limit of the triplets distribution function.

The last expression is rewritten, making use of $\boldsymbol{\mu} \cdot \boldsymbol{\mu}' = \sum_\nu \mu_\nu^* \cdot \mu_\nu'$,

$$\gamma_0^{(aab)\prime\prime} = \frac{4\pi^2}{3\hbar c} \frac{N_{a_a}^2 N_{a_b}}{4\pi\varepsilon_0} 8\pi^2 \int_0^\infty R^2 \, \mathrm{d}R \int_0^\infty R'^2 \, \mathrm{d}R' \int_0^\pi \sin\vartheta \, \mathrm{d}\vartheta$$

$$\times \mu(R)\, \mu(R')\, P_1(\cos\vartheta)\, g_0^{(aab)}(R, R', \cos\vartheta)\,. \tag{5.50}$$

We have used the relationship

$$\sum_\nu Y_{1\nu}^*(\widehat{\boldsymbol{R}})\, Y_{1\nu}(\widehat{\boldsymbol{R}}') = \frac{3}{4\pi} P_1(\cos\vartheta)\,;$$

P_1 is the Legendre polynomial of order 1. We have written $g_0^{(1a)}$ in terms of R, R' and $\cos\vartheta$, where ϑ is the angle subtended by $\boldsymbol{R}$ and $\boldsymbol{R}'$.

For the first moment, we find similarly [318]

$$\gamma_1^{(aab)\prime} = \frac{\pi}{3m_{ab}c^2} \frac{N_{a_a}^2 N_{a_b}}{4\pi\varepsilon_0} \int d^3R\, (\nabla\boldsymbol{\mu}(\boldsymbol{R})) : (\nabla\boldsymbol{\mu}(\boldsymbol{R}))\, g^{(a)}(R)\,,$$

$$\gamma_1^{(aab)\prime\prime} = \frac{\pi}{3m_b c^2} \frac{N_{a_a}^2 N_{a_b}}{4\pi\varepsilon_0} \times \int d^3R\, d^3R'\, (\nabla\boldsymbol{\mu}(\boldsymbol{R})) : (\nabla\boldsymbol{\mu}(\boldsymbol{R}'))\, g_0^{(aab)}(\boldsymbol{R}, \boldsymbol{R}')\,,$$

where m_b is the mass of atom type b and m_{ab} the reduced mass of the pair a, b. (If good approximations of the irreducible three-body dipole component were known, one could also evaluate the triply primed moments in similar fashion.) With

$$(\nabla\boldsymbol{\mu}) : (\nabla\boldsymbol{\mu})' = \sum_{\alpha\nu} (\nabla_\alpha \mu_\nu)^* (\nabla_\alpha \mu_\nu)'\,,$$

where ∇_α is a spherical component of the gradient operator, and with the use of the gradient formulae [80], we obtain

$$\gamma_1^{(aab)\prime} = \frac{\pi}{3m_{ab}c^2} \frac{N_{a_a}^2 N_{a_b}}{4\pi\varepsilon_0} \times 4\pi \int_0^\infty R^2\, dR \left[\left(\frac{d\mu(R)}{dR} \right)^2 + \frac{2}{R^2} (\mu(R))^2 \right] g_1^{(a)}(R)\,, \tag{5.51}$$

$$\begin{aligned} \gamma_1^{(aab)\prime\prime} = {} & \frac{\pi}{3m_b c^2} \frac{N_{a_a}^2 N_{a_b}}{4\pi\varepsilon_0} \\ & \times 8\pi^2 \int_0^\infty R^2\, dR \int_0^\infty R'^2\, dR' \int_0^\pi \sin\vartheta\, d\vartheta\, g_0^{(aab)}(R, R', \cos\vartheta) \\ & \times \left\{ \frac{2}{3} \left[\frac{d\mu(R)}{dR} - \frac{\mu(R)}{R} \right] \left[\frac{d\mu(R')}{dR'} - \frac{\mu(R')}{R'} \right] P_2(\cos\vartheta) \right. \\ & \left. + \frac{1}{3} \left[\frac{d\mu(R)}{dR} + 2\frac{\mu(R)}{R} \right] \left[\frac{d\mu(R')}{dR'} + 2\frac{\mu(R')}{R'} \right] P_0(\cos\vartheta) \right\}, \end{aligned} \tag{5.52}$$

and similar for $\gamma_1^{(abb)\prime}$ and $\gamma_1^{(abb)\prime\prime}$.

Ternary moments consist of an 'intermolecular force' or 'finite volume' term (the singly primed terms), and an 'interference' or 'cancellation' term (the doubly primed terms) [400, 401, 402]. As was mentioned above, a third term arising from the non-additive ternary dipole components should also be added which we have, however, suppressed here for lack of detailed information.

Classical expressions. The expressions for the ternary moments given above are rigorous. However, exact quantum calculations do not exist and practical calculations are best based on the semi-classical formulae which we will briefly consider here.

The semiclassical distribution functions $g_1^{(a)}$, $g_1^{(b)}$, $g_0^{(aab)}$, $g_0^{(abb)}$, are given by Eqs. 2.37 and 2.38 on p. 38 ff. With the help of computers, these expressions are easily evaluated. Some caution is advised when numerical differentiation is employed; see the remarks on p. 218.

Results. The theory of ternary processes in collision-induced absorption was pioneered by van Kranendonk [402, 400]. He has pointed out the strong cancellations of the contributions arising from the density-dependent part of the pair distribution function (the 'intermolecular force effect') and the destructive interference effect of three-body complexes ('cancellation effect') that leads to a certain feebleness of the theoretical estimates of ternary effects.

Early numerical estimates of ternary moments [402] were based on the empirical 'exp–4' induced dipole model typical of collision-induced absorption in the fundamental band, which we will consider in Chapter 6, and hard-sphere interaction potentials. While the main conclusions are at least qualitatively supported by more detailed calculations, significant quantitative differences are observed that are related to three improvements that have been possible in recent work [296]: improved interaction potentials; the quantum corrections of the distribution functions; and new, accurate induced dipole functions. The force effect is by no means always positive, nor is it always stronger than the cancellation effect.

Formulae for the classical zeroth and second moments of collision-induced absorption of the rare gases [148, 165, 333] have been reported; the classical second moment γ_2 is related to the first quantum moment, M_1 [317, 177],

$$\gamma_2 \approx \frac{2kT}{\hbar} M_1$$

for classical systems; quantum corrections were not given [148]. Among others, moments were computed for the He–Ar–Ar and He–He–Ar complexes using a selection of widely different empirical [177, 79] and theoretical [256] dipole moment functions and potentials, including Lennard-Jones 6–12 models. The dipole models (as well as the Lennard-Jones interaction potentials [248]) must now be considered obsolete; see Chapters 2 and 4 [277, 39, 193]. For the better dipole and potential models, the results are similar to the classical ones shown below for the highest temperature.

Helium–argon mixtures. For the He–Ar pair, an accurate *ab initio* induced dipole surface exists, Table 4.3 which, with the help of line shape calculations, was shown to reproduce the binary collision-induced absorption spectra within the accuracy of the measurement [278]. For the ternary moments, the 'SPFD2' He–Ar [12] and the 'HFD-C' Ar–Ar [11] interaction potentials were input, along with this *ab initio* dipole surface.

Table 5.2 shows the results at four selected temperatures. For each temperature, two lines of results are given: the upper line shows the quantum-corrected results and the lower, starred lines are computed from classical distribution functions without Wigner–Kirkwood corrections. For the relatively light He–Ar system at low temperatures, quantum corrections are generally significant. Whereas quantum corrections of the binary moments amount to just a few percent, some ternary moments show remarkably strong quantum corrections, from 10% to over 100%. We note that the correction may even change the sign of M_n'.

The Wigner–Kirkwood corrected results (starred lines in the Table) for the lowest temperature may be uncertain by some fraction of the quantum correction and should be considered rough estimates.

The binary moments are positive and increasing with temperature as Table 5.2 shows (second column). The third and fourth columns specify the contributions due to He–Ar–Ar complexes and the fifth and sixth columns give those of the He–He–Ar complexes. We note generally strong cancellations of the singly and doubly primed components, especially for the zeroth moments, but the M_0' arising from the density-dependent part of the pair distribution function generally dominate unless the temperature is near a zero of the function $M_n(T)$. A change of sign happens twice, once at temperatures between 45 and 85 K and again at temperatures between 85 and 165 K. The doubly primed three-body moments show lesser variations with increasing temperature; they are generally monotonic functions of temperature. A comparison of the binary and ternary coefficients shows that the computed density effects are relatively strong at the low temperatures, especially for the zeroth moment.

Comparison of ternary moments with measurements. The density dependence of the helium–argon collision-induced absorption spectra has been studied at the temperature of 165 K, helium densities from 66 to 130 amagats, and argon densities from 156 to 280 amagats. Ternary moments of

$$\gamma_1^{(\mathrm{HeArAr})} = 1.0 \qquad \text{and} \qquad \gamma_1^{(\mathrm{HeHeAr})} = 0.7$$

were measured [95], in units of 10^{-7} cm^{-2}amagat^{-3}. In Table 5.2 we find

Table 5.2. Various computed binary and ternary moments M_n, with and without Wigner–Kirkwood corrections, for helium–argon mixtures at various temperatures. Units of M_0 and M_1 are 10^{-33} J amagat^{-N} and 10^{-20} W amagat^{-N}, where $N = 2$ and 3 for binary and ternary moments, respectively. The asterisk indicates that Wigner–Kirkwood corrections have not been made to the entries on that line [296].

T (K)	$M_0^{(\mathrm{HeAr})}$	$M_0^{(\mathrm{HeArAr})\prime}$	$M_0^{(\mathrm{HeArAr})\prime\prime}$	$M_0^{(\mathrm{HeHeAr})\prime}$	$M_0^{(\mathrm{HeHeAr})\prime\prime}$
45	550	1.220	0.573	0.012	−0.162
*45	520	1.035	0.745	0.013	−0.186
85	680	−0.538	−0.154	0.154	−0.172
*85	658	−0.704	−0.190	0.132	−0.182
165	1,020	0.085	−0.372	0.325	−0.227
*165	1,000	0.066	−0.400	0.306	−0.233
298	1,580	0.818	−0.590	0.535	−0.322
*298	1,570	0.811	−0.611	0.520	−0.326

T (K)	$M_1^{(\mathrm{HeAr})}$	$M_1^{(\mathrm{HeArAr})\prime}$	$M_1^{(\mathrm{HeArAr})\prime\prime}$	$M_1^{(\mathrm{HeHeAr})\prime}$	$M_1^{(\mathrm{HeHeAr})\prime\prime}$
45	437	1.010	−0.114	0.005	0.016
*45	430	0.878	−0.067	−0.014	0.022
85	523	-0.422	0.043	0.113	0.016
*85	515	-0.562	0.061	0.098	0.018
165	747	0.050	0.097	0.232	0.019
*165	742	0.036	0.107	0.220	0.020
298	1,110	0.557	0.147	0.367	0.026
*298	1,110	0.555	0.155	0.358	0.026

the values calculated from Eq. 5.8

$$\gamma_1^{(\mathrm{HeArAr})} = 0.33 \qquad \text{and} \qquad \gamma_1^{(\mathrm{HeHeAr})} = 0.56\,,$$

in the same units. Experimental uncertainties were not specified [95], but the possible errors of a virial analysis of this kind may be rather high. In view of presumably large uncertainties, the computed and measured ternary moments may actually be consistent; they are certainly of comparable magnitude and have the correct sign. However, we point out that the measured ternary moments both seem to actually *exceed the computed values* by some fraction of 10^{-7} cm^{-2}amagats^{-2}.

If this were true – and evidence to be presented for molecular systems in Chapter 6 below seems to corroborate this – the observed excess of ternary absorption suggests the presence of a *positive*, additional component not accounted for in the computation. That component may arise from the irreducible three-body dipoles. Since the classical (i.e., long-range, electrodynamic) irreducible dipole component suggests a weak, *negative* spectral comonent, the much stronger observed positive component likely arises from irreducible exchange contributions [296]. More measurements of this kind, at a variety of temperatures, are most desirable. *Ab initio* calculations of irreducible dipole moments of dissimilar atom pairs would also help to shed new light on the situation.

Relationship with the intercollisional dip. The cancellation effect described by the doubly primed spectral moments $\gamma_n^{(aab)''}$, $\gamma_n^{(abb)''}$, is of course related to the intercollisional interference process observed near zero frequency, Fig. 3.5. The important difference is that the spectral moments are ternary quantities by design while the intercollisional dip is affected by many-body processes.

5.3 Virial expansion of line shape

For some time it has been known that the spectral moments, which are *static* properties of the absorption spectra, may be written as a virial expansion in powers of density, ϱ^n, so that the nth virial coefficient describes the n-body contributions ($n = 2, 3 \ldots$) [400]. That *dynamical* properties like the spectral density, $J(\omega)$, may also be expanded in terms of powers of density has been tacitly assumed by a number of authors who have reported low-density absorption spectra as a sum of two components proportional to ϱ^2 and ϱ^3, respectively [100, 99, 140]. It has recently been shown by Moraldi (1990) that the spectral components proportional to ϱ^2 and ϱ^3 may indeed be related to the two- and three-body dynamical processes, *provided* a condition on time is satisfied [318, 297]. The proof resorts to an extension of the static pair and triplet distribution functions to describe the time evolution of the initial configurations; these permit an expansion in terms of powers of density that is analogous to that of the static distribution functions [135].

It was recently shown that a formal density expansion of space–time correlation functions of quantum mechanical many-body systems is possible in very general terms [297]. The formalism may be applied to collision-induced absorption to obtain the virial expansions of the dipole autocorrelation function and its Fourier transform, the spectral profile. While the formalism permits one to evaluate the expressions to arbitrary high order, we will write down here only the coefficients of the density squared and cubed terms, i.e., the two- and three-body expressions.

The virial expansion of the time correlation functions is possible for times smaller than the mean time τ between collisions. Accordingly, the spectral profiles may be expanded in powers of density, for angular frequencies much greater than the reciprocal mean time between collisions, $\omega \gg 1/\tau$. Since at low density the mean time between collisions is inversely proportional to density, lower densities permit a meaningful virial expansion for a greater portion of the spectral profiles.

The three-body spectra and their associated correlation functions may be considered to be a superposition of three components of different nature. One part arises from two-body dynamics where the third atom acts strictly as a perturbing field. The second part represents the contributions of the irreducible three-body dynamics to the pairwise-additive induction. The third part is due to the three-body induction mechanism and contains the irreducible dipole. These agents vary differently with temperature and could in principle be separated on that basis.

For mixed gases A and B, we define the functions $G^{(n)}(\boldsymbol{R}^{\{n\}}, \boldsymbol{R}'^{\{n\}}; t)$ representing the combined probability of having the configuration $\boldsymbol{R}^{\{n\}}$ at some time t_0, and $\boldsymbol{R}'^{\{n\}}$ at a later time, $t_0 + t$. We note that t_0 is not an argument of the function G, because under equilibrium conditions this function is independent of t_0 (Moraldi 1990),

$$
\begin{aligned}
G^{(nm)} &\left(\boldsymbol{R}_A^{\{n\}}; \boldsymbol{R}_B^{\{m\}}; \boldsymbol{R}_A'^{\{n\}}; \boldsymbol{R}_B'^{\{m\}}; t\right) = \\
&= \sum \Big\langle \delta\left(\boldsymbol{R}_{A1} - \boldsymbol{q}_{Ai_1}\right) \cdots \delta\left(\boldsymbol{R}_{An} - \boldsymbol{q}_{Ai_n}\right) \\
&\quad \times \delta\left(\boldsymbol{R}_{B1} - \boldsymbol{q}_{Bj_1}\right) \cdots \delta\left(\boldsymbol{R}_{Bm} - \boldsymbol{q}_{Bj_m}\right) \\
&\quad \times \delta\left(\boldsymbol{R}'_{A1} - \boldsymbol{q}_{Ak_1}(t)\right) \cdots \delta\left(\boldsymbol{R}'_{An} - \boldsymbol{q}_{Ak_n}(t)\right) \\
&\quad \times \delta\left(\boldsymbol{R}'_{B1} - \boldsymbol{q}_{Bl_1}(t)\right) \cdots \delta\left(\boldsymbol{R}'_{Bm} - \boldsymbol{q}_{Bl_m}(t)\right) \Big\rangle .
\end{aligned}
\tag{5.53}
$$

The summation is over all possible groups of n, m atoms A and B, $i_1 \cdots i_n$, $j_1 \cdots j_m$, $k_1 \cdots k_n$, $l_1 \cdots l_m$ that may be chosen of the total number of atoms, N_A and N_B. The $\boldsymbol{q}_{Ai}$, $\boldsymbol{q}_{Ak}(t)$ are the positions of atoms i, k of species A at the times t_0 and t, and similarly for species B. The $\boldsymbol{R}^{\{n\}}$ stand for the set of coordinates, $\boldsymbol{R}_1 \ldots \boldsymbol{R}_n$; $\boldsymbol{R}_A^{\{1\}}$ and $\boldsymbol{R}_B^{\{1\}}$ are equivalent to $\boldsymbol{R}_A$ and $\boldsymbol{R}_B$, respectively.

With the function G, the autocorrelation function may be written as

$$
\begin{aligned}
C(t) = & \frac{n_A n_B}{V} \frac{1}{4\pi\varepsilon_0} \int \mathrm{d}^3 R_A \, \mathrm{d}^3 R_B \, \mathrm{d}^3 R_A' \, \mathrm{d}^3 R_B' \\
& \times G^{(11)}(\boldsymbol{R}_A, \boldsymbol{R}_B; \boldsymbol{R}_A', \boldsymbol{R}_B'; t) \, \boldsymbol{\mu}_{AB}(\boldsymbol{R}_A, \boldsymbol{R}_B) \cdot \boldsymbol{\mu}_{AB}(\boldsymbol{R}_A', \boldsymbol{R}_B') \\
& + \frac{n_A^2 n_B}{V} \frac{1}{4\pi\varepsilon_0} \int \mathrm{d}^3 R_{A1} \, \mathrm{d}^3 R_{A2} \, \mathrm{d}^3 R_B \, \mathrm{d}^3 R_{A1}' \, \mathrm{d}^3 R_{A2}' \, \mathrm{d}^3 R_B' \\
& \times G^{(21)}(\boldsymbol{R}_{A1}, \boldsymbol{R}_{A2}, \boldsymbol{R}_B; \boldsymbol{R}_{A1}', \boldsymbol{R}_{A2}', \boldsymbol{R}_B'; t) \\
& \times \left\{ \frac{1}{4} \boldsymbol{\mu}_{AAB}(\boldsymbol{R}_{A1}, \boldsymbol{R}_{A2}, \boldsymbol{R}_B) \cdot \boldsymbol{\mu}_{AAB}(\boldsymbol{R}_{A1}', \boldsymbol{R}_{A2}', \boldsymbol{R}_B') \right. \\
& \quad + \frac{1}{2} \boldsymbol{\mu}_{AB}(\boldsymbol{R}_{A1}, \boldsymbol{R}_B) \cdot \boldsymbol{\mu}_{AAB}(\boldsymbol{R}_{A1}', \boldsymbol{R}_{A2}', \boldsymbol{R}_B') \\
& \quad \left. + \frac{1}{2} \boldsymbol{\mu}_{AAB}(\boldsymbol{R}_{A1}, \boldsymbol{R}_{A2}, \boldsymbol{R}_B) \cdot \boldsymbol{\mu}_{AB}(\boldsymbol{R}_{A1}', \boldsymbol{R}_B') \right\} \qquad (5.54) \\
& + \frac{n_A n_B^2}{V} \frac{1}{4\pi\varepsilon_0} \int \mathrm{d}^3 R_A \, \mathrm{d}^3 R_{B1} \, \mathrm{d}^3 R_{B2} \, \mathrm{d}^3 R_A' \, \mathrm{d}^3 R_{B1}' \, \mathrm{d}^3 R_{B2}' \\
& \times G^{(12)}(\boldsymbol{R}_A, \boldsymbol{R}_{B1}, \boldsymbol{R}_{B2}; \boldsymbol{R}_A', \boldsymbol{R}_{B1}', \boldsymbol{R}_{B2}'; t) \\
& \times \left\{ \frac{1}{4} \boldsymbol{\mu}_{ABB}(\boldsymbol{R}_A, \boldsymbol{R}_{B1}, \boldsymbol{R}_{B2}) \cdot \boldsymbol{\mu}_{ABB}(\boldsymbol{R}_A', \boldsymbol{R}_{B1}', \boldsymbol{R}_{B2}') \right. \\
& \quad + \frac{1}{2} \boldsymbol{\mu}_{AB}(\boldsymbol{R}_A, \boldsymbol{R}_{B1}) \cdot \boldsymbol{\mu}_{ABB}(\boldsymbol{R}_A', \boldsymbol{R}_{B1}', \boldsymbol{R}_{B2}') \\
& \quad \left. + \frac{1}{2} \boldsymbol{\mu}_{ABB}(\boldsymbol{R}_A, \boldsymbol{R}_{B1}, \boldsymbol{R}_{B2}) \cdot \boldsymbol{\mu}_{AB}(\boldsymbol{R}_A', \boldsymbol{R}_{B1}') \right\} \\
& + \mathcal{O}(\varrho^4) \ldots
\end{aligned}
$$

where V is the volume and n_A and n_B are number densities for atoms of species A and B. The $\boldsymbol{\mu}_{AB}$, $\boldsymbol{\mu}_{AAB}$ and $\boldsymbol{\mu}_{ABB}$ are the two- and three-body induced dipole moments, respectively.

We have neglected the (weak) dipole component arising from three identical atoms, i.e., $\boldsymbol{\mu}_{AAA}$ and $\boldsymbol{\mu}_{BBB} \approx 0$ and the irreducible components of order greater than 3. The 'dynamic' functions G play the same role as the analogous static functions in the expressions for the spectral moments.

Following the example of the static case, we define the dynamic functions W_{nm} and U_{nm} according to

$$
\begin{aligned}
W_{nm}\left(\boldsymbol{R}_A^{\{n\}}, \boldsymbol{R}_B^{\{m\}}; \boldsymbol{R}_A'^{\{n\}}, \boldsymbol{R}_B'^{\{m\}}; t\right) = & \frac{\lambda_A^{3n} \lambda_B^{3m}}{n!m!} \\
& \times \left\langle \boldsymbol{R}_A^{\{n\}} \boldsymbol{R}_B'^{\{m\}} \middle| \mathscr{T}_{nm}^{\dagger}(t) \middle| \boldsymbol{R}_A'^{\{n\}} \boldsymbol{R}_B'^{\{m\}} \right\rangle \\
& \times \left\langle \boldsymbol{R}_A'^{\{n\}} \boldsymbol{R}_B'^{\{m\}} \middle| \mathscr{T}_{nm}(t - \mathrm{i}\hbar/kT) \middle| \boldsymbol{R}_A^{\{n\}} \boldsymbol{R}_B^{\{m\}} \right\rangle ,
\end{aligned}
$$

where $\mathscr{T}_{nm}(t) = \exp\{-\mathrm{i}\mathscr{H}_{nm} t/\hbar\}$ is the time evolution operator of the system consisting of n atoms of species A and m atoms B. In the above equation, for reasons of brevity we have not explicitly accounted for nuclear spin; for a fuller treatment, see Eq. 4 of [297] where the spin eigenstates are fully accounted for.

The new *dynamic* functions U_{nm} are related to the W_{nm} by

$$\begin{aligned}
U_{10}(\mathbf{R}_A, \mathbf{R}'_A; t) &= W_{10}(\mathbf{R}_A, \mathbf{R}'_A; t) \\
U_{01}(\mathbf{R}_B, \mathbf{R}'_B; t) &= W_{01}(\mathbf{R}_B, \mathbf{R}'_B; t) \\
U_{11}(\mathbf{R}_A\mathbf{R}_B, \mathbf{R}'_A\mathbf{R}'_B; t) &= W_{11}(\mathbf{R}_A\mathbf{R}_B, \mathbf{R}'_A\mathbf{R}'_B; t) \\
&\quad - W_{10}(\mathbf{R}_A, \mathbf{R}'_A; t)\, W_{01}(\mathbf{R}_B, \mathbf{R}'_B; t) \\
U_{20}(\mathbf{R}_{A1}, \mathbf{R}_{A2}, \mathbf{R}'_{A1}, \mathbf{R}'_{A2}; t) &= W_{20}(\mathbf{R}_{A1}, \mathbf{R}_{A2}, \mathbf{R}'_{A1}, \mathbf{R}'_{A2}; t) \\
&\quad - \frac{1}{2}\sum_p W_{10}\left(\mathbf{R}_{A1}\mathbf{R}'_{Ap1}; t\right) W_{01}\left(\mathbf{R}_{A2}\mathbf{R}'_{Ap2}; t\right) \\
U_{21}(\mathbf{R}_{A1}\mathbf{R}_{A2}\mathbf{R}_B, \mathbf{R}'_{A1}\mathbf{R}'_{A2}\mathbf{R}'_B; t) &= W_{21}(\mathbf{R}_{A1}\mathbf{R}_{A2}\mathbf{R}_B, \mathbf{R}'_{A1}\mathbf{R}'_{A2}\mathbf{R}'_B; t) \\
&\quad - \frac{1}{2}\sum_p W_{20}\left(\mathbf{R}_{A1}\mathbf{R}_{A2}\mathbf{R}'_{Ap1}\mathbf{R}'_{Ap2}; t\right) W_{01}(\mathbf{R}_B\mathbf{R}'_B; t) \\
&\quad + \frac{1}{2}\sum_p W_{10}\left(\mathbf{R}_{A1}\mathbf{R}'_{A1p}; t\right) W_{10}\left(\mathbf{R}_{A2}\mathbf{R}'_{A2p}; t\right) W_{01}(\mathbf{R}_B\mathbf{R}'_B; t) \\
\cdots &= \cdots
\end{aligned}$$

In the static case, the functions U_{nm} differ from zero only if the $n+m$ atoms are interacting significantly, i.e., if at near range, $|\mathbf{R}_A - \mathbf{R}_B| \lesssim R_0$; R_0 is the range of the interatomic interaction. Moraldi (1990) points out that, as a consequence, the validity of a virial expansion is limited to densities for which the average number of atoms in the volume R_0^3 is much smaller than unity. In the time-dependent case, we see that the functions U are zero only if among the $n+m$ atoms a non-vanishing number will never interact during the time t. Consequently, a condition on time t must be satisfied for the dynamic density expansion to be valid.

In the static case, the condition on density is obtained by postulating that the significance of the U_n decreases with increasing n so that the role of clusters is less significant as n increases. The relative significance of the complex of n particles, when compared to that of $n-1$ particles, is measured by the quantity

$$\left| \varrho \int \mathrm{d}^{3n}R\; U_n\left(\mathbf{R}^{\{n\}}\right) / U_{n-1}\left(\mathbf{R}^{\{n-1\}}\right) \right| .$$

So, if $n = 2$, we must have

$$\left| \varrho \int \mathrm{d}^3R_A\, \mathrm{d}^3R_B\, \left(W_{11}(\mathbf{R}_A, \mathbf{R}_B) - 1\right) \right| \ll 1 ,$$

hence $\varrho R_0^3 \ll 1$. Similarly, for the dynamic functions, we must have

$$\left| \varrho \int \mathrm{d}^3 R_A \, \mathrm{d}^3 R_B \; U_{11}\left(\boldsymbol{R}_A \boldsymbol{R}_B \boldsymbol{R}_A' \boldsymbol{R}_B' ; t\right) / U_1\left(\boldsymbol{R}_A \boldsymbol{R}_A' ; t\right) \right| \ll 1 \,,$$

and similar for $U_{01}(\boldsymbol{R}_B \boldsymbol{R}_B'; t)$.

Moraldi gives an estimate of this integral for large times t, on the basis of a dimensional argument. The integral must diverge as t for $t \to \infty$; it must also be proportional to the cross section for binary interactions which is of the order of σ^2, the square of the zero of the intermolecular potential functions, $V(\sigma) = 0$. In other words, the factor of proportionality not specified as yet has units of speed, i.e., the root mean square speed, or

$$\varrho \, \sigma^2 \, v \, t \ll 1 \quad \text{or} \quad t \ll \tau \,, \tag{5.55}$$

where τ is the low-density limit of the time between collisions.

At this point, the analogy with the static case is complete and one may take advantage of the results obtained for the static case. Interesting quantal expressions for the dynamic pair and triplet distribution functions have been communicated [297]; the latter may be separated into a pairwise additive and an irreducible component, as in the static case.

Induced dipole autocorrelation functions of three-body systems have not yet been computed from first principles. Such work involves the solution of Schrödinger's equation of three interacting atoms. However, classical and semi-classical methods, especially molecular dynamics calculations, exist which offer some insight into three-body dynamics and interactions. Very useful expressions exist for the three-body spectral moments, with the lowest-order Wigner–Kirkwood quantum corrections which were discussed above.

Summarizing, it may be said that virial expansions of spectral line shapes of induced spectra exist for frequencies much greater than the reciprocal mean free time between collisions. The coefficients of the density squared and density cubed terms represent the effects of purely binary and ternary collisions, respectively. At the present time, computations of the spectral component do not exist except in the form of the spectral moments; see the previous Section for details.

It has been argued that, in the low-density limit, intercollisional interference results from correlations of the dipole moments induced in subsequent collisions (van Kranendonk 1980; Lewis 1980). Consequently, intercollisional interference takes place in times of the order of the mean time between collisions, τ. According to what was just stated, intercollisional interference cannot be described in terms of a virial expansion. Nevertheless, in the low-density limit, one may argue that intercollisional interference may be modeled as a sequence of two two-body collisions; in this approximation, any irreducible three-body contribution vanishes.

5.4 Dipole autocorrelation function

The dipole autocorrelation function is the expectation value over the equilibrium ensemble of an unperturbed system of atoms or molecules (i.e., without an applied field),

$$C(t) = \frac{1}{V} \frac{1}{4\pi\varepsilon_0} \langle \boldsymbol{\mu}(0) \cdot \boldsymbol{\mu}(t) \rangle \; . \tag{5.56}$$

The Heisenberg dipole operator is given by

$$\boldsymbol{\mu}(t) = \exp\left(\mathrm{i}\mathscr{H}_S t/\hbar\right) \; \boldsymbol{\mu}(\boldsymbol{R}) \; \exp\left(-\mathrm{i}\mathscr{H}_S t/\hbar\right) \; ,$$

where $\mathscr{H}_S$ designates the unperturbed Hamiltonian of the system. We may express the correlation function by its inverse Fourier transforn (see Eq. 2.53ff.) of the spectral function, $J(\omega)$, Eqs. 5.3,

$$C(t) = \frac{1}{2\pi} \int_{-\infty}^{\infty} \exp\left(\mathrm{i}\omega t\right) \; J(\omega) \; \mathrm{d}\omega \; . \tag{5.57}$$

Knowledge of the spectral density $J(\omega)$ is equivalent to knowledge of the dipole autocorrelation function $C(t)$, and vice versa.

Symmetries. From the definition of the spectral density, Eq. 5.3, it is clear that $J(\omega)$ is real and obeys detailed balance, Eq. 5.73. The correlation function, $C(t)$, on the other hand, is complex and satisfies [318]

$$C(t)^* = C(-t) \; ;$$

the asterisk denotes the conjugate complex as usual. In other words, the real part of $C(t)$ is even in time,

$$\Re\{C(t)\} = \frac{1}{2V} \frac{1}{4\pi\varepsilon_0} \langle \boldsymbol{\mu}(0) \cdot \boldsymbol{\mu}(t) + \boldsymbol{\mu}(t) \cdot \boldsymbol{\mu}(0) \rangle \; ,$$

while the imaginary part is odd,

$$\Im\{C(t)\} = \frac{1}{2\mathrm{i}V} \frac{1}{4\pi\varepsilon_0} \langle \boldsymbol{\mu}(0) \cdot \boldsymbol{\mu}(t) - \boldsymbol{\mu}(t) \cdot \boldsymbol{\mu}(0) \rangle \; .$$

We note that *classical* dipole autocorrelation functions are real and symmetric in time, $C(-t) = C(t)$.

Short-time expansion. The Taylor series expansion of the correlation function is a useful representation at small times,

$$C(t) = \sum_{n=0}^{\infty} \frac{1}{n!} \, t^n \, \frac{\mathrm{d}^n}{\mathrm{d}t^n} \, C(t) \bigg|_{t=0} \, ,$$

For quantum systems, the time derivatives may be evaluated by repeated use of Heisenberg's commutator expression, Eq. 2.89, namely

$$\langle \boldsymbol{\mu}(0) \cdot \boldsymbol{\mu}(t) \rangle = \sum_{n=0}^{\infty} \left(\frac{\mathrm{i}t}{\hbar} \right)^n \frac{1}{n!} \left\langle \boldsymbol{\mu}(0) \cdot \underbrace{[\mathscr{H}_S, [\mathscr{H}_S, \cdots [\mathscr{H}_S, \boldsymbol{\mu}] \cdots]]}_{n \text{ times}} \right\rangle ,$$

where in the nth term the Hamiltonian $\mathscr{H}_S$ appears n times.

This expansion may be compared with an expansion of the exponential function, $\exp \mathrm{i}\omega t$, in the inverse Fourier transform of the spectral density,

$$C(t) = \sum_{n=0}^{\infty} \frac{(\mathrm{i}t)^n}{n!} \int_{-\infty}^{\infty} \omega^n \, J(\omega) \, \mathrm{d}\omega \, .$$

The integrals appearing to the right-hand side are the spectral moments,

$$M_n = \int_{-\infty}^{\infty} \omega^n \, J(\omega) \, \mathrm{d}\omega \, .$$

The comparison shows that the spectral moments are given by

$$M_n = \hbar^{-n} \left\langle \boldsymbol{\mu}(0) \cdot \underbrace{[\mathscr{H}_S, [\mathscr{H}_S, \cdots [\mathscr{H}_S, \boldsymbol{\mu}] \cdots]]}_{n \text{ times}} \right\rangle , \tag{5.58}$$

or, in the case of classical systems,

$$M_n = (-\mathrm{i})^n \left. \frac{\mathrm{d}^n}{\mathrm{d}t^n} C(t) \right|_{t=0} .$$

Specifically, the zeroth and first moments are given by

$$\begin{aligned} M_0 &= \langle \boldsymbol{\mu}(0) \cdot \boldsymbol{\mu}(0) \rangle \\ M_1 &= \langle \boldsymbol{\mu}(0) \cdot [\mathscr{H}_S, \boldsymbol{\mu}] \rangle \, . \end{aligned}$$

Density expansion. The method of cluster expansions has been used to obtain the time-dependent correlation functions for a mixture of atomic gases. The particle dynamics was treated quantum mechanically. Expressions up to third order in density were given explicitly [331]. We have discussed similar work in the previous Section and simply state that one may talk about binary, ternary, etc., dipole autocorrelation functions.

Binary interactions. Dipole autocorrelation functions of binary systems are readily computed. For binary systems, it is convenient to obtain the dipole autocorrelation function, $C(t)$, from the spectral profile, $G(\omega)$. Figure 5.2 shows the complex correlation function of the quantum profile of He–Ar pairs (295 K) given in Figs. 5.5 and 5.6. The real part is an even function of time, $\Re\{C(-t)\} = \Re\{C(t)\}$ (solid upper curve). The imaginary part, on the other hand, is negative for positive times; it is also an odd function of time, $\Im\{C(-t)\} = -\Im\{C(t)\}$ (solid lower curve, Fig. 5.2). For comparison, the classical autocorrelation function is also shown. It is real, positive and symmetric in time (dotted curve). In the case considered, the

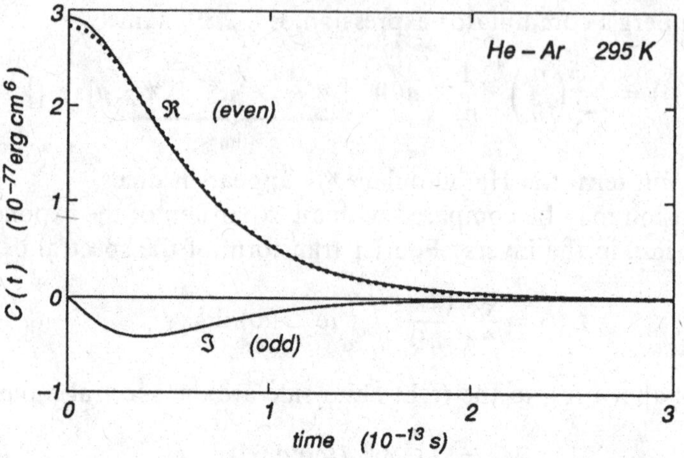

Fig. 5.2. The dipole autocorrelation function of He–Ar at 295 K, according to a quantal (solid lines) and a classical calculation (dotted). The quantum correlation function is complex; the real part is symmetric and positive ($\Re$) while the imaginary part ($\Im$) is anti-symmetric and negative at positive frequencies.

classical correlation function is nearly identical with the real part of the quantum correlation function, except at small times where the classical function is smaller than the quantal one. On a semi-logarithmic scale (not shown here) one does, however, notice that the quantal part is significantly greater than the classical function at large times ($> 1.5 \times 10^{-13}$ s).

We have seen above (Chapter 4) that on very general grounds the dipole moment induced in an interacting, dissimilar diatom pair is expected to change sign as we go from near range (where electron exchange dominates the interaction) to more distant interactions (where van der Waals attraction prevails). Consequently, the dipole autocorrelation function of the binary system may also change sign, certainly for the near central collisions (i.e., collisions with a very small impact parameter b) that are spectroscopically so significant; for central collisions, the angle between $\mu(0)$ and $\mu(t)$ changes very little so that the cosine of that angle does not change sign. In other words, at long times the real part of the dipole autocorrelation function may be expected to become negative if dissimilar atomic pairs are considered. The real part of the correlation function shown in Fig. 5.2 actually becomes negative at $t = 3.5 \times 10^{-13}$ seconds (not discernible in the figure). We shall see below that a more significant negative, long-time part arises through many-body interactions (intercollisional interference effect).

Classical correlation function. A system of N classical particles with a total of n degrees of freedom is fully specified at any time t by the set

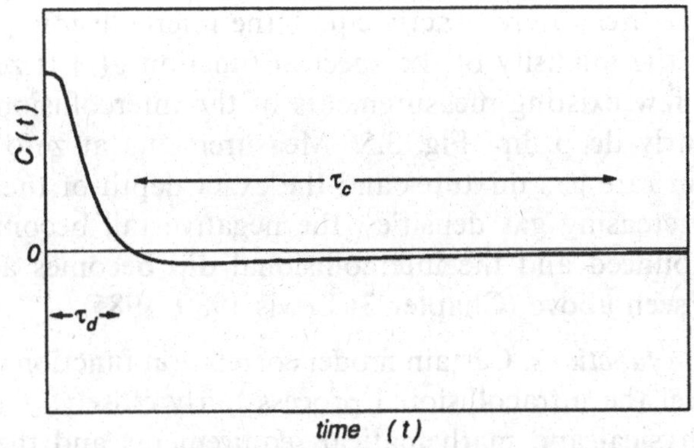

Fig. 5.3. The dipole autocorrelation function, long-time behavior (schematic). τ_d and τ_c are times of the order of the mean duration of a collision and the mean time between collisions, respectively.

of canonical position and momentum coordinates, $q_1, \ldots q_n$ and $p_1, \ldots p_n$, that is by a point in the $2n$ dimensional phase space. In the absence of external influences, the isolated system evolves in time t according to the laws of Hamiltonian mechanics. The state of the system at the time $\tau + t$ is uniquely determined by its state at the time τ. Time evolution takes place along a well defined trajectory in phase space.

Like many other properties of the N-particle system, the electric dipole moment, $\boldsymbol{\mu}(\boldsymbol{p}, \boldsymbol{q}, t)$, is a function of the canonical variables and time. We define the classical dipole autocorrelation function, according to

$$C(t) = \frac{1}{N} \sum_{i=1}^{N} \lim_{T \to \infty} \frac{1}{T} \int_0^T \frac{1}{4\pi\varepsilon_0} \boldsymbol{\mu}_i(\tau) \cdot \boldsymbol{\mu}_i(\tau + t) \, d\tau , \tag{5.59}$$

the average of $\boldsymbol{\mu}_i(\tau) \cdot \boldsymbol{\mu}_i(\tau + t)$ over time and over all particles. Knowledge of the correlation function is equivalent to knowledge of the (classical) spectrum, Eq. 2.90.

The classical dipole correlation function is symmetric in time, $C(-t) = C(t)$, as may be seen from Eq. 5.59 by replacing τ by $\tau - t$; the classical scalar product in Eq. 5.59 is, of course, commutative. Classical line shapes are, therefore, symmetric, $J(-\omega) = J(\omega)$, at all frequencies ω. Furthermore, classical dipole autocorrelation functions are real.

Long-time behavior of correlation functions. The dipoles induced in successive collisions are correlated as Fig. 3.4 on p. 70 suggests. As a consequence, the dipole autocorrelation function has a negative tail of a duration comparable to the mean time between collisions, Fig. 5.3. Furthermore, the area under the negative tail is of similar order of magnitude as the area under the positive (or intracollisional) part of $C(t)$. If the negative and positive areas were exactly equal, the intercollisional dip would be so deep that the intensity of the spectral function $g(\nu)$ is zero at zero frequency. The few existing measurements of the intercollisional process do suggest a fairly deep dip, Fig. 3.5. Measurements at zero frequency are impossible in rare gas mixtures and the exact depth of the dip is not known. With increasing gas densities, the negative tail becomes shorter and more pronounced and the intercollisional dip becomes accordingly broader as was seen above (Chapter 3; Lewis 1980, 1985).

Model correlation functions. Certain model correlation functions have been found that model the intracollisional process fairly closely. These satisfy a number of physical and mathematical requirements and their Fourier transforms provide a simple analytical model of the spectral profile. The model functions depend on the choice of two or three parameters which may be related to the physics (i.e., the spectral moments) of the system. Sears [363, 362] expanded the classical correlation function as a series in powers of time squared, assuming an exponential overlap-induced dipole moment as in Eq. 4.1. The series was truncated at the second term and the parameters of the dipole model were related to the spectral moments [79]. The spectral model profile was obtained by Fourier transform. Levine and Birnbaum [232] developed a classical line shape, assuming straight trajectories and a Gaussian dipole function. The model was successful in reproducing measured He–Ar [232] and other [189, 245] spectra. Moreover, the quantum effect associated with the straight path approximation could also be estimated. We will be interested in such three-parameter model correlation functions below whose Fourier transforms fit measured spectra and the computed quantum profiles closely; see Section 5.10. Intracollisional model correlation functions were discussed by Birnbaum *et al.*, (1982).

5.5 Line shapes of diatomic systems

When spectra of binary systems are considered, it is useful to introduce a normalized pair absorption coefficient, A, and a pair line shape function, G, according to

$$\begin{aligned} A(\omega) &= \alpha(\omega)/\varrho_1\varrho_2\,, \\ G(\omega) &= J(\omega)/N_{a_a}N_{a_b}\varrho_1\varrho_2\,, \end{aligned} \tag{5.60}$$

for mixtures of partial densities ϱ_1, ϱ_2. Instead of normalization by the densities expressed in amagat units, normalization by number densities is sometimes used in the literature. We note that one also finds in the literature the dimensionless quantity $\widetilde{A}$, usually as function of frequency ν in cm^{-1} units, which is related to the above absorption coefficient as

$\tilde{A}(\nu) = A(\nu)/\nu$. Whereas A describes the absorption of spectral intensity or energy, $\tilde{A}$ is proportional to the probability of absorbing a photon per unit path length. We will not make great use of the quantity $\tilde{A}$, because the spectral density G defined above is more closely related to the squared dipole transition matrix elements, even at low frequency; G is the preferred quantity.

The simplest systems of interest consist of two interacting, non-reactive atoms, such as He–Ar. In such cases (if electronic transitions are ignored), there is only the translational band to be considered. Line shape computations are straightforward but will in general require the use of digital computers if realistic intermolecular potentials are employed.

A treatment of the spectral line shape starts with the time-dependent Schrödinger perturbation theory, Eqs. 2.85 and 2.86. For absorbing pairs of dissimilar atoms, the number of absorbers, N_a, in the volume V is given by the number of dissimilar pairs, $N_a = N_1 N_2 = n_1 n_2 V^2$. In this expression, the number density, $n_i = N_i/V$, of species $i = 1$ or 2 is often expressed in amagat units, $\varrho_i \simeq N_{a_i}^{-1} n_i$ where N_a is the number density at 1 amagat of species i. For ideal gases, N_a equals Loschmidt's number, $N_L =$ 2.686763 $\times 10^{19}$ cm^{-3} amagat^{-1}. We thus compute the density invariant binary line shape, from Eqs. 5.2 and 5.3, as

$$\alpha(\omega)/\varrho_1 \varrho_2 = \frac{4\pi^2 N_{a_a} N_{a_b}}{3\hbar c n} \omega \left[1 - \exp(-\hbar\omega/kT)\right] V G(\omega) \quad (5.61)$$

$$V G(\omega) = \sum_{i,f} V P_i \frac{1}{4\pi\varepsilon_0} |\langle f | \boldsymbol{\mu} | i \rangle|^2 \, \delta(\omega_{fi} - \omega) \, . \quad (5.62)$$

The dipole operator is expressed in terms of spherical vector components, Eqs. 4.6 and 4.14, according to

$$\mu_\nu(R, \hat{R}) = B(R) \left(\frac{4\pi}{3}\right)^{1/2} Y_{1\nu}(\hat{R}) = B(R)\, C_\nu^{(1)} \, .$$

The factor $C_l^{(k)}$ of $B(R)$ is often called the lth component of the Racah spherical tensor of rank k; the three tensor components of rank 1 may be considered unit basis vectors spanning the (spherical) space.

Radial equation. In the laboratory frame, the two-atom Hamiltonian is given by Eq. 5.28. Transformation of $\mathscr{H}_{12}$ to center of mass and intermolecular separation coordinates, Eq. 5.29, allows the separation of these variables as was seen above, p. 207.

The Hamiltonian of relative motion is given by Eq. 5.30. The states $|i\rangle$, $|f\rangle$, of this Hamiltonian, $\mathscr{H}$, which we will call the initial and final states of a given spectroscopic transition, are associated with eigenenergies E_i, E_f. The wavefunctions of relative motion are obtained by introducing spherical coordinates R, ϑ, φ and the partial wave expansion, according

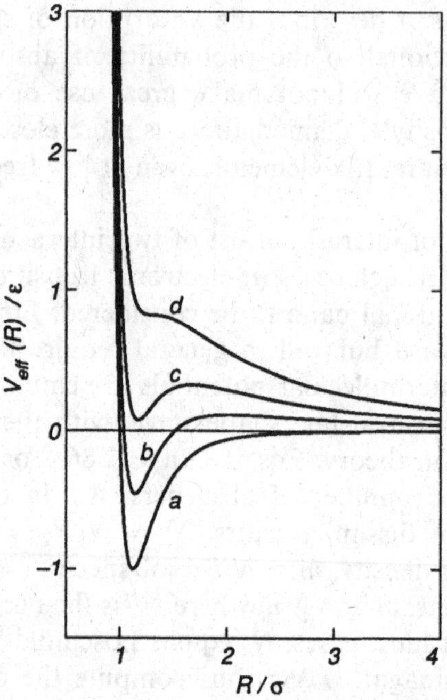

Fig. 5.4. The effective potential for four values of the angular momentum of relative motion: zero (a), small (b), intermediate (c) and large (d).

to Eq. 5.31. The radial wavefunctions ψ are solutions of the radial wave equation, Eq. 5.32.

For a given reduced mass and small angular momentum ℓ, this term is not significant and the effective potential is nearly the same as the potential, $V(R)$, (curve a of Fig. 5.4). Such is the case for 'head-on collisions' or non-rotating pairs. The potential has a well region of negative energies and, for positive energy, a repulsive core. For most molecular systems of interest, this well is deep and wide enough to support bound states (van der Waals molecules). Only a few of the more common systems form no bound states, for example H–He, He–He and H_2–He, because of the shallowness of the well and the small reduced mass which leads to de Broglie half wavelengths too large to fit inside the well. For more massive systems several vibrational bound states usually exist; these are numbered $v = 0,1,2, \ldots$

As a consequence of the attractive region of the interaction, several bound state solutions of the radial Schrödinger equation generally exist whose wavefunctions are localized in the well region. The bound state eigenenergies are negative, discrete and will be subscripted with the vibrational and rotational quantum numbers, v and ℓ; the normalization condition for bound states is

$$\int_0^\infty \psi^* (R; E_{\ell,v}) \; \psi (R; E_{\ell,v'}) \; \mathrm{d}R = \delta_{vv'}$$

as usual. In this expression, $\delta_{vv'}$ is the Kronecker delta symbol; it equals unity for $v = v'$ and else zero.

At large values of the angular momentum, $\ell \gg 1$, i.e., for distant 'fly-by' encounters at high speed or quickly rotating systems, the effective potential is a monotonically decreasing function of separation R (curve d of Fig. 5.4). Under such conditions, only free (i.e., 'collisional') states exist, with wavefunctions that are non-vanishing even at large separations of the pair. Free state energies are positive and continuous. For free state wavefunctions, energy density normalization is used,

$$\int_0^\infty \psi^* (R; E\ell) \; \psi (R; E'\ell) \; \mathrm{d}R = \delta(E - E') \tag{5.63}$$

for all fixed ℓ. In this expression, $\delta(x)$ designates Dirac's δ function. In other words, for $R \to \infty$, the energy normalized radial wavefunction is given by the right-hand side of

$$\psi(R; E, \ell) \sim \left(\frac{2m}{\pi \hbar^2 \kappa}\right)^{1/2} \sin\left(\kappa R - \frac{1}{2}\pi\ell + \eta_\ell\right),$$

where $\hbar^2\kappa^2 = 2mE$ and η_ℓ is the elastic scattering phase shift of the ℓth partial wave.

At intermediate values of the angular momentum, ℓ, a well region exists that may be associated with negative and positive energies (curve b of Fig. 5.4); or just positive energies (curve c of Fig. 5.4), depending on the details of the potential function and the magnitude of ℓ. States of negative energy (if these exist for $\ell > 0$) correspond to rotating van der Waals molecules; these are truly bound states. The states of positive energy are coupled more or less strongly to the continuum, across the centrifugal barrier. In scattering theory, such states coupled to the continuum are called a scattering resonance. Spectroscopists call them predissociating states. These are the continuation of the bound states into the continuum of energies with increasing ℓ. If the rotational excitation is increased beyond a certain point, centrifugal forces tear the van der Waals molecule apart (curve d of Fig. 5.4); the molecule ceases to exist.

Population factors. The probability of finding the state $|i\rangle = |E_i \ell_i m_{\ell_i}\rangle$ populated is given by a Boltzmann factor, which is a function of temperature T,

$$P_i = P_i(T) = \mathrm{e}^{-E_i/kT} / Z(T),$$

where $Z(T)$ is the pair partition function (with the center of mass factor

$V/\lambda_{\mathrm{CM}}^3$ suppressed, Eqs. 2.16 and 2.21)

$$Z(T) = \sum_{\ell,v}(2\ell+1)\,\mathrm{e}^{-E_{\ell v}/kT} + \sum_{\ell}(2\ell+1)\int_{E_\ell}^{\infty}\mathrm{e}^{-E/kT}\,\frac{\mathrm{d}n_\ell}{\mathrm{d}\kappa}\,\mathrm{d}\kappa\,. \qquad (5.64)$$

The first term to the right is the sum over bound states, $E_{v\ell} < 0$, which does not depend on the free volume. The second term is the sum over free states, $E > 0$, which is roughly proportional to volume V. The lower bound of the integral, $E_\ell \cong \hbar^2\ell(\ell+1)/2mR_c^2$, is the smallest free-state energy possible in the volume V of radius R_c when the centrifugal barrier becomes significant, as $\ell \to \infty$.

At small densities the first term to the right is usually negligible. In the sum over the free states, $\mathrm{d}n_\ell/\mathrm{d}\kappa$ is the density of states which may be expressed in terms of the radius R_c of the free volume and the scattering phase shift, η_ℓ, as

$$\frac{\mathrm{d}n_\ell}{\mathrm{d}\kappa} = \frac{1}{\pi}\left(R_c + \frac{\mathrm{d}\eta_\ell}{\mathrm{d}\kappa}\right).$$

The second term to the right of the equation for $Z(T)$ is orders of magnitude greater than the first one and is given by [185]

$$Z_{\mathrm{free}}(T) = \frac{1}{h^3}\int\cdots\int \exp\left(-\boldsymbol{p}^2/2mkT\right)\,\mathrm{d}^3p\,\mathrm{d}^3R = V\,\lambda_0^{-3}\,, \qquad (5.65)$$

where $\lambda_0 = \left(2\pi\hbar^2/mkT\right)^{1/2}$ is the thermal de Broglie wavelength of the pair; m is the reduced mass, $\boldsymbol{p}$ and $\boldsymbol{R}$ are the Cartesian momentum and position coordinates of relative motion. The volume V that appears in the partition sum of the free pairs cancels with the factor V of the spectral density G, Eq. 5.61; the products $V\,P_i$ and $V\,G(\omega)$ are independent of volume.

Matrix elements. The computation of the matrix elements $|\langle f|\mu_v|i\rangle|$ appearing in Eq. 5.62, for given E_i, E_f, ℓ_i, ℓ_f, starts with the summation (s) over the magnetic quantum numbers m_{ℓ_i}, m_{ℓ_f}, and the three components $v = 0, \pm 1$, of the dipole moment,

$$\sum_{(s)}\left|\left\langle E_f\ell_f m_{\ell_f}\,|\mu_v|\,E_i\ell_i m_{\ell_i}\right\rangle\right|^2 =$$

$$\left|\int_0^{\infty}\psi^*(R;E_f\ell_f)\,B(R)\,\psi(R;E_i\ell_i)\,\mathrm{d}R\right|^2\sum_{(s)}\left|\left\langle \ell_f m_{\ell_f}\left|C_v^{(1)}\right|\ell_i m_{\ell_i}\right\rangle\right|^2.$$

This expression factorizes into a radial integral and an angle-dependent part. The $|\ell m_\ell\rangle$ are of course given by the spherical harmonics,

$$|\ell m_\ell\rangle = Y_{\ell m_\ell}(\vartheta,\varphi)\,;$$

polar and azimuthal angles, ϑ and φ, have also been represented by the unit vector $\widehat{\boldsymbol{R}}$. The various terms under the sum to the right are squares of integrals over $Y_{\ell_f m_f}$, $Y_{1\nu}$, and $Y_{\ell_i m_i}$, familiar from the spectroscopy of ordinary atoms and molecules (Gaunt's coefficients).

From the Wigner–Eckart theorem, these matrix elements may be written in terms of 3–j symbols, as [350]

$$\left\langle \ell_f m_{\ell_f} \left| C_\nu^{(1)} \right| \ell_i m_{\ell_i} \right\rangle = \\ (-1)^{-m_f} \left[(2\ell_f+1)(2\ell_i+1)\right]^{1/2} \begin{pmatrix} \ell_f & 1 & \ell_i \\ -m_{\ell_f} & \nu & m_{\ell_i} \end{pmatrix} \begin{pmatrix} \ell_f & 1 & \ell_i \\ 0 & 0 & 0 \end{pmatrix} .$$

These vanish unless $\ell_f + \ell_i + 1$ equals an even integer and the triangular inequalities are satisfied,

$$|\ell_i - 1| \le \ell_f \le \ell_i + 1 .$$

These conditions imply the selection rules $\ell_f = \ell_i \pm 1$. The sum on the right-hand side of the equation thus becomes

$$\sum_{(s)} \left| \left\langle \ell_f m_{\ell_f} \left| C_\nu^{(1)} \right| \ell_i m_{\ell_i} \right\rangle \right|^2 \\ = (2\ell_f+1)(2\ell_i+1) \begin{pmatrix} \ell_f & 1 & \ell_i \\ 0 & 0 & 0 \end{pmatrix}^2 \sum_{(s)} \begin{pmatrix} \ell_f & 1 & \ell_i \\ -m_{\ell_f} & \nu & m_{\ell_i} \end{pmatrix}^2 \\ = \ell_> = \max\{\ell_i, \ell_f\} .$$

The sum to the right of this equation equals unity (orthonormality of 3–j symbols [350]). Hence, the sum to the left is readily shown to be equal to $\ell_>$, a number equal to the larger of ℓ_i and ℓ_f. In other words,

$$\sum_{(s)} \left| \left\langle E_f \ell_f m_{\ell_f} \left| \mu_\nu \right| E_i \ell_i m_{\ell_i} \right\rangle \right|^2 = \ell_> \; |\langle E_f \ell_f | B | E_i \ell_i \rangle|^2 ,$$

where $\ell_f = \ell_i \pm 1$ and

$$|\langle E_f \ell_f | B | E_i \ell_i \rangle| = \left| \int_0^\infty \psi^*(R; E_f \ell_f)\, B(R)\, \psi(R; E_i \ell_i)\, \mathrm{d}R \right| \tag{5.66}$$

is the magnitude of the radial matrix element. These expressions may be evaluated with the help of digital computers. For convenience, we have introduced the notation $\langle E_f \ell_f | B | E_i \ell_i \rangle$ for the radial matrix elements of the spherical dipole components $B(R)$.

Line shapes. Because of the energy normalization of the radial wavefunctions, Eq. 5.63, the summations over the free-state energies, Eq. 5.62, become integrations,

$$\sum_{E_f, E_i} \ldots = \int_0^\infty \int_0^\infty \ldots \mathrm{d}E_f \, \mathrm{d}E_i .$$

The integration over E_f replaces the δ function of frequency by $\hbar$ and fixes E_f at the value given by Bohr's frequency condition, Eq. 2.80. The integration over E_i must be done numerically. For each fixed value ℓ_i, the summation over ℓ_f is over only two terms (selection rules),

$$V\,G(\omega) = \lambda_0^3\hbar\sum_{\ell_i}\int_0^\infty VP_i\left\{\ell_i\,|\langle E_i+\hbar\omega,\ell_i-1|B|E_i\ell_i\rangle|^2 + (\ell_i+1)\,|\langle E_i+\hbar\omega,\ell_i+1|B|E_i\ell_i\rangle|^2\right\}dE_i\,.$$

The remaining summation over ℓ_i is also done numerically; the infinite sum may be conveniently truncated at some $\ell_{\max}$ as the matrix elements of large ℓ values fall off to zero.

If van der Waals molecules exist, one must account for bound-state to bound-state, and for bound-to-free and free-to-bound transitions, in addition to the free-to-free state transitions. The procedures are the same as explained above. The spectral density then becomes

$$\begin{aligned} VG(\omega) = \lambda_0^3\hbar\sum_{\ell_i,\ell_f}(2\ell_i+1)\,C(\ell_i\,1\,\ell_f;000)^2 \qquad (5.67)\\ \times\Bigg\{\int_0^\infty \exp(-E_i/kT)\;|\langle E_i+\hbar\omega,\ell_f|B|E_i\ell_i\rangle|^2\;\mathrm{d}E_i \\ +\sum_{v_i,v_f}\exp(-E_{v_i\ell_i}/kT)\left|\left\langle E_{v_f\ell_f}\ell_f|B|E_{v_i\ell_i}\ell_i\right\rangle\right|^2\delta(E_{v_f\ell_f}-E_{v_i\ell_i}-\hbar\omega) \\ +\sum_{v_i}\exp(-E_{v_i\ell_i}/kT)\,|\langle E_{v_i\ell_i}+\hbar\omega,\ell_f|B|E_{v_i\ell_i}\ell_i\rangle|^2 \\ +\sum_{v_f}\exp\Big(-(E_{v_f\ell_f}-\hbar\omega)/kT\Big)\left|\left\langle E_{v_f\ell_f}\ell_f|B|E_{v_f\ell_f}-\hbar\omega,\ell_i\right\rangle\right|^2\Bigg\}, \end{aligned}$$

with $\ell_f=\ell_i\pm1$. We note that the expression $(2\ell_i+1)\,C(\ell_i1\ell_f;000)^2$ is equal to $\ell_>$, the greater of ℓ_i and ℓ_f. This expression, Eq. 5.67, consists of four terms.

The first term will in general give the most substantial contribution which is from free-to-free transitions. The radial state vectors $|E_i+\hbar\omega,\ell_i\rangle$ correspond to free states, that is states of *positive* energy. This first term gives a relatively unstructured, diffuse spectral 'line'.

The second term describes the bound-to-bound contributions, that is the rotovibrational bands of the van der Waals dimer. If the system does not form dimers, this term and the following two terms all vanish. For practical use, the δ function in this term should be replaced by an instrumental slit function, or perhaps with some Lorentzian if pressure broadening affects the individual lines (as will often be the case). In any case, the δ function is symbolic for the relatively sharp dimer lines that must be contrasted with the very diffuse free-to-free contribution of the first term.

The third and fourth terms give the bound-to-free and free-to-bound contributions, respectively. These are diffuse continua, more or less structured if scattering resonances are important. As was noted above for the first term, the two free-state vectors of energy $E_{v\ell} \pm \hbar\omega$ seen in the third and fourth terms must again be true free-state energies, $E_{v_i\ell_i} + \hbar\omega > 0$ and $E_{v_f\ell_f} - \hbar\omega > 0$, respectively.

The spectral density may alternatively be expressed in terms of frequency f in Hertz, $VG(f)$, or frequency v in cm^{-1}, $VG(v)$, instead of angular frequency ω, as in $VG(\omega)$. In such a case, we must multiply the right-hand side, Eq. 5.67, by 2π or $2\pi c$, and replace all ω by $2\pi f$ or $2\pi cv$, respectively. Similarly, the δ function of energy seen in the bound-to-bound transition term may be converted to one of frequency v by multiplication by Planck's constant, h.

This concludes the theory of collision-induced line shapes of binary systems, that is the line shape that one might observe at gas densities that are not too high – with one exception: near zero frequency the intercollisional 'dip' will always be present, no matter how low the pressure may be. The absorption dip is a many-body effect and is not obtainable from a binary theory (Poll 1980). At low gas densities, the intercollisional process appears only over a very small frequency interval near zero, of the order of the mean collision frequency, and it can in general be readily distinguished from the binary profile which extends over a much greater range of frequencies.

Computations. If the induced dipole moment, $B(R)$, and the interaction potential, $V(R)$, are known as functions of the intermolecular separation, R, the computation of the spectral line shape is straightforward. For a meaningful comparison of the predictions of the fundamental theory with measurements one must select the best models for dipole and potential functions available. For a few simple systems advanced theoretical methods have provided reliable induced dipole surfaces. Good models of the intermolecular potentials, on the other hand, are generally semi-empirical, or at least thoroughly tested against the widest selection of empirical data, such as scattering cross sections, virial coefficients, diffusion coefficients, etc. (Maitland, Rigby, Smith and Wakeham 1981). Obsolete potential models, such as the Lennard-Jones model, have been shown to lead to spectral profiles which are not in agreement with the measurements [39].

Numerical integration of the radial Schrödinger equation is generally necessary if realistic interaction potentials are selected. For this purpose, efficient computer codes exist based, for example, on Cooley's method of integration. Other widely available codes written to compute scattering phase shifts are, however, not sufficient unless the radial wavefunctions can be extracted at near range where the spectroscopic interactions take place.

From the radial wavefunctions of initial and final states, the radial matrix elements are computed by numerical integration. Such computations should use a grid spacing that is small enough so that one has a sufficient number of grid points per de Broglie wavelength. For a modest accuracy, five grid points per wavelength may suffice but one might want to use several times that many grid points for better accuracy. Since the wavefunctions are rapidly oscillating functions, the radial integrals may be more accurate if the simple trapezoidal rule is employed; higher-order formulae such as Simpson's rule have been found to be less suited, especially when the de Broglie wavelengths of initial and final states are very different. Since the induced dipole operators fall off exponentially with separation, the radial integral may generally be truncated at some distance R_{max} without loss of accuracy. An essential part of a line shape calculation is the integration over initial and final state energies, Eq. 5.67. Normally, the radial wavefunctions, $\psi(R;E\ell)$, vary slowly with energy, except in the vicinity of scattering resonances where their amplitudes may vary by orders of magnitude over a relatively narrow energy band. It is important that the integration over predissociating states be done with an energy grid tailored to the width of the resonances where these exist. Extremely sharp resonances of a width much smaller than the energy step size may be treated like bound states. We note that for the integration over energy even the higher-order (20 to 40 point) Gauß–Hermite integration formulae were found inadequate because the scattering resonances could not be modeled. Third-order spline integration, on the other hand, gives the necessary flexibility. The sum over partial waves may be truncated at $\ell_{max} = 60$ for light systems like He–Ar, and at 180 for more massive systems like Ar–Kr, at temperatures near room temperature. The overall numerical precision of a line shape calculation may be kept at the 1% level. We note that spectral moments may be computed at much lower cost and may serve as a test of the accuracy of line shape calculations. A total of 20 000 to 50 000 radial matrix elements is generally obtained for a precision of the spectral profiles at the 1% level, at temperatures not exceeding $\approx$300 K.

Comparison with measurement. Measurements of the absorption of rare gas mixtures exist for some time. This fact has stimulated a good deal of theoretical research. A number of *ab initio* computations of the induced dipole moment of He–Ar are known, including an advanced treatment which accounts for configuration interaction to a high degree; see Chapter 4 for details. Figure 5.5 shows the spectral density profile computed

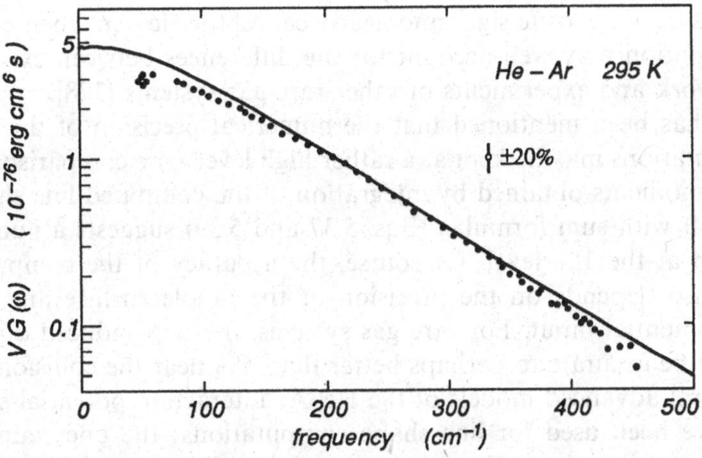

Fig. 5.5. The spectral function of He–Ar pairs at 295 K, computed from first principles, compared with the experiment (dots, [75]) [278].

from that dipole surface [278] and a refined potential [12]. No adjustable parameters are employed in the computation [278]. For comparison, Bosomworth and Gush's measurement is also shown (dots) on an absolute intensity scale. The agreement is good, especially in the region where the absorption is strong (around 200 cm^{-1}, see Fig. 3.1); at the extreme frequencies the measurement falls below theory by up to 20%, for an unknown reason. This drop exceeds perhaps the average estimated uncertainty of the measurement a little, but the increased scatter of the data points from 50 to 60 cm^{-1}, and also above 400 cm^{-1}, suggests a greater uncertainty of the measurement at these frequencies. The apparent noise is not seriously inconsistent with the differences observed between theory and measurement. In any case, we must remember that the direct result of such measurements is the absorption coefficient, α, not the spectral function, $G(\omega)$, and that measurements are difficult where the absorption is small, especially if the work is limited to low pressures to minimize three-body interference. Measurements of small absorption are necessarily somewhat more uncertain. The agreement of theory and measurement on an absolute intensity scale at frequencies where the absorption is strong is certainly reassuring. Nevertheless, new and accurate measurements of such binary profiles at varying temperature are desirable for a compelling test of the *ab initio* dipole model.

The *ab initio* He–Ar induced dipole surface that we have used here to compute the spectral profile suggests that exchange effects due to intra-atomic correlation and higher-order dispersion terms contribute significantly to the induced dipole. However, these exchange and dispersion contributions have opposite signs and nearly cancel for He–Ar. Their combined contribution may well account for the differences between existing theoretical work and experiments of other rare-gas systems [278].

Above, it has been mentioned that the numerical precision of the line shape computations may be kept at a rather high level. The comparison of the spectral moments obtained by integration of the computed line shape (as in Eq. 3.4) with sum formulae (Eqs. 5.37 and 5.39) suggests a numerical precision at the 1% level. Of course, the accuracy of the computed line shape also depends on the precision of the dipole surface and the interaction potential input. For rare gas systems, the best induced dipole surfaces may be accurate to perhaps better than 5% near the collision diameter. Several advanced models of the He–Ar interaction potential exist [13] and have been used for line shape computations; the uncertainties of the computed line shape due to the remaining differences of the best potential models amount to a few percent of the integrated intensity at most. Since the induced dipole moment is a function of separation that falls off rapidly to insignificantly small values with increasing separation, it is of special importance for line shape calculations that the interaction potential model the repulsive core of the interaction potential closely. The study of the variation of the absorption with temperature would be a discriminating test of the repulsive branch of a given potential model, but at present measurements of binary spectra of helium–argon mixtures at temperatures other than 295 K are scarce.

We note, moreover, that in a semi-logarithmic grid, for He–Ar, at frequencies above 85 cm^{-1}, theory suggests a nearly straight line of a very small *concavity* (i.e., a hanging chain-like curvature) while the measurement looks distinctly *convex*, Fig. 5.5. We have tried to modify within reasonable limits the dipole and potential models, to see if in this way one could obtain computationally the convex curvature of the measurement, to no avail. Only if one introduces modifications of the dipole model that are far beyond any existing uncertainties, for example by using a dispersion coefficient D_7 that is orders of magnitude greater than the theoretical estimates, are such convex line shapes computationally attainable, but they are hardly meaningful. One might think that the observed convexity of the experimental data, Fig. 5.5, is possibly an artifact. It would be interesting to see whether new measurements in helium–argon mixtures of low density confirm the convex line shape. The only other published absorption profile of helium–argon mixtures, by Tonkov and an associate [95], suggests no convexity in the profile, but the measurement is up to 20% *more intense* than theory predicts. Furthermore, it was obtained at much higher densities and might have been affected by ternary processes [278]. It is, therefore, probably not a valid test of a theory based on binary interactions.

Bound HeAr pairs, i.e., the van der Waals molecules, exist. The so-called SPFD2 model of the interaction potential [12], which we have used for the computation, suggests the existence of just one vibrational level, $v = 0$, with rotational structures at −7.05, −6.47, −5.30, −3.59, −1.39, +1.18, +4.10 cm^{-1} relative to the dissociation limit, for $\ell = 0$ through 6. The latter two levels correspond to predissociating states of a lifetime greater than the mean time between collisions. If the rotational excitation is increased beyond the $\ell = 6$ level, centrifugal forces become stronger than the van der Waals binding force and the dimer dissociates. The spectroscopic effects of the bound HeAr dimer are rather weak at room temperatures. Rotational states are all populated, but at 295 K the dimer concentrations amount to only 0.2% at 10 atmospheres, and to 2% at 100 atmospheres, so that significant spectral contributions are hardly expected. Selection rules are $\Delta\ell = \pm 1$; accordingly, the computed rotational dimer band consists of six lines between roughly 1 and 3 cm^{-1} and are difficult to access. Furthermore, at the gas densities commonly used in collision-induced absorption studies, these bound-to-bound state transitions must be expected to be strongly pressure broadened because of the anisotropy of the van der Waals molecule, so that it should be difficult, perhaps impossible, to actually observe the rotational band. The rotational dimer band intensity amounts to roughly 2% of the absorption between 2 and 3 cm^{-1} of the free-to-free state contributions, not a very striking spectral feature when pressure broadened. Besides the bound-to-bound state transitions, bound-to-free and free-to-bound induced transitions will generate a continuum at low frequencies, roughly from 0 to 60 cm^{-1} that amounts to a 1% enhancement of the total absorption. These enhancements involving bound dimer states are actually included in the computed profile, Fig. 5.5, but are hardly discernible.

Summarizing, it is clear that in helium–argon mixtures at 295 K roughly 98% of the observed absorption is due to free-to-free transitions. However, at lower temperatures, and for the more massive systems (like Kr–Xe), a greater spectroscopic effect of the dimers must be expected. Massive systems show, in general, stronger dimer concentrations than the light systems, for two reasons. First, a greater reduced mass is associated with a smaller de Broglie wavelength, thereby allowing a greater number of bound states to exist in a given potential well; this in turn increases the number of dimer states which makes them statistically more significant. Second, the more massive systems show generally deeper wells than the light systems, so that the number of bound states is further increased.

Other quantum line shape computations of absorption by dissimilar atomic pairs based on empirical or *ab initio* dipole models have been known for some time [39, 44, 76, 251, 330, 332, 361, 365, 386]. Such studies are of interest for the analysis of measurements, for predicting absorption at temperatures that differ from those of the measurements, and for the study of the influences of the remaining uncertainties of the theoretical line shape arising from the imperfect knowledge of potential and dipole function. Although it has been stated above that variations of the potential function reflecting the existing uncertainties of the potential models may affect the results of moment calculations less than similar variations of the induced dipole model, the former, nevertheless, may have a pronounced effect on the computed profile. It is important to select the most advanced interaction potential models for such calculations; the formerly widely used Lennard-Jones 6–12 potential models have given rise to markedly inferior fits to the experimental spectra [68, 58].

For line shape calculations, purely exponential dipole models, and exponential models combined with a dispersion term, D_7/R^{-7}, have been used, see Chapter 4 for details. The variation of the line shape with these dipole models has been studied. In all cases, the purely exponential dipole model gives inferior results when compared with the exp–7 models; the influence of the dispersion term, while small, is nevertheless significant in line shape analyses; moment analyses, in contrast, have reportedly not been able to demonstrate the significance of the dispersion term. It would seem that the accuracy of quantum line shape calculations of the absorption by rare gas pairs has reached the point where further progress must await more accurate experiments at low gas densities and over a wider range of frequencies and temperatures.

Collision-induced profiles of pairs involving one or more molecules will be considered in the next Chapter. Classical line shape calculations of rare gas pairs will be considered next.

5.6 Classical diatom line shapes

For classical line shape calculations one needs the induced dipole moment as function of time, $\boldsymbol{\mu}(R(t))$, averaged over angular momenta and speeds of relative motion. In other words, one solves Newton's equation of motion, or one of its integrals, of the two-particle system. After suitable averaging, one obtains the spectral profile by Fourier transform.

Suppose we have two particles with masses m_1 and m_2, and position vectors $\boldsymbol{R}_1$ and $\boldsymbol{R}_2$ in the laboratory frame, respectively, that interact through the isotropic intermolecular potentential $V(R)$, with $R = |\boldsymbol{R}_2 - \boldsymbol{R}_1|$, with no other forces acting on them. For each particle, an equation of motion may be written down,

$$\boldsymbol{F}_{12} = m_1 \ddot{\boldsymbol{R}}_1 \qquad \boldsymbol{F}_{21} = m_2 \ddot{\boldsymbol{R}}_2 ,$$

with $\boldsymbol{F}_{12} = -\boldsymbol{F}_{21} = -\nabla_1 V(R)$. Differentiation with respect to time is indicated by dots. We introduce as usual center of mass and relative coordinates, Eq. 5.29. Dividing these equations by m_i ($i = 1,2$) and subtracting the second from the first, we obtain Newton's equation of the relative motion of the pair,

$$m\ddot{\boldsymbol{R}} = -\nabla V(R) ; \tag{5.68}$$

$m = m_1 m_2/(m_1 + m_2)$ designates the reduced mass. For a derivation of the equation of motion of the center of mass, we simply add the equations of motion of the two particles and get force-free motion of the center of mass, which is of no great interest here; therefore, it is suppressed.

Since the direction of the interaction force, $-\nabla V$, is along the relative position vector, $\boldsymbol{R}$, the vector product of $\boldsymbol{R}$ with $\ddot{\boldsymbol{R}}$, Eq. 5.68, vanishes, $\boldsymbol{R}\times\ddot{\boldsymbol{R}} = 0$. In other words, $\boldsymbol{R}\times m\dot{\boldsymbol{R}} = \boldsymbol{R}\times m\boldsymbol{v} = \boldsymbol{L}$, the angular momentum of relative motion, is a constant of motion. Its magnitude may be expressed in polar coordinates R, ϑ, according to

$$L = mR^2\dot{\vartheta} , \tag{5.69}$$

by integrating the angular part of Eq. 5.68 with respect to time.

The equation of relative motion, Eq. 5.68, may also be integrated with respect to position, dR, to give the law of energy conservation,

$$\frac{1}{2}\, mv_0^2 = \frac{1}{2}\, m\left(\dot{R}^2 + R^2\dot{\vartheta}^2\right) + V(R) ;$$

v_0 is the relative initial speed (at infinite separation). In this expression, we may substitute $\dot{\vartheta}$ from Eq. 5.69 so that

$$\frac{1}{2}\, mv_0^2 = \frac{1}{2}\, m\dot{R}^2 + V_{\text{eff}}(L;R) ,$$

with $V_{\text{eff}}(L;R) = V(R) + L^2/(2mR^2)$; V_{eff} is the classical analog of the effective potential, Eq. 5.33. The above equation is a differential equation of first order,

$$\dot{R} = \left[v_0^2 - \frac{2}{m}\left(V(R) + \frac{L^2}{2mR^2}\right)\right]^{1/2} ,$$

which is easier to solve than the second order equation. Separating the variables, one gets at once

$$t = \int_{R_0}^{R} \frac{\mathrm{d}R'}{\{v_0^2 - (2/m)\,[V(R') + L^2/(2mR'^2)]\}^{1/2}} . \tag{5.70}$$

In this expression, R_0 designates the greatest root of the argument of the square root appearing in the denominator; R_0 is the outermost classical turning point. Equation 5.70 represents the inverse function, $t(R)$, of the desired solution $R(v_0, b; t)$. With the knowledge of the time dependence of

R, Eq. 5.69 may be integrated directly; the solution is $\vartheta(v_0, b; t)$. The two constants of integration v_0, b (or, alternatively, $E_0 = mv_0^2/2$ and $L = bmv_0$) are parameters of the solution.

One often needs the classical scattering angle, $\chi = \pi - 2\vartheta(\infty)$. It is obtained from

$$\chi = \pi - 2\vartheta(\infty) = \pi - 2b \int_{R_0}^{\infty} \left(1 - \frac{2V(R)}{mv_0^2} - \frac{b^2}{R^2}\right)^{-1/2} \frac{\mathrm{d}R}{R^2} . \tag{5.71}$$

For a certain range of values of impact parameter b and initial speed v_0, the radicand in Eq. 5.70 may actually have three roots. By choosing the largest classical turning point R_0, we have limited our considerations to free, i.e., colliding particles. The other two turning points (if existent) are characteristic of bound states which may be similarly treated [168].

The solutions $R(v_0, b; t)$ and $\vartheta(v_0, b; t)$ give the dipole moment $\boldsymbol{\mu}(R) = \mu(R)\hat{\boldsymbol{R}}$ as function of time. This expression may be Fourier transformed to get the emission profile (as in Eq. 1ff., with $q\boldsymbol{r} = \boldsymbol{\mu}$ and $\omega = 2\pi\nu$, after integration over $\mathrm{d}\Omega$)

$$\frac{\mathrm{d}E_{\mathrm{rad}}(v_0, b; \omega)}{\mathrm{d}\omega} = \frac{2\omega^4}{3c^3\pi} \frac{1}{4\pi\varepsilon_0} \left| \int_{-\infty}^{\infty} \mathrm{d}t \, \exp(-i\omega t) \, \boldsymbol{\mu}(v_0, b; t) \right|^2$$

(Larmor's formula), the energy radiated per unit angular frequency per encounter with a given v_0 and b. The classical line shape is then given by

$$I(\omega) = \int \int N(v_0, b) \frac{\mathrm{d}E_{\mathrm{rad}}(v_0, b; \omega)}{\mathrm{d}\omega} \, \mathrm{d}b \, \mathrm{d}v_0 , \tag{5.72}$$

where $N(v_0, b)\, \mathrm{d}b\, \mathrm{d}v_0$ is the number of collisions of dissimilar pairs, per unit volume, per unit time, with impact parameter in the range between b and $b + \mathrm{d}b$, and with relative initial speeds between v_0 and $v_0 + \mathrm{d}v_0$,

$$N(v_0, b) \, \mathrm{d}v_0 \, \mathrm{d}b = \frac{N_a N_b}{V} \, 2\pi \, v_0 \, b \, \mathrm{d}b \left(\frac{m}{2\pi kT}\right)^{3/2} \exp\left(\frac{-mv_0^2}{2kT}\right) 4\pi v_0^2 \, \mathrm{d}v_0 .$$

We note that $b\,\mathrm{d}b$ is related to the classical differential scattering cross section. It is a short-hand notation for $b(\chi)\, |\mathrm{d}b/\mathrm{d}\chi| \, \mathrm{d}\chi$; the magnitude of the derivative $\mathrm{d}b/\mathrm{d}\chi$ falls off to zero quickly so that the integral over $b\,\mathrm{d}b$ is actually finite.

The classical emission profile, Eq. 5.72, may be converted to an absorption profile with the help of Kirchhoff's law, Eq. 2.70, which relates the absorption coefficient α to the emitted power per unit frequency interval per unit volume, with the help of Planck's law, Eq. 2.71, according to

$$\alpha(\omega)/\varrho_a \varrho_b = \frac{4\pi^2}{3\hbar c} N_L^2 \omega \left[1 - \exp\left(-\hbar\omega/kT\right)\right] G(\omega) .$$

The comparison of Eqs. 2.74 and 2.86 suggests a relationship between the emission profile, $I(\omega)$, and the spectral density introduced above, $G(\omega)$ of Eq. 5.60, according to $I(\omega) = (4\omega^4/3c^3)\, VG(\omega)$; see also Eq. 5.4.

It may look strange to have Planck's constant appear in a 'classical' formula of the absorption; in classical physics Planck's constant vanishes. At low enough frequencies, $\omega \ll kT/\hbar$, the appearance of $\hbar$ may be avoided by substituting the classical Rayleigh–Jeans law for Planck's law, Eq. 2.72. However, in that case, as frequency approaches or exceeds a certain limit, the use of Rayleigh–Jeans' law must lead to coarse disagreement with observation ('radiation catastrophe'). Therefore, a classical line shape computation is best limited to the spectral density profile, $G(\omega)$; once the profile is obtained, the absorption coefficient should probably always be obtained from an expression like that for the absorption coefficient, $\alpha(\nu)$, which contains Planck's constant. Purely classical absorption breaks down at frequencies ω that are comparable to $kT/\hbar$, that is roughly 200 cm^{-1} at 300 K. Stimulated emission is absent in the classical formulae but is generally significant in the translational spectra of rare gas mixtures.

Computations of classical line shapes are known [167, 168, 232, 271]; see also Fig. 5.6 below. Simplifying assumptions have frequently been invoked, such as hard-sphere interaction [303, 379], but the resulting spectral profiles have generally been disappointing; collisions of hard spheres overemphasize the high frequencies. We note that in [167] the bound states are accounted for in the framework of classical theory; moreover, an asymptotic formula is given for the far wings of the spectral density function. The discussions on the desymmetrization of classical line shapes (pp. 251ff., see also Fig. 5.6 and Table 5.1) suggest, of course, that quantum dynamics and detailed balance generally have a strong effect on the far wings of the profile; classical treatments of the wings must, therefore, be used with caution.

Computation. Classical line shapes may be computed in a fraction of the time required for the computation of quantum profiles. Special care is required to match the time increment Δt and the total time of integration, t_{max}, to the frequency band of interest, $f_{min} \cdots f_{max}$, so that $\Delta t f_{max} < 1$ and $t_{max} f_{min} < 1$, and at the same time have a reasonable radial step size for the integration of the equation of motion; the latter is best entirely separated from the Fourier transform. Furthermore, special attention must be devoted to the treatment of orbiting collisions, i.e., collisions at fixed relative speed v_0 for which with increasing impact parameter b the scattering angle goes through infinity; bound states require likewise special consideration.

Classical trajectories are even in time, $R(v_0, b; -t) = R(v_0, b; t)$, and $\vartheta(v_0, b; -t) = -\vartheta(v_0, b; t)$. Classical profiles are, therefore, even functions of frequency. We note that for the translational absorption profile of rare

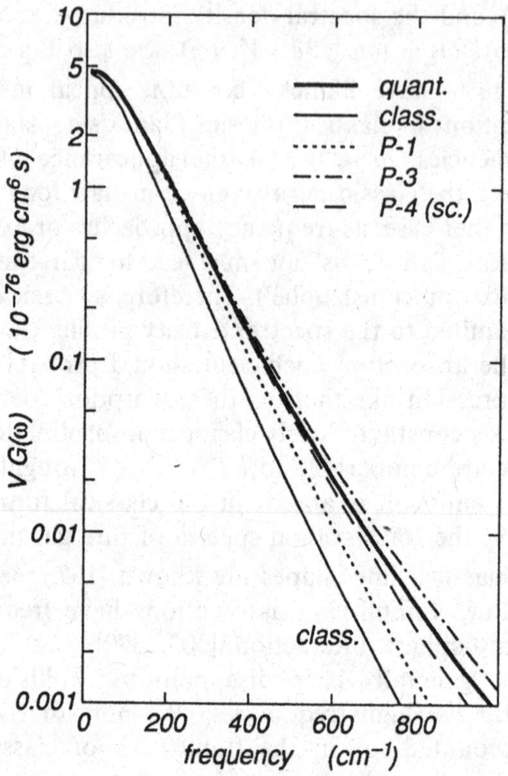

Fig. 5.6. Comparison of classical and quantal profiles computed from the same dipole and potential function, for He–Ar at 295 K (thin and thick solid lines, respectively). Classical profiles desymmetrized according to the procedures P-1 through P-4 are also shown.

gas mixtures knowledge of the spectral function is required for the positive frequencies only. However, an equation like Eq. 5.72 allows computation of the spectral density profile at *negative* frequencies as well; obviously, we have $G(-\omega) = G(\omega)$ for the classical profiles.

Results. Figure 5.6 compares the classical and quantum profiles of the spectral function, $VG(\omega)$, of He–Ar pairs at 295 K (light and heavy solid curves, respectively), over a wide frequency band. Whereas at the lowest frequencies the profiles are quite similar, at the higher frequencies we observe increasing differences which amount up to an order of magnitude. The induced dipole [278] and the potential function [12] are the same for both computations. Bound state contributions have been suppressed; we have seen above that for He–Ar at 295 K, the spectroscopic effects involving van der Waals molecules amount to only ≈2% at the lowest frequencies, and to much less than that at higher frequencies.

If bound state effects are suppressed, the classical profile peaks at zero frequency where it has a zero slope; the classical profile is symmetric in frequency. The quantum profile, on the other hand, peaks at somewhat higher frequencies and has a logarithmic slope of $\hbar/2kT$ near zero frequency. At positive frequencies, the quantum profile is more intense than the classical profile, but at not too small negative frequencies the opposite is true. These facts are related to the different symmetries of these profiles, which we examine in the next subsection. We note that various procedures have been proposed to 'correct' classical profiles somehow so that these simulate the symmetry of quantum profiles.

5.7 Symmetry of line profiles

Detailed balance. If in the lower part of Eq. 2.86 we interchange the arbitrary subscripts i and f, with the help of Eq. 2.82, we have

$$J(-\omega) = e^{-\hbar\omega/kT} \, J(\omega) \,, \tag{5.73}$$

which is often referred to as the detailed balance condition. Equation 5.73 is a consequence of quantum mechanics and must be contrasted with the classical result, $J(-\omega) = J(\omega)$. Analogous equations exist for the binary spectral function, $VG(\omega)$. Equation 5.73 is important because spectral profiles of collision-induced absorption are generally so diffuse that in the wings $|\hbar\omega|$ is comparable to kT. This fact leads to a striking asymmetry in the observed line profiles; compare quantum and classical profiles, Fig. 5.6 which differ increasingly with increasing frequency. We note that ω designates absolute frequencies.

The condition, Eq. 5.73, is also often quoted using frequency *shifts* instead of absolute frequencies; in that case, the symmetry may be altered, see Chapter 6.

Welsh and his associates have pointed out early on that the observed spectral profiles are strikingly asymmetric [422]. Of course, line shapes computed on the basis of a quantum formalism will always have the proper asymmetry so that measurement and theory may be directly compared. Problems may arise, however, if classical profiles are employed for analysis of a measurement, or if classical expressions for computation of spectral moments are used for a comparison with the measurement.

It was widely believed that the main defect of classical line shape can approximately be corrected with the help of one of the various 'desymmetrization' procedures proposed in the literature that formally satisfy Eq. 5.73. However, it has been pointed out that the various procedures give rise to profiles that differ greatly in the wings [70]. While they are *sufficient* to generate the asymmetry, Eq. 5.73, the resulting desymmetrized classical profiles will not necessarily approximate the quantum profile obtained from an *identical input* (dipole and potential function). The symmetrization procedures are not uniquely defined by the detailed balance condition alone. It has been argued that infinitely many different procedures exist to satisfy Eq. 5.73. One wants reasonable consistency of the desymmetrized, classical profile with the quantum profile obtained with an identical input, especially in the far wings.

As a quick illustration, we mention that the two functions,

$$\exp(-\hbar\varpi/2kT) \qquad \text{and} \qquad 2/[1+\exp(-\hbar\varpi/kT)]$$

when used as factors of a quantum profile satisfying Eq. 5.73, each create a symmetric function; their inverse expressions could be used to desymmetrize a classical profile so that it satisfies Eq. 5.73. These functions obviously differ, however, especially at large frequency shifts. Given the fact that the computation of the quantum profile was based on a certain induced dipole and potential function, the classical profile is well defined and cannot possibly agree with both even functions; the classical line shape may even be inconsistent with both symmetric profiles. Desymmetrization does not necessarily make a quantum line shape out of a classical one.

For any given potential and dipole function, at a fixed temperature, the classical and quantum profiles (and their spectral moments) are uniquely defined. If a desymmetrization procedure applied to the classical profile is to be meaningful, it must result in a close approximation of the quantum profile over the required frequency band, or the procedure is a dangerous one to use. On the other hand, if a procedure can be identified which will approximate the quantum profile closely, one may be able to use classical line shapes (which are inexpensive to compute), even in the far wings of induced spectral lines; a computation of quantum line shapes may then be unnecessary.

We will briefly consider several desymmetrization procedures that have been mentioned in the literature. These may simply employ various factors applied to the symmetric, classical profile, $G(\omega)$, or alternatively attempt to correct the classical dipole autocorrelation functions in the time domain.

P-1 Procedure #1:

$$G_D^{(1)}(\omega) = \frac{2}{1+\exp(-\hbar\omega/kT)}\, G_{\text{cl}}(\omega)$$

P-2 Procedure #2:

$$G_D^{(2)}(\omega) = \frac{\hbar\omega}{kT}\,\frac{1}{1-\exp(-\hbar\omega/kT)}\, G_{\text{cl}}(\omega)$$

P-3 Schofield's procedure [360]:

$$G_D^{(3)}(\omega) = \exp(\hbar\omega/2kT)\; G_{\mathrm{cl}}(\omega)$$

P-4 Egelstaff's procedure [143]:

$$G_D^{(4)}(\omega) = \frac{1}{2\pi}\int \mathrm{d}t\, \mathrm{e}^{-\mathrm{i}\omega t}\, C_{\mathrm{cl}}\left(\left[t\,(t-(\mathrm{i}\hbar/kT))\right]^{1/2}\right).$$

In these formulae, G_D is the desymmetrized profile, G_{cl} is the classical (symmetric) line profile; ω and T are angular frequency and temperature. In all cases, the desymmetrized function $G_D(\omega)$ obeys Eq. 5.73 exactly. We note that at low frequencies, $|\hbar\omega| \ll kT$, the four expressions are practically equivalent. However, at high frequencies the results of these desymmetrizations differ strikingly. One needs only to compare the magnitude of the factors of G_{cl} of the upper three defining equations, for $\omega \to \pm\infty$, to realize enormous differences among these. Hence, the question arises as to which one (if any) of these procedures approximates the exact quantum profile, $G(\omega)$. We note that in Egelstaff's procedure the desymmetrization is accomplished in the *time* domain rather than the frequency domain. The classical correlation function, $C_{\mathrm{cl}}(t)$, and spectral function, $G_{\mathrm{cl}}(\omega)$, are related by Fourier transform.

The reasoning behind the desymmetrizations, P-1 through P-4, will be described next. If the quantal spectrum is decomposed into a symmetric and antisymmetric part (which is always possible),

$$E(\omega) = (G(\omega) + G(-\omega))/2 \qquad \text{and} \qquad O(\omega) = (G(\omega) - G(-\omega))/2$$

the odd and even functions must be related by [318, 28]

$$O(\omega) = \tanh(\hbar\omega/2kT)\, E(\omega)\,. \tag{5.74}$$

This is just another way of stating the detailed balance condition, Eq. 5.73.

On the other hand, $O(\omega)$ and $E(\omega)$ are proportional to the Fourier transforms of two real functions of time, the real and imaginary parts of the autocorrelation function,

$$E(\omega) = \frac{1}{2\pi}\int_{-\infty}^{\infty} \mathrm{d}t\, \exp(-\mathrm{i}\omega t)\, C_R(t) \tag{5.75}$$

$$O(\omega) = \frac{\mathrm{i}\hbar}{4\pi kT}\int_{-\infty}^{\infty} \mathrm{d}t\, \exp(-\mathrm{i}\omega t)\, C_I(t)\,.$$

The expansions of $C_R(t)$ and $C_I(t)$ in terms of a power series of Planck's constant show a dependence on even powers of $\hbar$ only, but $C_R(t)$ is symmetric in t and $C_I(t)$ is antisymmetric.

In the classical limit, ($\hbar \to 0$), we have

$$C_R(t) \;\to\; C_{\mathrm{cl}}(t) \tag{5.76}$$

$$C_I(t) \quad \rightarrow \quad -\frac{\mathrm{d}C_{\mathrm{cl}}(t)}{\mathrm{d}t}\,, \tag{5.77}$$

so that the spectrum $G_{\mathrm{cl}}(\omega)$ is the Fourier transform of a function $C(t)$ whose first two terms in the expansion of powers of $\hbar$ are

$$C(t) \simeq C_{\mathrm{cl}}(t) - \frac{i\hbar}{2kT}\,\frac{\mathrm{d}C_{\mathrm{cl}}(t)}{\mathrm{d}t}\,. \tag{5.78}$$

If one were able to find an exact relationship of the classical correlation function, $C_{\mathrm{cl}}(t)$, with $C(t)$ (or $C_R(t)$, $C_I(t)$), no assumptions need to be made and an exact desymmetrization procedure could be defined. This is, however, not possible. We will briefly discuss the assumptions made in the procedures P-1 through P-4 mentioned above.

P-1 Desymmetrization is obtained if we arbitrarily identify $C_R(t)$ with $C_{\mathrm{cl}}(t)$, which of course is correct only in the classical limit, Eq. 5.76; from Eqs. 5.75 and 5.74 one then calculates $E(\omega)$ and $O(\omega)$.

P-2 In this case, we set instead $C_I(t)$ equal to $-\mathrm{d}C_{\mathrm{cl}}(t)/\mathrm{d}t$ as if Eq. 5.77 were valid only in the classical limit, calculate $O(\omega)$ and, with the help of Eq. 5.74, we get $E(\omega)$.

P-3 Schofield's procedure assumes that Eq. 5.78 represents the first two terms of an exact Taylor expansion of the function $C(t)$, with $C_{\mathrm{cl}}(t - i\hbar/2kT)$; by Fourier transform we get at once P-3.

P-4 Egelstaff's procedure may be obtained as a product of two transformations,

$$C_{\mathrm{cl}}(t) \rightarrow C_{\mathrm{cl}}\left(\left[t^2 + \hbar^2/(2(kT)^2)\right]^{1/2}\right) = \widetilde{C}(t) \rightarrow \widetilde{C}\left(t - i\hbar/(2kT)\right)\,. \tag{5.79}$$

The second transformation is again Schofield's, and the first one has been added to assure the exact reproduction of the quantum mechanical ideal gas behavior if particles exchange symmetry effects may be neglected [318, 286].

P-4 (scaled) For reasons that will be apparent below, we also list here a slight variant of Egelstaff's procedure which is identical with Eq. 5.79, except for the rescaling of $\widetilde{C}(t)$ by the constant factor

$$F = C_{\mathrm{cl}}(0)/C_{\mathrm{cl}}(\hbar/\sqrt{2}kT)\,, \tag{5.80}$$

so that $\widetilde{C}(t) = C_{\mathrm{cl}}\left([t^2 + \hbar^2/2(kT)^2]^{1/2}\right) F$. The factor F is greater than unity by a few percent.

For the procedure P-4 one needs to transform from frequency domain to time domain, and back to frequency domain, once the first time transform

Table 5.3. Desymmetrized classical profiles of He–Ar at 295 K for comparison with the quantum profile (last column) obtained from an identical input (dipole [278] and potential [12] functions).

freq.	class.	P-1	P-2	P-3	P-4	P-4(sc)	quan.
(cm^{-1})	(10^{-79} erg cm^6 s)						
3	4571	4604	4605	4605	4309	4709	4673
10	4548	4659	4660	4660	4359	4763	4711
30	4290	4603	4611	4616	4314	4715	4621
60	3525	4037	4066	4080	3791	4143	4048
100	2393	2965	3024	3054	2803	3063	3002
150	1334	1801	1881	1923	1731	1891	1871
200	710.0	1031	1112	1156	1017	1111	1112
250	373.1	576.0	645.7	686.4	588.3	642.8	652.2
300	196.8	319.6	374.7	409.0	340.8	372.4	383.2
350	105.0	177.8	219.0	246.5	199.2	217.7	227.3
400	56.85	99.55	129.3	150.8	117.9	128.8	136.6
450	31.31	56.34	77.33	93.81	70.85	77.42	83.33
500	17.53	32.25	46.84	59.34	43.18	47.19	51.59
600	5.752	10.92	17.79	24.85	16.56	18.09	20.66
700	1.994	3.861	7.039	10.99	6.558	7.167	8.728
800	0.724	1.420	2.884	5.094	2.569	2.807	3.864
900	0.274	0.5403	1.216	2.455	0.9349	1.022	1.783
1000	0.107	0.2124	0.5259	1.226	0.2874	0.3140	0.8519
1100	0.043	0.0861	0.2331	0.6322	0.1140	0.1246	0.4200
1200	0.018	0.0361	0.1061	0.3373			0.2128
1300	0.0079	0.0157	0.0500	0.1873			0.1104
1400	0.0036	0.0072	0.0245	0.1088			0.0585
1500	0.0017	0.0034	0.0126	0.0665			0.0316
1600	0.0009	0.0017	0.0068	0.0432			0.0174

is executed. These Fourier transforms may be done with the help of widely available computer programs, or else, in closed form, through the use of simple transfer functions.

Efficient computer codes exist for transforming from the frequency to the time domain, and vice versa. (We have used the program FOUR1 described in the book *Numerical Recipes* by Press *et al.*, Cambridge 1986.) With the help of such programs, it is straightforward to generate a numerical representation of the classical correlation function from the known classical spectral function. With the help of another standard code, such as a third-order, natural spline interpolation program, we generate the classical correlation function, $C_{cl}(t)$, as a continuous function of time, in a form suitable for the time substitution, $t \rightarrow [t^2 + \hbar^2/2(kT)^2]^{1/2}$. In this way the symmetric, quasi-classical correlation function, $\tilde{C}(t)$, is obtained which, with or without scaling (Eq. 5.80), is converted back to the frequency domain. As the final step, Schofield's desymmetrization P-3 is readily applied in frequency space to complete the P-4 and P-4 (scaled) procedures. The results of the various desymmetrizations are compared in Table 5.3 and Fig. 5.6.

Before looking at the results we mention that, as an alternative to the Fourier transforms just described, one may take advantage of the fact that both the classical line shape, $G_{cl}(\omega)$, and the correlation function, $C_{cl}(t)$, may be represented very closely by an expression as in Eq. 5.110 [70]. The parameters $\tau_1 \cdots \tau_4$, ε and S of these functions are adjusted to match the classical line shape. These six parameter model functions have Fourier transforms that may be expressed in closed form so that the inverse and forward transforms are obtained directly in closed form. We note that the use of transfer functions is merely a convenience, certainly not a necessity as the above discussion has shown.

The desymmetrization procedures are compared in Fig. 5.6 and Table 5.3, using the classical He–Ar profile at 295 K as an example (lowermost curve, solid thin line in the figure; column 2 in the Table). The quantum profile is also shown for comparison (heavy solid line; last column). At positive frequencies, all four procedures mentioned enhance the wing of the classical line shape toward that of the quantum profile. Specifically,

P-1 (dotted line; column 3) approximates the quantum profile at the 2% level up to frequencies around 100 cm^{-1}, but deteriorates quickly above roughly 200 cm^{-1} beyond the 10% level, rendering it less desirable for the analyses of measurements taken over a broad region of frequencies. For the positive frequency wing, the logarithmic slope of the P-1 profile is that of the classical profile, which is very different from that of the quantum profile.

P-2 (not shown in the figure; column 4) provides, together with P-4 (scaled) from which it differs by generally much less than 10%; one of the better approximations to the quantum profile.

P-3 (dash-dot-dash pattern; column 5) is the only one that overestimates the corrections at the higher frequencies; for positive frequencies, it might be a useful upper bound for an estimate of quantum effects of classical calculations. It differs by generally more than 10% from the quantum profile at frequencies greater than 400 cm^{-1}.

P-4 (only the scaled version is shown in the Figure, but columns 6 and 7 of the Table present both the scaled and unscaled P-4 results) provides the closest approach at small frequencies; even at high frequencies it is clearly the best approximation of the quantum profile, especially the scaled version.

Of the four procedures considered, the Egelstaff procedure P-4 (scaled) is clearly the best approximation to the exact quantum profile. It results in the almost exact quantum profile from the classical profile, even far in the wing where intensities have fallen off to a small fraction of the peak intensity. In other words, quantum corrections based on the ideal gas approximation are the leading ones. Next are the extremely simple procedures P-2 and P-3. The widely used procedure P-1, on the other hand, leads to rapidly deteriorating wings and should be avoided, unless limited to a narrow frequency band near the line center.

5.8 Intercollisional interference

A small dip of the absorption in the *Q* branch of hydrogen was recognized as a genuine feature of the collision-induced spectra in the earliest recordings of collision-induced absorption spectra (Welsh 1972). Systematic experimental studies showed that the dip becomes more pronounced at high densities; see Chapters 3 and 6 for details. Similar dips were subsequently discovered at zero frequency in rare gas mixtures [249, 130, 101, 103]; we will be concerned with a theoretical analysis of this interesting feature in this Chapter. The correct interpretation in terms of destructive interference of the dipole moments induced in successive collisions of a given atom of type A with the surrounding atoms B_1, B_2, ... was given by van Kranendonk and Lewis years after the discovery (van Kranendonk 1980; Lewis 1980; 1985). The process is now commonly referred to as *intercollisional interference*, and the dip is called the *intercollisional spectrum*, in contrast to the intracollisional spectra. At low enough densities, the intracollisional spectra are identical with the binary collision-induced spectra discussed in great detail above.

Intercollisional interference is a many-body process. Poll (1980) has pointed out that, no matter how low the gas densities actually are, this many-body effect will always have to be reckoned with, for principal reasons. In more practical terms, at low densities intercollisional dips are generally reasonably well separable from the intracollisional profiles, because intercollisional profiles are relatively sharp while intracollisional ones are rather diffuse. In other words, a reasonably clear distinction between binary and many-body profiles is straightforward in low-density recordings. For this reason, separate theoretical discussions of the intra- and intercollisional processes are convenient and quite natural.

The kinetic theory of intercollisional interference has been worked out by Lewis and van Kranendonk [234, 404, 236, 235]. Being a many-body theory, it is rather complex. We will, therefore, consider here a somewhat simplified version of the theory which includes the essential features and should describe the observations well if densities are not too high and if the mass ratio of atoms A and B is small, $m_A/m_B \ll 1$. More precisely, we will assume i) an infinite mass for the atoms of species B, ii) small concentrations of atoms of species A so that A–A collisions may be neglected, and iii) low enough concentrations of atoms of type B so that the mean duration of the interaction of A with B is small compared to the mean time between collisions A–B ('Lorentz gas approximation'); the effects of ternary and higher-order collisions are here totally ignored. The atoms of type B are considered to form a cloud of randomly distributed, stationary scatterers for the atoms of type A. We want to derive an expression for the intercollisional line shape, based on these three assumptions [236], along with the kinetic theory of gases.

The kinetic theory of gases asserts that the probability of finding an atom in a small volume element, d^3R, with a velocity between $\boldsymbol{v}$ and $\boldsymbol{v} + d\boldsymbol{v}$, is independent of whether or not there is another molecule with known velocity in its immediate vicinity ('molecular chaos' assumption). One can then deduce in the usual way that the number of collisions, per unit volume and per unit time, between molecules with velocities in the ranges d^3v_A and d^3v_B and collision parameters in the range specified by db is equal to $f_A f_B \, b \, db \, d^3v_A \, d^3v_B$, where the f_A, f_B are Maxwellian distributions of velocities. The factor of $f_A d^3v_A$ is interpreted as the probability that a molecule with velocity $\boldsymbol{v}_A$ will undergo a collision specified by a range of impact parameters db and velocities, $d\boldsymbol{v}_B$, independent of the previous history of atom B. In that sense, successive collisions of a given molecule are assumed to be uncorrelated in the standard kinetic theory of gases. However, this does not mean that certain correlation functions, such as the induced dipole autocorrelation function, show no intercollisional correlation; see Fig. 3.4 (p. 70) for an illustration of this important fact. It is clear that the kinetic theory tends to break down for times much longer than the mean time between collisions because of collective effects which are ignored in the kinetic theory. This breakdown is, however, spectroscopically insignificant because the induced dipole moments generally fall off to small values rapidly with increasing separations; dipole correlation functions fall off to zero on a time scale much longer than the mean free time between collisions at low densities (Lewis 1980).

Line shape function. Given a mixture of two monatomic gases of densities N_A/V and N_B/V in the volume V, with $N_A \ll N_B$. The dipole moment autocorrelation function is given by

$$C(t) = \frac{1}{2}\frac{1}{4\pi\varepsilon_0}\langle \boldsymbol{\mu}(0)\cdot\boldsymbol{\mu}(t) + \boldsymbol{\mu}(t)\cdot\boldsymbol{\mu}(0)\rangle .$$

The brackets denote an ensemble average and $\boldsymbol{\mu}(t)$ is the total dipole moment. The translational motion of the atoms is treated classically and, in the spirit of the kinetic theory, the ensemble average is replaced by a time average, $C(\tau) = \frac{1}{4\pi\varepsilon_0}\langle \boldsymbol{\mu}(t)\cdot\boldsymbol{\mu}(t+\tau)\rangle$.

With the assumption that the atomic masses are very different, $m_A/m_B \ll 1$, a significant simplification is introduced. The atoms of type B may be assumed to be at rest in the laboratory frame and the relative velocities of the A–B encounters are equal to the velocities of A measured in the laboratory frame. The density of atoms of type B is assumed small enough so that the mean time τ_c between optical collisions of the type A–B is much longer than the mean duration τ_d of these collisions; the density N_A/V of atoms A is assumed small and collisions of the type A–A are neglegible. The range of the induced dipole falls off to insignificant values rapidly with increasing separation so that $\tau_c \gg \tau_d$.

Since elastic collisions conserve the relative speed throughout the history of each atom A, its speed, v_A, remains constant. For the computation of the correlation function, we consider the time average,

$$C(\tau; v_A) = \frac{1}{4\pi\varepsilon_0}\lim_{T\to\infty}\frac{1}{T}\int_0^T \boldsymbol{\mu}(t)\cdot\boldsymbol{\mu}(t+\tau)\,\mathrm{d}t \tag{5.81}$$

and the velocity average

$$C(\tau) = \int_0^\infty P_A(v_A)\; C(\tau; v_A)\; \mathrm{d}v_A\,, \tag{5.82}$$

where $P_A(v_A)$ is the probability distribution of speeds v_A (as in Eq. 2.10).

Next we follow the history of one atom of type A over a time $T \gg \tau_c$ in which the atom makes N collisions at times t_i, with $i = 1, \cdots N$, where t_i is the time of closest approach of a collision. The dipole moment $\boldsymbol{\mu}(t)$ may be written as the superposition of the moments induced in subsequent binary collisions, namely

$$\boldsymbol{\mu}(t) = \sum_{i=1}^{N} \boldsymbol{\mu}_i\,(t - t_i)\,,$$

where $\boldsymbol{\mu}_i$ is the dipole moment induced in the ith collision; ternary collisions are neglected. Accordingly, we may write

$$\boldsymbol{\mu}(t)\cdot\boldsymbol{\mu}(t+\tau) \;=\; \sum_{i,j}\boldsymbol{\mu}_i(t-t_i)\cdot\boldsymbol{\mu}_j(t+\tau-t_j)$$

$$= \sum_i \boldsymbol{\mu}_i(t - t_i) \cdot \boldsymbol{\mu}_i(t + \tau - t_i)$$
$$+ \sum_i [\boldsymbol{\mu}_i(t - t_i) \cdot \boldsymbol{\mu}_{i+1}(t + \tau - t_i - x_i)$$
$$+ \boldsymbol{\mu}_{i+1}(t - t_i - x_i) \cdot \boldsymbol{\mu}_i(t + \tau - t_i)]$$
$$+ \cdots ,$$

where $x_i = t_{i+1} - t_i > 0$ is the time interval between successive collisions, i and $i + 1$. The $(n + 1)$th term in this series expresses the correlation between the ith and $(i + n)$th collisions.

We substitute this expression into Eq. 5.81 and get

$$C(\tau; v_A) = \sum_{n=0}^{\infty} C_n(\tau; v_A) , \tag{5.83}$$

where

$$C_0(\tau; v_A) = \lim_{T \to \infty} \frac{1}{T} \sum_i C_{i,i}(\tau; v_A)$$

and

$$C_n(\tau; v_A) = \lim_{T \to \infty} \frac{1}{T} \sum_i [C_{i,i+n}(x_i + \cdots + x_{i+n-1} - \tau; v_A) + C_{i,i+n}(x_i + \cdots + x_{i+n-1} + \tau; v_A)] , \tag{5.84}$$

for $n = 1, 2, \ldots$, with

$$C_{i,i+n}(\tau; v_A) = \frac{1}{4\pi\varepsilon_0} \int_{-\infty}^{\infty} \boldsymbol{\mu}_i(t) \cdot \boldsymbol{\mu}_{i+n}(t - \tau)\, dt .$$

These expressions $C_{i,i+n}$ depend not only on v_A, but also on the collision parameters of collisions i and $i + n$.

The expression on the right-hand side of Eq. 5.84 is equal to the collision frequency, $v_c(v_A) = \lim N/T$ for $T \to \infty$, times the average value of the quantity in square brackets averaged over the time intervals, $x_i, \ldots x_{i+n-1}$, and over the collision variables of collisions $i, i+1, \ldots i+n$. These collision variables are the speed v_A and the scattering angle χ subtended by the directions of the velocity unit vectors, $\hat{\boldsymbol{v}}_i = \boldsymbol{v}_A/v_A$, before and after the respective collision. Since intermolecular forces are repulsive at near range and attractive at distant range, the impact parameter $b(\chi)$ is in general a multi-valued function of the scattering angle, so that we must be careful to sum over the various branches of the function $b(\chi)$ where these exist. The collision sequence i, $i + 1$, ..., $i + n$ can be specified by the speed v_A, the time intervals x_i, ..., x_{i+n-1}, and the collision variables $\hat{\boldsymbol{v}}_i$, ..., $\hat{\boldsymbol{v}}_{i+n+1}$.

It is assumed that the distribution of the collision variables $\hat{\boldsymbol{v}}_i, \cdots \hat{\boldsymbol{v}}_{i+n+1}$ is independent of the time intervals $x_i, \ldots x_{i+n-1}$. The average value of the

$C_{i,i+n}(x_i + \cdots + x_{i+n-1} - \tau; v_A)$ over the collision variables $\hat{v}_i, \cdots \hat{v}_{i+n+1}$ is written as

$$\langle C_{i,i+n}(x_i + \cdots + x_{i+n-1} - \tau; v_A)\rangle_i \, .$$

In this way, we may rewrite the above equations in the form

$$C_0(\tau; v_A) = v_c \langle C_{i,i}(\tau; v_A)\rangle_i \tag{5.85}$$

$$\begin{aligned} C_n(\tau; v_A) = {} & v_c \int_0^\infty \cdots \int_0^\infty p(x_i, \cdots x_{i+n-1}) \, \mathrm{d}x_i \cdots \mathrm{d}x_{i+n-1} \\ & \times \Big[\langle C_{i,i+n}(x_i + \cdots x_{i+n-1} - \tau; v_A)\rangle_i \\ & + \langle C_{i,i+n}(x_i + \cdots x_{i+n-1} + \tau; v_A)\rangle_i\Big] \end{aligned} \tag{5.86}$$

for $n = 1, 2, \ldots$, where $p(x_i, \cdots x_{i+n-1})$ is the distribution function of the intervals x_m, with uncorrelated x_m, according to

$$p(x_m) = v_c \exp(-v_c \, x_m) \, . \tag{5.87}$$

The line shape is the Fourier transform of the dipole autocorrelation function,

$$J(\omega) = \int_{-\infty}^{\infty} \exp(-\mathrm{i}\omega\tau) \, C(\tau) \, \mathrm{d}\tau \, .$$

Using Eqs. 5.82 and 5.83, we may write $J(\omega) = \sum_{n=0}^{\infty} J_n(\omega)$, where

$$J_n(\omega) = \int_{-\infty}^{\infty} \mathrm{d}\tau \exp(\mathrm{i}\omega\tau) \int_0^\infty \mathrm{d}v_A \, P_A(v_A) \; C_n(\tau; v_A) \, .$$

With the help of Eqs. 5.85, 5.86 and 5.87 we thus get

$$\begin{aligned} J(\omega) = {} & \int_0^\infty P_A(v_A) \, \frac{1}{4\pi\varepsilon_0} \, v_c \Big[\langle \boldsymbol{\mu}_i(\omega) \cdot \boldsymbol{\mu}_i(-\omega)\rangle_i \\ & + 2\Re\Big\{\sum_{n=1}^{\infty} \left(\frac{v_c}{v_c + \mathrm{i}\omega}\right)^n \langle \boldsymbol{\mu}_i(\omega) \cdot \boldsymbol{\mu}_{i+n}(-\omega)\rangle_i\Big\}\Big] \, \mathrm{d}v_A \, , \end{aligned} \tag{5.88}$$

where $\boldsymbol{\mu}(\omega)$ is the Fourier transform of the induced dipole moment as function of time,

$$\boldsymbol{\mu}_i(\omega) = \int_{-\infty}^{\infty} \exp(\mathrm{i}\omega\tau) \; \boldsymbol{\mu}_i(\tau) \; \mathrm{d}\tau \, .$$

The leading term on the right-hand side of Eq. 5.88 is, of course, the intracollisional profile, $g(\omega)$, that is the spectrum that would be observed if no correlations existed between the dipoles induced in successive collisions. The remaining term describes the intercollisional process, which is the central theme of this subsection.

For a collision (which we will arbitrarily label 1) the directions of initial and final velocities are given by the unit vectors $\hat{v}_1$ and $\hat{v}_2$. The scattering

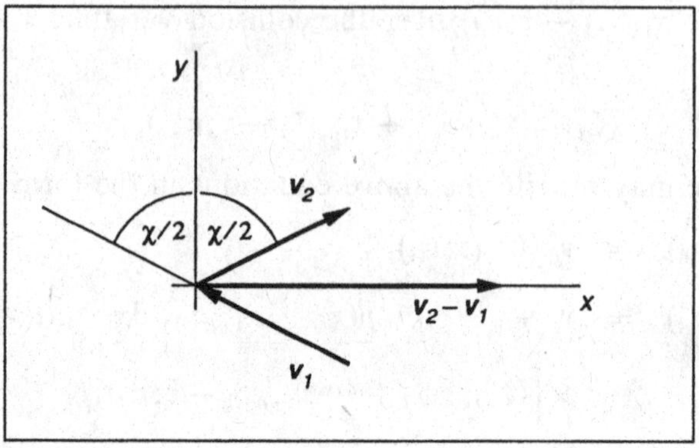

Fig. 5.7. The Cartesian x, y frame in the collision plane. Velocities of the atom of type A before and after the collision labeled 1 are $\boldsymbol{v}_1$ and $\boldsymbol{v}_2$; χ is the scattering angle. The x-axis is aligned with the direction of the velocity change, $\boldsymbol{v}_2 - \boldsymbol{v}_1$.

angle χ is the angle subtended by these vectors and φ is the azimuth of $\widehat{\boldsymbol{v}}_2$ relative to $\widehat{\boldsymbol{v}}_1$ (z-axis) and some plane. For central forces we describe the time evolution of the encounter by the magnitude $R = R(t - t_1)$ and the polar angle $\vartheta = \vartheta(t - t_1)$ of the radius arm vector in the collision plane, with $\vartheta(0) = 0$ and $\vartheta < 0$ for $t < t_1$. Hence, for classical trajectories,

$$R(-t) = R(t) \qquad \text{and} \qquad \vartheta(-t) = -\vartheta(t)\,. \tag{5.89}$$

We note that the scattering angle χ and the polar angle at $t \to \infty$ are related according to $\chi + 2\vartheta(\infty) = \pi$. The dipole moment $\boldsymbol{\mu}(\boldsymbol{R})$ is parallel to the intermolecular axis. Introducing a Cartesian frame with its x-axis along the apse line and the y-axis in the collision plane as illustrated in Fig. 5.7, we get from the definition of $\boldsymbol{\mu}_i(\omega)$ for $i = 1$,

$$\begin{aligned} \boldsymbol{\mu}_1(\omega) &= \int_{-\infty}^{\infty} \exp(\mathrm{i}\omega\tau)\, \mu_0(R(t)) \begin{Bmatrix} \cos\vartheta(t) \\ \sin\vartheta(t) \\ 0 \end{Bmatrix} dt \\ &= \begin{Bmatrix} \mu_x(\omega;\chi,v_A) \\ i\mu_y(\omega;\chi,v_A) \\ 0 \end{Bmatrix}, \end{aligned} \tag{5.90}$$

with $\mu_0 = \boldsymbol{\mu} \cdot \widehat{\boldsymbol{R}}$. Making use of Eq. 5.89, we have

$$\begin{Bmatrix} \mu_x(\omega;\chi,v_A) \\ \mu_y(\omega;\chi,v_A) \end{Bmatrix} = 2\int_0^{\infty} \mu_0(R(t))\, dt \begin{Bmatrix} \cos\omega t \, \cos\vartheta(t) \\ \sin\omega t \, \sin\vartheta(t) \end{Bmatrix}. \tag{5.91}$$

The vectors $\boldsymbol{\mu}_m(t)$ and $\boldsymbol{\mu}_m(\omega)$ lie in the plane of the mth collision. Accordingly, $\boldsymbol{\mu}_m(\omega)$ is a random vector uniformly distributed in its azimuth about $\widehat{\boldsymbol{v}}_m$, the direction of the atom of type A as it approaches an atom B during an early phase of the mth collision. In order to compute the scalar products $\boldsymbol{\mu}_i(\omega) \cdot \boldsymbol{\mu}_{i+n}(-\omega)$ appearing in Eq. 5.88, we must average $\boldsymbol{\mu}_{i+n}(-\omega)$ over $\widehat{\boldsymbol{v}}_{i+n+1}$ to give a vector parallel to $\widehat{\boldsymbol{v}}_{i+n}$,

$$\langle \boldsymbol{\mu}_{i+n}(-\omega) \rangle_{\widehat{\boldsymbol{v}}_{i+n+1}} = A\,(-\omega, v_A)\ \widehat{\boldsymbol{v}}_{i+n}\ ,$$

where $A(-\omega, v_A)$ is independent of $\widehat{\boldsymbol{v}}_{i+n}$. Next we average the right-hand side over $\widehat{\boldsymbol{v}}_{i+n}$, keeping $\widehat{\boldsymbol{v}}_{i+n-1}$ fixed to give a vector parallel to $\widehat{\boldsymbol{v}}_{i+n-1}$,

$$\langle \boldsymbol{\mu}_{i+n}(-\omega) \rangle_{\widehat{\boldsymbol{v}}_{i+n+1}, \widehat{\boldsymbol{v}}_{i+n}} = A(-\omega, v_A)\ \Delta\,(v_A)\ \widehat{\boldsymbol{v}}_{i+n-1}\ ,$$

where Δ is independent of $\widehat{\boldsymbol{v}}_{i+n-1}$ and ω. Averaging the right-hand side over $\widehat{\boldsymbol{v}}_{i+n-1}, \widehat{\boldsymbol{v}}_{i+n-2}, \ldots$, one finally obtains

$$\langle \boldsymbol{\mu}_i(\omega) \cdot \boldsymbol{\mu}_{i+n}(-\omega) \rangle_i = A\,(-\omega, v_A)\ B\,(\omega, v_A)\ \Delta(v_A)^{n-1}\ ,$$

where

$$\begin{aligned} B\,(\omega, v_A) &= \langle \boldsymbol{\mu}_i(\omega) \cdot \widehat{\boldsymbol{v}}_{i+1} \rangle_{\widehat{\boldsymbol{v}}_{i+1}} \\ &= \langle \boldsymbol{\mu}_1\,(\omega; \chi, v_A)\ (\boldsymbol{v}_2 - \boldsymbol{v}_1) \cdot \boldsymbol{v}_2 \rangle_{\chi,\varphi} \\ &= -A^*\,(\omega, v_A)\ . \end{aligned}$$

A and B are both complex functions of ω.

Using Eq. 5.90, we obtain the real and imaginary parts of A as

$$\begin{aligned} \Re\left\{A\,(\omega, v_A)\right\} &= -\int_0^\pi \sin\frac{1}{2}\chi\ \mu_x\,(\omega; \chi, v_A)\ P\,(\chi; v_A)\ \sin\chi\ d\chi \\ \Im\left\{A\,(\omega, v_A)\right\} &= \int_0^\pi \cos\frac{1}{2}\chi\ \mu_y\,(\omega; \chi, v_A)\ P\,(\chi; v_A)\ \sin\chi\ d\chi\ . \end{aligned}$$

According to Eq. 5.91, these are even and odd functions of ω, respectively. Since the average of $\widehat{\boldsymbol{v}}_{i+n}$ over $\widehat{\boldsymbol{v}}_{i+n}$ for fixed $\widehat{\boldsymbol{v}}_{i+n-1}$ is given by

$$\langle \widehat{\boldsymbol{v}}_{i+n} \rangle_{\widehat{\boldsymbol{v}}_{i+n}} = \Delta\,(v_A)\ \widehat{\boldsymbol{v}}_{i+n-1}\ ,$$

with a function $\Delta(v_A)$ that is independent of $\widehat{\boldsymbol{v}}_{i+n-1}$, namely

$$\begin{aligned} \Delta\,(v_A)\,\widehat{\boldsymbol{v}}_1 &= \langle \widehat{\boldsymbol{v}}_2 \rangle_{\chi,\varphi} \\ &= \langle \widehat{\boldsymbol{v}}_1 \cdot \widehat{\boldsymbol{v}}_2 \rangle_\chi\ \widehat{\boldsymbol{v}}_1 \\ &= \int_0^\pi \cos\chi\ P\,(\chi; v_A)\ \sin\chi\ d\chi\ \widehat{\boldsymbol{v}}_1\ , \end{aligned}$$

which defines the function $\Delta(v_A)$; it is obviously independent of $\boldsymbol{\mu}$.

Putting it all together, we have

$$\langle \boldsymbol{\mu}_i(\omega) \cdot \boldsymbol{\mu}_{i+n}(-\omega) \rangle_i = -\Delta\,(v_A)^{n-1} \left(\Re\left\{A(\omega, v_A)\right\} - \mathrm{i}\Im\left\{A(\omega, v_A)\right\} \right)^2 .$$

Finally, with the help of Eq. 5.88, by summing the geometric series over n, we get the line shape according to

$$J(\omega) = g(\omega)\, n_A\, n_B \tag{5.92}$$

$$- 2\frac{1}{4\pi\varepsilon_0}\int_0^\infty \frac{dv_A\, P_A(v_A)\, v_c(v_A)}{(1-\Delta(v_A))^2 + (\omega/v_c(v_A))^2}$$
$$\times \left[(1-\Delta(v_A))\left(\Re\left\{A(\omega,v_A)\right\} + \Im\left\{A(\omega,v_A)\right\}\right)^2 + \left(\frac{\omega}{v_c(v_A)}\right)\Re\left\{A(\omega,v_A)\right\}\Im\left\{A(\omega,v_A)\right\}\right].$$

In this expression, $g(\omega)$ is the intracollisional part,

$$g(\omega)\, n_A n_B = \int_0^\infty dv_A\, P_A(v_A)\; v_c(v_A) \times \int_0^\pi \frac{1}{4\pi\varepsilon_0}\mu(\omega;\chi,v_A)^2\; P(\chi,v_A)\,\sin\chi\, d\chi\,,$$

where $\mu^2 = \mu_x^2 + \mu_y^2$.

In the Lorentz gas approximation, this term is proportional to the number densities of atoms of type A and B, n_A and n_B, because the probability of finding an atom of the light species with a speed between v_A and $v_A = dv_A$ is given by the Maxwellian distribution function,

$$P_A(v_A)\, dv_A = N_A\left(\frac{m_A}{2\pi kT}\right)^{3/2} \exp\left(-m_A v_A^2/2kT\right)\, 4\pi\, v_A^2\, dv_A$$

with $n_A = N_A/V$, and the frequency v_c of optical collisions is proportional to n_B,

$$v_c(v_A) = n_B\, v_A\, \sigma_{AB}\;; \tag{5.93}$$

the cross section for elastic collision of dissimilar atoms is called σ_{AB}; it is a function of the speed v_A.

The integral in Eq. 5.92, the other term besides $g(\omega)$, is of course the intercollisional profile. Its density dependence is more complex; it is proportional to the product of densities $n_A n_B$ because of the appearance of P_A and v_c in the numerator of the first fraction under the integral. Furthermore, n_B appears in the form of the collision frequency once more in the denominator of that fraction, as well as in the denominator of the last term. In deriving Eq. 5.92, we have summed over many collisions of a selected atom of type A with the various atoms B. This summation is responsible for the complex density dependence observed here.

The spectral profile, $J(\omega)$ Eq. 5.92, is the average of the speed v_A of the superposition of the intracollisional and intercollisional profile. The sign of the intercollisional term depends on the product of real and imaginary parts of A, $\Re\{A\}$ and $\Im\{A\}$, Eq. 5.92, which will be positive at low frequencies if both, the dipole function and the interatomic potential are short range [236]. In other words, the intercollisional absorption profile is *negative* under such conditions; an absorption dip is obtained, in agreement with the observations.

Lewis gives several alternative forms of the line shape function, $J(\omega)$, for example for the simplified case of zero duration of the interatomic interactions [236], and for finite mass ratio m_A/m_B [235].

Computations. The kinetic theory of intercollisional interference is a classical theory. We briefly summarize the basic relationships needed for the computation of a line profile from Newtonian mechanics.

For every value of the impact parameter b, at a given speed of the encounter, v_A, Newton's equation of motion predicts a scattering angle χ from the knowledge of the A–B interaction potential (assumed to be isotropic). Conversely, for every scattering angle χ, one or more values of the impact parameter exist that will produce scattering at that angle. In that case, $b(\chi)$ is a multi-valued function. Intermolecular forces are attractive at long range and repulsive at near range which gives rise to such behavior; Fig. 2.4. We define a differential scattering cross section of elastic scattering of an atom A by B in terms of the function $b(\chi)$ by the relationship,

$$\frac{\partial^2 \sigma_{AB}(v_A)}{\partial^2 \Omega} \, \mathrm{d}\Omega = \sum_{\chi} b(\chi) \left| \frac{\mathrm{d}b}{\mathrm{d}\chi} \right| \mathrm{d}\chi \, \mathrm{d}\varphi \, ,$$

where $\mathrm{d}^2\Omega = \sin\chi \, \mathrm{d}\chi \, \mathrm{d}\varphi$ is the solid angle element defined by the polar angle χ and azimuth φ. (For central forces, the integration over the azimuth can be immediately performed and results in the factor 2π.) The sum is over all values $b(\chi)$ with the same scattering angle χ, in the case more than one b value exist which give rise to the same scattering angle χ. The right-hand side of this relationship is often written as $b \, \mathrm{d}b \, \mathrm{d}\varphi$. The total scattering cross section for A–B collisions is given by

$$\sigma_{AB}\,(v_A) = \int\!\!\int \frac{\partial^2 \sigma_{AB}(v_A)}{\partial^2 \Omega} \, \mathrm{d}^2\Omega = 2\pi \int b \left| \frac{\mathrm{d}b}{\mathrm{d}\chi} \right| \mathrm{d}\chi \, , \qquad (5.94)$$

The last integral is often abbreviated as $\int b \, \mathrm{d}b$ which is, of course, finite when read correctly. The collision frequency may be computed in terms of the function $b(\chi)$ by combining Eqs. 5.94 with 5.93. The probability $P(\chi; v_A)$ of scattering by the angle χ needed for the computation of the quantities A, B and Δ introduced above is similarly expressed in terms of $b(\chi)$ as

$$P\,(\chi; v_A) = \frac{2\pi}{\sigma_{AB}} \frac{\partial^2 \sigma_{AB}}{\partial^2 \Omega} \, .$$

In other words, all required quantities may be computed from classical mechanics if the interaction potential and dipole function are known. Greater detail may be found elsewhere, see, for example, Hirschfelder, Curtiss and Bird (1964), or Chapman and Cowling (1960).

5.9 Other approaches to line shape

To the extent that the dipole induced in a cluster of atoms may be represented by a sum of pair dipoles induced in dissimilar atoms, i and i', $\boldsymbol{\mu}_{ii'} = \boldsymbol{\mu}(\boldsymbol{R}_{ii'})$, we may write for the total dipole $\boldsymbol{\mu} = \sum_{ii'} \boldsymbol{\mu}_{ii'}$. The summation is over all atoms i and all atoms i'. The dipole correlation function can then be written as a sum of three parts (Poll 1980),

$$C(t) = \frac{1}{V}\left\langle \sum_{ii'} \boldsymbol{\mu}_{ii'}(0) \cdot \sum_{jj'} \boldsymbol{\mu}_{jj'}(t) \right\rangle = C_2(t) + C_3(t) + C_4(t) , \qquad (5.95)$$

where

$$\begin{aligned} C_2(t) &= \frac{1}{V}\left\langle \sum_{ii'} \boldsymbol{\mu}_{ii'}(0) \cdot \boldsymbol{\mu}_{ii'}(t) \right\rangle \\ C_3(t) &= \frac{1}{V}\left\langle \sum_{i} \sum_{i' \neq j'} \boldsymbol{\mu}_{ii'}(0) \cdot \boldsymbol{\mu}_{ij'}(t) \right\rangle + \frac{1}{V}\left\langle \sum_{i'} \sum_{i \neq j} \boldsymbol{\mu}_{ii'}(0) \cdot \boldsymbol{\mu}_{ji'}(t) \right\rangle \\ C_4(t) &= \frac{1}{V}\left\langle \sum_{i \neq j} \sum_{i' \neq j'} \boldsymbol{\mu}_{ii'}(0) \cdot \boldsymbol{\mu}_{jj'}(t) \right\rangle . \end{aligned}$$

The correlation function $C_2(t)$ models the binary interactions. The three-particle function $C_3(t)$ describes the interactions of one atom of one kind with two atoms of the other kind; it includes the intercollisional effect. The four-particle function $C_4(t)$ models the correlations between distinct pairs of dissimilar atoms.

The pair dipole may be expressed according to

$$\boldsymbol{\mu}_{ii'}(t) = \boldsymbol{\mu}(\boldsymbol{R}_{ii'}(t)) = \int \mathrm{d}^3R' \, \boldsymbol{\mu}(\boldsymbol{R}') \, \delta(\boldsymbol{R}' - \boldsymbol{R}_{ii'}(t)) ,$$

where $\boldsymbol{R}_{ii'}(t)$ is the Heisenberg operator corresponding to the separation of the atoms i and i' and $\boldsymbol{R}'$ is the integration variable. The dipole correlation function may thus be written

$$C(t) = \int \mathrm{d}^3R' \int \mathrm{d}^3R'' \boldsymbol{\mu}(\boldsymbol{R}') \cdot \boldsymbol{\mu}(\boldsymbol{R}'') \; G(\boldsymbol{R}', \boldsymbol{R}''; t) . \qquad (5.96)$$

The function $G(\boldsymbol{R}', \boldsymbol{R}''; t)$ can always be split into two parts that are symmetric and antisymmetric under interchange of $\boldsymbol{R}'$ and $\boldsymbol{R}''$, respectively. The antisymmetric part does not contribute to $C(t)$ so that we may write

$$G(\boldsymbol{R}', \boldsymbol{R}''; t) = \frac{1}{2V}\left\langle \sum_{ii'} \delta(\boldsymbol{R}' - \boldsymbol{R}_{ii'}(0)) \sum_{jj''} \delta(\boldsymbol{R}'' - \boldsymbol{R}_{jj'}(t)) + \mathrm{cycl} \right\rangle , \qquad (5.97)$$

cycl refers to a term which is identical to the preceding one, except for the interchange of $\boldsymbol{R}'$ and $\boldsymbol{R}''$. The function G can be written out in three parts, corresponding to the pair-, three-particle and four-particle functions introduced above.

The correlation functions G are the same functions which Oppenheimer and Bloom introduced in their theory of nuclear spin relaxation [304]. The function has been called the time-dependent pair distribution function; it should not, however, be confused with the van Hove correlation function which is often referred to by the same name. The name time-dependent intermolecular correlation function (TDICF) has, therefore, been proposed for G [71].

Constant acceleration approximation. An approximation introduced to the time-dependent intermolecular correlation function G, which was commonly referred to as the constant acceleration approximation (CAA), was used to compute the line shapes of collision-induced absorption spectra of rare gas mixtures, but the computed profiles were found to be unsatisfactory [286]. It does not give the correct first spectral moment.

A simple extension of the constant acceleration approximation was later introduced which gave results that agree rather well with the measured spectral profiles and moments [71]. The model has no free parameters although the required value of the derivative of the potential may be used as an adjustable parameter if desired. The computational efforts are minor and the extended constant acceleration approximation should be useful for all types of short-range induction components.

We start with a simple model of the two-particle intracollisional correlation function in the semiclassical limit [324],

$$\begin{aligned} G_2(\boldsymbol{R}', \boldsymbol{R}''; t) &= n_a n_b \frac{V}{2} \langle \delta(\boldsymbol{R}' - \boldsymbol{R}(0))\, \delta(\boldsymbol{R}'' - \boldsymbol{R}(t)) \\ &\quad + \delta(\boldsymbol{R}'' - \boldsymbol{R}(0))\, \delta(\boldsymbol{R}' - \boldsymbol{R}(t)) \rangle \\ &= n_a n_b\, g(\boldsymbol{R}', \boldsymbol{R}''; t)\ , \end{aligned}$$

where $\langle \cdots \rangle$ denotes an ensemble average over the relative motion of the pair. The separation of the pair is $\boldsymbol{R}(t)$.

The correlation function g thus defined can be written in terms of the propagator of relative motion, K, as

$$\begin{aligned} g(\boldsymbol{R}', \boldsymbol{R}''; t) &= \frac{1}{2} \lambda^3 \Big[K(\boldsymbol{R}', \boldsymbol{R}''; t)^* \, K(\boldsymbol{R}', \boldsymbol{R}''; t - \mathrm{i}(\hbar/kT)) \\ &\quad + K(\boldsymbol{R}'', \boldsymbol{R}'; t)^* \, K(\boldsymbol{R}'', \boldsymbol{R}'; t - \mathrm{i}(\hbar/kT)) \Big]\ ; \end{aligned}$$

λ designates the de Broglie wavelength. The propagator is given by the Hamiltonian $\mathscr{H}$ of relative motion as

$$K(\boldsymbol{R}', \boldsymbol{R}''; t) = \langle \boldsymbol{R}'' | \exp(-\mathrm{i}\mathscr{H} t/\hbar) | \boldsymbol{R}' \rangle$$

and can be written as a Feynman path integral,

$$K(\boldsymbol{R}',\boldsymbol{R}'';t) = \int \mathrm{d}\boldsymbol{R}_t \exp\left[\mathrm{i}S(\boldsymbol{R}_t)/\hbar\right] . \tag{5.98}$$

The integral is over all possible paths $\boldsymbol{R}_t$ which connect the points $\boldsymbol{R}'$ and $\boldsymbol{R}''$ in time t, and $S(\boldsymbol{R}_t)$ is the action along the path.

In the semiclassical limit, only the classically allowed paths are taken into account, and we get

$$\begin{aligned} K(\boldsymbol{R}',\boldsymbol{R}'';t) &= \sum_p \left(\frac{\mathrm{i}}{2\pi\hbar}\right)^{3/2} \left[\mathrm{Det}\frac{\partial^2 S^p_{\mathrm{cl}}(\boldsymbol{R}',\boldsymbol{R}'';t)}{\partial\boldsymbol{R}'\,\partial\boldsymbol{R}''}\right]^{1/2} \\ &\quad \times \exp\left[\mathrm{i}S^p_{\mathrm{cl}}(\boldsymbol{R}',\boldsymbol{R}'';t)/\hbar\right] . \end{aligned} \tag{5.99}$$

The sum is over all classical paths connecting $\boldsymbol{R}'$ and $\boldsymbol{R}''$ in time t, S^p_{cl} is the classical action along such paths, and Det denotes the determinant which ensures unitarity of the propagator K up to second-order variations in S. The classical action is a solution of the Hamilton–Jacobi equation,

$$\frac{1}{2m}(\nabla''S)^2 + V(\boldsymbol{R}'') = -\frac{\partial S}{\partial t} .$$

We assume that for short times we can write

$$S = S_0 - t\,S_1 - t^3\,S_2 + \cdots ,$$

where $S_0 = m(\boldsymbol{R}'' - \boldsymbol{R}')^2/(2t)$ is the classical action for free particles, and S_1, S_2, ..., are time independent.

We assume, furthermore, that the interaction takes place during very short intervals of time. In this approximation, there is only one classical path for which we solve the Hamilton–Jacobi equation. In the vicinity of the collision diameter (where the spectroscopically significant interactions take place) we have, to a good approximation (Poll 1980)

$$\begin{aligned} S_1(\boldsymbol{R}',\boldsymbol{R}'') &= \frac{1}{|\boldsymbol{R}''-\boldsymbol{R}'|}\int_{\boldsymbol{R}'}^{\boldsymbol{R}''} V(R)\,\mathrm{d}\boldsymbol{R} , \\ S_2(\boldsymbol{R}',\boldsymbol{R}'') &= \frac{1}{|\boldsymbol{R}''-\boldsymbol{R}'|^3}\int_{\boldsymbol{R}'}^{\boldsymbol{R}''} \frac{1}{2m}\left[\nabla S_1(\boldsymbol{R}',\boldsymbol{R})\right]^2 (\boldsymbol{R}-\boldsymbol{R}')^2\,\mathrm{d}\boldsymbol{R} . \end{aligned} \tag{5.100}$$

Since we are dealing with a small range of intermolecular separations, in first approximation, we can write

$$\begin{aligned} S_1(\boldsymbol{R}',\boldsymbol{R}'') &\approx \frac{1}{2}\left(V(R'') + V(R')\right) , \\ S_2(\boldsymbol{R}',\boldsymbol{R}'') &\approx \frac{1}{24m}\left(\frac{\mathrm{d}V(R)}{\mathrm{d}R}\right)^2\Bigg|_{R=R_0} , \end{aligned} \tag{5.101}$$

with $R' \le R_0 \le R''$, provided that the derivative of the potential does not vary too much.

Combining the above results, we write the intramolecular correlation function as

$$\begin{aligned} g\,(\boldsymbol{R}',\boldsymbol{R}'';t) &= g_0\,(\boldsymbol{R}',\boldsymbol{R}'';t)\exp\left[-V(R'')/(2kT)\right]\,\exp\left[-V(R')/(2kT)\right] \\ &\quad\times\exp\left(-\frac{y^2}{8mkT}\,V'^2(R_0)+\frac{\hbar^2}{24m(kT)^3}\,V'^2(R_0)\right). \end{aligned} \tag{5.102}$$

In this expression,

$$g_0\,(\boldsymbol{R}',\boldsymbol{R}'';t)=\left(\frac{m}{2\pi y^2kT}\right)^{3/2}\exp\left[-m\left(\boldsymbol{R}'-\boldsymbol{R}''\right)^2/(2y^2kT)\right]$$

is the quantum mechanical free-particle correlation function. It has been obtained from the classical one by substituting for t the Egelstaff time, y, defined by $y^2=t^2-i\hbar t/(kT)$.

By setting the derivative of the potential equal to zero in the lower Eq. 5.101, Oppenheim and Bloom's constant acceleration approximation is obtained; a more appropriate name would be 'zero acceleration approximation'. By avoiding neglect of the derivative of the potential, one has a simple and certainly more accurate approximation [71].

Specifically, an exact quantum profile of Ne–Ar pairs (which was in agreement with the measurement) was compared with an extended CAA profile computation, using the same input [71]. The approximation to the profile did not differ from the exact profile by more than 10% even in the far wing, where the intensities have dropped by nearly two orders of magnitude relative to the peak. Considering the extreme simplicity of the model, this agreement is remarkable.

In practical terms, the quality of approximation obtained with the extended constant acceleration approximation is comparable to that of the best *ad hoc* model profiles to be discussed at the end of this Chapter. This approximation does not make any *ad hoc* assumptions concerning the line shapes. It may be considered a one-parameter profile (when the derivative of the potential is replaced by an adjustable parameter) and is as such somewhat inferior to the alternative three-parameter profiles mentioned in the cases where a direct comparison is possible. It is noteworthy that the CAA theory can be further refined in a number of ways that will doubtlessly be investigated in the future.

Memory function. Spectral profiles may be computed from the memory function, $K(t)$, which is related to the dipole autocorrelation function, $C(t)$, according to

$$\frac{\mathrm{d}C(t)}{\mathrm{d}t}=-\int_0^t K\,(t-t')\;C\,(t')\;\mathrm{d}t'\,.$$

The spectral profile is given by a simple algebraic relationship involving the Laplace transform of the memory function. The spectrum may be written in the form of a continued fraction representation (Mori theory; Birnbaum, Guillot and Bratos (1982)). The formalism has been used mainly for condensed states and is, therefore, of lesser interest here.

Molecular dynamics. The spectroscopy of dense fluids is closely related to the dynamics of the many-body system. For that reason, molecular dynamics and other computer simulations of many-body dynamics have been used to compute spectral moments and line shapes of dense fluids. Such simulations are important for liquids or highly compressed gases (Frenkel 1980).

5.10 Model line profiles

Despite the obvious power of quantum line shape calculations for the analysis of measured collision-induced absorption spectra, a need persists for simple but accurate model line profiles, especially for extrapolating experimental spectra to both low and high frequencies for an accurate determination of the spectral moments. Reliable model profiles are also useful for line shape analyses, i.e., for representing complex spectra as a superposition of lines (where this is possible).

A historical sketch, as well as theoretical and practical aspects of the numerous attempts to obtain reliable line shape models for the collision-induced spectroscopies has been given by Birnbaum, Guillot and Bratos (1982). Early empirical studies were based on the familiar models of spectroscopy, such as the Lorentzian, Gaussian and exponential profiles that were more or less modified (symmetry!) for the purpose. Empirical attempts are briefly sketched in Section 3.7 above. Theoretical efforts, on the other hand, usually start by simplifying the treatment of the molecular dynamics, for example, by assuming classical trajectories – often straight paths – for the molecular encounter, or by writing down a correlation function whose time behavior at short and long times approximates that of real molecular encounters. We will consider here only the most successful of these and discuss the criteria useful for determining the various degrees of utility of such functions for the modeling of collision-induced absorption spectra.

The LB model. Levine and Birnbaum have developed a classical line shape theory, assuming straight paths for the molecular encounters and a dipole model of the form [232]

$$\mu(R) = \mu_\gamma \, \gamma \, \exp\left(-\gamma^2 R^2\right) .$$

The factor γR was included to simulate the repulsive intermolecular interactions in this potential-free model calculation. The quantum formalism based on the same assumptions was developed by Levine [231], who gives the absorption coefficient as

$$\alpha(\omega) = \left(\frac{2}{\pi}\right)^{1/2} \frac{\mu_\gamma^2 \pi^3 n_a n_b}{6\hbar c \gamma^3} \sinh(\tau_0 \omega) \frac{x^3 K_2\left(x\left[1+\theta^2\right]^{1/2}\right)}{1+\theta^2},$$

with

$$\tau_0 = \frac{\hbar}{2kT} \qquad \theta = \frac{\gamma \hbar}{2(mkT)^{1/2}} \qquad x = \frac{\omega}{\gamma}\left(\frac{m}{kT}\right)^{1/2}.$$

$K_2(z)$ is the second-order, modified Bessel function of the second kind. For $\hbar \to 0$, this expression for α reduces to the classical result [232].

The autocorrelation function (normalized to unity at $t = 0$) is given by

$$C(t) = 2^{-1} \tau_\alpha^5 \left[\left(\tau_\alpha^2 + y_+^2\right)^{-5/2} + \left(\tau_\alpha^2 + y_-^2\right)^{-5/2}\right], \tag{5.103}$$

with $\tau_\alpha = (m/kT)^{1/2}/\gamma$ and $y_\pm^2 = t^2 \pm 2i\tau_0 t$; y_- is Egelstaff's complex time. The parameters may be related to the spectral moments according to Eq. 5.58ff.

Asymptotically, the Bessel function K_2 is near exponential,

$$K_2(z) \sim (\pi/2z)^{1/2} \exp -z$$

for $z \to \infty$. In other words, the spectral function G (related to α by Eq. 5.61) shows a near exponential fall-off. The fit of Bosomworth and Gush's He–Ar spectrum was satisfactory when the γ was properly adjusted (Birnbaum, Guillot and Bratos 1982). Interesting comparisons of the quantum effects on the classical line shape could also be made.

The BC model. Birnbaum and Cohen start from a two-parameter time correlation function [36, 38]

$$\hat{C}(t) = \exp\left(\frac{1}{\tau_1}\left[\tau_2 - \left(t^2 + \tau_2^2\right)^{1/2}\right]\right). \tag{5.104}$$

This classical model is symmetric in time t and leads to symmetric spectral profiles. However, by introducing the (complex) Egelstaff time, y, by substituting y^2 for t^2, where y^2 is given by [143]

$$y^2 = t^2 - i\hbar t/kT,$$

the real and imaginary parts of the correlation function become even and odd, respectively, in time and spectral profiles are derived by Fourier transform that satisfy the quantum principle of detailed balance, Eq. 5.73. The Birnbaum–Cohen (BC) model profile thus becomes

$$g_{BC}(\overline{\omega}) = \frac{S\tau_1}{\pi} \frac{1}{1+(\tau_1 \omega)^2} \exp\left(\frac{\tau_2}{\tau_1} + \tau_0 \overline{\omega}\right) z K_1(z), \tag{5.105}$$

with $z = \left[\left(1+(\tau_1\varpi)^2\right)\left(\tau_2^2+\tau_0^2\right)\right]^{1/2}/\tau_1$ and $\tau_0 = \hbar/2kT$. In this expression, $K_1(z)$ is a modified first-order Bessel function of the second kind which for numerical evaluations may be approximated by polynomials and exponential functions (Abramowitz and Stegun 1964). The time parameter $\tau = (\tau_2^2+\tau_0^2)^{1/2}$, which equals τ_2 in the high-temperature limit, controls the exponential decay in the wings as frequency shifts $|\varpi|$ go to infinity. Near the line center, for small frequency shifts, if $\tau_1 \gg \tau_2$, the shape of the BC model resembles a Lorentzian with an approximate half-width $\Delta\varpi \approx 1/\tau_1$. However, if the parameters τ_1 and τ_2 are of comparable magnitude, as is true for most collision-induced profiles studied to date, the BC model is never Lorentzian and $\Delta\varpi$ differs appreciably from $1/\tau_1$. This model is capable of representing various computed and measured line shapes closely and is perhaps the most useful model profile of the collision-induced spectroscopies known today [37, 68].

The spectral moments of the BC profile may be expressed in terms of three parameters, S, τ_1, and τ_2. They are obtained as the coefficients of t^n in a Taylor series expansion of the correlation function as

$$\begin{aligned} M_0 &= S \\ M_1 &= \frac{S\tau_0}{\tau_1\tau_2} \\ M_2 &= \frac{S}{\tau_1\tau_2}\left[1+\frac{\tau_0^2}{\tau_2}\left(\frac{1}{\tau_1}+\frac{1}{\tau_2}\right)\right]. \end{aligned} \tag{5.106}$$

The inverse system of equations which permits the computation of the parameters of the BC model from the moments, M_n for $n = 0\ldots2$, is readily derived from these expressions [52, 290].

The K0 line profile. The K0 model is a three-parameter analytical expression which has been shown to approximate closely a wide selection of profiles associated with overlap induction [70, 69]. We start with the correlation function

$$\widehat{C}(t) = \frac{\tau_3}{\left(y^2+\tau_3^2\right)^{1/2}} \exp\left(\frac{1}{\tau_4}\left[\tau_3 - \left(y^2+\tau_3^2\right)^{1/2}\right]\right), \tag{5.107}$$

where y is again the Egelstaff time [143]. A 'classical' correlation function is obtained by setting $\hbar = 0$. The Fourier transform of Eq. 5.107 gives the so-called K0 profile,

$$g_{K0}(\varpi) = \frac{S'\tau_3}{\pi} \exp\left(\frac{\tau_3}{\tau_4} + \tau_0\,\varpi\right) K_0(z'), \tag{5.108}$$

with $z' = \left[\left(\tau_3^2+\tau_0^2\right)\left(1+\tau_4^2\varpi^2\right)\right]^{1/2}/\tau_4$ and $\tau_0 = \hbar/2kT$; $K_0(z')$ is a modified zeroth-order Bessel function of the second kind (Abramowitz and

Stegun 1964) which has given the model its name. The three adjustable parameters are S', τ_3 and τ_4. The lowest three spectral moments may be obtained as functions of these parameters from the Taylor series expansion of the correlation function, according to

$$\begin{aligned} M_0 &= S', \\ M_1 &= S'\frac{\tau_0}{\tau_3}\left(\frac{1}{\tau_3}+\frac{1}{\tau_4}\right), \\ M_2 &= S'\left\{\frac{1}{\tau_3}\left(\frac{1}{\tau_3}+\frac{1}{\tau_4}\right)+\left(\frac{\tau_0}{\tau_3}\right)^2\left(\frac{1}{\tau_4^2}+\frac{3}{\tau_3\tau_4}+\frac{3}{\tau_3^2}\right)\right\}. \end{aligned} \tag{5.109}$$

When plotted in a semi-logarithmic grid, the K0 model features gentle curvature at low frequencies. At the higher frequencies, a nearly exponential degradation of intensity is observed; in a semi-logarithmic grid, the wings of the K0 function appear nearly as straight lines, in contrast to those of the BC model which show slightly more concave curvature (like a hanging chain).

Related to this near absence of logarithmic curvature of the wings is the fact that the K0 model is superior in describing line profiles resulting from overlap induction. The BC shape, on the other hand, shows more curvature in the wing, as this is needed for the modeling of profiles generated by low-order multipolar induction. Purely quadrupole-induced components are closely modeled by the BC shape.

The dipole components of the multipole induction terms are in general more or less affected by an overlap component. Consequently, an extended BC (EBC) model, a combination of the BC and K0 models, was found to be useful for such applications where the resulting large number (six) of free parameters is no impediment,

$$\begin{aligned} g(\overline{\omega}) = \frac{S}{1+\varepsilon}\Bigg[&\frac{\tau_1}{\pi}\frac{zK_1(z)}{1+(\tau_1\overline{\omega})^2}\exp\left(\frac{\tau_2}{\tau_1}+\tau_0\overline{\omega}\right) \\ &+\varepsilon\frac{\tau_3}{\pi}K_0(z')\exp\left(\frac{\tau_3}{\tau_4}+\tau_0\overline{\omega}\right)\Bigg]. \end{aligned} \tag{5.110}$$

Very good representations of computed quantum profiles have thus been possible. Since the inverse Fourier transform of this function is known, it may serve as a very useful transfer function. However, for line shape analyses of measured spectra, many-parameter models are not acceptable and the usefulness of the EBC model is therefore limited.

IT model. A line shape based on information theory (IT) has been proposed for the collision-induced spectra [159, 203]. The profile was really never intended to be used to represent collision-induced spectra with the best accuracy possible. Rather, the emphasis is on simplicity; it is the result of a qualitative theory of the line shape in situations where only a few spectral moments of low order are known. Furthermore, for liquids, where most of the approaches discussed above do not work, information theory was suggested as a perhaps semi-quantitative line shape theory. Our ideas of model profiles are, of course, different: while the models considered above are still simple and free from adjustable parameters (if a sufficient number of spectral moments is known), certain simple facts concerning the shape of spectral profiles of low-pressure gases have been readily incorporated so that a close modeling of real profiles, be these measured or computed from an exact quantum formulation, is possible. That has been our main concern. For these reasons, a discussion of merits of the information-theoretical line shape is somewhat out of place here. Nevertheless, it is of interest to consider some of the implications and properties.

Starting from the assumption that the *only* known information concerning the line shape is a finite number of spectral moments, a quantity called information is computed from the probability for finding a given spectral component at a given frequency; this quantity is then minimized. Alternatively, one may maximize the number of configurations of the various spectral components. This process yields an expression for the spectral density as function of frequency which contains a small number of parameters which are then related to the known spectral moments. If classical (i.e., Boltzmann) statistics are employed, information theory predicts a line shape of the form

$$g(\omega) = A_0 \exp\left(-A_2\omega^2 - A_4\omega^4 - \cdots - A_{n_{\max}}\omega^{n_{\max}}\right) ,$$

with the parameters A_0, A_2, ... For $n_{\max} = 2$ the profile is Gaussian, which falls off too fast when compared with real profiles. For $n_{\max} = 4$, profiles exist only if a certain relationship among the moments is satisfied [159, 274], namely

$$1 \leq M_4 M_0 / M_2^2 \leq 3 ;$$

these fall off even faster in the wings. For the collision-induced profiles, ratios of $M_4 M_0 / M_2^2$ amount typically to 10 so that the classical (Boltzmann) profile above must be replaced by a model employing Bose–Einstein statistics [326],

$$g(\omega) = \frac{A}{\exp\left(A_0 + A_2\omega^2 + \cdots + A_{n_{\max}}\omega^{n_{\max}}\right) - 1} ,$$

The numerator A represents the density of states function, which in practice is set equal to a constant [326]. In this way, profiles may be obtained for the collision-induced spectra. However, one should not expect the IT shape to approximate real line shapes well [83, 274]. The basic assumption of IT mentioned above ignores the wealth of information

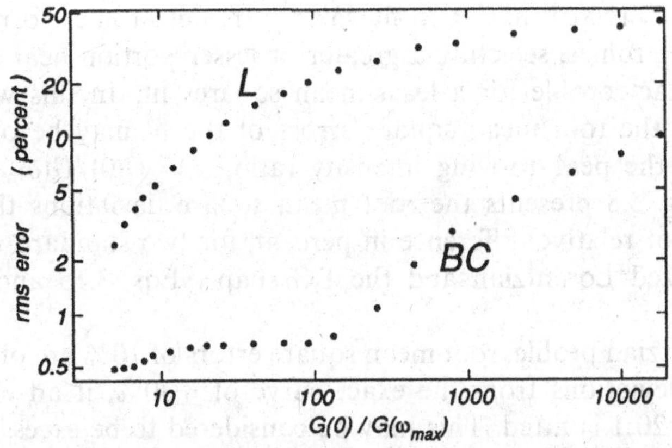

Fig. 5.8. Root mean square relative errors of model line shapes fitted to a quantum profile, the quadrupole-induced ($\lambda L = 23$) component [69]. The abscissa gives the ratio of peak intensity and wing intensity of the fitted portion of the exact profile. The superiority of the BC model (lower set of data points) over the desymmetrized Lorentzian (upper set) is evident.

that is actually known about the shapes of collision-induced, individual lines.

Criteria for choosing specific models. We have seen above that spectral profiles of binary complexes can be computed from a rigorous quantum formalism. These describe the measurements well. 'Long' profiles, i.e., profiles with a peak-to-wing intensity of several orders of magnitude, can be readily obtained. As an illustration of how well the various available model profiles approximate the exact computations, we show results for one basic profile type characteristic of absorption by H_2–H_2 pairs [69], a pure quadrupole-induced profile which accounts for roughly 90% of the total intensity of the rototranslational H_2–H_2 collision-induced absorption spectra at 77 K.

Model profiles may be expected to mimic exact quantum profiles over a certain range of intensities that is more or less limited; beyond that range, in the far wings, model and exact profiles may rapidly diverge. If the divergence occurs far enough in the wing where intensities have fallen off to insignificant values, it may not be objectionable. For quantitative work, acceptable model profiles must have their low-order spectral moments in close agreement with the exact profile; this requirement defines a practical limit for the onset of the divergence. Normally, agreement of the zeroth through second moments will be sufficient, but if higher moment expressions are to be employed, model profiles that agree with the exact profiles over an even greater range must be used for consistency of the derived data.

It is a straightforward matter to fit various model profiles to realistic, exact computed profiles, selecting a greater or lesser portion near the line center of the exact profile for a least mean squares fit. In this way, the parameters and the root mean square errors of the fit may be obtained as functions of the peak-to-wing intensity ratio, $x = G(0)/G(\overline{\omega}_{\max})$. As an example, Fig. 5.8 presents the root mean square deviations thus obtained, in units of relative difference in percent, for two standard models, the desymmetrized Lorentzian and the BC shape, Eqs. 3.15 and 5.105, respectively.

For the Lorentzian profile, root mean square errors of 10% are obtained, with maximal deviations from the exact curve of $\pm 30\%$, if an intensity range x of only 20:1 is fitted. This may be considered to be excessive; for good measurements experimental uncertainties do not normally exceed 10%. Errors increase monotonically if greater peak-to-wing intensity ranges x are fitted as is necessary for agreement of the second moments, Fig. 5.8, upper curves. Even more disturbing is the fact that for the Lorentzian, the values of the fitting parameters vary significantly with the intensity range fitted. Specifically, the parameter p, Eq. 3.19, is ill defined for some of the most important profile types of interest. While nearly constant S and τ values were obtained when the intensity range x was increased from roughly 3 to 100 for quadrupole-induced profiles, the p values of the Lorentzian varied exponentially from about 1 to 5 for x varying from 3 to 1000, showing no region where it might be considered independent of x; the shape of the Lorentzian never 'locks on' to the quantum shape. The variation of p has a strong influence on the logarithmic slopes, especially in the wings. If the parameter p is fixed at its average value, $p = 1.7$, the quality of the fit suffers substantially. Only the overlap component was modeled reasonably well by the Lorentzian but the fit could still be improved if an exponential desymmetrization function, $\exp(-\hbar\overline{\omega}/2kT)$, were employed instead of the leading factor shown to the right of Eq. 3.19 [69].

Numerical problems arising from the use of the Lorentzian for fitting spectra have also been reported [69]. These are related to the non-differentiable profile at the points where the exponential wings are attached to the Lorentzian core. Partial derivatives with respect to the line shape parameters are usually needed in least mean squares fitting procedures.

The BC and K0 models, on the other hand, show much smaller root mean square errors, typically in the 1% range, over an amazingly substantial range x of intensities fitted, Fig. 5.8, lower set of data points. Maximal deviations from the exact profiles amount to no more than twice the root mean square errors shown, that is well within the experimental uncertainties of the best measurements. The BC model is especially well suited to approximate quadrupole-induced profiles. The K0 model, on the other hand, represents overlap-induced profiles with root mean square errors of typically well below 1% over intensity ranges x as big as 10 000. The parameters of the fit are nearly independent of the range x fitted, just as this is seen for the root mean square relative error, Fig. 5.8. Consequently, the spectral moments of fits based on the BC and K0 model are likewise independent of x. Indeed, the exact spectral moments and the moments of the fit agreed to within 1% – which is a remarkable, most desirable property of these model profiles not found in many other model profiles known today [69].

Modeling of bound states. Bound state contributions cannot be modeled very well with any of the profiles mentioned. If an accurate modeling is necessary; bound-to-bound and bound-to-free contributions must be treated separately [58].

Symmetry. For the translational and rototranslational bands, Eq. 5.73 is usually a good description of the observed asymmetry. In that case, the use of the BC and K0 profiles is straightforward because these have the same symmetry, Eq. 5.73. In the next Chapter we will see that profiles of the rotovibrational bands are of a symmetry which is different from that relationship [295]. In that case, new line profiles must be constructed that satisfy the correct symmetry relationship [295, 62, 48]. These will be discussed next Chapter.

5.11 General references

M. Abramowitz and I. A. Stegun, *Handbook of Mathematical Functions*, National Bureau of Standards, U.S.A., 1964.

G. Birnbaum, B. Guillot and S. Bratos, Theory of Collision-Induced Line Shapes – Absorption and Light Scattering at Low Density, *Adv. Chem. Phys.*, **51**, (1982) 49.

J. Borysow and L. Frommhold, The Infrared and Raman Line Shapes of Pairs of Interacting Molecules, in *Phenomena Induced by Intermolecular Interactions*, G. Birnbaum, ed., Plenum, New York, 1985.

S. Chapman and T. G. Cowling, *The Mathematical Theory of Nonuniform Gases*, Cambridge University Press 1960.

J. H. R. Clarke, Bandshapes and Molecular Dynamics in Liquids, in *Advances in Infrared and Raman Spectroscopy*, **4** (1978) 109, R. J. H. Clarke and R. E. Hesters, eds.

D. Frenkel, Intermolecular Spectroscopy and Computer Simulation, in *Intermolecular Spectroscopy and Dynamical Properties of Dense Systems*, J. van Kranendonk, ed., p. 156, North-Holland, Amsterdam 1980.

R. G. Gordon, Correlation Functions for Molecular Motion, *Adv. Magnetic Resonance*, vol. **3**, 1 (1968).

J. O. Hirschfelder, C. F. Curtiss and R. B. Bird, *Molecular Theory of Gases and Liquids*, John Wiley & Sons, New York, 1964.

J. C. Lewis, Intercollisional Interference Effects, p. 91, in *Intermolecular Spectroscopy and Dynamical Properties of Dense Systems*, J. van Kranendonk, ed., North-Holland, Amsterdam 1980.

J. C. Lewis, Intercollisional Interference – Theory and Experiment, p. 215, in *Phenomena Induced by Intermolecular Interactions*, G. Birnbaum, ed., Plenum, New York, 1985.

G. C. Maitland, M. Rigby, E. B. Smith, and W. A. Wakeham. *Intermolecular Forces*, Clarendon Press, Oxford, 1981.

M. Moraldi, Virial expansion of correlation functions for collision-induced spectroscopies, in *Spectral Line Shapes* **6**, L. Frommhold and J. W. Keto, eds., American Institute of Physics 1990.

J. D. Poll, Intermolecular Spectroscopy of Gases, in *Intermolecular Spectroscopy and Dynamical Properties of Dense Systems*, J. van Kranendonk, ed., North-Holland, Amsterdam, 1980.

W. H. Press, B. P. Flannery, S. A. Teukolsky and W. T. Vetterling. *Numerical Recipes*, Cambridge University Press 1986.

J. van Kranendonk, Intermolecular Spectroscopy, *Physica*, **73** (1974) 156.

J. van Kranendonk, Intercollisional Interference Effects, p. 77, in *Intermolecular Spectroscopy and Dynamical Properties of Dense Systems*, J. van Kranendonk, ed., North-Holland, Amsterdam 1980.

6

Theory: molecular gases

In Chapter 5 the absorption spectra of complexes of interacting *atoms* were considered. If some or all of the interacting members of a complex are *molecular*, additional degrees of freedom exist and may be excited in the presence of radiation. As a result, besides the translational profiles discussed in Chapter 5, new spectral bands appear at the rotovibrational transition frequencies of the molecules involved, and at sums and differences of such frequencies – even if the non-interacting molecules are infrared inactive. The theory of absorption by small complexes involving molecules is considered in the present Chapter.

We will be concerned with the spectral bands in the microwave and infrared regions. The translational and the purely rotational bands appear both at low frequencies and form in general one composite band, especially at the higher temperatures where individual lines tend to overlap ('rototranslational band'). Moreover, various rotovibrational bands in the near infrared will be considered, such as the fundamental and the overtone bands. Even high overtone bands in the visible are of interest, e.g., of H_2. We have seen in Chapter 3 that induced spectra of the kind are readily discernible in gases whose (non-interacting) molecules are infrared inactive, but evidence exists that suggests the presence of induced absorption in the allowed molecular bands as well. Induced absorption involving electronic transitions will be briefly considered in Chapter 7.

We note that even the unmixed molecular gases absorb infrared radiation by collisional induction, in contrast to the unmixed monatomic gases which do not. This is so because pairs of like atoms possess an inversion symmetry that is inconsistent with a dipole moment; pairs of like molecules, on the other hand, may be found in various orientations and rotovibrational excitations that are consistent with a dipole moment of the complex. We will, therefore, consider in this Chapter mostly the unmixed molecular gases (without neglecting mixtures unduly).

Besides the isotropic overlap-induced dipole component familiar from the rare gas pairs, we will now in general have other significant induced dipole components if molecules are present, namely multipole-induced and distorted frame-induced dipole components, see Chapter 4 for details. Moreover, these anisotropic dipole components couple with the polarizability tensor and thus give rise to simultaneous transitions in two (or perhaps more) molecules. Furthermore, molecules in general interact with more or less anisotropic forces which to some extent does also affect the spectra of molecular systems.

6.1 Spectral moments

The motion of the electrons is treated in the adiabatic approximation. For absorption in the infrared, the electrons remain in the ground state. The electric dipole moment of the complex of two molecules is the expectation value of the dipole moment operator over the ground electronic state, which is a function of the nuclear coordinates only. Specifically, the dipole moment of a complex of n molecules is dependent on the vibrational (r_i) and orientational ($\widehat{\boldsymbol{r}}_i$) coordinates, and on the position ($\boldsymbol{R}_i$) of the mass centers of the molecules i, for $1 \leq i \leq n$,

$$\boldsymbol{\mu} = \boldsymbol{\mu}(\boldsymbol{R}_1, \boldsymbol{r}_1, \boldsymbol{R}_2, \boldsymbol{r}_2, \cdots \boldsymbol{R}_n, \boldsymbol{r}_n) \ .$$

For simplicity, we may assume diatomic molecules (characterized by their vibrational and rotational coordinates, $\boldsymbol{r}_i = r_i \widehat{\boldsymbol{r}}_i$). For polyatomic molecules, $\boldsymbol{r}_i$ stands for the set of normal and rotational coordinates.

Molecules generally interact with anisotropic forces. The accounting for the anisotropy of intermolecular interactions introduces substantial complexity, especially for the quantum mechanical treatment. We will, therefore, use as much as possible the *isotropic* interactions isotropic interaction approximation (IIA), where the Hamiltonian is given by a sum of two independent terms representing rotovibrational and translational motion. The total energy of the complex is then given by the sum of rotovibrational and translational energies. The state of the supermolecule is described by the product of rotovibrational and translational wavefunctions, with an associated set of quantum numbers r and t, respectively.

A theory of collision-induced absorption by molecular systems starts again with Eq. 5.1,

$$\alpha(\omega) = \frac{4\pi^2}{3\hbar^2 cV} \frac{1}{4\pi\varepsilon_0} \sum_{i<f} (P_i - P_f) \ |\langle i | \boldsymbol{\mu} | f \rangle|^2 \ (E_f - E_i) \ .$$

The notation $i < f$ indicates the restricted summation over all states with $E_i < E_f$. Similarly as was done in the previous Chapter, the integrals,

Eq. 5.6, of this expression are now rewritten as ensemble averages of certain operators. If the refractive index n differs significantly from unity or is strongly frequency-dependent at the frequencies of interest, the Polo–Wilson factor should be added to the right-hand side of this equation (and of other equations below; see p. 197).

For the rotovibrational spectra of molecules interacting through purely isotropic forces, the Hamiltonian may be written as the sum of two independent terms. One term describes the rotational motion of the molecules, the other the translational motion of the pair. The total energy of the system is then equal to the sum of the rotovibrational and the translational energies. At the same time, the supermolecular wavefunctions are products of rotovibrational and translational functions. Let r designate the set of the rotovibrational quantum numbers and t the set of translational quantum numbers, the equation for γ_0 may be written [314]

$$\gamma_0 = \frac{4\pi^2}{3\hbar c V}\frac{1}{4\pi\varepsilon_0}\sum_{rt<r't'}(P_r P_t - P_{r'} P_{t'}) \coth\frac{hc\nu}{2kT}\left|\langle t\,|\boldsymbol{\mu}_{rr'}|\,t'\rangle\right|^2 ,$$

where $\boldsymbol{\mu}_{rr'} = \langle r|\boldsymbol{\mu}|r'\rangle$ is the dipole matrix element for transitions between initial and final rotovibrational states; it is still a function of the positions of the centers of mass of the molecules. The restriction $rt < r't'$ of the summation is short for $E_r + E_t < E_{r'} + E_{t'}$. Using manipulations similar to those employed above, pp. 199ff., we obtain

$$\gamma_0 = \frac{4\pi^2}{3\hbar c V}\frac{1}{4\pi\varepsilon_0}\sum_{r,r'} P_r \sum_{tt'} P_t\left|\langle t\,|\boldsymbol{\mu}_{rr'}|\,t'\rangle\right|^2 \tag{6.1}$$

where $\sum_{tt'} P_t|\langle t|\boldsymbol{\mu}|t'\rangle|^2$ may be replaced by the trace $\mathrm{Tr}\,\{P_{\mathrm{tr}}\boldsymbol{\mu}_{rr'}\}$ of the density operator of translational motion,

$$P_{\mathrm{tr}} = \exp\left(-\mathscr{H}_{\mathrm{tr}}/kT\right)\Big/\mathrm{Tr}\left\{\exp -\mathscr{H}_{\mathrm{tr}}/kT\right\} .$$

Similarly, an expression for γ_1 may be obtained,

$$\gamma_1 = \frac{2\pi}{3\hbar^2 c^2 V}\frac{1}{4\pi\varepsilon_0}\mathrm{Tr}\left\{P\boldsymbol{\mu}\cdot[\mathscr{H},\boldsymbol{\mu}]\right\}, \tag{6.2}$$

where the trace now includes a summation over all rotational states. This may be written in the form

$$\gamma_1 = \sum_{r\neq r'}\gamma_{1,\mathrm{rot}}(r,r') + \sum_{r=r'}\gamma_{1,\mathrm{tr}}(r,r') , \tag{6.3}$$

where

$$\gamma_{1,\mathrm{rot}} = \frac{2\pi}{3\hbar^2 c^2 V}\frac{1}{4\pi\varepsilon_0} P_r\sum_{tt'} P_t\left|\langle t\,|\boldsymbol{\mu}_{rr'}|\,t'\rangle\right|^2 (E_{r'} - E_r + E_{t'} - E_t)$$

$$\begin{aligned} \gamma_{1,\mathrm{tr}} &= \frac{2\pi}{3\hbar^2 c^2 V} \frac{1}{4\pi\varepsilon_0} P_r \sum_{tt'} P_t \left|\langle t\,|\boldsymbol{\mu}_{rr'}|\,t'\rangle\right|^2 (E_{t'} - E_t) \\ &= \frac{2\pi}{3\hbar^2 c^2 V} \frac{1}{4\pi\varepsilon_0} P_r \,\mathrm{Tr}\left\{P_{\mathrm{tr}}\, \boldsymbol{\mu}_{rr'} \cdot [\mathscr{H}_{\mathrm{tr}}, \boldsymbol{\mu}_{rr'}]\right\} \end{aligned}$$

and $\mathscr{H}_{\mathrm{tr}}$ designates the Hamiltonian of translational motion.

1 Virial expansion

Using the same reasoning as in the previous Chapter, for the monatomic mixtures, one arrives at the virial expansion, Eq. 5.18,

$$\gamma_n = \gamma_n^{(2)} \varrho^2 + \gamma_n^{(3)} \varrho^3 + \cdots$$

which we have written here for an unmixed, molecular gas. In that case, the binary and ternary first moments, Eqs. 5.19, 5.20, 5.25, 5.26, assume the form [314, 296]

$$\gamma_1^{(2)} = \frac{1}{2} \frac{2\pi V}{3\hbar^2 c^2} \frac{1}{4\pi\varepsilon_0} \mathrm{Tr}\left\{P_2^0(12)\boldsymbol{\mu}(12) \cdot [\mathscr{H}(12), \boldsymbol{\mu}(12)]\right\} \tag{6.4}$$

$$\begin{aligned} \gamma_1^{(3)} &= \frac{1}{2} \frac{2\pi V}{3\hbar^2 c^2} \frac{1}{4\pi\varepsilon_0} \mathrm{Tr}\left\{P_2^1(12)\boldsymbol{\mu}(12) \cdot [\mathscr{H}(12), \boldsymbol{\mu}(12)]\right\} \\ &\quad + \frac{2\pi V}{3\hbar^2 c^2} \frac{1}{4\pi\varepsilon_0} \mathrm{Tr}\Big\{P_3^0(123)\Big(2\boldsymbol{\mu}(12) \cdot [\mathscr{H}(13), \boldsymbol{\mu}(13)] \\ &\quad + \boldsymbol{\mu}(12) \cdot [\mathscr{H}(123), \boldsymbol{U}_3(123)] + \boldsymbol{U}_3(123) \cdot [\mathscr{H}(12), \boldsymbol{\mu}(12)] \\ &\quad + \frac{1}{3} \boldsymbol{U}_3(123) \cdot [\mathscr{H}(123), \boldsymbol{U}_3(123)]\Big)\Big\}. \end{aligned} \tag{6.5}$$

For the zeroth moments, we have similarly

$$\gamma_0^{(2)} = \frac{4\pi^2 V}{3\hbar c} \frac{1}{4\pi\varepsilon_0} \mathrm{Tr}\left\{P_2^0(12)\mu(12)^2\right\} \tag{6.6}$$

$$\begin{aligned} \gamma_0^{(3)} &= \frac{4\pi^2 V}{3\hbar c} \frac{1}{4\pi\varepsilon_0} \mathrm{Tr}\left\{P_2^1(12)\mu(12)^2\right\} \\ &\quad + \frac{4\pi^2 V}{3\hbar c} \frac{1}{4\pi\varepsilon_0} \mathrm{Tr}\Big\{P_3^0(123)\Big(\boldsymbol{\mu}(12) \cdot \boldsymbol{\mu}(23) \\ &\quad + \boldsymbol{\mu}(12) \cdot \boldsymbol{U}_3(123) + \boldsymbol{U}_3(123) \cdot \boldsymbol{\mu}(12) + \frac{1}{2} \boldsymbol{U}_3(123) \cdot \boldsymbol{U}_3(123)\Big)\Big\}. \end{aligned} \tag{6.7}$$

For molecular pairs, the low-density limit P_2^0 of the pair density operator is given by

$$P_2^0(12) = \frac{\lambda_0^6}{Z_1 Z_2} \exp\left(-\mathscr{H}(12)/kT\right), \tag{6.8}$$

where $\lambda_0^2 = 2\pi\hbar^2/(mkT)$ is the square of the thermal de Broglie wavelength of the pair, m is the reduced mass and the Z_i are the rotational partition functions of molecules $i = 1$ and 2. $\mathscr{H}(12)$ designates the Hamiltonian of the pair 1,2.

The dipole moment of a molecular pair depends also on the molecular orientations, $\widehat{r}_1$ and $\widehat{r}_2$, and vibrational coordinates, r_1, r_2. For linear molecules we have Eq. 4.18,

$$\mu_\nu = \frac{(4\pi)^{3/2}}{\sqrt{3}} \sum_{\lambda_1\lambda_2\Lambda L} A_{\lambda_1\lambda_2\Lambda L}(r_1, r_2, R) \sum_{m_1 m_2 M m} C(\lambda_1\lambda_2\Lambda; m_1 m_2 m)$$
$$\times C(\Lambda L 1; m M \nu)\, Y_{\lambda_1}^{m_1}(\widehat{r}_1)\, Y_{\lambda_2}^{m_2}(\widehat{r}_2)\, Y_L^M(\widehat{R}) .$$

The vibrational averages of the dipole components, $\langle v_1 j_1 v_2 j_2 | \mu_\nu | v_1' j_1' v_2' j_2' \rangle$, are given by a similar expression, with the $A_{\lambda_1\lambda_2\Lambda L}$ coefficients replaced by Eq. 4.20,

$$B_{\lambda_1\lambda_2\Lambda L}(R) = \langle v_1 j_1 v_2 j_2 \,| A_{\lambda_1\lambda_2\Lambda L}(r_1, r_2, R) |\, v_1' j_1' v_2' j_2' \rangle .$$

The dependence of the vibrational wavefunctions $|v_1 j_1 v_2 j_2\rangle$ on the rotational quantum numbers is sometimes a weak one and may in that case be suppressed, in particular in the rototranslational bands, $|v_1 j_1 v_2 j_2\rangle \approx |v_1 0 v_2 0\rangle$, or short $|v_1 v_2\rangle$.

It is well known that in n-body complexes rotational transitions of the order n may occur [400]. However, we will assume here that the interaction forces are pairwise additive and isotropic so that rotations and translations are uncorrelated. In this case, at most double transitions occur [400] and the correlation function of the total dipole moment can be written as

$$C(t) = \frac{4\pi^2}{3\hbar c} \varrho^2 \sum_{\lambda_1\lambda_2\Lambda L} \sum_{j_1 j_1' j_2 j_2'} P(j_1)\,(2j_1+1)\,P(j_2)\,(2j_2+1) \tag{6.9}$$
$$\times C(j_1\lambda_1 j_1'; 000)^2\, C(j_2\lambda_2 j_2'; 000)^2\, C_{\lambda_1\lambda_2\Lambda L}(t)\, \exp\left[it(\omega_{j_1 j_1'} + \omega_{j_2 j_2'})\right] ,$$

where ϱ is the density of the diatomic gas in amagat units; $P(j)$ the population probability of the rotational state with the principal quantum numbers j, m_j; $\omega_{jj'}$ is a rotational transition frequency; and $C_{\lambda_1\lambda_2\Lambda L}(t)$ is the correlation function of the $\lambda_1\lambda_2\Lambda L$ dipole component [291, 292],

$$C_{(c)}(t) = \frac{1}{4\pi\varepsilon_0} \operatorname{Tr}\left\{ e^{-\mathscr{H}_{\rm tr}/kT}\, B_c(R)\, e^{i\mathscr{H}_{\rm tr}t/\hbar}\, B_c(R)\, e^{-i\mathscr{H}_{\rm tr}t/\hbar} \right\} , \tag{6.10}$$

the subscript (c) is short for $\lambda_1\lambda_2\Lambda L$. In this equation, $\mathscr{H}_{\rm tr} = T + V_0$ designates the Hamiltonian of relative motion. The function $C_{(c)}(t)$ depends only on the motion of the centers of mass of the molecules, but not on their orientation. We next define the translational spectral moments with the help of the correlation function, according to

$$M_n^{(c)} = (-\mathrm{i})^n \left. \frac{\mathrm{d}^n C_{(c)}(t)}{\mathrm{d}t^n} \right|_{t=0}, \tag{6.11}$$

These moments permit a virial expansion that can be written as

$$M_n^{(c)} = M_n^{(c)}(2) + \varrho \left(M_n^{(c)}(3)' + M_n^{(c)}(3)'' + M_n^{(3)'''} \right) + \mathcal{O}(\varrho^2) \tag{6.12}$$

up to ternary contributions; the arguments (2) and (3) of M_n mean binary and ternary contributions, respectively.

Expressions for the moments exist that are analogous to Eqs. 5.37, 5.38, 5.49, 5.51, and 5.52, except for a replacement of the spherical harmonics Y_1^m by Y_L^M. With

$$\sum_M Y_L^M(\widehat{\boldsymbol{R}})^* \, Y_L^M(\widehat{\boldsymbol{R}}') = \frac{2L+1}{4\pi} P_L(\cos\vartheta) ,$$

and the help of the gradient formula, we compute

$$\sum_M \left(\nabla B_{(c)}(R)\, Y_L^M(\widehat{\boldsymbol{R}}) \right)^* \left(\nabla B_{(c)}(R)\, Y_L^M(\widehat{\boldsymbol{R}}) \right)$$

to get expressions for the binary and ternary translational moments,

$$M_0^{(c)}(2) = \frac{1}{2}\, 4\pi N_L^2 \int_0^\infty R^2\, \mathrm{d}R\, \frac{1}{4\pi\varepsilon_0}\, [B_c(R)]^2\, g_0^{(2)}(R) \tag{6.13}$$

$$M_0^{(c)}(3)' = \frac{1}{2}\, 4\pi N_L^3 \int_0^\infty R^2\, \mathrm{d}R\, \frac{1}{4\pi\varepsilon_0}\, [B_c(R)]^2\, g_1^{(2)}(R) \tag{6.14}$$

$$\begin{aligned} M_0^{(c)}(3)'' = {} & 8\pi^2 N_L^3\, \delta_{\lambda 2 0} \int_0^\infty R^2\, \mathrm{d}R \int_0^\infty R'^2\, \mathrm{d}R' \int_0^\pi \sin\vartheta\, \mathrm{d}\vartheta \\ & \times g_0^{(3)}(R, R', \cos\vartheta)\, \frac{1}{4\pi\varepsilon_0}\, B_c(R)\, B_c(R')\, P_L(\cos\vartheta) \end{aligned} \tag{6.15}$$

$$\begin{aligned} M_1^{(c)}(2) = {} & \frac{1}{2} \frac{\hbar}{2m_r}\, 4\pi N_L^2 \int_0^\infty R^2\, \mathrm{d}R\, g_0^{(2)}(R) \\ & \times \frac{1}{4\pi\varepsilon_0} \left\{ \left(\frac{\mathrm{d}}{\mathrm{d}R} B_c(R) \right)^2 + \frac{L(L+1)}{R^2}\, (B_c(R))^2 \right\} \end{aligned} \tag{6.16}$$

$$\begin{aligned} M_1^{(c)}(3)' = {} & \frac{1}{2} \frac{\hbar}{2m_r}\, 4\pi N_L^3 \int_0^\infty R^2\, \mathrm{d}R\, g_1^{(2)}(R) \\ & \times \frac{1}{4\pi\varepsilon_0} \left\{ \left(\frac{\mathrm{d}}{\mathrm{d}R} B_c(R) \right)^2 + \frac{L(L+1)}{R^2}\, (B_c(R))^2 \right\} \end{aligned} \tag{6.17}$$

$$\begin{aligned} M_1^{(c)}(3)'' = {} & \frac{\hbar}{2m}\, 8\pi^2 N_L^3\, \delta_{\lambda 2 0} \int_0^\infty R^2\, \mathrm{d}R \int_0^\infty R'^2\, \mathrm{d}R' \\ & \times \int_0^\pi \sin\vartheta\, \mathrm{d}\vartheta\, g_0^{(3)}(R, R', \cos\vartheta) \end{aligned} \tag{6.18}$$

$$\times \frac{1}{4\pi\varepsilon_0} \left\{ \frac{L+1}{2L+1} \left[\frac{\mathrm{d}}{\mathrm{d}R} B_c(R) - \frac{L}{R} B_c(R) \right] \right.$$
$$\times \left[\frac{\mathrm{d}}{\mathrm{d}R'} B_c)(R') - \frac{L}{R'} B_c(R') \right] P_{L+1}(\cos \vartheta)$$
$$+ \frac{L}{2L+1} \left[\frac{\mathrm{d}}{\mathrm{d}R} B_c(R) + \frac{L+1}{R} B_c(R) \right]$$
$$\left. \times \left[\frac{\mathrm{d}}{\mathrm{d}R'} B_c(R') + \frac{L+1}{R'} B_c(R') \right] P_{L-1}(\cos \vartheta) \right\} ,$$

where $g_0^{(2)}$ is the zero density limit of the pair distribution function; $g_1^{(2)}$ is the coefficient of the density ϱ of its virial expansion; $g_0^{(3)}(R, R', \cos \vartheta)$ is the zero density limit of the triplet distribution function, and m and m_r are the mass of one molecule and the reduced mass of two molecules, respectively. We note that the triply primed terms, Eq. 6.12, arise from the irreducible (i.e., not pairwise-additive) ternary dipole components which are poorly known.

With the help of Eq. 5.6, the spectral moments γ_0 and γ_1 can be approximated in terms of the translational moments [314], M_n,

$$\gamma_0 = \frac{4\pi^2}{3\hbar c} \sum_{(c)} M_0^{(c)} \tag{6.19}$$

$$\gamma_1 \cong \frac{4\pi^2}{3\hbar c} \sum_{(c)} \left\{ \frac{1}{2\pi c} M_1^{(c)} \right. \tag{6.20}$$
$$\left. + \left[\nu_{v_1 v_1'} + \nu_{v_2 v_2'} + \lambda_1(\lambda_1+1) B^{(1)}_{v'-v} + \lambda_2(\lambda_2+1) B^{(2)}_{v'-v} \right] M_0^{(c)} \right\} ,$$

where $B^{(i)}_{v'-v} = \hbar/4\pi c I$ designates the rotational constant of molecule i and the ν are the vibrational transition frequencies. The expression for γ_1 is exact for the rototranslational band.

For the rotovibrational bands, the induced dipole components B_c of Eqs. 6.13 through 6.18 become the vibrational matrix element [281], Eq. 4.20. Furthermore, each $\lambda_1\lambda_2\Lambda L$ component now occurs twice, once with molecule 1 vibrating and once with molecule 2 vibrating in the final state. For like molecular pairs, this may be taken into account by removing the factors of 1/2 in Eqs. 6.13, 6.14, 6.16, and 6.17. For dissimilar molecular pairs, the factors 1/2 are absent from the equations quoted.

Atom–diatom mixtures. The treatment of this case is very similar to the above if one remembers that the induced dipole component is of the form of Eq. 4.15. The moment expressions are obtained from Eqs. 6.13 through 6.18 by setting $\lambda_2 = 0$ and removing the factors of 1/2 from Eqs. 6.13, 6.14, 6.16, and 6.17.

2 Second and third binary moment

For the rototranslational spectra, within the framework of the isotropic interaction approximation, the expressions for the zeroth and first moments, Eqs. 6.13 and 6.16, are exact provided the quantal pair distribution function (Eq. 5.36) is used [314]. A similar expression for the binary second translational moment has been reported [291],

$$\begin{aligned}
M_2^{(c)}(2) = \frac{1}{4\pi\varepsilon_0}\Bigg[\hbar^2 \int_0^\infty g^{(0)}(R)\Bigg\{ & -\frac{1}{4}B_c(R)B_c^{IV}(R) - \frac{1}{2}B_c^I(R)\, B_c^{III}(R) \\
& + \frac{L(L+1)}{2R^2}\left(B_c^I(R)\right)^2 - \frac{L(L+1)}{R^3}B_c(R)B_c^I(R) \\
& + \frac{L^2(L+1)^2}{4R^2}B_c(R)^2\Bigg\}\, R^2\, \mathrm{d}R \\
& + m\int_0^\infty g^{(0)}(R)\{B_c(R)B_c^I(R)V_0^I(R) \\
& + 2B_c(R)B_c^{II}(R)V_0(R)\}\, R^2\, \mathrm{d}R \\
& - 2m\int_0^\infty g_E(R)\, B_c(R)B_c^{II}(R)\, R^2\, \mathrm{d}R \qquad (6.21)\\
& + \int_0^\infty g_M(R)\Bigg\{\frac{1}{R^2}B_c(R)B_c^{II}(R) - \frac{1}{R^3}B_c(R)B_c^I(R) \\
& + \frac{L(L+1)}{2R^4}B_c(R)^2\Bigg\}\, R^2\, \mathrm{d}R\Bigg] .
\end{aligned}$$

The superscripts I···IV in this expression designate the first through fourth derivatives of potential and dipole function with respect to R.

The pair distribution function $g^{(0)}$ is similar to Eq. 5.36, except that we now have to consider the possibility of like pairs. Specifically, we write

$$g^{(0)}(R) = \lambda_0^3 \int_0^\infty \mathrm{d}E \sum_\ell \frac{2\ell+1}{4\pi}\, \mathrm{e}^{-E/kT}\, \frac{1+(-1)^\ell \gamma_S}{R^2}\, |\psi(R;E,\ell)|^2 \, . \qquad (6.22)$$

For dissimilar pairs, the parameter γ_S equals zero and we have Eq. 5.36. Like pairs of zero spin are bosons and all odd-numbered partial waves are ruled out by the requirement of even wavefunctions of the pair; this calls for $\gamma_S = 1$. In general, for like pairs, the symmetry parameter γ_S will be between –1 and 1, depending on the monomer spins (fermions or bosons) and the various total spin functions of the pair. A simple example is considered below (p. 288ff.). If vibrational states are excited, the radial wavefunctions ψ must be obtained from the vibrationally averaged potential, $V_0^v(R)$. The functions $g_E(R)$ and $g_M(R)$ are similar to the pair distribution function, namely [294]

$$g_E(R) = \tag{6.23}$$
$$\lambda_0^3 \int_0^\infty \mathrm{d}E \sum_{\ell=0}^{\infty} \frac{2\ell+1}{4\pi} \, \mathrm{e}^{-E/kT} \, \frac{E}{R^2} \left[1 + (-1)^\ell \gamma_S\right] |\psi(R;E,\ell)|^2$$

$$g_M(R) =$$
$$\lambda_0^3 \int_0^\infty \mathrm{d}E \sum_{\ell=0}^{\infty} \frac{2\ell+1}{4\pi} \, \mathrm{e}^{-E/kT} \, \frac{\hbar^2 \ell(\ell+1)}{R^2} \left[1 + (-1)^\ell \gamma_S\right] |\psi(R;E,\ell)|^2 .$$

The translational wavefunctions ψ are given by Eq. 5.32. Extensive use of Eq. 6.21 has been made [52, 58], especially for the modeling of spectral profiles [53, 54, 55, 57, 61, 62, 64, 65, 66], and for tests of line shape calculations.

For the rototranslational band, the spectral moment γ_2 has been shown to be related to the translational moments M_n, according to [292]

$$\begin{aligned} \gamma_2 \;=\; & \frac{4\pi^2}{3\hbar c} \sum_{(c)} \left\{ \frac{M_2^{(c)}}{(2\pi c)^2} + 2\left[B_1\lambda_1(\lambda_1+1) + B_2\lambda_2(\lambda_2+1)\right] \frac{M_1^{(c)}}{2\pi c} \right. \\ & + \Big[\left[B_1\lambda_1(\lambda_1+1) + B_2\lambda_2(\lambda_2+1)\right]^2 \\ & \left. + 2B_1\lambda_1(\lambda_1+1) \, \langle E_{\mathrm{rot};1}\rangle + 2B_2\lambda_2(\lambda_2+1) \, \langle E_{\mathrm{rot};2}\rangle \Big] M_0^{(c)} \right\} , \end{aligned} \tag{6.24}$$

with the mean rotational energy (in wavenumber units) being given by

$$\langle E_{\mathrm{rot;i}}\rangle = \sum_\ell P_\ell \, (2\ell+1) \, B_i \, \ell(\ell+1) \, .$$

The B_i are the rotational constants of molecule i in the vibrational ground state in wavenumber units. P_ℓ is the normalized Boltzmann factor.

The third moment has also been reported [319]; the expression for the translational moment is given by Eq. 5.40, with $L_1 = L(L+1)$ and $\mu(R)$ replaced by $B_c(R)$, with $c = \lambda_1\lambda_2\Lambda L$, Eq. 4.17; see pp. 210ff. for details.

About exchange symmetry. We compute the symmetry parameter γ_S appearing in Eq. 6.22 for given para-H_2 and ortho-H_2 densities, n_p, n_o. We assume low temperatures so that all para-H_2 is in the $j = 0$, and all ortho-H_2 in the $j = 1$ state. (Symmetry should not matter at temperatures above roughly 20 K, unless spectral dimer features are considered.)

For para-H_2 pairs, the spins and rotational states of the monomers are zero, $s_1 = s_2 = 0$ and $j_1 = j_2 = 0$. Accordingly, for the pair, we have just a single state with a total spin $S = 0$ and total angular momentum $J = 0$ which are both even. For the Bose symmetry required in this case, only even partial waves occur. The multiplicity equals $(2S+1)(2J+1) = 1$.

For ortho-H_2, $s_1 = s_2 = 1$ and $j_1 = j_2 = 1$. So, both for the total spin of the pair and the rotational wavefunctions, $|SM_S\rangle$, $|JM_J\rangle$, singlets ($S = 0$), triplets ($S = 1$) and quintuplets ($S = 2$) occur of which states with even S, J are even, and odd ones are odd. For bosons, the total wavefunction must be even. Hence even ℓ are associated with those states whose products of $|SM\rangle|JM_J\rangle$ are even, and odd ℓ if the products are odd. The sum of multiplicities of all even states, $S, J = 00, 02, 20, 11$ and 22, thus amounts to 45, and those of the odd states, $S, J = 01, 10, 12$ and 21, amounts to 36. The ratio of even to odd states of ortho-H_2 pairs is, therefore, 5:4.

For the treatment of a mixture of para-H_2 and ortho-H_2, we next compute r_{even}, the fraction of H_2 pairs associated with *even* ℓ, and r_{odd}, the fraction associated with *odd* ℓ, and r_{no}, the fraction without exchange symmetry (para-H_2–ortho-H_2 pairs). Particle conservation requires

$$r_{\text{even}} + r_{\text{odd}} + r_{\text{no}} = 1 \,.$$

These fractions are related to the symmetry parameter γ_S according to

$$2r_{\text{even}} + r_{\text{no}} = 1 + \gamma_S \qquad \text{and} \qquad 2r_{\text{odd}} + r_{\text{no}} = 1 - \gamma_S \tag{6.25}$$

for even and odd ℓ, respectively. These equations arise from the fact that the square of the wavefunction, $|\Psi_{\text{total}}|^2$, is (upon integration over configuration space) the probability of occurence of that state. Subtracting the second from the first of the above equations, we get at once

$$\gamma_S = r_{\text{even}} - r_{\text{odd}} \,. \tag{6.26}$$

We note that $|\gamma_S| \leq 1$.

The computation of the fractions r is straightforward. For equilibrium, the para-H_2 and ortho-H_2 densities, n_p, n_o, are obtained from the partition sum of rotational states, but if an excess of para-H_2 is to be accounted for (as is often the case in astrophysics), then the n_p, n_o are input in some form. The number of para-H_2 pairs in the volume V is $\frac{1}{2}n_p^2V^2$ which enter the expression for r_{even} with a multiplicity of unity. The number of ortho-H_2 pairs is $\frac{1}{2}n_o^2V^2$ of which the fraction 5/9 enters r_{even}, and the fraction 4/9 enters r_{odd}. The number of para-H_2–ortho-H_2 pairs without exchange symmetry is $n_pn_oV^2$, and the total number of H_2–H_2 pairs equals $\frac{1}{2}(n_p + n_o)^2V^2$. Hence, we have

$$\begin{aligned} r_{\text{even}} &= \frac{n_p^2 + \frac{5}{9}n_o^2}{(n_p + n_o)^2} \,; \\ r_{\text{odd}} &= \frac{\frac{4}{9}n_o^2}{(n_p + n_o)^2} \,; \\ r_{\text{no}} &= \frac{2\, n_p\, n_o}{(n_p + n_o)^2} \,; \end{aligned} \tag{6.27}$$

and hence, at low temperature,

$$\gamma_S = \frac{n_p^2 + \frac{1}{9} n_o^2}{(n_p + n_o)^2}. \tag{6.28}$$

At elevated temperatures, γ_S falls off to near zero as a consequence of rotational excitation of the H_2 molecules (which is here not accounted for), and the effect of the summation over many partial waves will further reduce the effects of symmetry.

Anisotropic potentials. The anisotropy of the interaction potentials may be taken into account in the computation of spectral moments. For the zeroth and first moments, the anisotropy of the interaction affects the pair distribution function, $g(R, \Omega_1, \Omega_2)$, which thus becomes dependent on the orientations of molecules 1 and 2. A perturbation treatment based on the assumption of small anisotropy was given later for an estimate of the effects of the anisotropy of H_2–He and H_2–H_2 pairs [293]. Moments of more strongly anisotropic molecules (N_2, CO_2) were recently considered [67, 122].

Classical and semiclassical moment expressions. The expressions for the spectral moments can be made classical by substituting the classical distribution function, $g(R) = \exp(-V(R)/kT)$, for the quantum expressions. Wigner–Kirkwood corrections are known which account to lowest order for the static quantum corrections, Eq. 5.44 [177, 292]. For the second and higher moments, dynamic quantum corrections must also be made [177]. As was mentioned in the previous Chapter, such semiclassical corrections are useful in supplementing quantum computations of the spectral moments at large separations where the quantum effects are small; the computational effort of quantum calculations, which is substantial at large separations, may thus be avoided.

Higher-order classical moments have also been reported. We mention the classical expressions for the translational spectral moments M_n, with $n = 0, 2, 4$, and 6, for pairs of linear molecules given in an appendix of [204]. Spectral moments of spherical top molecules have been similarly considered [163, 205]. We note that for $n > 1$, spectral moments show dynamic as well as static quantum correction, which become more important as the order n of the spectral moments is increased. The discussions on pp. 219, and Table 5.1, suggest that, even for the near-classical systems, quantum corrections may be substantial and can rarely be ignored.

3 *Results: binary rototranslational moments*

A selection of computational results of binary, rototranslational moments as function of temperature is shown in Fig. 3.27, p. 100 (solid lines). For the systems H_2–He and H_2–H_2, the dipole moment surfaces are obtained from first principles [277]; for the other systems shown, the classical multipole-induction model was employed, sometimes supplemented by an empirical overlap-induced dipole of low order (λL = 01 or 21, for example); see [58] for details and references. The agreement of computed moments with the measurements (dots, circles, squares, ...) is generally well within the experimental uncertainties.

Second moments have also been computed, both from first principles and on the basis of the classical multipole-induction model. These are found to be in close agreement with measurements where these exist. Second moments are of a special interest in connection with modeling of three-parameter line profiles from three spectral moments [52]. In analyses based on classical expressions, the second moment is expressible in terms of the first moment specified above, multiplied by $2kT/\hbar$.

Spectral moments can also be computed from classical expressions with Wigner–Kirkwood quantum corrections [177, 189, 317] of the order $\mathcal{O}(\hbar^2)$. For the quadrupole-induced 0223 and 2023 components of H_2–H_2, at the temperature of 40 K, such results differ from the exact zeroth, first and second moments by –10%, –10%, and +30% respectively. For the leading overlap-induced 0221 and 2021 components, we get similarly +14%, +12%, and –56%. These numbers illustrate the significance of a quantum treatment of the hydrogen pair at low temperatures. At room temperature, the semiclassical and quantum moments of low order differ by a few percent at most. Quantum calculations of higher-order moments differ, however, more strongly from their classical counterparts.

For moment calculations, the accounting for the spectral moments hardly affects the spectral moments at the temperatures considered; the extreme variation of the symmetry parameter γ_S from 0 to 1 modifies the moments M_n by less than 1% at 40 K, and much less at higher temperatures. It has been previously reported [315] that exchange symmetry matters at temperatures of less than 10 K. At the higher temperatures, one may often neglect symmetry ($\gamma_S = 0$ in Eq. 6.22) unless dimer features are considered; the dimer structures of like pairs are in general dependent on symmetry at any temperature.

4 *Results: binary rotovibrational moments*

The binary moment relations quoted above have been successfully used for the rototranslational bands and, to some extent, also for the rotovibrational bands [342]. However, it has been noted [151] that for the roto*vibrational* bands, all but the zeroth spectral moment are affected by the variation of the interaction potential with the vibrational excitation, an effect not accounted for in Eqs. 6.13 through 6.18, 6.21.

It is clear that the interaction potential is an essential part of the two-particle Hamiltonian and thus of the translational state of the supermolecule; the interaction potentials of initial and final state differ if a molecule undergoes vibrational excitation. For example, for a diatom–diatom pair like H_2–H_2, the translational state is determined by the vibrational average of the potential,

$$V_0^{v_1 v_2}(R) = \langle v_1 j_1 v_2 j_2 \,|\, V_0(R; r_1, r_2) |\, v_1 j_1 v_2 j_2 \rangle \,, \qquad (6.29)$$

if the interacting molecules are in the vibrational states $|v_1 j_1\rangle$ and $|v_2 j_2\rangle$. We note that for small rotational quantum numbers j_i, the dependence of the vibrational wavefunctions on j_i is generally a weak one which can in such a case be suppressed. Vibrations and, to some extent, the centrifugal forces arising from rotation, increase the size of molecules relative to the ground state.

For H_2 and, presumably, for most other molecules, significant differences between $V_0^{00}(R)$ and $V_0^{01}(R)$, etc., exist, especially at near range which is so important for the spectroscopic interactions. Table 6.1 shows as an example various vibrational matrix elements of vibrating H_2 pairs which were computed for a number of separations R from first principles [281]. The data show how the interaction energy (i.e., the apparent size of the interacting pair) increases with increasing vibrational excitation. We note that the anisotropy of vibrating diatomic molecules is also greater than that of non-vibrating molecules, but this fact has been often neglected in the past.

Spectral moments were obtained from the autocorrelation function of the $\lambda_1 \lambda_2 \Lambda L$ component of the induced dipole component [294],

$$C_c^{vv'}(t) = \frac{1}{4\pi\varepsilon_0} \operatorname{Tr} \left\{ \mathrm{e}^{-\mathscr{H}_v/kT} \, B_c^{vv'}(R) \, Y \, \mathrm{e}^{\mathrm{i}\mathscr{H}_{v'} t/\hbar} \, B_c^{v'v}(R) \, Y \, \mathrm{e}^{-\mathrm{i}\mathscr{H}_v t/\hbar} \right\} . \qquad (6.30)$$

The zeroth translational moment is obtained for $t = 0$, according to

$$\begin{aligned} M_0^{vv'}(c;T) &= C_c^{vv'}(t)\Big|_{t=0} \qquad (6.31) \\ &= \frac{1}{4\pi\varepsilon_0} \operatorname{Tr} \left\{ \exp\left(-\mathscr{H}_v/kT\right) \left| B_c^{vv'} \right|^2 \right\} \\ &= 4\pi \, \frac{1}{4\pi\varepsilon_0} \int_0^\infty \left| B_c^{vv'} \right|^2 g_v(R) \, R^2 \, \mathrm{d}R \,, \end{aligned}$$

where the subscript c stands for λL or $\lambda_1 \lambda_2 \Lambda L$, respectively. Apart from an obvious generalization to 'hot bands' ($v > 0$), for $v = 0$ this result agrees with earlier work, Eq. 6.13. The zeroth moment equals the total intensity which is seen not to depend on the vibrational excitation of the final state.

Table 6.1. Various vibrational averages, $\langle v_1 v_2 | V_0(R; r_1, r_2) | v_1 v_2 \rangle$, of the H_2–H_2 interaction potential [281]; the dependences on the rotational quantum numbers j_1, j_2 are here suppressed.

R	$\langle 00\|V\|00\rangle$	$\langle 00\|V\|00\rangle$	$\langle 01\|V\|01\rangle$	$\langle 11\|V\|11\rangle$	$\langle 02\|V\|02\rangle$
(a.u.)	(10^{-6} hartree)				
3.5	17883.3	18872.3	19425.3	19965.5	19978.8
4.0	7117.0	7608.6	7925.3	8238.5	8252.8
4.5	2518.6	2810.5	2972.8	3135.0	3144.6
5.0	771.5	880.2	952.2	1024.6	1030.1
5.5	125.4	166.7	192.7	218.9	221.7
6.0	−73.4	−60.2	−54.7	−49.3	−47.9
6.5	−111.6	−107.7	−109.9	−112.4	−111.8
7.0	−98.7	−98.1	−102.3	−106.8	−106.5
8.5	−39.2	−39.3	−41.7	−44.3	−44.3
10.0	−14.5	−14.5	−15.4	−16.3	−16.3

The first translational moment is given by

$$\begin{aligned} M_1^{vv'}(c;T) &= -\mathrm{i} \left. \frac{\mathrm{d}C_c^{vv'}}{\mathrm{d}t} \right|_{t=0} \qquad (6.32) \\ &= \frac{1}{\hbar} \frac{1}{4\pi\varepsilon_0} \operatorname{Tr} \left\{ \exp\left(-\mathscr{H}_v/kT\right) B_c^{vv'} \left(\mathscr{H}_{v'} B_c^{v'v} - B_c^{vv'} \mathscr{H}_v \right) \right\} \\ &= \frac{1}{\hbar} \frac{1}{4\pi\varepsilon_0} \operatorname{Tr} \left\{ \exp\left(-\mathscr{H}_v/kT\right) B_c^{vv'} \right. \\ &\qquad \left. \times \left(\left[\mathscr{H}_{.v}, B_c^{v'v} \right] + \left(V_0^{v'}(R) - V_0^{v}(R) \right) B_c^{vv'} \right) \right\} . \end{aligned}$$

The comparison with Eq. 6.4 shows that the first of the two terms given here is the familiar first moment, $M_1(\lambda L;T)$, evaluated for the proper vibrational state. The second term is new and may be considered a correction; it vanishes for v-independent interaction potentials and $v = v'$. The first translational moment thus becomes

$$\begin{aligned} M_1^{vv'}(c;T) &= \left[M_1(c;T) \right]_v \qquad (6.33) \\ &+ \frac{4\pi}{\hbar} \frac{1}{4\pi\varepsilon_0} \int_0^\infty \left(B_c^{vv'} \right)^2 \left(V_0^{v'}(R) - V_0^{v}(R) \right) g_v(R)\, R^2 \, \mathrm{d}R . \end{aligned}$$

Skipping over lengthy algebra, for the second translational moment we get similarly [296]

$$M_2^{vv'}(c;T) = \left[M_2(c;T) \right]_v \qquad (6.34)$$

Table 6.2. The zeroth, first, and second translational moments of the fundamental band of H_2–He, of the three main induced dipole components, at four temperatures. The M_n are the uncorrected values of the nth moment, obtained with the assumption $V_0^v(R) \equiv V_0^{v'}(R)$, and the M_n^{01} are values corrected for the vibrational dependences of the interaction potential; the M_n^* are obtained from line shape calculations which account for the v' dependence of the interaction.

T (K)	M_0^{01} (10^{-64} erg cm^6)	M_0^*	M_1 (10^{-51} erg cm^6 s^{-1})	M_1^{01}	M_1^*	M_2 (10^{-37} erg cm^6 s^{-2})	M_2^{01}	M_2^*
$\lambda L = 01$								
26.8	1.80	1.81	3.84	4.20	4.24	1.75	2.07	2.11
40.0	2.17	2.18	4.58	5.05	5.07	2.20	2.61	2.66
155.	5.92	6.04	11.85	13.89	13.68	8.78	10.54	10.74
298.	11.20	11.57	21.52	27.02	26.47	23.59	28.28	29.25
$\lambda L = 23$								
26.8	0.472	0.476	0.514	0.559	0.560	0.184	0.217	0.220
40.0	0.524	0.528	0.593	0.650	0.650	0.233	0.274	0.278
140.	0.918	0.932	1.222	1.415	1.384	0.814	0.959	0.986
298.	1.509	1.552	2.254	2.800	2.668	2.471	2.899	2.951
$\lambda L = 21$								
26.8	0.117	0.118	0.276	0.301	0.305	0.141	0.167	0.171
40.0	0.141	0.142	0.333	0.366	0.369	0.179	0.212	0.218
140.	0.372	0.379	0.861	0.996	0.992	0.662	0.795	0.801

$$+ \frac{4\pi}{\hbar^2} \frac{1}{4\pi\varepsilon_0} \int_0^\infty \left(B_c^{vv'}\right)^2 \left(V_0^{v'}(R) - V_0^v(R)\right)^2 g_v(R)\, R^2\, dR$$

$$+ \frac{4\pi}{m} \frac{1}{4\pi\varepsilon_0} \int_0^\infty \left\{ \left[\left(\frac{dB_c^{vv'}}{dR} \right)^2 + \frac{L(L+1)}{R^2} \left(B_c^{vv'}\right)^2 \right] \right.$$

$$\left. \times \left(V_0^{v'}(R) - V_0^v(R)\right) + B_c^{vv'} \frac{dB_c^{vv'}}{dR} \left(\frac{dV_0^{v'}}{dR} - \frac{dV_0^v}{dR} \right) \right\} g_v(R)\, R^2\, dR\,.$$

This is the familiar result, $M_2(c;T)$ of Eq. 6.21, computed with the appropriate potential $V_0^v(R)$, plus several correction terms which vanish in the limit of a v-independent interaction and for $v = v'$. The radial pair distribution function, $g_v(R)$, may be obtained by quantal or semiclassical methods.

For the case of induced absorption of H_2–He pairs, the vibrational matrix elements of potential and dipole function are well known [151, 294]. The spectral moments $M_0 \cdots M_2$ have been computed for the main induction components with the corrections for the vibrational dependences of the interaction potential, Table 6.2. While the interaction of H_2–He pairs is not dramatically different if the H_2 molecule is vibrationally excited [276], the vibrational corrections to the translational spectral moments are significant, even for the lowest vibrational excitations ($v = 0 \rightarrow 1$). The zeroth moment does not require vibrational corrections for any v', Eq. 6.31. For the first moments, the corrections amount to $\approx 10\%$ at the lower temperatures, and to $\approx 20\%$ at the higher temperatures shown. For the second moment, even greater corrections are observed which increase with increasing temperature.

The comparison of the corrected translational moments, M_n^{01}, with the moments M_n^* obtained by integration of the computed line shapes (to be discussed below) shows agreement in the 1% range, the estimated numerical precision of the computational work. At the higher temperatures, the second moments differ by up to 5% which is, however, due to inadequate extrapolation of the computed line shapes to high frequencies; this error could have been largely avoided by extending the line shape computations to much higher frequency shifts.

Figures 6.21 and 6.22 below (p. 345) show calculations of the three lowest moments of the H_2–He enhancement spectra of the fundamental band (left panels). The range of temperatures is a much wider one than those of Table 6.2. We notice that with increasing temperature, the differences between the moments computed with and without the corrections for the vibrational dependences of the interaction potential (solid and dashed lines, respectively) increase substantially for the first moments and decrease for the second moments to the point of insignificance.

Figure 6.1 shows similar results for the H_2–H_2 system. One notices again significant corrections for the vibrational dependences of the interaction potential (the differences between the solid and the dashed lines), especially for the first moments at the higher temperatures. Since in the classical limit the first moments vanish, one would think that these corrections are largely of a quantum nature. We note that the more familiar quantum corrections, e.g., of the Wigner–Kirkwood type, Chapter 5, tend to be less significant at elevated temperatures. This fact suggests a non-classical behavior of so-called classical systems in the vibrational bands [295].

The data presented in Table 6.2 and Fig. 6.1 show that accounting for the vibrational dependences of the interaction potential is significant, especially at the higher temperatures, certainly for systems like H_2–He and H_2–H_2.

Figures 3.42 through 3.44 (pp. 122 and 123) compare the spectral moments γ_0, γ_1, computed from first principles (solid curves) for the fundamental band of hydrogen, of the systems H_2–H_2 and H_2–He, with the existing measurements (dots, circles, squares, etc.). The agreement is well within the experimental uncertainties of such measurements.

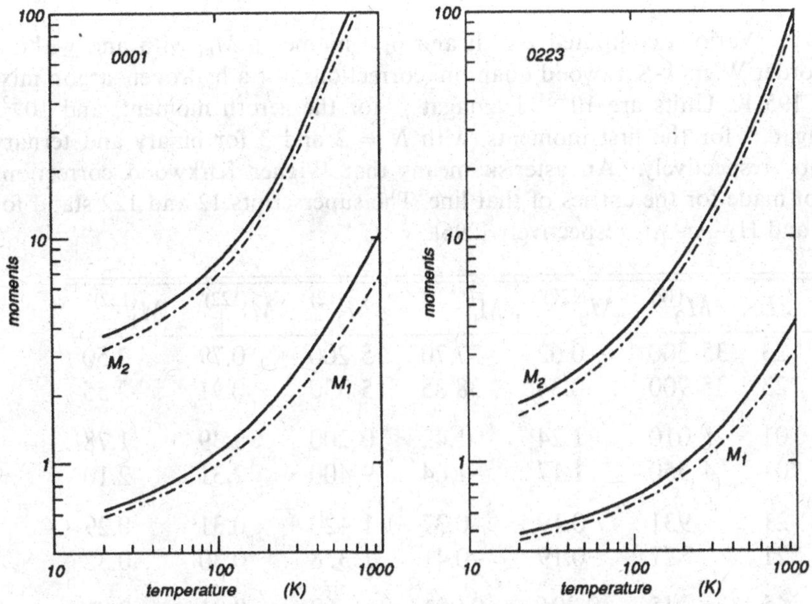

Fig. 6.1. First and second moments of the main dipole components, $\lambda_1\lambda_2\Lambda L =$ 0001 (left) and 0223 (right) of the fundamental band of hydrogen in unmixed hydrogen, as function of temperature; units are 10^{-50} erg cm^6 s^{-1} and 10^{-37} erg cm^6 s^{-2} for the first and second moments, respectively. The dot-dashed curves represent the moments obtained without accounting for the vibrational state dependence of the interaction potential [281].

5 Ternary moments

The theory of ternary processes in collision-induced absorption was pioneered by van Kranendonk [402, 400]. He has pointed out the strong cancellations of the contributions arising from the density-dependent part of the pair distribution function (the 'intermolecular force effect') and the destructive interference effects of three-body complexes ('cancellation effect') that lead to a certain feebleness of the theoretical estimates of ternary effects in collision-induced absorption spectra.

Ternary moments have been computed for several systems of practical interest [314, 422]. Recent studies are based on accurate *ab initio* pair dipole surfaces obtained with highly correlated wavefunctions. Because not much is presently known about the irreducible ternary components, it is important to determine to what extent the measured three-body spectral components arise from pairwise-additive contributions [296, 299].

H_2–Ar–Ar rototranslational spectra. Tables 6.3 and 6.4 show the rototranslational spectral moments computed from first principles, with the advanced 'TT3' H_2–Ar [228] and the 'HFD-C' Ar–Ar interaction poten-

Table 6.3. Various computed zeroth and first moments M_n, with and without lowest-order Wigner–Kirkwood quantum corrections, for a hydrogen–argon mixture at 195 K. Units are 10^{-34} J amagat^{-N} for the zeroth moment, and 10^{-21} W amagat^{-N} for the first moments, with $N = 2$ and 3 for binary and ternary moments, respectively. An asterisk means that Wigner–Kirkwood corrections were not made for the entries of that line. The superscripts 12 and 122 stand for H_2–Ar and H_2–Ar–Ar, respectively [296].

λL	$M_0^{(12)}$	$M_0^{(122)'}$	$M_0^{(122)''}$	$M_1^{(12)}$	$M_1^{(122)'}$	$M_1^{(122)''}$
23	35 300	−0.62	−37.70	15 200	0.79	−2.50
*23	35 900	−0.34	−38.85	15 330	0.91	−2.55
01	5 010	1.24	−1.42	10 200	2.39	1.78
*01	4 550	1.17	−1.64	9 400	2.31	2.10
21	931	0.19	−0.37	1 420	0.31	0.29
*21	877	0.19	−0.41	1 328	0.30	0.32
45	45	0.006	0.001	60	0.01	0.06
*45	45	0.006	0.001	58	0.01	0.07

tial [11]. The He–Ar pair dipole surface is believed to be one of the best known induced dipole surfaces; line shape computations based on it describe the measured binary, rototranslational spectra closely [280, 139]. The second and fifth columns of Table 6.3 show the binary zeroth and first spectral moments for the four most important dipole components, at a temperature of 195 K, with (upper lines) and without (lower, starred lines) quantum corrections. For the quadrupole-induced component ($\lambda L = 23$), quantum corrections are insignificant at 195 K, but for the other components shown, especially for the overlap-induced components ($\lambda L = 01$ and 21), these amount to a non-negligible $\approx$10%. For some three-body moments, on the other hand, the quantum corrections are more substantial, Table 6.3. (In Table 6.4, only the Wigner–Kirkwood quantum corrected data are shown for four temperatures.)

For the quadrupole-induced component ($\lambda L = 23$), ternary zeroth moments, $M_0^{(T)} = M_0' + M_0''$, are negative except at the lowest temperature, Table 6.4. Negative zeroth moments are well known from related work with nitrogen [375] and have been observed in some other systems where quadrupole induction is significant. The related first moments may be negative or positive, depending on the temperature, and vary rapidly with temperature, even changing sign at a relatively high temperature, somewhere between 195 and 298 K.

Table 6.4. Temperature variation of the two- and three-body moments of the spectral function for the H_2–Ar–Ar complex: the superscripts (12) and (122) stand for H_2–Ar and H_2–Ar–Ar; $M^{(T)}$ is actually the sum of $M^{(122)'}$ and $M^{(122)''}$. Units are 10^{-34} J amagat^{-N} for the zeroth and 10^{-21} W amagat^{-N} for the first moments; $N = 2$ for binary and $N = 3$ for ternary moments [296].

λL	T (K)	$M_0^{(12)}$ ($N = 2$)	$M_0^{(T)}$ ($N = 3$)	$M_1^{(12)}$ ($N = 2$)	$M_1^{(T)}$ ($N = 3$)
23:	45	53 200	243	21 000	272
	85	39 200	−84.9	15 700	−0.30
	165	35 200	−46.5	14 900	−3.54
	298	37 000	−22.1	16 800	3.15
01:	45	4 280	124	9 090	227
	85	3 490	4.61	7 350	10.4
	165	4 470	−0.39	9 160	3.20
	298	7 180	1.23	14 200	8.68
21:	45	903	25.2	1 350	31.5
	85	723	0.917	1 080	1.47
	165	851	−0.201	1 290	0.46
	298	1 250	0.036	1 930	1.18
45:	45	54.2	1.27	66.0	1.58
	85	41.8	0.072	51.7	0.080
	165	43.3	0.014	56.2	0.013
	298	54.3	0.027	74.8	0.033

At the lowest temperatures, the three-body moments are relatively strong, Table 6.4. A density of only ≈10 amagat will modify the observed moments by roughly 10%. The strong temperature dependence of the three-body moments at low temperature may be quite important for some applications, for example for the spectroscopic modeling of planetary atmospheres. It seems to be related to the formation of dimers and, consequently, to monomer–dimer interactions which are three-body processes by our definition. Of course, at 45 K, quantum corrections are substantial and the numbers quoted must be considered rough estimates. Nevertheless, the general trend of the temperature dependence seems clear.

Hydrogen–Helium Mixtures. Accurate *ab initio* dipole surfaces for both the rototranslational collision-induced absorption spectrum in the far infrared [280], and the rotovibrational collision-induced absorption spectrum in the near infrared [151], have been obtained that could have

Table 6.5. Temperature dependence of the moment of the enhancement spectra of hydrogen–helium mixtures in the fundamental band of H_2. The superscripts 12 and 122 stand for H_2–He and H_2–He–He; the term $M_n^{(122)} = M_n^{(H_2-He-He)\prime} + M_n^{(H_2-He-He)\prime\prime}$; and similar for $M_n^{(112)}$. Units are 10^{-35} J amagat^{-N} and 10^{-22} W amagat^{-N} for the zeroth and first moments, with $N = 2$ for the binary and $N = 3$ for the ternary moments [296].

λL	T (K)	$M_0^{(12)}$ ($N = 2$)	$M_0^{(122)}$ ($N = 3$)	$M_0^{(112)}$	$M_1^{(12)}$ ($N = 2$)	$M_1^{(122)}$ ($N = 3$)	$M_1^{(112)}$
23:	45	393	−0.060	−0.018	459	0.087	0.029
	85	510	−0.002	−0.036	636	0.177	0.215
	165	732	0.072	0.137	1000	0.321	0.486
	298	1090	0.016	0.334	1630	0.533	0.861
01:	45	1750	0.35	−0.543	3670	1.18	−0.447
	85	2590	0.58	0.287	5340	2.39	1.93
	165	4510	1.02	1.24	9000	4.32	4.89
	298	8030	1.71	2.55	15400	7.14	8.96
21:	45	117	0.024	−0.035	274	0.09	−0.031
	85	176	0.042	0.023	411	0.18	0.151
	165	320	0.079	0.097	738	0.35	0.410
	298	602	0.144	0.212	1370	0.64	0.774
45:	45	5.6	0.0016	−0.0004	16.4	0.005	−0.0022
	85	8.2	0.0030	0.0031	24.7	0.009	0.0084
	165	14.4	0.0058	0.0080	44.8	0.018	0.024
	298	26.2	0.0103	0.0154	84.5	0.034	0.048

been used to compute the pairwise-additive ternary contributions for both bands. However, measurements of the density-dependence exist only for the fundamental band. For that reason, computational results are reported only for that band, Table 6.5.

Hydrogen. Accurate *ab initio* dipole surfaces for both the rototranslational collision-induced absorption spectrum in the far infrared [282], and the rotovibrational collision-induced absorption spectrum in the near infrared [281] have been obtained that may be used to compute the pairwise-additive ternary contributions for both bands. However, measurements of the density-dependence exist only for the fundamental band, Table 6.6. We note that according to the nomenclature adopted, molecule 1 undergoes the vibrational transition.

Of all gases considered, hydrogen molecules form the lightest complexes. At any given temperature, hydrogen will, therefore, show the most substantial quantum corrections.

Table 6.6. Temperature dependence of the moments of collision-induced absorption of pure hydrogen, in the fundamental band. The superscripts 12 and 123 stand for H_2–H_2 and H_2–H_2–H_2. Units are 10^{-35} J amagat^{-N} and 10^{-22} W amagat^{-N} for the zeroth and first moments, respectively, with $N = 2$ for the binary and $N = 3$ for the ternary moments [296].

$\lambda_1\lambda_2\Lambda L$	T (K)	$M_0^{(12)}$ (N=2)	$M_0^{(123)'}$ ($N = 3$)	$M_0^{(123)''}$	$M_1^{(12)}$ (N=2)	$M_1^{(123)'}$ ($N = 3$)	$M_1^{(123)''}$
0223:	40	825	−0.508	0	600	−0.343	0
	50	847	−0.179	0	622	−0.105	0
	60	855	−0.077	0	635	−0.030	0
	78	865	0.014	0	650	0.040	0
	165	950	0.214	0	760	0.212	0
	298	1100	0.338	0	930	0.338	0
2023:	40	2180	−1.14	−2.76	2255	−0.88	0.129
	50	2235	−0.366	−3.40	2307	−0.22	−0.168
	60	2263	−0.114	−3.34	2355	0.022	−0.237
	78	2320	0.128	−3.07	2460	0.278	−0.254
	165	2730	0.750	−2.55	3180	1.08	−0.225
	298	3432	1.265	−2.46	4440	1.91	−0.217
0001:	40	1340	−0.46	−0.103	3600	−1.48	−0.173
	50	1323	−0.201	−0.290	3570	−0.63	0.257
	60	1350	−0.045	−0.421	3640	−0.165	0.435
	78	1445	0.167	−0.583	3885	0.416	0.588
	165	2215	0.89	−1.10	5780	2.27	0.927
	298	3675	1.84	−1.81	9250	4.52	1.34
0221:	40	13.8	−0.007	0	35.9	−0.017	0
	50	13.8	−0.004	0	35.6	−0.010	0
	60	13.8	−0.002	0	35.6	−0.007	0
	78	14.1	−0.001	0	35.8	−0.004	0
	165	16.2	0.003	0	39.2	0.002	0
	298	20.0	0.005	0	45.6	0.008	0
2021:	40	238	−0.074	−0.042	670	−0.240	−0.034
	50	235	−0.033	−0.067	661	−0.106	0.037
	60	240	−0.006	−0.087	673	−0.025	0.068
	78	257	0.030	−0.114	720	0.081	0.096
	165	399	0.162	−0.205	1090	0.436	0.161
	298	675	0.340	−0.335	1790	0.875	0.243

We note that the intermolecular force effect was once thought to lead in general to positive M_0', but Table 6.6 shows that at low temperatures the M_n' are negative. The interference terms M_0', on the other hand, are generally negative (unless they are zero) and may actually be more significant than the force terms for the zeroth moments.

H_2–Ar–Ar rototranslational band. Experimental studies of the density variation of the rototranslational collision-induced absorption spectra of argon gas with a small admixture of hydrogen or deuterium have been reported [140, 108, 109, 106]. Since there is no induced dipole component associated with Ar–Ar interactions, the spectroscopically dominant three-body interactions involve one hydrogen molecule and two argon atoms, H_2–Ar–Ar. These spectra consist mainly of the quadrupole-induced rotational $S_0(J)$ lines arising from the $\lambda L = 23$ component.

Collision-induced absorption spectra obtained at a number of argon densities from 42 to 182 amagats, with 3% hydrogen admixtures, show subtle changes of shape and intensity that indicate the presence of many-body effects beyond binary ones. Evaluations in terms of line shape models and correlation functions [108, 109, 106] and virial expansions of the line shape [140] have been attempted. The zeroth moment, normalized by the product of densities, $\gamma_0/\varrho_1\varrho_2$, shows a slight linear decrease with increasing argon density, with a slope that agrees well with theory, Table 6.7 [140]. The experimental uncertainties should be somewhere between 10 and 25%, but are hard to estimate dependably. The measurement is taken at 195 K. The measured first moment, $\gamma_1/\varrho_1\varrho_2$, on the other hand, is much greater than the theoretical value, Table 6.7, by about 4×10^{-7} cm^{-2} amagat^{-2}; even the signs differ from the theoretical value.

H_2–He–He rotovibrational band. The density dependence of the H_2–He enhancement spectrum in the fundamental band of hydrogen has been measured previously, using a trace of hydrogen in helium of thousands of amagats [121, 175, 142]: ternary moments were measured at room temperature. The measurements suggest again greater values of the spectral moments, Table 6.7.

H_2–H_2–H_2 rotovibrational band. Probably the most substantial measurements of the density dependence of collision-induced absorption spectra in the fundamental band near the onset of three-body effects, at densities from 15 to 400 amagat, are due to Hunt [187]. He shows a plot of γ_1/ϱ^2 as function of the hydrogen density at six temperatures, from 40 to 300 K (Fig. 3.45). Especially at the low temperatures, these dependences are well represented by straight lines of well defined slopes, which in turn define the three-body moments. Inferred slopes are reported for both γ_0 and γ_1 in that work.

It had been suggested at one point that virial expansions of this kind are possibly meaningless on the grounds of the very high densities necessary

Table 6.7. Comparison of the computed third virial coefficients of the induced spectra of various gases and mixtures with the existing measurements. The three-body coefficients were computed with the assumption of pairwise-additive dipole components [48].

T	source	$\gamma_1^{(112)}$		$\gamma_0^{(122)}$		$\gamma_1^{(122)}$	
	(expt.)	$10^{-6}\text{cm}^{-2}\text{am}^{-3}$		$10^{-8}\text{cm}^{-1}\text{am}^{-3}$		$10^{-6}\text{cm}^{-2}\text{am}^{-3}$	
(K)	Ref.	meas.	calc.	meas.	calc.	meas.	calc.
helium–argon mixture, translational band:							
165	[95]	0.07	0.056			0.10	0.033
hydrogen–argon mixture, rotational lines S_0:							
195	[139]			−1.60	−1.65	0.44	−0.036
hydrogen–helium, rototranslational band:							
298	[121]					0.55	0.37
298	[175]					0.55	0.37
300	[142]					0.41	0.37

for such measurements [402] which may necessitate accounting for many-body interactions. However, the evidence considered suggests that virial expansions of spectral moments are probably valid procedures at densities up to a few hundred amagats.

Figures 3.46 (p. 126) and 6.2 show the measured three-body moments (squares, dots, etc.) as function of temperature. A visual average of the data presented is sketched (thin line). For comparison, the calculated theoretical predictions are also plotted (heavy curves). Both the measured and computed moments γ_n increase with increasing temperature. However, the measurements increase faster with temperature than the calculations. Moments are negative at low temperature; measured moments turn positive at some intermediate temperature. Theory and measurements never coincide, the measured values always being greater if $T > 50$ K.

Error bars of the experimental data were not specified. However, Hunt provided measurements at two different ortho- to para-H_2 concentration ratios, 3:1 (solid squares and dots) and 1:1 (open squares and circles). According to the theory developed above, variation of this ratio should not affect the results. While symmetry considerations of the interacting H_2 molecules are important at low temperatures ($T < 40$ K), the semiclassical approach does not distinguish para- and ortho-H_2; we think that the differences of the data taken at different ortho–para ratios may, in essence, reflect the uncertainties of the measurement. We note that earlier works by Chisholm and Welsh [121] and by Hare and Welsh [175] gave values

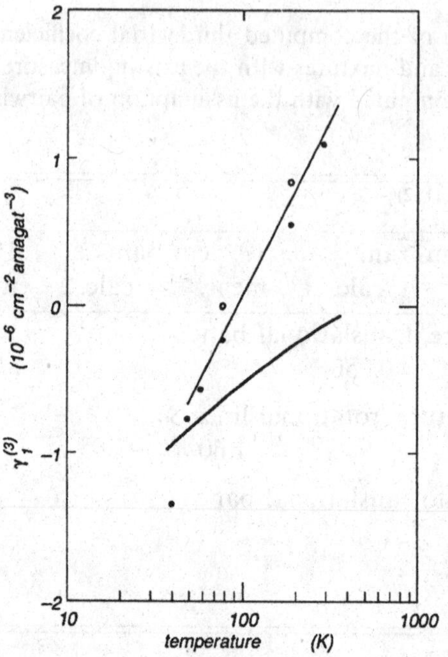

Fig. 6.2. Comparison of measured three-body γ_1 moments with theory (heavy curve) for hydrogen and the fundamental band; $\gamma_1^{(3)}$: dots and circles from [187]; the thin line is a visual average through the experimental points [295].

of the three-body moments that differ from Hunt's measurements roughly by the uncertainties suggested here, Table 6.7.

Theory, on the other hand, should be accurate to within roughly ten percent as far as the pairwise-additive contributions are concerned. Theory neglects the slightly different interaction potentials of two H_2 molecules when one is vibrationally excited [151, 294, 295], which may introduce errors of this magnitude.

It seems, therefore, not unreasonable to assume that the differences observed in Figs. 3.46 and 6.2 between theory and measurement are due, in essence, to the pairwise non-additive three-body contributions. These amount to roughly $\approx 2 \times 10^{-10}$ cm^{-1} amagat^{-3} for the moment γ_0, and to $\approx 5 \times 10^{-7}$ cm^{-2} amagat^{-3} for the moment γ_1; the differences increase with temperature as the diverging experimental and theoretical curves suggest, Figs. 3.46 and 6.2. Such a conclusion is to be taken with some caution, but the difference between theory and measurements shown may be the best evidence obtained to date for the existence and the approximate magnitude of the non-additive dipole contributions.

Early work. Previous numerical estimates [402] were based on the empirical 'exp–4' induced dipole model typical of collision-induced absorption in the fundamental band, and hard-sphere interaction potentials. While the main conclusions are at least qualitatively supported by more detailed, recent calculations, significant quantitative differences are observed that are related to the use of much improved interaction potentials now available; to quantum corrections; and to accurate induced dipole functions that include significant overlap components in the quadrupole-induced terms. Reliable estimates of the $\lambda_1\lambda_2\Lambda L = 0001$ and 0221, 2021 components are also included for the first time. As a consequence, at 298 K, the more recent three-body moment $\gamma_0^{(3)}$ is negative, –0.34, but the previous estimate [402] is positive, 0.4, in units of 10^{-10} cm^{-1} am^{-3}. If the previous data were plotted in Fig. 6.2, this difference would not be too striking. Moreover, the force effect is not always positive, nor is it always stronger than the cancellation effect, but for hydrogen near room temperature, for the main induction components, previous conclusions to that effect [402] are confirmed.

A computational study of the density dependence of the rototranslational collision-induced absorption spectrum of nitrogen, oxygen and carbon dioxyde was reported by Steele and Birnbaum [375] on the basis of the classical quadrupole induction model.

Conclusions. The existing, accurate *ab initio* induced dipole surfaces permit the computation of binary collision-induced absorption spectra in close agreement with existing measurements when these pair-dipole functions are used to compute the pairwise additive components of the ternary spectral moments for such systems and conditions for which experimental data are available for comparison, it is evident that the measured ternary moments are consistently greater than calculations, Figs. 3.46 and 6.2 and Table 6.7. Specifically, measured $\gamma_0^{(3)}$ moments exceed the calculated values significantly, roughly by 10^{-10} cm^{-1} amagat^{-3}, and the $\gamma_1^{(3)}$ by $\approx 10^{-7}$ cm^{-2} amagat^{-3}, in all examples considered above. The excess absorption is increasing rapidly with temperature in hydrogen gas. It is almost certainly related to the non-additive components of the collision-induced dipoles of molecular triplets that were neglected in the theory.

Very little is known about the irreducible ternary dipole components. An early estimate based on classical electrodynamics, hard-sphere interaction and other simplifying assumptions suggests very small, *negative* contributions to the zeroth spectral moment [402], namely -0.13×10^{-10} cm^{-1} amagat^{-3}.

It is, therefore, interesting to point out that in a recent molecular dynamics study, shapes of intercollisional dips of collision-induced absorption were obtained. These line shapes are considered a particularly sensitive probe of intermolecular interactions [301]. Using recent pair potentials and empirical pair dipole functions, for certain rare-gas mixtures spectral profiles were obtained that differ significantly from what is observed experimentally; high densities around 1000 amagat are considered. This inconsistency of pairwise-additive and measured shapes was considered compelling evidence for the presence of irreducible three-body forces and dipole moments, in qualitative support of the above results.

The classical long-range components of the ternary irreducible induced dipole do probably not generate a significant portion of the irreducible effects seen in the figures. The non-additive close-range contributions may be the main mechanism. Guillot, Mountain and Birnbaum have developed a 'one effective electron model' of the latter, for a first theoretical estimate of the magnitude of the spectroscopic effects arising from the ternary, irreducible dipole component [171, 173]. For systems like H_2–He–He at liquid state densities, this model suggests also a significant *enhancement* of the absorption due to overlap-induced irreducible dipole components. The strong temperature dependence shown in Fig. 6.2 is consistent with the assumption of overlap induction of the irreducible components.

Ternary moments are generally associated with greater quantum corrections than binary moments, Tables 5.2 and 6.3. Quantum corrections are most significant near the repulsive core of the interaction potential. Apparently, for three-body interactions, core penetration is more significant spectroscopically than for two-body interactions.

Theory suggests that ternary moments vary substantially with temperature; even sign changes occur with modest temperature variations. This fact offers intriguing possibilities for an experimental separation of the pairwise-additive and the irreducible three-body effects and, perhaps, for a critical search for irreducible ternary contributions.

6.2 Line shapes of pairs of linear molecules

The theory of line shapes of systems involving one or more molecules starts from the same relationships mentioned in Chapter 5. We will not repeat here the basic developments, e.g., the virial expansion, and proceed directly to the discussion of binary molecular systems. It has been amply demonstrated that at not too high gas densities the intensities of most parts of the induced absorption spectra vary as density squared, which suggests a binary origin. However, in certain narrow frequency bands, especially in the Q branches, this intensity variation with density ϱ differs from the ϱ^2 behavior (intercollisional effect); the binary line shape theory does not describe the observed spectra where many-body processes are significant. In the absence of a workable theory that covers *all* frequencies at once, even in the low-density limit one has to treat the intercollisional parts of the spectra separately and remember that the binary theory fails at certain narrow frequency bands [318].

For the binary spectra, a density normalized absorption coefficient, A, and a spectral density, function, G, are widely used. These are related to the absorption coefficient, $\alpha(\nu)$, and spectral density, $J(\nu)$, according to

$$A(\omega) = \alpha(\omega)/\varrho_1\varrho_2 \quad \text{or} \quad \alpha(\omega)/\varrho^2, \tag{6.35}$$

$$G(\omega) = J(\omega)/N_L^2\varrho_1\varrho_2 \quad \text{or} \quad J(\omega)\, 2/\varrho^2, \tag{6.36}$$

for mixtures of partial densities ϱ_1, ϱ_2, and like pairs of molecules of density ϱ, respectively. We note that the factor 2 for like pairs appears only in the expression of the spectral density J, but not for the absorption coefficient α; this factor accounts for the identity of like pairs of the type A–B and B–A. Instead of normalization by the densities expressed in amagat units, normalization by number densities is sometimes used in the literature. We note that elsewhere the dimensionless quantity $\tilde{A}$, is used, usually as function of frequency ν in cm^{-1} units, which is related to the normalized absorption coefficient as $\tilde{A}(\nu) = A(\nu)/\nu = A(\nu)2\pi c/\omega$. The coefficient A describes the absorption of spectral intensity or energy, $\tilde{A}$ is proportional to the probability of absorbing a photon per unit path length. We will not make great use of the quantity $\tilde{A}$, because the spectral density G defined above is more closely related to the squared dipole transition matrix elemements, even at low frequency; it is therefore the preferred quantity.

If collisional systems involving one or more molecules are considered, the internal degrees of freedom of the molecule(s) (e.g., rotation, vibration) have to be taken into account. This often leads to cumbersome notations and other complications. Furthermore, we now have to deal with anisotropic intermolecular interactions which again calls for a significant modification of the formal theory. In that sense, this Chapter differs from the previous one; but otherwise the reader will find here much the same material, techniques, etc., as discussed in Chapter 5.

We will begin with the important case of pairs of linear molecules, such as diatomic molecules; the formalism to be developed may be reduced to the case of atom–diatom collisions, or to the case of molecules of arbitrary geometry.

The Schrödinger equation for the interacting pair of linear molecules is again separated into the equations of center-of-mass motion and relative motion, exactly as this was done in Chapter 5. The equation of the center-of-mass motion is of little interest and will be ignored. The Schrödinger equation of relative motion is given by [354]

$$\left\{-\frac{\hbar^2}{2m}\nabla_{\boldsymbol{R}}^2 + \mathscr{H}_1(\boldsymbol{r}_1) + \mathscr{H}_2(\boldsymbol{r}_2) + V(\boldsymbol{r}_1, \boldsymbol{r}_2, \boldsymbol{R}) - E\right\}\Psi(\boldsymbol{r}_1, \boldsymbol{r}_2, \boldsymbol{R}) = 0\,, \tag{6.37}$$

where the $\boldsymbol{r}_i$ designate the vibrational coordinates of the linear molecule $i = 1, 2$, and $\boldsymbol{R}$ connects the centers of mass of the two interacting molecules; it is the vector of the relative position. The term in braces $\{\cdots\}$ (aside from E) may be regarded the two-particle, or supermolecular Hamiltonian. The internal motion of the molecules is described by the Hamiltonians $\mathscr{H}_i$,

$$\left\{\mathscr{H}_i(r_i) - \varepsilon_i(vj)\right\} \frac{1}{r_i} \chi(r_i; vj)\; Y_{jm_j}(\widehat{\boldsymbol{r}}_i) = 0\,. \tag{6.38}$$

The interaction potential is expanded as

$$V(\boldsymbol{r}_1, \boldsymbol{r}_2, \boldsymbol{R}) = \sum_{\lambda_1\lambda_2\lambda} V(r_1, r_2, R; \lambda_1\lambda_2\lambda) \times \tag{6.39}$$
$$\sum_{m_{\lambda_1} m_{\lambda_2} m_\lambda} C(\lambda_1\lambda_2\lambda; m_{\lambda_1} m_{\lambda_2} m_\lambda)\; Y_{\lambda_1 m_{\lambda_1}}(\widehat{\boldsymbol{r}}_1)\; Y_{\lambda_2 m_{\lambda_2}}(\widehat{\boldsymbol{r}}_2)\; Y^*_{\lambda m_\lambda}\left(\widehat{\boldsymbol{R}}\right),$$

where the $C(j_1j_2J; m_1m_2M)$ are Clebsch–Gordan coefficients.

The eigenvectors of Eq. 6.37 are represented by the product of the eigenfunctions of the free molecules and a function of the internuclear separation. The angular momenta $\boldsymbol{j}_i$ of the molecules are coupled to form $\boldsymbol{j}$, and $\boldsymbol{j}$ couples with the orbital angular momentum $\boldsymbol{\ell}$ to the total momentum $\boldsymbol{J}$,

$$\boldsymbol{j}_1 + \boldsymbol{j}_2 = \boldsymbol{j} \quad \text{and} \quad \boldsymbol{j} + \boldsymbol{\ell} = \boldsymbol{J}\,.$$

The wavefunction Ψ is expanded in terms of the total J eigenfunctions which must satisfy the boundary conditions for scattering,

$$\Psi_{v_1j_1v_2j_2}(\boldsymbol{r}_1, \boldsymbol{r}_2, \boldsymbol{R}) \sim \sum_{J,M,\ell} \psi^{JM}_{v_1j_1v_2j_2\ell}(\boldsymbol{r}_1, \boldsymbol{r}_2, \boldsymbol{R})\,. \tag{6.40}$$

The v_1j_1, v_2j_2 refer to the initial states of the two molecules and

$$\psi^{JM}_{v_1j_1v_2j_2\ell} = \sum (r_1r_2R)^{-1}\, f^{JM}(R; v_1'j_1'v_2'j_2'j'\ell' \leftarrow v_1j_1v_2j_2j\ell) \tag{6.41}$$
$$\times \chi(r_1; v_1'j_1')\; \chi(r_2, v_2'j_2')\; Y^{JM}\left(\widehat{\boldsymbol{r}}_1, \widehat{\boldsymbol{r}}_2, \widehat{\boldsymbol{R}}; j_1'j_2'j'\ell'\right),$$

where the sum is over all $v_1'j_1'v_2'j_2'j'\ell'$. The angular dependences are represented by

$$Y^{JM}\left(\widehat{\boldsymbol{r}}_1, \widehat{\boldsymbol{r}}_2, \widehat{\boldsymbol{R}}; j_1'j_2'j'\ell'\right) = \sum_{m_1m;m_2m_\ell} C(j_1j_2j; m_1m_2m)\; C(j\ell J; m\,m_\ell\,M)$$
$$\times\, Y_{j_1m_1}(\widehat{\boldsymbol{r}}_1)\; Y_{j_2m_2}(\widehat{\boldsymbol{r}}_2)\; Y_{\ell m_\ell}\left(\widehat{\boldsymbol{R}}\right). \tag{6.42}$$

A set of coupled equations for the Jth partial wave is obtained by combining Eqs. 6.37, 6.41 and 6.42 and integrating over the angular dependences and the vibrational coordinates, according to

$$\frac{\hbar^2}{2m}\left(\frac{d^2}{dR^2} - \frac{\ell(\ell+1)}{R^2} + k_{\alpha'}^2\right) f^J(R;\alpha' \leftarrow \alpha) = \sum_{\alpha''} V^J_{\alpha'\alpha''}(R)\, f^J(R;\alpha'' \leftarrow \alpha)\,. \tag{6.43}$$

Each equation ('channel') is defined by the set of six quantum numbers, $\alpha = v_1 j_1 v_2 j_2 j \ell$, and the squared wave vector is given by

$$\hbar^2 k_{\alpha'}^2 = 2m\,(K - E_{\alpha'} + E_\alpha)\,, \tag{6.44}$$

with K designating the initial kinetic energy of relative motion, and

$$E_\alpha = \varepsilon_1(v_1 j_1) + \varepsilon_2(v_2 j_2)$$

is the sum of the molecular rotovibrational energies.

The M independent coupling matrix elements are functions of the intermolecular distance only,

$$\begin{aligned} V^J_{\alpha\alpha'} &= (-1)^{J+j_1+j_2+j'}\,(4\pi)^{-3/2} \sum_{\lambda_1\lambda_2\lambda} \int dr_1\, dr_2\, \chi(r_1;v_1 j_1)\ \chi(r_2;v_2 j_2) \\ &\quad \times V(r_1 r_2 R;\lambda_1\lambda_2\lambda)\ \chi(r_1;v_1' j_1')\ \chi(r_2;v_2' j_2') \\ &\quad \times \left([\lambda]^2\,[\lambda_1]\,[\lambda_2]\,[j_1]\,[j_2]\,[j]\,[\ell]\,[j_1']\,[j_2']\,[\ell']\right)^{1/2} \\ &\quad \times \begin{pmatrix} \lambda & \ell' & \ell \\ 0 & 0 & 0 \end{pmatrix} \begin{pmatrix} \lambda_1 & j_1 & j_1' \\ 0 & 0 & 0 \end{pmatrix} \begin{pmatrix} \lambda_2 & j_2 & j_2' \\ 0 & 0 & 0 \end{pmatrix} \\ &\quad \times \begin{Bmatrix} \ell' & \ell & \lambda \\ j & j' & J \end{Bmatrix} \begin{Bmatrix} j' & j_2' & j_1' \\ j & j_2 & j_1 \\ \lambda & \lambda_2 & \lambda_1 \end{Bmatrix}, \end{aligned} \tag{6.45}$$

with $[n] = (2n+1)$. The terms in parentheses are Wigner's 3–j symbols and the ones in braces are 6–j and 9–j symbols, respectively.

The solution vector $f^J(R;\alpha \leftarrow \alpha')$ is obtained by numerical integration; Gordon's method has been employed successfully (with certain modifications of the original code) [162, 354].

So far, we have assumed dissimilar pairs of linear molecules, such as H_2–He or H_2–N_2. If like pairs are considered, we must include symmetry and spin. The parity of each channel is given by

$$P = (-1)^{J+j_1+j_2+\ell}\,. \tag{6.46}$$

Only states of the same parity couple. The set of Eqs. 6.43 separates therefore into two uncoupled sets; solutions will be marked by a superscript + or – depending on the sign of the parity.

For like pairs, the angular part of the wavefunction, Eq. 6.42, must be symmetrized, according to

$$\begin{aligned} Y^{JM\pm}\left(\widehat{\boldsymbol{r}}_1,\widehat{\boldsymbol{r}}_2,\widehat{\boldsymbol{R}};j_1j_2j\ell\right) & \\ = N\left[Y^{JM}\left(\widehat{\boldsymbol{r}}_1,\widehat{\boldsymbol{r}}_2,\widehat{\boldsymbol{R}};j_1j_2j\ell\right) \pm Y^{JM}\left(\widehat{\boldsymbol{r}}_1,\widehat{\boldsymbol{r}}_2,-\widehat{\boldsymbol{R}};j_1j_2j\ell\right)\right] & \\ = N\left[Y^{JM}\left(\widehat{\boldsymbol{r}}_1,\widehat{\boldsymbol{r}}_2,\widehat{\boldsymbol{R}};j_1j_2j\ell\right)\; Y^{JM}\left(-\widehat{\boldsymbol{r}}_1,-\widehat{\boldsymbol{r}}_2,-\widehat{\boldsymbol{R}};j_2j_1j\ell\right)\right] & \end{aligned} \tag{6.47}$$

with $N = [2(1+\delta_{j_1j_2})]^{-1/2}$ being the normalization factor. The angular function $Y^{JM\pm}$ is multiplied by the spin function giving the total spin $\boldsymbol{S} = \boldsymbol{s}_1 + \boldsymbol{s}_2$ of the pair,

$$|s_1s_2SM_S\rangle = \sum_{m_{s1}m_{s2}} C\left(s_1s_2S;m_{s1}m_{s2}M_S\right)\, |s_1m_{s1}\rangle\; |s_2m_{s2}\rangle\ . \tag{6.48}$$

The product of angular and spin function must be even under exchange for bosons, and odd for fermions. Moreover, for like pairs the matrix elements that couple the rotovibrational states may be written as

$$\begin{aligned} V^{J\pm}_{\alpha\alpha'} &= \left[\left(1+\delta_{v_1v_2}\,\delta_{j_1j_2}\right)\left(1+\delta_{v_1'v_2'}\,\delta_{j_1'j_2'}\right)\right]^{-1/2} \\ &\quad\times\left(V^{J}_{\alpha\alpha'} \pm (-1)^{j_1+j_2-j+\ell}\,\mathscr{T}_{j_1j_2}\,V^{J}_{\alpha\alpha'}\right)\,, \end{aligned} \tag{6.49}$$

where $\mathscr{T}_{j_1j_2}$ is the transposing operator for j_1 and j_2. The interaction potential must also be symmetric for identical molecules. Further details concerning such computations may be found elsewhere [31, 354].

1 Isotropic potential approximation

We will consider the simplifications possible with the assumption of isotropic interactions. As this was done above, we assume for simplicity that we are dealing with an unmixed gas of diatomic molecules in the low-density limit. We start with Eq. 5.2,

$$\alpha(\nu) = \frac{1}{2}\,\frac{4\pi^2}{3\hbar}\,N_a^2\,\varrho^2\,\nu\,\left[1-\exp\left(-hc\nu/kT\right)\right]\,V\,g(\nu;T)\,. \tag{6.50}$$

In this equation, the energies E_i and E_f of the initial and final states, $|i\rangle$ and $|f\rangle$, and the dipole moment all refer to a pair of diatomic molecules; $hc\nu_{i,f} = E_f - E_i$ is Bohr's frequency condition. With isotropic interaction, rotation and translation may be assumed to be independent so that the rotational and translational wavefuntions, population factors, etc., factorize. Furthermore, we express the position coordinates of the pair in terms of center-of-mass and relative coordinates as this was done in Chapter 5.

The spectral density is defined in terms of the matrix elements of the induced dipole moment $\boldsymbol{\mu}$ by the 'golden rule', Eq. 5.3,

$$V\,g(\omega;T) = \sum_{ss'} P_s \sum_{tt'} V\,P_t \frac{1}{4\pi\varepsilon_0} \left|\langle t|\boldsymbol{\mu}_{ss'}|t'\rangle\right|^2 \,\delta(\omega_{ss'} + \omega_{tt'} - \omega)\,. \quad (6.51)$$

The subscripts $s = \{j_1, m_1, v_1, j_2, m_2, v_2\}$ and $t = \{\ell, m_\ell, E_t\}$ denote molecular and translational states, respectively; the j, m refer to rotational states and the v to vibrational states, the subscripts 1 and 2 refer to molecules 1 and 2, and a prime denotes final states. The normalized translational Boltzmann factor may be written as

$$P_t(T) = \frac{\lambda_0^3}{V}\,\mathrm{e}^{-E_t/kT} \quad \text{with} \quad \lambda_0 = \left(\frac{2\pi\hbar^2}{mkT}\right)^{1/2}, \quad (6.52)$$

where λ_0 is the thermal de Broglie wavelength of the pair. The rotational Boltzmann factor is written as $P_s(T) = P_{j_1}(T)\,P_{j_2}(T)$, with

$$P_j(T) = \frac{g_j\,\exp(-E_j/kT)}{\sum_J g_J\,(2J+1)\,\exp(-E_J/kT)}\,. \quad (6.53)$$

At high temperatures, vibrational states must also be included in the partition sum above. The nuclear weights are g_j; for hydrogen we have, for example, $g_j = 1$ for even j, and $g_j = 3$ for odd j. However, we mention that in low-temperature laboratory measurements as well as in astrophysical applications, para-H_2 and ortho-H_2 abundances may actually differ from the proportions characteristic of thermal equilibrium (Eq. 6.53). In such a case, at any fixed temperature T, one may account for non-equilibrium proportions by assuming g_j values so that the ratio g_0/g_1 reflects the actual para to ortho abundance ratio. Positive frequencies correspond to absorption, but the spectral function $g(\omega;T)$ is also defined for negative frequencies which correspond to emission. We note that the product $V\,g(\omega;T)$ actually does not depend on V because of the reciprocal V-dependence of P_t, Eq. 6.52.

The induced dipole moment $\boldsymbol{\mu}$ is expressed in the form of Eq. 4.18 in spherical components [314, 317]. As was seen in [43], we obtain the spectral function as a multiple sum of incoherent components,

$$\begin{aligned} g(\omega;T) \;=\; & \sum_{\lambda_1\lambda_2\Lambda L}\; \sum_{j_1 j_1' j_2 j_2'} (2j_1+1)\,P_{j_1}\,C(j_1\,\lambda_1\,j_1';000)^2\,(2j_2+1)\,P_{j_2} \\ & \times\, C(j_2\,\lambda_2\,j_2';000)^2\,G_{\lambda_1\lambda_2\Lambda L}(\omega - \omega_{j_1 j_1'} - \omega_{j_2 j_2'};T)\,. \end{aligned} \quad (6.54)$$

In the isotropic potential approximation, the complete binary spectrum is obtained by superimposing basic line profiles, $G_{\lambda_1\lambda_2\Lambda L}(\omega;T)$, shifted by sums of molecular rotational frequencies $\omega_{j_1 j_1'}, \omega_{j_2 j_2'}$ which may be positive, zero and negative. If vibrational bands are considered, rotovibrational frequencies, $\omega_{v_i j_i v_i' j_i'}$, must be substituted, with $i = 1$ and 2. The translational profiles (which may depend on the molecular quantum number j) are given by Eq. 5.67,

$$V\,G_c(\omega;T) = \lambda_0^3 \hbar \sum_{\ell,\ell'} (2\ell+1)\, C(\ell\, L\, \ell';000)^2\, w(\ell\ell' j_1 j_1' j_2 j_2' v v') \tag{6.55}$$

$$\times \Big\{ \int_0^\infty \exp(-E_t/kT)\, \mathrm{d}E_t \left| \langle \ell, E_t, v | B_c(R) | \ell', E_t + \hbar\omega, v' \rangle \right|^2$$

$$+ \sum_{n,n'} \exp(-E_{n\ell}/kT) \left| \langle \ell, E_{n\ell}, v | B_c(R) | \ell', E_{n'\ell', v'} \rangle \right|^2 \times \delta(E_{n'\ell'} - E_{n\ell} - \hbar\omega)$$

$$+ \sum_{n} \exp(-E_{n\ell}/kT) \left| \langle \ell, E_{n\ell}, v | B_c(R) | \ell', E_{n\ell} + \hbar\omega, v' \rangle \right|^2$$

$$+ \sum_{n'} \exp\left[-(E_{n'\ell'} - \hbar\omega)/kT\right] \left| \langle \ell, E_{n'\ell'} - \hbar\omega, v | B_c(R) | \ell', E_{n'\ell'}, v' \rangle \right|^2 \Big\} ,$$

$c = \lambda_1 \lambda_2 \Lambda L$ designates the expansion parameters and $v = (v_1, v_2)$ represents the vibrational quantum numbers of molecules 1 and 2.

In this expression, the terms $\langle \cdots \rangle$ represent the radial matrix elements,

$$\langle \ell, E, v\, | B_c |\, \ell', E', v' \rangle = \int_0^\infty \psi^*(R; \ell', E', v')\, B_c^{vv'}(R)\, \psi(R; \ell, E, v')\, \mathrm{d}R\,, \tag{6.56}$$

with the radial wavefunctions ψ defined below. The weights $w(\cdots)$ account for the molecular exchange symmetry if line shapes of like molecules are considered. We will see below that for the low-resolution, collision-induced rototranslational spectra normally considered, this symmetry only matters at temperatures much lower than 40 K; only the generally weak dimer structures are affected by symmetry (even at higher temperature). At temperatures above 40 K and when dimer structures are not resolved, one may usually assume $w = 1$.

As was explained on pp. 240ff., the expression to the right of Eq. 6.55 consists of a sum of four terms. The first one, the integral, represents the free → free transitions of the collisional pair and is usually the dominant term in this expression. The second term, a sum, gives the bound → bound transitions of the van der Waals dimers. The vibrational and rotational quantum numbers of the dimer are designated by n and ℓ; ℓ also numbers the partial waves. For practical purposes, one will convolute the delta function with some suitable instrumental or line profile for a more realistic simulation of the observation. The last two terms to the right of Eq. 6.55 account for bound → free and free → bound transitions of the molecular pair, respectively. We note that the energies $E_{n\ell} + \hbar\omega$ and $E_{n'\ell'} - \hbar\omega$ appearing in these terms must be legitimate free state energies, i.e., they must be positive or else the radial matrix elements vanish.

The radial wavefunctions $\psi(R;\ell,E_t)/R$ needed for the computation of Eq. 6.55 are solutions of the Schrödinger equation of relative motion,

$$-\frac{\hbar^2}{2m}\frac{d^2\psi}{dR^2}+\left(V_0(R)+\frac{\hbar^2\ell(\ell+1)}{2mR^2}-E_t\right)\psi=0\,. \qquad (6.57)$$

$V_0(R)$ is the vibrational average of the interaction potential, Eq. 6.29. Free-state wavefunctions may be energy normalized,

$$\int_0^\infty \psi^*(R;\ell,E')\,\psi(R;\ell,E)\,dR=\delta(E'-E) \qquad (6.58)$$

for any fixed ℓ. For bound states, the δ-function in Eq. 6.58 is replaced by the Kronecker delta symbol, $\delta_{nn'}$, as usual; E is then replaced by the negative $E_{n\ell}$. We note that the integration over predissociating states in the first term to the right of Eq. 6.55 must be done with carefully tailored energy grids. Third-order spline integration gives the necessary flexibility, and satisfactory results are obtained for a truncated integration interval, from $E_t \approx kT/100$ (instead of 0) to $\approx 15kT$ (instead of ∞). Similarly, the radial integrals $\langle \ell E|B_c|\ell' E'\rangle$ may be truncated at an R_{max} of 1.5 or 2 nm.

The spectral function, Eq. 6.55, and thus the absorption coefficient, Eqs. 6.50 and 6.54, can be computed if an interaction potential and the significant induction operator components are known. The spectral function, Eq. 6.54, consists of a number of lines at rotational transition frequencies of the monomers ($j_1 \neq j_1'$ and $j_2 = j_2'$; or $j_2 \neq j_2'$ and $j_1 = j_1'$), lines at sums and differences of such rotation frequencies (both $j_1 \neq j_1'$ and $j_2 \neq j_2'$), and a translational component ($j_1 = j_1'$ and $j_2 = j_2'$; or $j_1 = j_2'$ and $j_2 = j_1'$). Equation 6.54 shows how the λ_1, λ_2 impose selection rules on the molecular transitions, and Eq. 6.55 shows how the expansion parameter L controls translational transitions through basic properties of the Clebsch–Gordan coefficients; $C(j\,\lambda\,j';000) = 0$ unless the value of the sum of j, λ and j' is an even integer and the triangular inequalities $j+\lambda \geq j'$, $j+j' \geq \lambda$, and $\lambda + j' \geq j$ are satisfied; similar conditions exist for $C(\ell L \ell';000)$. The evaluations of the radial functions ψ and the translational profile, Eq. 6.55, represent the main computational effort of line shape calculations.

Detailed balance – preliminary remarks. Induced spectral lines $G(\omega)$ are strikingly asymmetric. The intensities of the high-frequency (or 'blue') wing are related to the intensities of the low-frequency (or 'red') wing through *detailed balance.* In essence, we have

$$G(-\omega) \approx \exp(-\hbar\omega/kT)\,G(\omega)\,, \qquad (6.59)$$

which has been referred to as the detailed balance 'condition'. Under certain circumstances this relation is exact and the $\approx$ sign may be replaced by the sign of equality (=). The Boltzmann factor reflects the population ratios of the (many) initial states undergoing the 'down' transition $E_+ \to E_-$ and their counterparts which undergo the inverse ('up') transitions, $E_- \to E_+$, with $E_+ - E_- = \hbar\omega$. Indeed, experimental case studies [422] and quite general theoretical treatments [313, 318] have confirmed an asymmetry of the line profiles which is consistent with Eq. 6.59 for several important cases. Much use of Eq. 6.59 has been made, for example in modeling attempts of the spectra, and for line shape computations which could thus be limited to the computations of just one wing of the profile.

It is, therefore, of interest to point out that Eq. 6.59 may not be exact in important cases. Rather, a few more or less striking violations of Eq. 6.59 are known, both from experiment and from computations of the profiles, Eq. 6.55. For simplicity, we will briefly consider two such cases. We note that line shapes calculated from Eq. 6.55 are consistent with the measurements even even if these do not support Eq. 6.59.

Case I: *dissimilar pairs.* In that case, the weights w appearing in Eq. 6.55 equal unity, for all ℓ, ℓ', j, j', v, v'. Moreover, if we temporarily suppress the dimer states for simplicity, only the first term of Eq. 6.55 (the integral representing free-to-free transitions) survives,

$$\begin{aligned} V\,G_c(\omega) &= \lambda_0^3\hbar\sum_{\ell\ell'}(2\ell+1)\;C(\ell L\ell';000)^2 \\ &\times\int_0^\infty \exp\left(-E/kT\right)\;\mathrm{d}E\;\left|\langle \ell,E,v\,|B_c|\,\ell'E+\hbar\omega,v'\rangle\right|^2 . \end{aligned}$$

The states $|\ell E v\rangle$ and $|\ell',E+\hbar\omega,v'\rangle$ must be genuine free states of *positive* energy, $E>0$ and $E+\hbar\omega>0$. For $\omega>0$ we compute the integral in the expression for $VG(\omega)$ above for a negative frequency shift $-\omega$. In that case, the integration $\int\cdots\mathrm{d}E$ starts at $E=\hbar\omega$ and goes to ∞. We substitute E' for $E-\hbar\omega$ and drop the prime. In this way, we get

$$\begin{aligned} V\,G_c(-\omega) &= \lambda_0^3\hbar\sum_{\ell\ell'}(2\ell+1)\;C(\ell L\ell';000)^2 \\ &\times\int_0^\infty \exp\left[-(E+\hbar\omega)/kT\right]\;\mathrm{d}E\;\left|\langle \ell,E+\hbar\omega,v\,|B_c|\,\ell'E,v'\rangle\right|^2 . \end{aligned}$$

Since $(2\ell+1)\,C(\ell L\ell',000)^2 = (2\ell'+1)\,C(\ell' L\ell,000)^2$, we may formally interchange ℓ and ℓ'. Furthermore, the radial integrals (Eq. 6.50)

$$\left|\langle \ell',E+\hbar\omega,v\,|B_c|\,\ell,E,v'\rangle\right| \quad\text{and}\quad \left|\langle \ell,E,v\,|B_c|\,\ell',E+\hbar\omega,v'\rangle\right| \tag{6.60}$$

are equal, *provided* $V_0^{v'}(R)\equiv V_0^{v}(R)$. In that case, we have Eq. 6.59, with the $\approx$ sign replaced by an equal sign. This will in general be the case for the rototranslational spectra.

For the rotovibrational bands, on the other hand, the vibrational quantum numbers of initial and final states differ, $v\neq v'$. In that case, the radial integrals, Eq. 6.60, differ and Eq. 6.59 is not exact if one accounts for the dependence of the interaction potential on the vibrational excitation $\langle v|V_0^v|v\rangle \neq \langle v'|V_0^{v'}|v'\rangle$ [151, 281]. A modified detailed balance relation describing the symmetry of induced lines has been obtained for the rotovibrational bands [295]; details may be found below, pp. 338ff. The asymmetry of the line profiles computed from Eq. 6.55 for positive and negative frequency shifts differ strikingly from Eq. 6.59 (see pp. 341) and give improved fits of the existing measurements [63, 151, 281] if compared to profiles which obey Eq. 6.59 exactly. These results were obtained for the rotovibrational bands of H_2–He and H_2–H_2 pairs, but they may well be of a broad validity.

The above limitation to pairs not forming bound dimers is not essential and one may readily include the bound/ dimer contributions, thereby extending the above consideration to any dissimilar pair with isotropic interactions. For dissimilar pairs, the conclusions just obtained are not affected by the presence or absence of dimer components.

Case II: *like pairs*, such as H_2–H_2. If like pairs are considered, the wavefunction must reflect the various exchange symmetries. In other words, the weights w appearing to the right of Eq. 6.55 (first line) are now dependent on the list of arguments shown on p. 310 and may affect detailed balance, especially when one or more of the last three of the four terms given in that equation are significant. We shall discuss below the case of the $(H_2)_2$ dimer structures of the H_2 $S_0(0)$ line (p. 318). These discussions show that for hydrogen at low temperature, the weight w is small ($w_e \approx 0.35$) for even partial wave quantum numbers ℓ and big ($w_o \approx 0.65$) for odd ℓ; the sum of these weights equals unity, $w_e + w_o = 1$.

As a consequence, in Eq. 6.55 the summation over ℓ may be divided into a sum over even ℓ and a sum over odd ℓ. If the sums are over a sufficiently large number of partial waves ℓ, the sum of the sums over even and odd ℓ will not depend much on the weights w_e, w_o and we expect again Eq. 6.59 to be more or less satisfied (unless one accounts for the vibrational dependences of the interaction). However, the last three of the four terms of Eq. 6.55 which involve bound dimer contributions may have prominent features, such as the $\ell = 1 \leftrightarrow \ell' = 2$ spike seen in Fig. 6.5 which will be much stronger in the red wing ($\ell = 2 \rightarrow 1$) than in the blue wing ($\ell = 1 \rightarrow 2$), because of the difference in the weights w_e, w_o. This observation is strikingly inconsistent with Eq. 6.59 which for the red wing would have suggested a slightly smaller spike than for the blue wing, see Figs. 6.4 and 6.5 below. In other words, whenever dimer features of like pairs arise from just one quantum state (as opposed to a sum of many superimposed structures), the weight factors w may drastically change the symmetry of induced lines, even in the rototranslational bands. This computational fact is fully supported by measurements of dimer spectra of hydrogen [268]; see also Fig. 6.19.

2 Computations and results

H_2–H_2 rototranslational spectra. For the significant $\lambda_1\lambda_2\Lambda L$ induction components, Table 4.11, values of the various spectral functions have been computed at frequencies from 0 to 1800 cm^{-1} and for temperatures from 40 to 300 K, Fig. 6.3 [282]. As a test of these line shape computations, the zeroth, first and second spectral moments have been computed in two independent ways: by integration of the spectral functions with respect to frequency, Eq. 3.4, and also from the quantum sum formulae, Eqs. 6.13, 6.16, and 6.21. Agreement of the numerical results within 0.3% is observed for the 0223, 2023 components, and 1% for the other less important components. This agreement indicates that the line shape computations are as accurate as numerical tests with varying grid widths, etc., have indicated, namely about 1%; see Table 6.2 as an example (p. 293).

The quadrupole-induced components $\lambda_1\lambda_2\Lambda L = 0223, 2023$ are the most important components of the spectrum. Figure 6.3 compares the positive-frequency wings of the various spectral functions, Eq. 6.55, at the temperature of 77 K. These consist of free → free and bound → free components, the first and third terms to the right of Eq. 6.55. The dimer structures were suppressed in Fig. 6.3 but are shown (as obtained in the isotropic interaction approximation) in Fig. 6.4. The low-frequency end of the profiles, Fig. 6.3, may be considered a low-resolution rendition (as may be obtained with a monochromator of low resolution, $\approx$ 10 cm^{-1}; pressure broadening would similarly flatten the spectral dimer structures).

At 77 K, the positive-frequency wings of the quadrupole-induced components 0223, 2023 have a half-width of about 75 cm^{-1}, Fig. 6.3. This relatively small value is typical of the long-range R^{-4} dependence of the quadrupole induction operator that dominates this interaction. The quadrupole interacting with the anisotropy $\beta = \alpha_\parallel - \alpha_\perp$ of the polarizability tensor (2233) is seen to be only slightly broader, because it has somewhat more overlap mixed in with the long-range term. However, the pure overlap components 0221, 2021 show a halfwidth about three times greater. The 0445,4045 components are very broad, on account of the high-frequency content of the R^{-6} dependence characteristic of the hexadecapolar induction. The far wings of the quadrupolar induction term have intensities and logarithmic slopes approaching those of the overlap components, despite the substantial differences at low frequencies. We mention that the red wings ($\omega < 0$) of the profiles are obtainable from the blue wing by the principle of detailed balance, Eq. 6.59, as long as the striking dimer features (to be discussed below) are suppressed.

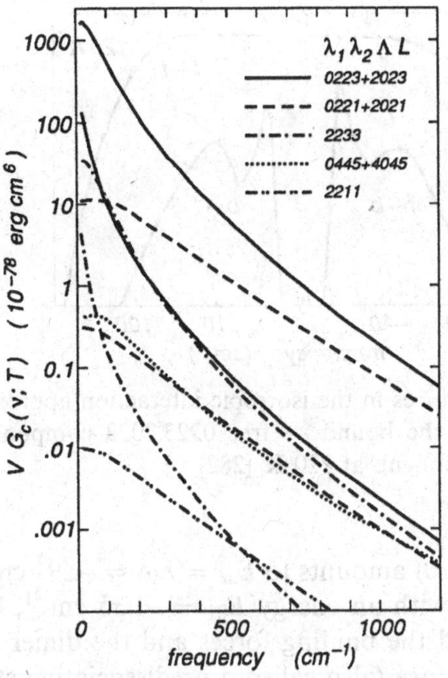

Fig. 6.3. The significant free → free components of the spectral functions of molecular hydrogen pairs at 77 K. For a given set of expansion parameters $\lambda_1\lambda_2\Lambda L$, a different line type is chosen. When two curves of the same type are shown, the upper one represents the free → free, the lower the bound → free contributions; their sum is the total $VG_{\lambda_1\lambda_2\Lambda L}(\omega; T)$. The extreme low-frequency portion of the bound → free contributions with the dimer fine structures is here suppressed [282].

Where in Fig. 6.3 two curves are drawn with the same type of line, the upper curve describes the free → free transitions of the collisional pair, and the lower one the bound → free contributions. Asymptotically, at high frequencies, the bound → free contributions amount to only about 1% of the free → free components. At the lower frequencies, however, the bound → free components are relatively more significant. In fact, the bound → free components which must be superimposed with the free → free components to obtain the spectral function, Eq. 6.54, affect the shapes of the profiles near the line centers. We note that the quadrupole-induced components 0223, 2023 do not feature a bound → bound spectrum, but the less important overlap components 0221, 2021 do. However, absorption due to this overlap component is insignificant, a few percent of the total absorption.

$(H_2)_2$ dimers. The isotropic interaction potential [257] supports just one vibrational dimer level, the ground state ($n = 0$). The energy level of the

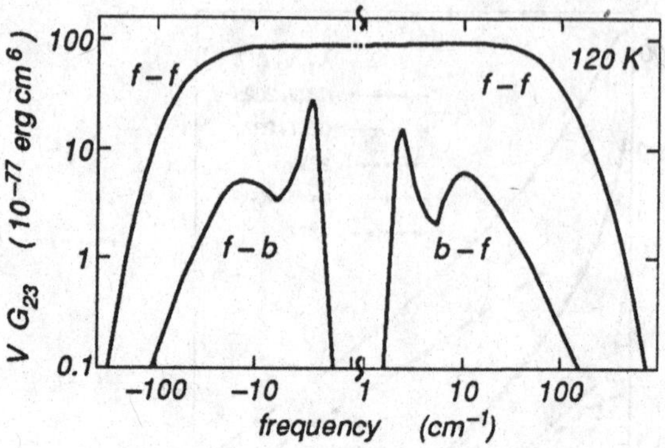

Fig. 6.4. The dimer structures in the isotropic interaction approximation, in the low-frequency portion of the bound → free 0223,2023 components, compared with the free → free component, at 120 K [282].

non-rotating dimer ($\ell = 0$) amounts to $E_{n\ell} = E_{00} = -2.91$ cm^{-1} [282]. The $\ell = 1$ state also exists, with an energy $E_{01} = -1.35$ cm^{-1}, but for $\ell > 1$, centrifugal forces exceed the binding forces and the dimer dissociates. A narrow scattering resonance (also called a predissociating state) exists for $\ell = 2$ at $E = 1.15 \pm 0.7$ cm^{-1} and a much broader one for $\ell = 3$ at $E \approx 5$ cm^{-1} [282]. Slightly better isotropic potential functions now exist, but the dimer levels and widths are similar to the numbers just quoted. In fact, similar dimer energies based on isotropic model potentials were known for some time [161]. The energy spacings are, furthermore, in reasonable agreement with measurements of the main dimer structures [259, 268]. We note that dimer levels [133, 134] and dimer emission spectra [355, 269] were recently computed from first principles, taking the anisotropy of the interaction into account. The computed spectra are in agreement with the measurements and are much richer than the spectra computed from an isotropic potential, Figs. 6.15 and 6.19. Nevertheless, the main spectral features are reasonably consistent with the coarse structures computed from the isotropic interaction approximation which thus offers a simple, 'low-resolution' treatment of these interesting dimer features.

While the free → free spectral components are relatively unstructured, the bound → free contributions have interesting 'line' structures, Fig. 6.4 (H_2 $S_0(0)$ line). For a study of such dimer structures which appear near the line centers, the ℓ, j dependence of the weights $w(\cdots)$ in Eq. 6.55 must be computed. For processes contributing to the induced $S_0(0)$ line intensity, at least one of the collisional partners must be para-H_2 in the $j_1 = 0$ state. The colliding partner, with a probability P_0, may also be para-H_2 in the ($j_2 = 0$) state. In that case, boson symmetry requires that all odd ℓ-values vanish but even ℓ are enhanced. No symmetry restrictions exist for distinguishable collisional pairs and hence $w = 1 + P_0$ if ℓ is even, and $w = 1 - P_0$ if ℓ is odd. At the temperature of 120 K, P_0 amounts to about 0.31. The consideration of the symmetries associated with the induced $S_0(1)$ transitions leads similarly to weights w which are more nearly equal to unity.

In Fig. 6.4, we notice vanishingly small bound → free intensities from 0 to about ± 2 cm^{-1} relative to the H_2 rotational transition frequencies. Near ± 2.6 cm^{-1}, peaks that at 120 K amount to roughly 40% of the free → free intensities, are seen. The one to the right arises mainly from transitions of the ℓ=1 bound state to the ℓ'=2 predissociating state. It is smaller than the one to the left which arises from the inverse $\ell = 2 \rightarrow \ell' = 1$ transitions. Their intensity ratio is given by the factor $(1 - P_0)/(1 + P_0)$. This ratio of the weights differs from unity and violates strikingly the detailed balance relation, Eq. 6.59. The widths of the line structure reflect the lifetime of the predissociating state. One sees another, much broader structure (20 cm^{-1}) that peaks near 12 cm^{-1}. A smooth roll-off begins at frequencies greater than about 30 cm^{-1}. At positive frequencies this structure is due to transitions from the other bound state (ℓ=0) to the other predissociating state at ℓ'=3. These structures are generated by the third term to the right of Eq. 6.55; the fourth term at positive frequencies is here an empty sum of zero value.

Structures similar to those seen in Fig. 6.4 are obtained for the other induction components (not shown). The anisotropic overlap components 0221, 2021 even show the ($\ell = 0 \rightarrow 1$) bound → bound transition near 1.6 cm^{-1} which, however, is weak and difficult to observe.

Since in laboratory measurements of hydrogen collision-induced roto-translational spectra the instrumental widths have generally been high, roughly 10 cm^{-1}, and the pressures in the range of 50 amagat [15], the structures shown in Fig. 6.5 were only recently observed using high resolution and low temperatures [266]. It is noteworthy that previously unexplained structures near the $S_0(0)$ and $S_0(1)$ rotational transition frequencies were recorded by the *IRIS* infrared spectrometer aboard the Voyager space probes in the upper atmospheres of Jupiter and Saturn [174, 157]. These structures have been explained in terms of the hydrogen dimer fine structure [262, 150, 266]. Similar dimer structures are familiar from laboratory measurements of the induced fundamental band of hydrogen [259].

Comparison of theory and measurement. For a comparison of theory with measurements, rototranslational absorption spectra were computed in the isotropic interaction approximation and compared with low-resolution ($\approx$ 10 cm^{-1}) spectra, dimer structures are not discernible in the measurement. The frequencies range from 0 to 2250 cm^{-1}. Temperatures were chosen

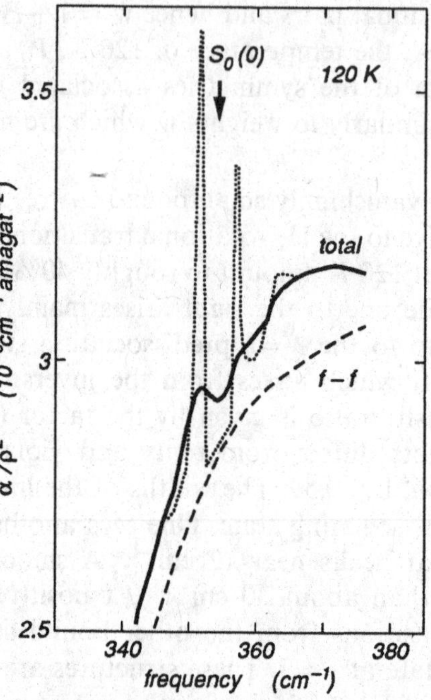

Fig. 6.5. Computed structures due to the hydrogen dimer, in the quadrupole-induced (0223,2023) components near the $S_0(0)$ line center at 120 K (the temperature of Jupiter's upper atmosphere). Superimposed with the smooth free → free continuum (dashes) are structures arising from bound → free (below 354 cm^{-1}) and free → bound (above 354 cm^{-1}) transitions of the hydrogen pair (dotted). The convolution of the spectrum with a 4.3 cm^{-1} slit function (approximating the instrumental profile of the Voyager infrared spectrometer) is also shown (heavy line) [282].

[15] for which measurements exist, Figs. 6.6 and 6.7. The computations use the *ab initio* induced dipole components. The spectra are given in a semi-logarithmic grid so that regions of high and low absorption are rendered with constant relative precision. The 2211 and the 0445, 4045 components are not shown in Figs. 6.6 and 6.7 because their intensities are below 10^{-8} cm^{-1} amagat^{-2}; they were, however, computed and are actually included in the curves marked total absorption. Near the rotational lines, the dominating contributions are from the 0223, 2023 components. Near the $S_0(0)$ and $S_0(1)$ lines, these quadrupole-induced intensities are in fact nearly identical with the total intensity (heavy curve) and differ from it only at high frequencies (short dashed line), where the quadrupole interacting with the anisotropy of the polarizability (2233) and some other components (0221, 2021, 0445, 4045) add significantly to the total. At 77 K, Fig. 6.6, the 2233 component shows five broad structures which from left

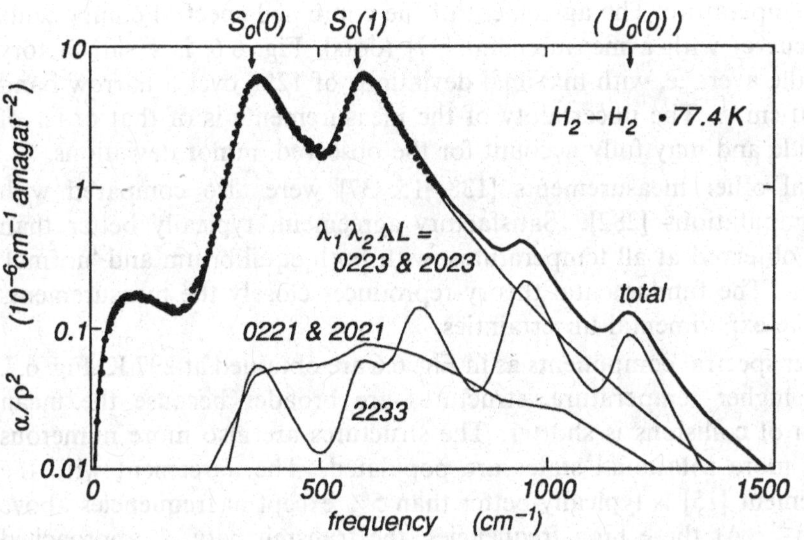

Fig. 6.6. Absorption by H_2–H_2 in the far infrared at 77.4 K. Agreement between theory (heavy line) and measurements • is observed; from [282].

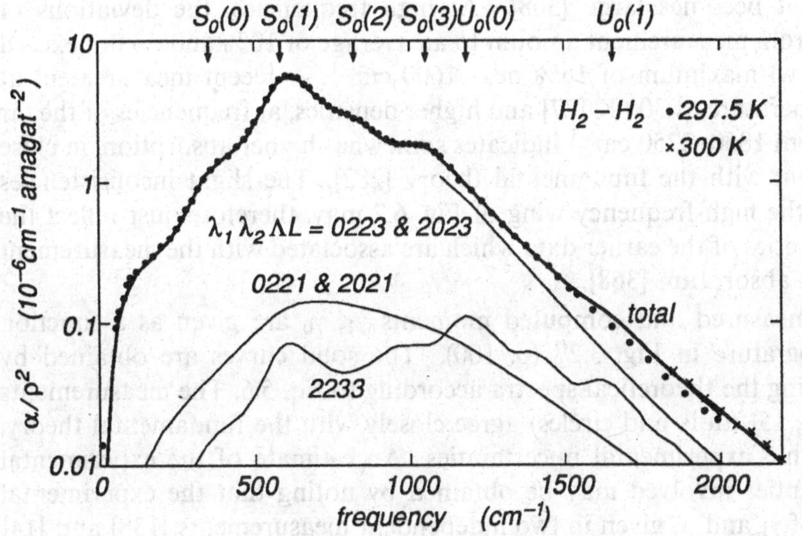

Fig. 6.7. Same as Fig. 6.6, but at the higher temperatures of 297.5 K (•) and 300 K (×) [282].

to right are labeled $S_0(0)$, $S_0(1)$, $2S_0(0)$, $S_0(0)+S_0(1)$, $2S_0(1)$. These are due to simultaneous rotational transitions; the two labels mentioned first stand for combinations $S_0(j_1) + Q_0(j_2)$ involving orientational transitions Q_0 of zero frequency shift. Some of these structures at high frequencies should be clearly visible but measurements above 800 cm^{-1} do not exist at this temperature. The agreement of the combined spectral components (heavy curve) with a measurement [37] (dots), Fig. 6.6, is a satisfactory 5% on the average, with maximal deviations of 12% over a narrow band near 230 cm^{-1}. The uncertainty of the measurements is of that order of magnitude and may fully account for the observed, minor deviations.

Several other measurements [138, 15, 37] were also compared with such computations [282]. Satisfactory agreement, typically better than 10%, is observed at all temperatures, with both equilibrium and 'normal' hydrogen. The fundamental theory reproduces closely the measurements within the experimental uncertainties.

Similar spectral components as in Fig. 6.6 are obtained at 297 K, Fig. 6.7. At this higher temperature, structures are broader because the mean duration of collisions is shorter. The structures are also more numerous because more rotational states are populated. The agreement with the measurement [15] is typically better than 5% except at frequencies above 1500 cm^{-1}. At these high frequencies, the transmittivity $\mathcal{T}$ approached unity at the pressure employed, and the absorption dropped to 1% or less. For values of $\mathcal{T}$ near unity, the fractional error in the absorption coefficient becomes large [368]. At high frequencies, the deviations of theory from measurement amount to an average of 10% and do not exceed a (narrow) maximum of 16% near 1600 cm^{-1}. A recent measurement at the temperature of 300 K [77] and higher densities, at frequencies of the far wing from 1860–2250 cm^{-1} indicates somewhat higher absorption, in close agreement with the fundamental theory [282]. The slight inconsistencies seen in the high-frequency wing of Fig. 6.7 may, therefore, just reflect the uncertainties of the earlier data which are associated with the measurement of small absorption [368].

The measured and computed moments γ_1, γ_0 are given as a function of temperature in Fig. 3.27 (p. 100). The solid curves are obtained by integrating the theoretical spectra according to Eq. 5.6. The measurements [138, 37, 15] (dots and circles) agree closely with the fundamental theory, well within experimental uncertainties. An estimate of the experimental uncertainties involved may be obtained by noting that the experimental values of γ_1 and γ_0 given in two independent measurements [138] and [14] agree to within 4% or less.

H_2–H_2 fundamental band. For the significant $\lambda_1\lambda_2\Lambda L$ induction components, values of the various spectral functions were computed over a frequency band of ± 1800 cm^{-1}, relative to the line center, and for temperatures from 20 to 300 K where measurements exist.

The spectra of bound dimers, $(H_2)_2$, have been recorded in the fundamental band [422, 259, 334, 412]. However, for the low-resolution spectra, dimer features are not discernible (albeit they are present) except at low temperatures and low pressures [281]. Free–bound transitions contribute several percent to the observable spectra in the form of relatively unstructured, broad continua near the $S_1(j)$ etc., line centers. While theory is capable of modeling the major features involving dimers [161, 150], no attempt was made to model the weak features that are discernible at higher resolution; these are pressure broadened. The diffuse bound–free contributions are, however, included in the line shape calculations.

We note that a computational study of the dimer features is involved. It must account for the anisotropy of the interaction as this was done for the pure rotational bands of hydrogen pairs [355, 357]. Whereas a treatment based on the isotropic potential approximation may be expected to predict nearly correct total intensities of the free–bound, bound–free, and bound–bound transitions involving the $(H_2)_2$ van der Waals molecule (and, of course, the free–free transitions which make up more than 90% of the observed intensities), the anisotropy of the interaction causes elaborate fine structure that is of considerable interest for the measurement of the anisotropy [248].

Recent reviews [342] suggest that the effect of molecular vibrations has not been studied in the rotovibrational collision-induced absorption spectra of H_2 pairs, presumably due to the previous lack of a reliable interaction potential. Such data for hydrogen pairs do now exist and the influence of molecular vibrations on the collision-induced absorption spectra has recently been studied. Similar work on the H_2–He system indicated significant effects of vibration on the spectral moments and the symmetry of the lines [151, 295, 294].

As an example, Fig. 6.20 below compares the $\lambda_1\lambda_2\Lambda L = 0001$ and 2023 line profiles at 195 K which were computed with and without (solid and dashed curves, respectively) accounting for the vibrational dependences of the interaction potential. The correct profiles (solid curves) are more intense in the 'blue' wing, and less intense in the 'red' wing by up to 25% relative to the approximation (dashed), over the range of frequencies shown. Whereas the dashed profiles satisfy the detailed balance relation, Eq. 6.59, if ω is taken to be the frequency shift relative to the line center, the exact profiles deviate by up to a factor of 2 from that equation over the range of frequencies shown. In a comparison of theory and measurement the different symmetries are quite striking; use of the correct symmetry clearly improves the quality of the fits attainable.

Figures 6.8 and 6.9 compare the computational results with the existing measurements. All measurements shown agree very closely with the fundamental theory except for the dips discernible in the measurements near 4160 cm^{-1}, the so-called intercollisional dips [404, 238]. That feature arises from correlations of the induced dipoles in subsequent collisions. It cannot be described by a theory that considers binary interactions only.

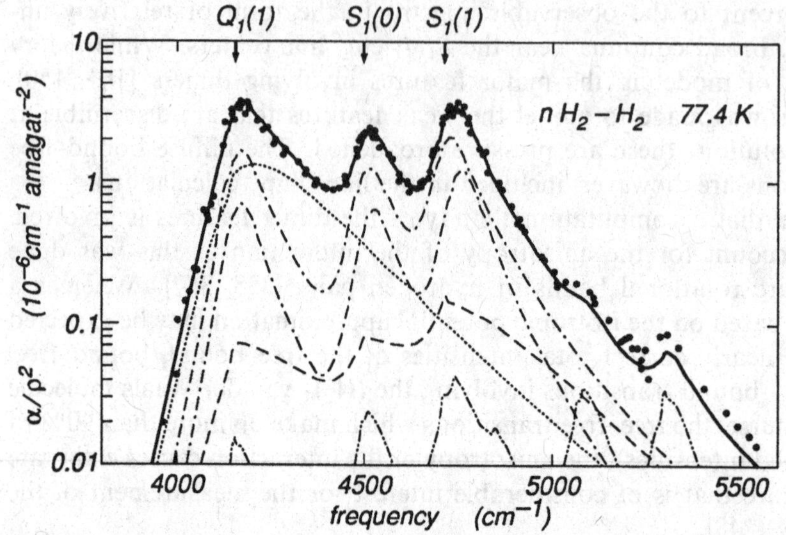

Fig. 6.8. The binary absorption spectrum of H_2–H_2 in the fundamental band of H_2, at 77 K. Computation: heavy line; measurements: dots, circles. Also shown are intermediate computational results: profiles labeled $\lambda_1\lambda_2\Lambda L$ =0001 (dotted); 0223 and 2023 (dash-dot); 0221 (dashes); 4045 and 0445 (dot–dash); and 2233 (thin solid line) [281].

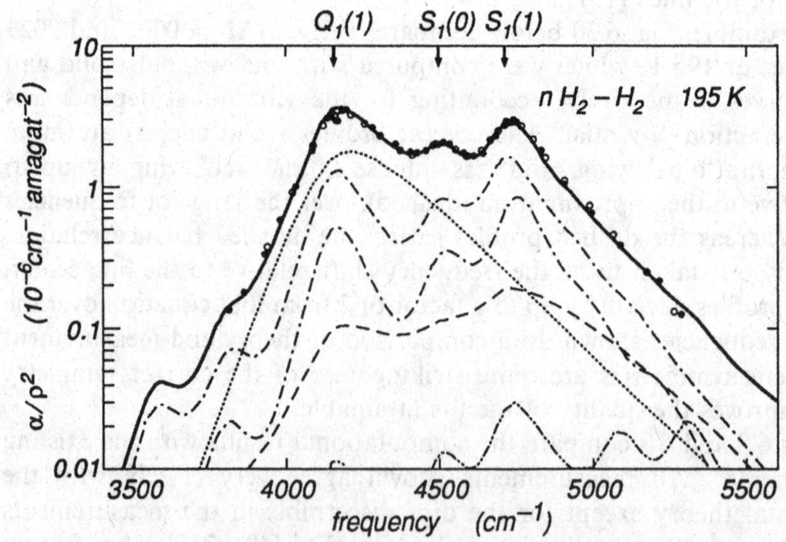

Fig. 6.9. Same as Fig. 6.8, but at the temperature of 195 K [281].

Apart from that feature, the agreement at the higher temperatures, Fig. 6.9, is indeed a very close one. In the figures, the dot diameter is consistent with an uncertainty of about 7%. In most cases, the agreement of theory and measurement is within that range. We note that no adjustable parameters were used anywhere in the theory. Rather, the agreement is in terms of absolute intensities.

For the H_2–H_2 rototranslational band the collision-induced absorption spectrum was seen to be composed of essentially eight components labeled $\lambda_1\lambda_2\Lambda L = 2023$, 0223, 2021, 0221, 2233, 2211, 4045, and 0445 [282]. Of these, the three pairs 2023 and 0223; 2021 and 0221; 4045 and 0445; are identical pairs. For the fundamental band, we have another important component, namely $\lambda_1\lambda_2\Lambda L = 0001$ (dotted in the figures) arising from overlap induction. That component is familiar from the CIA spectra of gas mixtures. It occurs here because a vibrating H_2 molecules differs from a non-vibrating one. Furthermore, the pairs like 2023,0223, etc., are not identical any more, for the same reason.

At lower temperature, striking features arise from van der Waals dimers that have been studied experimentally [422, 259, 334, 412] and theoretically [161, 150] by a number of investigators.

We have not attempted to exhibit in great detail the effects of the rotational excitations on the induced dipole components B and those of vibrational excitation on the interaction potential because this was done elsewhere for similar systems [151, 63, 295, 294]. The significance of the j, j' corrections is readily seen in the Tables and need not be displayed beyond that. The vibrational influence is displayed in Fig. 6.20; first and second spectral moments are strongly affected, especially at high temperatures, similar to that which was seen earlier for H_2–He [294], Fig. 6.23. The close agreement of the measurements of the rotovibrational collision-induced absorption bands of hydrogen with the fundamental theory shown above certainly depends on proper accounting for the rotational dependences of the induced dipole moment, and of the vibrational dependences of the final translational states of the molecular pair.

In conclusion, we mention that theoretical line shapes were compared with measurements of the hydrogen fundamental absorption for some time. The agreement was generally good; see the review articles [422, 342] for examples. Given the impressive consistency that was observed in earlier efforts, one may wonder what the point of the *ab initio* work is. A brief comparison with the previous efforts seems in order.

Previous work of the kind was generally based on empirical induced dipole models whose parameters were adjusted to fit measured spectra. For molecular systems like hydrogen pairs, empirical dipole models are highly simplified, for example, by either suppressing the anisotropic overlap terms, the $\Lambda L = 21$ components, in favor of an overlap term in the quadrupole-induced $\Lambda L = 23$ components, or else suppressing the overlap term in the 23 component and using instead excessive overlap in the 21 components. The *ab initio* work is an advanced treatment of all significant induced dipole components; no adjustable parameters are used.

Previous work used empirical model profiles, such as Lorentzians with exponential wings attached which were desymmetrized to satisfy Eq. 6.72. These profiles were often matched to theoretical spectral moments, thus providing a 'theoretical' spectrum. While such profiles represent the *cores* of induced lines fairly well, they are deficient in the wings [69]. Furthermore, as was mentioned above, Eq. 6.72 is not the correct symmetry of vibrational profiles; especially the wings of real line profiles deviate significantly as we have seen above. The linear plots usually used for the comparison emphasize the line centers, that is regions of strong absorption, but a poor representation of the wings (like a 50% discrepancy between theoretical and measured far wings) is hardly noticeable in such plots. On the other hand, an exact quantum line shape is used, based on the most reliable input available. In the semi-logarithmic plots shown above, a 10% difference between theory and measurement is as discernible near peak absorption as it is in the far wings. In that sense, in the hydrogen fundamental band, theory has not been compared as closely with the measurements in most of the earlier studies.

H_2–H_2 overtone band. Figure 6.10 shows the results of similar calculations for the first overtone band of H_2 (heavy line) [284]. Two measurements (dots and circles) are also given for comparison. The light dotted and dashed lines give intermediate results, corresponding to single and double transitions; the latter are even more numerous in the overtone band than they were in the fundamental band. The comparison of such calculations with measurements at a variety of temperatures shows satisfactory agreement [284].

H_2–He rototranslational spectra. The H_2–He system does not exist as a bound van der Waals molecule nor does any striking scattering resonance occur, the interaction potential is too shallow. Line shape computations for that system are, therefore, simpler than for almost any other system [43, 279].

The computed profiles are shown in Figs. 6.11 and 6.12. The various components labeled $\lambda L = 01$, 21, 23, and 45 are sketched lightly. Their sum is given by the heavy curve marked 'total.' The spectra consist of a broad, purely translational part that is dominated at the low frequencies by the isotropic component ($\lambda L = 01$). Other, generally smaller contributions are noticable, the most significant of which is the quadrupole-induced component ($\lambda L = 23$) which shapes the rotational, induced lines, $S_0(J)$ with $J = 0, 1, \ldots$, of H_2; this component arises from a dipole component

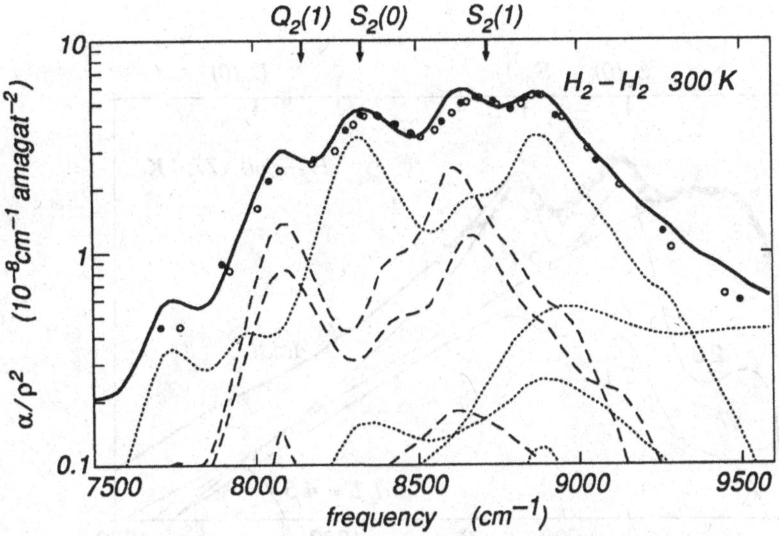

Fig. 6.10. Absorption in the first overtone band of H_2 by H_2–H_2 at 300 K. Comparison of theory (heavy line) and measurements (dots) [284].

which is significantly modified from the pure quadrupole-induced dipole form by a strong overlap contribution, Fig. 4.1. Another component ($\lambda L = 21$) of quadrupolar symmetry, which is 100% overlap-induced, gives rise to very diffuse 'lines' at the $S_0(J)$ rotational transition frequencies. The hexadecapole-induced component ($\lambda L = 45$) is important only at the highest frequencies where the intensities of the other components have fallen off to rather small values; few measurements exist at these high frequencies. The high-frequency presence of the hexadecapole-induced component is due mainly to the H_2 $U_0(J)$ lines, which correspond to transitions with $\Delta J = 4$. For the $\lambda L = 01$ component we get $\Delta J = 0$, which explains the absence of the rotational lines in that component; it consists of a purely translational component whose spectral profile $g(\nu)$ is centered at zero frequency. For the $\lambda L = 23$ and 21 components we have $\Delta J = 0, \pm 2$; these generate the Q_0 and S_0 lines, respectively, and the $\lambda L = 45$ component is associated with $\Delta J = 0, \pm 2, \pm 4$, adding the U_0 lines of H_2.

Close agreement with Birnbaum's measurements (dots) [37] is observed. We note that both theory and measurement are given on an absolute intensity scale; no adjustable parameters of any kind have been used. The agreement between theory and measurement is generally at the 10% level, except at the lowest temperature and near some peaks of the S_0 lines where the margin is slightly greater ($\pm 20\%$). This reflects the greater uncertainties of measurements of the *enhancement* spectra of mixtures

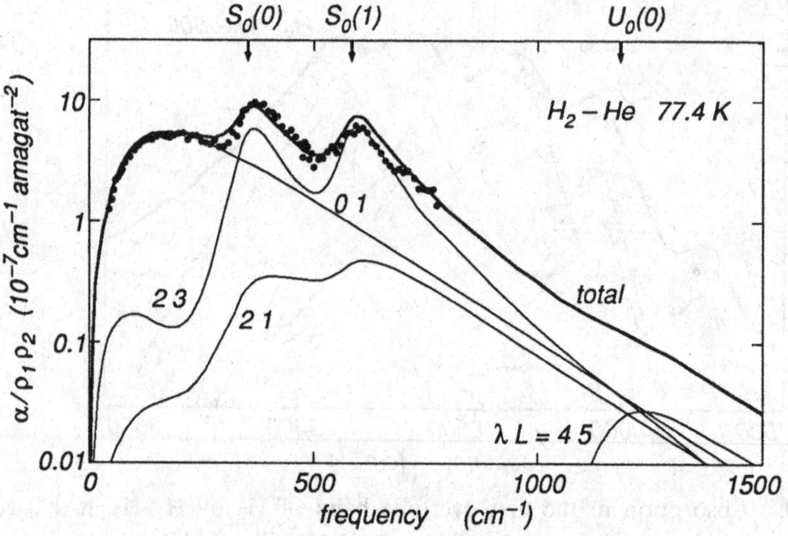

Fig. 6.11. Absorption by H_2–He in the far infrared, at the temperature of 77.4 K – comparison of theory (heavy line) and measurement • [279].

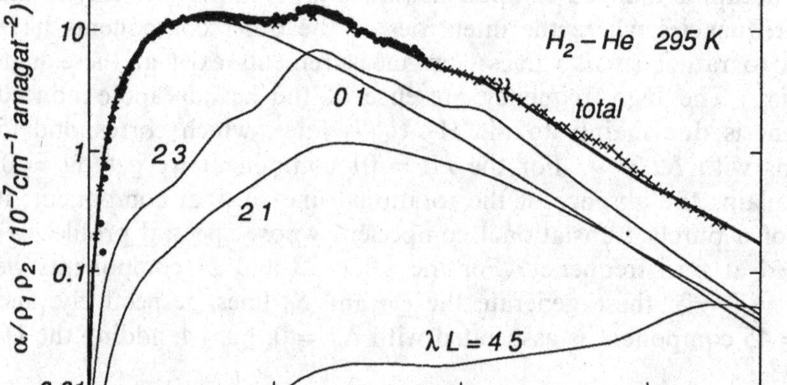

Fig. 6.12. Same as Fig. 6.11, but at the higher temperature of 292.4 K (•) for low frequencies < 1100 cm^{-1}, and 297 K at high frequencies > 1100 cm^{-1} (×) [279].

like hydrogen–helium, which must be obtained as a difference of the absorption spectra of the mixture and of unmixed hydrogen; subtraction of measured data of comparable intensities increases the uncertainty of the data.

H_2–He rotovibrational spectra. The dependence of the interaction potential of H_2–He on the vibrational coordinate of H_2 is well known [276] and has been accounted for in line shape calculations of collision-induced absorption spectra of the fundamental band of hydrogen [151]. That study has demonstrated the significance of the vibrational dependence of the interaction potential for collision-induced absorption.

Figure 6.14 compares the results of line shape computations based on the isotropic interaction approximation with the measurement by Hunt [187]. This spectrum does not have many striking features because of the relatively high temperature of 300 K. We notice only a broad, unresolved Q branch and a diffuse $S_1(1)$ line of H_2 is seen; other lines such as $S_1(J)$ with $J = 0, 2, 3, \ldots$ are barely discernible. Various dips of the absorption at 4126, 4154 and 4712 cm^{-1} are caused by intercollisional interference, a many-body effect which is not accounted for in a binary theory. Roughly 90% of the Q branch (in the broad vicinity of 4150 cm^{-1}) arises from the isotropic overlap induced dipole component ($\lambda L = 01$). The anisotropic overlap component ($\lambda L = 21$) is a little less than one-half as intense as the quadrupole induced term ($\lambda L = 23$). These two components with $\lambda = 2$ are responsible for the S_1 line structures superimposed on the broad isotropic induction component which is of roughly comparable intensity near the S_1 line center.

Figure 6.13 compares the low-temperature measurement of McKellar *et al.* [261] with theory. A mixture of para-hydrogen and helium has been used. At temperatures as low as 18 K, only the rotational ground state ($J = 0$) is populated which is isotropic so that the choice of an isotropic interaction potential is fully justified. The very sharp intercollisional dips seen in the original measurement have been suppressed in the figure. Agreement of measurement and theory is obtained, except near the peak of the H_2 $S_1(0)$ line where the measurement falls below theory by roughly 10%. An excess intensity is noticeable between 4650 and 4900 cm^{-1}, the $S_1(1)$ line of ortho-H_2 which was present with roughly a 10% abundance; the presence of ortho-H_2 also explains the defect at $S_1(0)$ because for the computation a 100% para-H_2 concentration was assumed. Proper accounting for the residual ortho-H_2 abundance would doubtlessly improve the agreement.

Apart from the intercollisional dips (which are not accounted for in a binary theory) and the effects of the ortho-H_2 contamination, the agreement between theory and measurement is generally better than 10%.

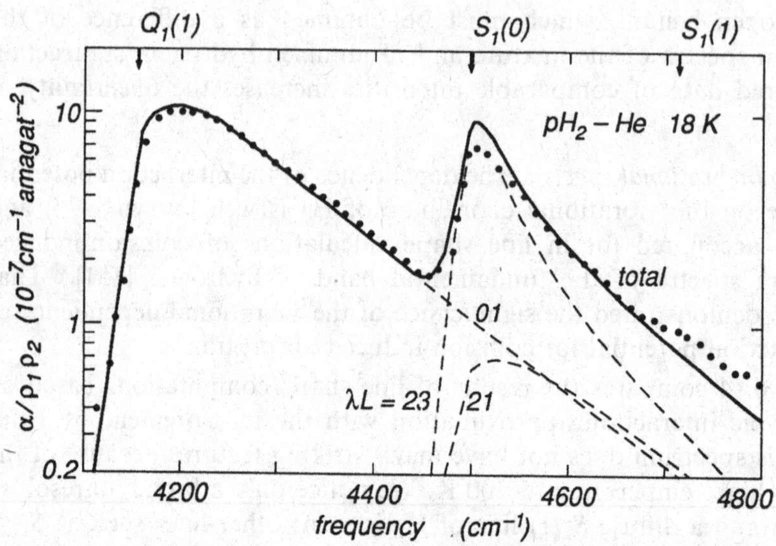

Fig. 6.13. The absorption spectrum of H_2–He in the fundamental band of H_2 at 18 K. Computation (heavy line); measurements (dots); intermediate results are also shown [63].

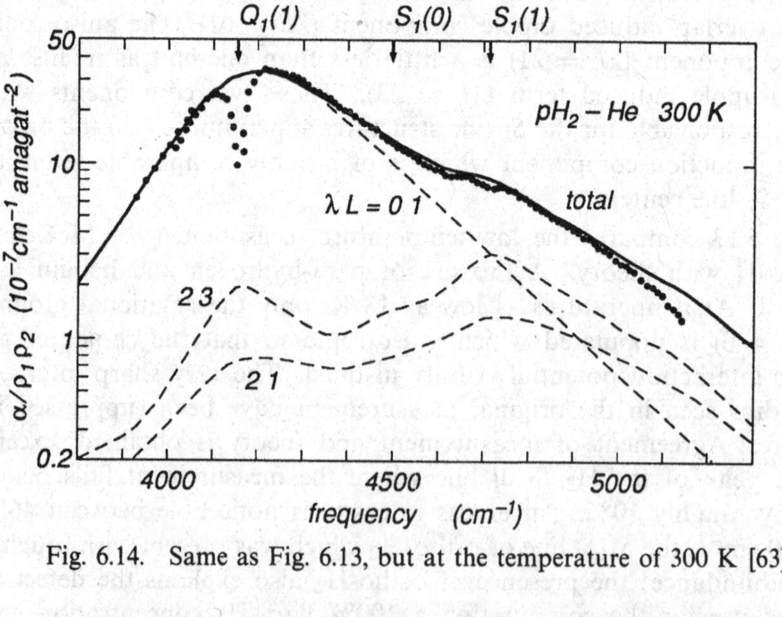

Fig. 6.14. Same as Fig. 6.13, but at the temperature of 300 K [63].

Measurements at several other temperatures were similarly compared with the theory and were found to be in agreement [151, 63].

It is well known [190] that most properties of the H_2 molecule, for example the multipole moments, q_2, q_4, and the polarizability invariants, $a = \alpha_\parallel + 2\alpha_\perp$ and $\beta = \alpha_\parallel - \alpha_\perp$, depend on the vibrational coordinate.

While for an accurate ($\pm 1\%$) treatment of the rototranslational spectra ($v = v' = 0$) the matrix elements $\langle vj|\mathcal{O}|v'j'\rangle$ of the lower rotational states do not much depend on the rotational transitions (j, j'), for the vibrational bands ($v' > 0$), for $v \neq v'$, relatively strong j, j' dependences are usually observed; $\mathcal{O}$ designates the multipole and polarizability operator. Similar j dependences are also obtained for the dipole components B_c that are significant for line shape computations [63]. The accounting for the j dependences is relatively easy because the main effect of the j dependence is on the integrated intensity, but not so much on the shape of the profile. The main effect of neglecting the j dependence in the low-temperature spectra is an excess intensity of the $S_1(1)$ lines.

The enhancement overtone spectra of hydrogen–helium mixtures have not been recorded. Theoretical estimates indicate that these are weak [61].

3 Line shapes of systems with anisotropic interaction

Practically all computations shown above were undertaken in the framework of the isotropic interaction approximation. For the examples considered, agreement of calculated and observed spectra was found. The most critical comparisons between theory and measurement were made for the H_2–X systems whose anisotropy is relatively mild. Nevertheless, some understanding is desirable of what the spectroscopic effects of the anisotropy are. Furthermore, other important systems like N_2–N_2 and CO_2–CO_2 are more anisotropic than H_2–X. The question thus remains as to what the spectroscopic significance of anisotropic interaction might be. In this Section, an attempt is made to focus on the known spectroscopic manifestations of the anisotropy of the intermolecular interaction.

The anisotropy of the interaction couples the translational and rotational states of collisional systems. This in turn couples the various dipole components. Instead of computing for each set of expansion parameters $\lambda_1\lambda_2\Lambda L$ one general profile for all rotational components associated with that set, one now has a much more complex computational task to compute the induced absorption continua. Moreover, the energy level diagrams as well as the spectra of van der Waals dimers are much more complex when the anisotropy of the interaction is accounted for.

Theory. The theory of collision-induced absorption profiles of systems with anisotropic interaction [43, 269] is based on Arthurs and Dalgarno's close coupled rigid rotor approximation [10]. Dipole and potential functions are approximated as rigid rotor functions, thus neglecting vibrational and centrifugal stretching effects. Only the H_2–He and H_2–H_2 systems have been considered to date, because these have relatively few channels (i.e., rotational levels of H_2 to be accounted for in the calculations). The computations must be done numerically and are costly; actual calculations are not practical for systems with greater numbers of channels, such as N_2–N_2 and CO_2–CO_2. The most advanced induced dipole and interaction surfaces have been used [277, 269, 354, 358].

The close coupled scheme is described on pp. 306 through 308. Specifically, the intermolecular potential of H_2–H_2 is given by an expression like Eq. 6.39 [354, 358]; the potential matrix elements are computed according to Eq. 6.45ff. The dipole function is given by Eq. 4.18. Vibration, i.e., the dependences on the H_2 vibrational quantum numbers v_i, will be suppressed here so that the formalism describes the rototranslational band only. For like pairs, the angular part of the wavefunction, Eq. 6.42, must be symmetrized, according to Eq. 6.47.

The wave vector of the initial free state $|j_1'm_1'j_2'm_2'\rangle$ is called $\boldsymbol{k}_{j_1'j_2'}$. For the transitions between initial ($k_{j_1'j_2'}$) and final ($k_{j_1j_2}$) scattering states, the spectral profile of free–free transitions becomes [358]

$$g_{ff}(\omega, T; j_1'j_2'j_1j_2) \tag{6.61}$$
$$= \frac{\lambda_0^3}{V} P_{j_1'} P_{j_2'} \int \frac{\mathrm{d}k_{j_1'j_2'}}{(2\pi)^3} \exp\left\{-\hbar^2 k_{j_1'j_2'}^2/2mkT\right\}$$
$$\times \int \frac{\mathrm{d}k_{j_1j_2}}{(2\pi)^3} \frac{(4\pi)^4}{4\pi\varepsilon_0} \sum_{J'\alpha' J\alpha} \left| \sum_{\alpha''\alpha'''} \langle \alpha'' J \,|\boldsymbol{\mu}|\, \alpha''' J' \rangle \right|^2 \delta(\omega_f - \omega_i - \omega) ,$$

where the rotational quantum numbers $j_1 j_2$ of the final state are kept fixed in α and α', respectively; the difference between final and initial energy, $E_f - E_i$, equals the absorbed photon energy as the δ function requires. λ_0 designates the thermal de Broglie wavelength, Eq. 5.65.

The transition matrix elements are given by [358]

$$\langle \alpha J \,|\boldsymbol{\mu}|\, \alpha' J' \rangle = \tag{6.62}$$
$$\sum_{(c)} \int \mathrm{d}R \, f_\alpha^J(KR) \, f_{\alpha'}^{J'}(K'R) \, B_c^{j_1'j_2'j_1j_2}(R)$$
$$\times \left([\lambda_1][\lambda_2][\Lambda][L][j_1][j_2][j][\ell][J][j_1'][j_2'][j'][\ell'][J'] \right)^{1/2}$$
$$\times (-1)^{j_1+j_2+\ell} \begin{pmatrix} j_1 & \lambda_1 & j_1' \\ 0 & 0 & 0 \end{pmatrix} \begin{pmatrix} j_2 & \lambda_2 & j_2' \\ 0 & 0 & 0 \end{pmatrix} \begin{pmatrix} \ell & L & \ell' \\ 0 & 0 & 0 \end{pmatrix}$$
$$\times \begin{pmatrix} j_1 & j_1' & \lambda_1 \\ j_2 & j_2' & \lambda_2 \\ j & j' & \Lambda \end{pmatrix} \begin{pmatrix} j & j' & \lambda \\ \ell & \ell' & L \\ J & J' & 1 \end{pmatrix}$$

with

$$B_c^{j_1'j_2'j_1j_2}(R) = \int \mathrm{d}r_1 \, \mathrm{d}r_2 \, \chi_{j_1}(r_1) \, \chi_{j_2}(r_2) \, A_c(r_1 r_2 R) \, \chi_{j_1'}(r_1) \, \chi_{j_2'}(r_2) ,$$

$(c) = (\lambda_1\lambda_2\Lambda L)$ and $[l] = 2l+1$. The centrifugal stretching effects are thus included in the integrations over r_1 and r_2. The formalism accounts for rotationally inelastic components in Eq. 6.61, in the sums over α'' and α''', the substitutions of the wave vectors must be done so that energy is always conserved.

Bound states are readily included in the line shape formalism either as initial or final state, or both. In Eq. 6.61 the plane wave expression(s) are then replaced by the dimer bound state wavefunction(s) and the integration(s) over $k_{j_1'j_2'}$ and/or $k_{j_1j_2}$ are replaced by a summation over the n bound state levels with total angular momentum J_n' or J_n. The kinetic energy is then also replaced by the appropriate eigen energy. In this way the bound–free spectral component is expressed as [358]

$$g_{bf}\,(\omega, T; j_1'j_2'j_1j_2) = \frac{\lambda_0^3}{V} P_{j_1'} P_{j_2'} \frac{(4\pi)^4}{4\pi\varepsilon_0} \sum_{J_n'} \int \frac{\mathrm{d}k_{j_1j_2}}{(2\pi)^3} \exp\left(-E_{J_n'}/kT\right) \times \sum_{J\alpha} \left| \sum_{\alpha''\alpha_n'} \langle \alpha'' J \,|\boldsymbol{\mu}|\, \alpha_n' J_n' \rangle \right|^2 \delta\left(\omega_f - \omega_i - \omega\right). \tag{6.63}$$

For the free–bound components, we get similarly

$$g_{fb}\,(\omega, T; j_1'j_2'j_1j_2) = \frac{\lambda_0^3}{V} P_{j_1'} P_{j_2'} \int \frac{\mathrm{d}k_{j_1'j_2'}}{(2\pi)^3} \exp\left(-\hbar^2 k_{j_1'j_2'}^2/2mkT\right) \times \frac{(4\pi)^2}{4\pi\varepsilon_0} \sum_{J'\alpha' J_n} \left| \sum_{\alpha_n\alpha'''} \langle \alpha_n J_n \,|\boldsymbol{\mu}|\, \alpha''' J' \rangle \right|^2 \delta\left(\omega_f - \omega_i - \omega\right), \tag{6.64}$$

and for the bound–bound contributions we get

$$g_{bb}\,(\omega, T; j_1'j_2'j_1j_2) = \frac{\lambda_0^3}{V\,4\pi\varepsilon_0} P_{j_1'} P_{j_2'} \sum_{J_n' J_m} \exp\left(-E_{J_n'}/kT\right) \times \left| \sum_{\alpha_m\alpha_n'} \langle \alpha_m J_m \,|\boldsymbol{\mu}|\, \alpha_n' J_n' \rangle \right|^2 \delta\left(\omega_f - \omega_i - E_{\mathrm{rad}}\right). \tag{6.65}$$

Continua. The wavefunctions of scattering and bound states have been calculated numerically in the close coupled approximation [358]. Converged partial wave expansions of the elastic scattering solutions have been calculated for pairs of angular momenta $j_1j_2 = 00$, 02, 22, 10, 30, 12, 11, and 13 at several hundred energy points. Rotationally inelastic

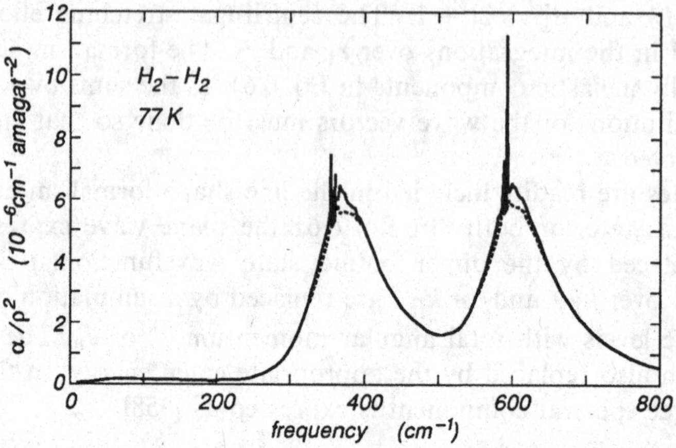

Fig. 6.15. Theoretical collision-induced absorption spectrum of equilibrium hydrogen at 77 K, obtained in the close coupled scheme. The lower curve (dotted) shows the free–free component and the upper curve (heavy solid line) includes the dimer contribution [358].

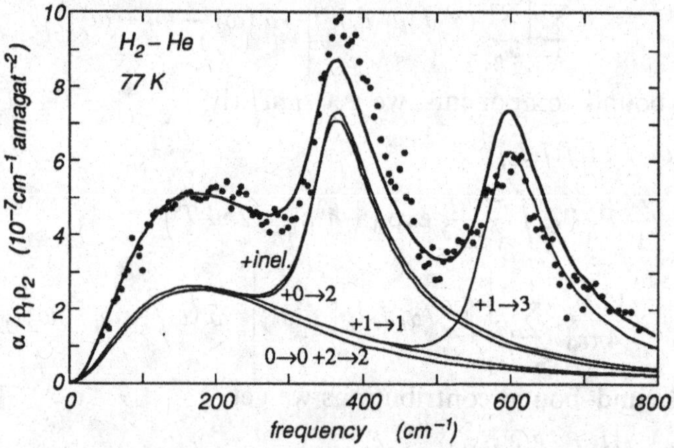

Fig. 6.16. Theoretical collision-induced absorption enhancement spectrum of H_2–He at 77 K, equilibrium H_2 obtained in the close coupled scheme; after J. Schäfer, unpublished. The labels are explained in the text.

wavefunctions have been neglected. The energy grids are chosen to be consistent with the scattering resonances [354, 357, 358, 269].

Figure 6.15 shows the rototranslational spectrum of H_2–H_2 at 77 K as an example. The diffuse lines are nearly indistinguishible from those obtained in the isotropic interaction approximation, Fig. 6.6. Strikingly different are the various dimer structures near the H_2 $S_0(0)$ and $S_0(1)$ rotational transition frequencies.

Figure 6.16 shows for comparison similar calculations for the H_2–He system which does not form bound dimer states. Consequently, the line structures near the H_2 rotational transition frequencies seen in Fig. 6.15 are totally absent. Here again, the profiles obtained in the isotropic interaction approximation, Fig. 6.11, are nearly indistinguishable from the much more involved calculations shown in Fig. 6.16.

Some intermediate results (thin lines, lower curves) are also shown in Fig. 6.16. The relatively unstructured lowest curve represents the $j = 0 \to 0$ and $2 \to 2$ contributions. If the ($j = 0 \to 2$) component is added, the $S_0(0)$ line appears; addition of all inelastic contributions gives the just slightly larger $S_0(0)$ line structure seen in the figure. Inclusion of the $j = 1 \to 3$ component leads to the appearance of the $S_0(1)$ line. The sum of all components is shown as a heavy line (uppermost curve). The dots represent Birnbaum's measurement [37].

Dimers. It is well known that H_2 pairs form bound states which are called van der Waals molecules. The discussions above based on the isotropic interaction approximation have shown that for the $(H_2)_2$ dimer a single vibrational state, the ground state ($n = 0$), exists which has two rotational levels ($\ell = 0$ and 1). If the van der Waals molecule rotates faster ($\ell > 1$), centrifugal forces tear the molecule apart so that bound states no longer exist. However, two prominent predissociating states exist which may be considered rotational dimer states in the continuum ($\ell = 2$ and 3). The effect of the anisotropy of the interaction is to split these levels into a number of sublevels.

It is instructive to look at the $(D_2)_2$ dimer, an isotope of hydrogen which has roughly twice the reduced mass of $(H_2)_2$. Consequently, the de Broglie wavelength of D_2 pairs is significantly shorter than that of H_2 pairs. Since the interaction potentials of H_2 and D_2 pairs are virtually the same, more bound states exist in the given well: in the isotropic potential approximation, one has four rotational states ($\ell = 0 \cdots 3$) that are true bound states, i.e., states of negative energy. Since molecules in the $j_1 = j_2 = 0$ rotational states interact like isotropic systems, Fig. 6.17 shows how the anisotropy (for $j = 2$) of the interaction splits the observable dimer energies into multiple sublevels. For the other rotational transitions of D_2, and of course also for the H_2 dimer, similar splittings are seen [133, 134, 269].

Spectral transitions occur between states of different parity. Figure 6.18 shows a few examples of theoretically predicted levels and transitions between bound states of $(H_2)_2$ [277, 358]. The measurement of dimer bands is useful for refining potential functions [268].

The most striking hydrogen dimer features are, however, not due to bound–bound transitions, but to bound–free and free–bound transitions, just as this was concluded above, pp. 316ff. Figure 6.19 shows a comparison

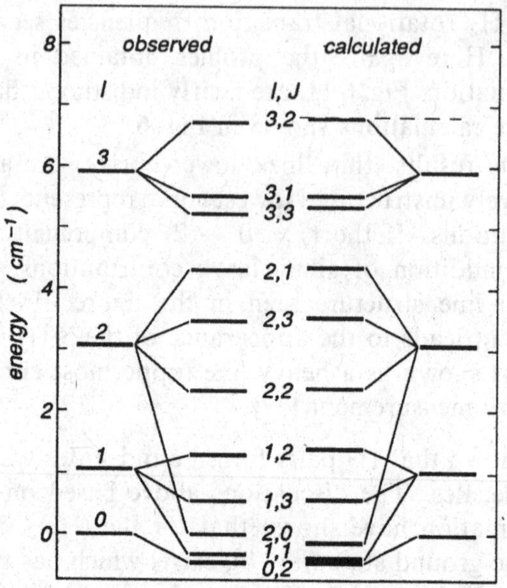

Fig. 6.17. Calculated and observed energy levels of $(D_2)_2$. The outer groups of four levels labeled with just one quantum number ℓ represent the dimer levels for $v_1 = v_2 = j_1 = j_2 = 0$. In the center, we see the effect of the anisotropy; the quantum numbers ℓ, J are used as labels and $j_1, j_2 = 2{,}0$ and 0,2, the initial and final states of the $S_0(0)$ transition ($v_1 = v_2 = 0$). The zero of energy is chosen to coincide with the ground state energy of the non-rotating molecules. The dashed lines indicate the dissociation limit which is, however, not accurately known for the observed levels. Only those final states are shown which are experimentally observed [269].

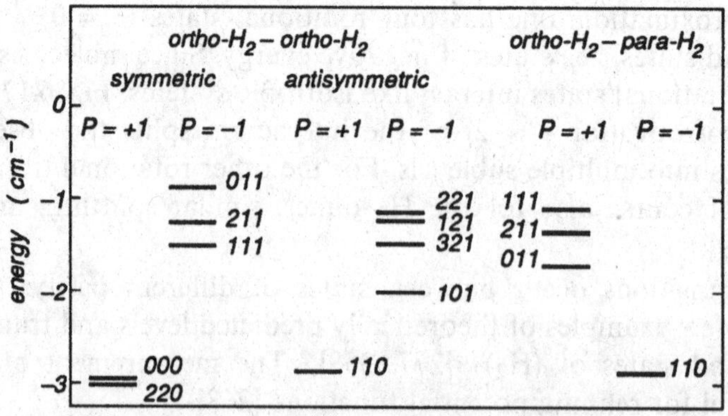

Fig. 6.18. Theoretical $(H_2)_2$ dimer levels and main transition frequencies [277]. The labels are J, j, ℓ; spectral transitions are as indicated.

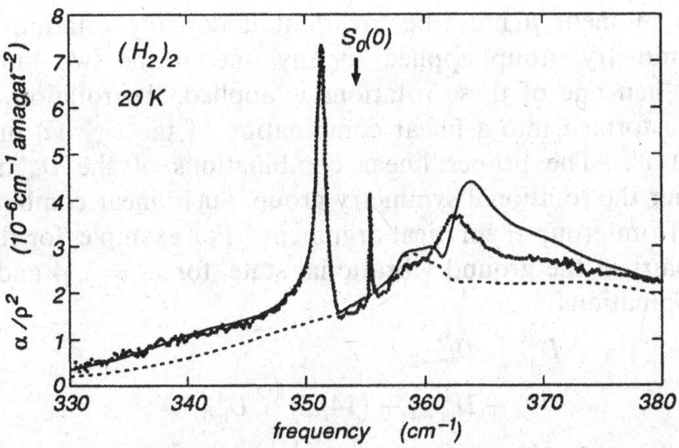

Fig. 6.19. Comparison of the theoretical and experimental spectra of para-H_2 in the $S_0(0)$ region at 20 K. There is good agreement of the overall level of absorption and in the sharper dimer features at 351.3 and 355.4 cm^{-1}. However, the broad dimer feature in the 360–370 cm^{-1} region does not agree well. The dashed curve shows the free–free spectral component [269].

of the calculated and observed dimer features near the H_2 $S_0(0)$ transition frequency of ≈ 354 cm^{-1}. While the agreement of the state of the art computation and a recent observation [269] is not fully satisfactory, several of the main observed features are clearly reproduced from theory on an absolute intensity scale, i.e., without using any adjustable parameters for the 'fit'. The few remaining areas of disagreement may be due to certain approximations that are still necessary. For example, it is clear that the dimer lines are affected by pressure broadening (McKellar 1990) which could be partially included in the theory.

Other dimers have been investigated to a greater or lesser extent. The work of Le Roy and associates [229] must be mentioned, along with other work presented elsewhere (McKellar 1990; Weber 1987; Halberstadt and Janda 1990).

4 Spherical molecules

The induced dipole moment is given by Eq. 4.8 [141],

$$\mu_\nu = \sum_{(c)} A_c \left(r_1^N r_2^N ; R\right) \ Y_{(c)}^{1\nu} \left(\Omega_1, \Omega_2, \widehat{\boldsymbol{R}}\right) ,$$

for spherical molecules. The r_i^N are the vibrational coordinates of molecule $i = 1$ or 2 and the Ω_i describe their orientations; the Ω_i may be the Euler angles of a molecule fixed frame with respect to a space fixed frame. The summation index (c) is short for $(c) = (\lambda_1\lambda_2\Lambda L; \upsilon_1\upsilon_2)$ and $Y_{(c)}^{1\nu}$ is given by Eq. 4.8.

The dipole moment $\boldsymbol{\mu}$ must be invariant under any rotation of the molecular symmetry group applied to any one of the two molecules [166, 374]. When one of these rotations is applied, the rotation matrix $D^{\lambda *}_{m\upsilon}(\Omega)$ is transformed into a linear combination of the $D^{\lambda *}_{m\upsilon'}(\Omega)$ matrices with different υ'. The proper linear combinations of the $D^{\lambda *}_{m\upsilon}(\Omega)$ are invariant under the rotational symmetry group. Such linear combinations are obtained from group-theoretical arguments. For example, for the case of methane pairs in the ground vibrational state, for $\lambda_1 = 3$, 4 and 6, we have the combinations

$$\begin{gathered} D^{3*}_{m2} - D^{3*}_{m-2} \\ D^{4*}_{m4} + D^{4*}_{m-4} + (14/5)^{1/2} D^{4*}_{m0} \\ D^{6*}_{m4} + D^{6*}_{m-4} - (2/7)^{1/2} D^{6*}_{m0} \,. \end{gathered} \tag{6.66}$$

These combinations represent the case when the body fixed coordinate frame has its axes parallel to the cube faces [374]. Since only the linear combinations just mentioned may appear in the expansion of the dipole moment, relations between the B components of the dipole moment which differ in their υ values are seen seen to be [141]

$$\begin{aligned} B_{3\lambda_2\Lambda L;2\upsilon_2}(R) &= -B_{3\lambda_2\Lambda L;-2\,\upsilon_2}(R) \\ B_{4\lambda_2\Lambda L;4\upsilon_2}(R) &= B_{4\lambda_2\Lambda L;-4\,\upsilon_2}(R) = (5/14)^{1/2} B_{4\lambda_2\Lambda L;0\upsilon_2}(R) \\ B_{6\lambda_2\Lambda L;4\upsilon_2}(R) &= B_{6\lambda_2\Lambda L;-4\,\upsilon_2}(R) = (7/2)^{1/2} B_{6\lambda_2\Lambda L;0\,\upsilon_2}(R) \,. \end{aligned} \tag{6.67}$$

B coefficients with different υ_1 vanish. Similar relations hold when λ_2 is considered instead of λ_1.

Pursuing the case of tetrahedral molecules (like CH_4–CH_4) a little further [141], we note that the induced dipole components μ_ν are in general complex. However, since $\boldsymbol{\mu}$ is a Hermitian operator, we have $\mu_\nu^* = (-1)^\nu \mu_{-\nu}$. Furthermore, it is seen that the functions $\mathscr{Y}$ satisfy the relationship

$$\mathscr{Y}^{(c)*}_{1\nu} = (-1)^{\nu+\upsilon_1+\upsilon_2+\lambda_1+\lambda_2+L+1} \mathscr{Y}^{(\bar{c})}_{1-\nu} \,,$$

with $\bar{c} = \lambda_1\lambda_2\Lambda L; -\upsilon_1 - \upsilon_2$. In other words,

$$(-1)^{\upsilon_1+\upsilon_2+\lambda_1+\lambda_2+L+1} A^{(c)*} = A^{(\bar{c})} \,.$$

Furthermore, when dealing with two identical molecules (CH_4 in this case), the interchange of the coordinates $r_i^N \Omega_i$, together with the transformation $\widehat{\boldsymbol{R}} \to -\widehat{\boldsymbol{R}}$ leads to the same dipole. Hence

$$A_{\lambda_2\lambda_1\Lambda L;\upsilon_2\upsilon_1} = (-1)^{\lambda_1+\lambda_2-\Lambda+L} A_{\lambda_1\lambda_2\Lambda L;\upsilon_1\upsilon_2} \,.$$

From these relationships, we conclude that the B coefficients are real for $\lambda = 4$ and 6, and purely imaginary if $\lambda = 3$ (octopolar induction).

In other words, for tetrahedral molecules, these relationships differ from the ones used for the linear molecules, especially Eq. 4.18. As a consequence, we must rederive the relationships for the spectral line shape and spectral moments. If the intermolecular interaction potential may be assumed to be isotropic, the line shape function $Vg(\omega;T)$, Eq. 6.49, which appears in the expression for the absorption coefficient α, Eq. 6.50, may still be written as a superposition of individual profiles,

$$g(\omega;T) = \sum_{(c)} g_c(\omega;T) .$$

But, instead of Eq. 6.54, we write the spectral component g_c now as

$$g_c(\omega;T) = \sum_{ss'} a_{ss'}^{(c)} \, G_c\left(\omega - \omega_{ss'}\right) ,$$

the s, s' label initial and final molecular states of molecules 1 and 2, excluding the degeneracies. The quantities $a_{ss'}^{(c)}$ measure the relative line strengths; they are normalized according to

$$\sum_{ss'} a_{ss'}^{(c)} = 1 .$$

The translational profiles are now expressed as [141]

$$G_c(\omega;T) = \frac{4\pi^2 n_L^2}{3\hbar c} \frac{1}{4\pi\varepsilon_0} \sum_{tt'} P_t \times \frac{4\pi}{2L+1} \sum_M \left| \left\langle t \left| B_c(R;ss') Y_{LM}\left(\widehat{\mathbf{R}}\right) \right| t' \right\rangle \right|^2 \delta\left(\omega - \omega_{tt'}\right) , \tag{6.68}$$

where the t, t' refer to initial and final translational states, respectively. The quantity $B_c\left(R;ss'\right) = \langle s \,|A_c|\, s'\rangle$ is the matrix element between states s and s'.

The profiles of the rototranslational absorption of CH_4–CH_4 in the far infrared have been reported [56]; see Fig. 3.22 for an example. The treatment of the spectra is based on the multipolar induction model and an advanced isotropic potential; empirical overlap-induced dipole components have also been included for fitting the experimental data at several temperatures (126 through 300 K). At the lower temperatures, satisfactory fits of the measurements are possible. The analysis seems to suggest that at temperatures near room temperature a significant rotation-induced distortion of the tetrahedral frames occurs which affects the properties of the individual molecules (multipole strengths, molecular symmetry, polarizabilities, and perhaps the interaction).

The rototranslational absorption band of the system He–CH_4 has also been considered [387], Fig. 3.24. The induced dipole of the He–CH_4

system is largely overlap-induced; multipolar induction was seen to be almost negligible, a fact related to the small polarizability of He.

6.3 Asymmetry of line profiles

On pp. 311ff., a preliminary discussion of the symmetry of induced line profiles was given. The spectral lines encountered in collision-induced absorption show a striking asymmetry which is described roughly by a Boltzmann factor, Eq. 6.59. However, it is clear that at any fixed frequency shift, the intensity ratio of red and blue wings is not always given exactly by a Boltzmann factor, for example if dimer structures of like pairs shape the profile, or more generally in the vibrational bands. We will next consider the latter case in some detail.

The computation of the translational wavefunctions of interacting molecular pairs requires knowledge of the interaction potential. It is well known that interaction potentials depend not only on the separation R of the pair (and the orientations of the molecules), but also on the vibrational coordinates $\boldsymbol{r}_i$ of the molecules $i = 1, \cdots$ (Maitland *et al.*, 1982). As a consequence, the translational wavefunctions of the final state must be computed with an interaction potential which differs from that of the initial state if vibrational excitation occurs. For H_2–H_2 and H_2–He the vibrational averages of the interaction potentials of initial and final state differ enough to affect the rotovibrational spectra; for other systems similar effects may be expected. The dynamics of initial and final translational states differ, therefore, especially if higher vibrational excitations are involved (overtone bands). Line shapes computed with and without accounting for these vibrational dependences of the interaction differ strikingly, especially at high temperatures.

In systems involving molecules like H_2, the vibrational coordinates may be considered as changing in time much faster than the translational ones, so that the motion of the latter can be determined by an isotropic potential that is the vibrational average, e.g., $V_0^v(R) = \langle vj|V(R;r)|vj\rangle$.

We define the spectral density according to Eq. 6.51,

$$Vg(\omega) = \frac{1}{4\pi\varepsilon_0} \sum_{s,s'} P_s \, |\langle s|\boldsymbol{\mu}|s'\rangle|^2 \, \delta(\omega - \omega_{ss'}) \, .$$

In this expression, the s, s' indicate the initial and final states of the supermolecule, P_s is the population probability of the state s, $\boldsymbol{\mu}$ is the dipole moment induced by intermolecular interactions, $\omega_{ss'} = (E_{s'} - E_s)/\hbar$ where E_s is the energy of the state $|s\rangle$. For simplicity, we consider an atom–diatom system, such as H_2–He, or a diatom–diatom system like H_2–H_2, with the absorption coefficient given by Eq. 6.50. Since we will regard the molecules as rotating freely, it is convenient to expand the spherical components of the induced dipole moment μ in terms of angular functions, Eq. 4.18. If we assume, furthermore, that the rotational motion does not couple with the translational motion, the states $|s\rangle$ of the supermolecule may be written as the product of a state representing the free rotation of the molecules, $|r, m_r\rangle$, and one describing both vibrational and translational variables, $|\phi\rangle$, according to $|s\rangle = |r, m_r\rangle\ |\phi\rangle$. Here, m_r designates a set of quantum numbers which distinguishes among states with the same rotational energy, E_r. For the computation of the spectral density, the sum over r, m_r, r', m_r' may be computed as

$$V\,g(\omega) = \sum_{(c),r,r'} a_{r,r'}^{(c)}\, F^{(c)}(\omega - \omega_{r,r'}) \,. \tag{6.69}$$

In other words, the spectral function is written as a sum over rotational line profiles centered at the rotational transition frequencies $\omega_{r,r'}$. For each set of expansion parameters $(c) = \lambda_1\lambda_2\Lambda L$, the quantities $a_{r,r'}^{(c)}$ satisfy the selection rules appropriate for the (c) component, and are chosen such that

$$\sum_{r'} a_{r,r'}^{(c)} = P_r \qquad \sum_{r,r'} a_{r,r'}^{(c)} = 1 \,,$$

where P_r is the population probability of the rotational state $|r\rangle$. Hence,

$$V \int_{-\infty}^{\infty} g(\omega)\, d\omega = \sum_{(c)} \int_{-\infty}^{\infty} F^{(c)}(\omega)\, d\omega \,.$$

The line profiles $F^{(c)}(\omega)$ may be written as

$$\begin{aligned} F^{(c)}(\omega) \;\; &= \sum_{\phi,\phi'} V\, P_\phi \frac{4\pi}{2L+1} \\ &\times \sum_M \left| \left\langle \phi \middle| A^{(c)}(R; r_1; r_2)\, Y_{LM}(\Omega) \middle| \phi' \right\rangle \right|^2 \delta(\omega - \omega_{\phi\phi'}) \,. \end{aligned} \tag{6.70}$$

Next, we assume that the translational states $|\phi\rangle$ and $|\phi'\rangle$ are of the form

$$|\phi\rangle = |v\rangle\, |r\rangle_v \qquad |\phi'\rangle = |v'\rangle\, |r'\rangle_{v'} \,,$$

with $|v\rangle$ and $|\phi'\rangle$ designating the initial and final vibrational states of the pair, $|\tau\rangle_v$ and $|\tau'\rangle_{v'}$ the translational eigenstates of the Hamiltonians $\langle v|\mathcal{H}|v\rangle$ and $\langle v'|\mathcal{H}|v'\rangle$ whose eigenvalues are the energies E_τ and $E_{\tau'}$, respectively. The Hamiltonian $\mathcal{H}$ is the total Hamiltonian of the pair *minus* the rotational and vibrational kinetic energies of the molecules.

If a transition from the initial state v to the final state v' takes place, we get from Eq. 6.70,

$$F^{(c)}_{vv'}(\omega) = P_v \sum_{\tau,\tau'} V \frac{4\pi}{2L+1} P_\tau \times \sum_M \left| \left\langle \tau^{(v)} \left| B^{(c)}_{vv'}(R) Y_{LM}(\Omega) \right| \tau'^{v'} \right\rangle \right|^2 \delta(\omega - \omega_{rr'}), \qquad (6.71)$$

where ω now designates the frequency shift $\omega - \omega_{vv'}$ relative to the vibrational transition frequency, and $B^{(c)}_{vv'}$ is the vibrational matrix element of the induced dipole component, Eq. 4.13. P_v and P_τ designate population probabilities of the initial vibrational and translational states, $|v\rangle$ and $|\tau^{(v)}\rangle$, respectively. Equation 6.71 is formally the same as Eq. 6.50, but the radial part of the translational wavefunction of initial and final states is now computed with the vibrational average of the interaction potential. Computations of such profiles were found to be in agreement with the existing measurements at temperatures from 18 to 300 K [151, 281].

For some time, the line profiles used to construct collision-induced spectra were believed to satisfy 'detailed balance', Eq. 5.73,

$$\Gamma(-\omega) = \exp(-\hbar\omega/kT)\,\Gamma(\omega)\,, \qquad (6.72)$$

where ω represents the frequency shift relative to the transition frequency. Indeed, for the rototranslational spectra, this relation accurately describes the observed line symmetries as long as dimer features of like pairs are of no great concern. It is therefore noteworthy that for the rotovibrational bands, the exact profiles $F_{vv'}(\omega)$ do *not* satisfy Eq. 6.72 unless $v = v'$. In fact, because of $\delta(-\omega - \omega_{\tau\tau'}) = \delta(\omega - \omega_{\tau'\tau})$, replacing ω by $-\omega$ is equivalent to exchanging the translational states τ and τ' in Eq. 6.71, except in the δ function. $P_{\tau'}$ equals $\exp(-\hbar\omega/kT)\, P_\tau$, but the square modulus of the matrix elements remains the same only if the initial and final vibrational states are the same (or if the translational motion does not depend on the vibrational state). The fact is that the $F(\omega)$ satisfy the detailed balance condition for inversion of the *absolute* frequency, ϖ, but not of the frequency shift, ω. The inversion of the frequency shift in Eq. 6.71 simply means that we go from

$$\varpi = \omega_{vv'} + \omega \qquad \text{to} \qquad \varpi = \omega_{vv'} - \omega\,.$$

The inversion of absolute frequency, ϖ, on the other hand, is equivalent to inverting both ω and $\omega_{vv'}$,

$$-\varpi = \omega_{v'v} - \omega\,.$$

In fact, starting from Eq. 6.71, it is apparent from the symmetry condition

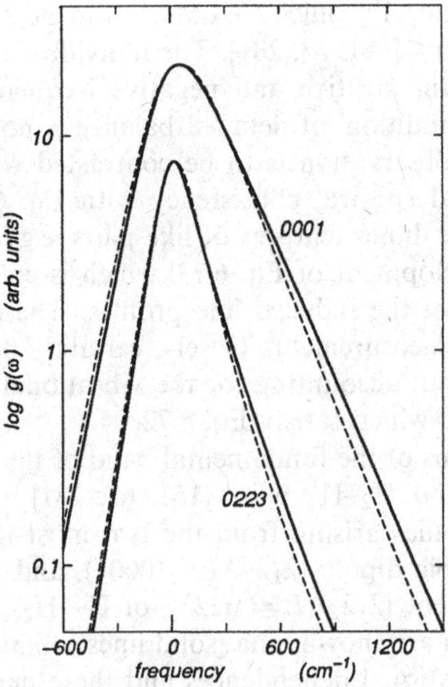

Fig. 6.20. Symmetry of the line profiles of the two main components of unmixed hydrogen in the fundamental band at 297 K. Higher curve: isotropic overlap component ($\lambda_1\lambda_2\Lambda L = 0001$); lower curve: quadrupole-induced component ($\lambda_1\lambda_2\Lambda L = 0223$). For comparison, the dashed curves represent the profiles computed without accounting for the vibrational dependence of the interaction potential [281].

that the vibrational profile F satisfies is given by

$$\frac{1}{P_{v'}} F_{v'v}(-\omega) = \frac{\exp(-\hbar\omega/kT)}{P_v} F_{vv'}(\omega) \,, \tag{6.73}$$

which differs from Eq. 6.72 if $v \neq v'$. The profile $F_{vv'}$ to the right of Eq. 6.73 is associated with the transition $v \rightarrow v'$, while the $F_{v'v}$ to the left is associated with the inverse transition, $v' \rightarrow v$. Rotovibrational spectra should be modeled with the help of profiles which satisfy Eq. 6.73, not Eq. 6.72. Profiles computed from Eq. 6.55 satisfy Eq. 6.73.

1 Numerical computations and results

For the H_2–He and H_2–H_2 pairs, accurate interaction potentials exist that account for the vibrational dependences. We will discuss some of the results obtained for these systems.

If the vibrational averages of the potential, $\langle v|V_0(R;r_1,r_2)|v\rangle$ with $v = 0$ and $v = 1$, are used in computations of the initial and final translational state wavefunctions, spectral profiles are obtained in close agreement with the existing measurements [151, 63, 281]. The individual line profiles have been computed for both, positive and negative frequency shifts and it was noticed that the 'condition' of detailed balance is not satisfied in the form of Eq. 6.72; this observation is to be contrasted with the fact that for the rototranslational spectra, consistency with Eq. 6.72 was always observed (except for the dimer features of like pairs, e.g., pp. 316ff.). This fact has led to the development of Eq. 6.73, which is consistent with the numerical calculations of the induced line profiles. These computed line profiles fit the existing measurements closely, certainly better than other profiles computed without accounting for the vibrational dependences of the interaction potential which satisfy Eq. 6.72.

Line shape calculations of the fundamental band of the hydrogen molecule exist for H_2–He and H_2–H_2 pairs [151, 63, 281]. As an example, Fig. 6.20 shows the profiles arising from the two most important dipole components, the isotropic dipole ($\lambda_1\lambda_2\Lambda L = 0001$), and the quadrupole-induced dipole component ($\lambda_1\lambda_2\Lambda L = 0223$), of H_2–H_2. For each dipole component, two profiles are shown, one (solid lines) obtained with proper accounting for the vibrational dependences and the other (dashed curve) by using identical potentials (with $v = 0$) for initial and final translational states. The correct profiles (solid curves) are more intense in the high-frequency or 'blue' wing, and less intense in the 'red' wing by up to 25% relative to the approximation (dashed curves), over the frequency band shown. Since the accuracy of measurements taken under favorable conditions is in the 10% range, or perhaps slightly better; deviations of that magnitude are significant for detailed comparisons between measurements and theory. We note that the dashed profiles satisfy the detailed balance condition, Eq. 6.72. The red and blue wings of the exact profiles differ from Eq. 6.72 by more than a factor of two over the frequency band shown and differ increasingly with increasing frequency shift (far wings – not shown). The exact profiles satisfy the more general detailed balance relationship of Eq. 6.73 (except for certain dimer features of like pairs which require separate consideration).

2 Model line shapes

The rototranslational profiles of molecular systems are surprisingly closely approximated by certain model profiles, which are functions of frequency, temperature, and three parameters, see Eqs. 5.105ff. The BC profiles have been found to approximate most closely, and for a maximal peak-to-wing intensity ratio, the multipole-induced profiles of low order while the overlap-induced components are best represented by the K0 model [69]. Analytical models of the rototranslational spectra of H_2–He pairs have been constructed that match the exact quantum computations almost perfectly [72]. It is important to note, however, that the H_2–He system is nearly unique in that it does not form bound van der Waals pairs. One reason the H_2–He rototranslational spectra can be modeled so accurately by the analytical model profiles is the absence of bound dimers for that system.

Other systems like H_2–H_2 feature a small number of bound states. Whenever molecular pairs form bound dimers, spectroscopic structures appear. First and usually most importantly, the continuum of the purely rotational band appears, but various other structures associated with bound-to-free transition usually show up that are harder to model closely. As a rule, the rototranslational absorption spectra of most molecular systems are not as easily modeled as that of H_2–He, because of the dimer structures. Of course, in the typical high-pressure laboratory measurements, dimer structures may be broadened to the point where these are hardly discernible. In such a case, the BC and K0 model profiles may become adequate again. In any case, the rototranslational spectra of a number of binary systems have been modeled closely over a broad range of temperatures [58], including the (coarse) dimer structures.

For the *rotovibrational* spectra, certain model profiles, such as the Lorentzian (preferably modified to satisfy Eq. 6.72) and the BC profiles have previously been used successfully [422, 342]. However, significant improvements are possible if model profiles are chosen that mimic the symmetry of the rotovibrational line profiles, Eq. 6.73.

Spectral profiles $K_{vv'}^{(c)}(\omega) = F_{vv'}^{(c)}(\omega)/P_v$ suitable for the modeling of rotovibrational profiles should be consistent with Eq. 6.73. For the rototranslational bands, the highly successful Γ profiles, e.g., the BC and K0 models, satisfy Eq. 6.72, not Eq. 6.73. In order to benefit from these trusted profiles, we describe a method to construct the profile $K(\omega)$ with the proper symmetry (Eq. 6.73) by combining the familiar Γ models in an *ad hoc* way which satisfies Eq. 6.73.

We define functions $X(\omega)$ and $Y(\omega)$, such that

$$\begin{aligned} X^{(c)}(-\omega) &= \exp(-\hbar\omega/kT)\, X^{(c)}(\omega) \\ Y^{(c)}(-\omega) &= -\exp(-\hbar\omega/kT)\, Y^{(c)}(\omega)\,, \end{aligned} \tag{6.74}$$

and construct a rotovibrational model according to

$$K_{vv'}^{(c)}(\omega) = X^{(c)}(\omega) + Y^{(c)}(\omega)\,. \tag{6.75}$$

This is always possible. It is sufficient to choose

$$\begin{aligned} X^{(c)}(\omega) &= 0.5\left[K_{vv'}^{(c)}(\omega) + \exp(\hbar\omega/kT)\, K_{vv'}^{(c)}(-\omega)\right] \\ Y^{(c)}(\omega) &= 0.5\left[K_{vv'}^{(c)}(\omega) - \exp(\hbar\omega/kT)\, K_{vv'}^{(c)}(-\omega)\right]\,. \end{aligned} \tag{6.76}$$

With the help of Eq. 6.73, we write these as

$$X^{(c)}(\omega) = 0.5\left(K^{(c)}_{vv'}(\omega) + K^{(c)}_{v'v}(-\omega)\right) \qquad (6.77)$$
$$Y^{(c)}(\omega) = 0.5\left(K^{(c)}_{vv'}(\omega) - K^{(c)}_{v'v}(-\omega)\right) .$$

The X and Y are thus expressed in terms of the profiles of the 'up' and their inverse 'down' ($v \leftrightarrow v'$) transitions. The spectral moments can thus be written as a simple combination of the moments of the up and down transitions, which may be computed from the induced dipole components and the interaction potential. Furthermore, the function X satisfies the (old) detailed balance condition, Eq. 6.72, and is conveniently represented by the successful BC or K0 models. A simple choice for Y could be $(\omega/\Delta)\,\Gamma'(\omega)$ where Δ is a constant to be specified and $\Gamma'(\omega)$ is another model function Γ which satisfies Eq. 6.72. In other words, according to Eq. 6.75, the rotovibrational profiles K can be represented by the familiar model functions whose parameters may be defined with the help of the associated moment expressions.

The parameters $\tau_1 \cdots \tau_N$ of the function Γ, and $\tau'_1, \cdots \tau'_N$, of Γ', can be determined by matching the spectral moments of the model functions to those of the $X^{(c)}$, $Y^{(c)}$, Eq. 6.77. The moments of X and Y are called $M^{(c)}_{Xn}$ and $M^{(c)}_{Yn}$, respectively. We give here the first three moments of X, obtained from Eq. 6.77 and [291, 294],

$$M^{(c)}_{X0} = \int |B|^2 \, \frac{1}{2}\,(g_v + g_{v'})\, \mathrm{d}^3R\,, \qquad (6.78)$$

$$M^{(c)}_{X1} = \frac{\hbar}{2m}\int \left[\left(B^I\right)^2 + L(L+1)\,\frac{|B|^2}{R^2}\right] \frac{1}{2}\,(g_v + g_{v'})\, \mathrm{d}^3R \qquad (6.79)$$
$$+ \frac{1}{\hbar}\int |B|^2\,(V_{v'} - V_v)\,\frac{1}{2}\,(g_v - g_{v'})\,\mathrm{d}^3R$$

$$M^{(c)}_{X2} = \frac{\hbar^2}{m^2}\int \Bigg[-\frac{1}{4}BB^{IV} - \frac{1}{2}\,B^I B^{III} + \frac{L(L+1)}{2R^2}\,(B^I)^2 \qquad (6.80)$$
$$- \frac{L(L+1)}{R^3}\,BB^I + \frac{L^2(L+1)^2}{4R^4}\,B^2\Bigg]\,\frac{1}{2}\,(g_v + g_{v'})\,\mathrm{d}^3R$$
$$+ \frac{1}{m}\frac{1}{2}\int \left[BB^I\,(g_v V^I_v + g_{v'} V^I_{v'}) + 2BB^{II}\,(g_v V_v + g_{v'} V_{v'}\right]\mathrm{d}^3R$$
$$- \frac{2}{m}\int BB^{II}\,\frac{1}{2}\,(g^{(E)}_v + g^{(E)}_{v'})\,\mathrm{d}^3R$$
$$+ \frac{1}{m^2}\int \left[\frac{BB^{II}}{R^2} - \frac{BB^i}{R^3} + \frac{L(L+1)}{2R^4}\,B^2\right]\frac{1}{2}(g^{(M)}_v + g^{(M)}_{v'})\,\mathrm{d}^3R$$
$$+ \frac{1}{\hbar^2}\int B^2(V_{v'} - V_v)\frac{1}{2}\,(g_v + g_{v'})\,\mathrm{d}^3R$$

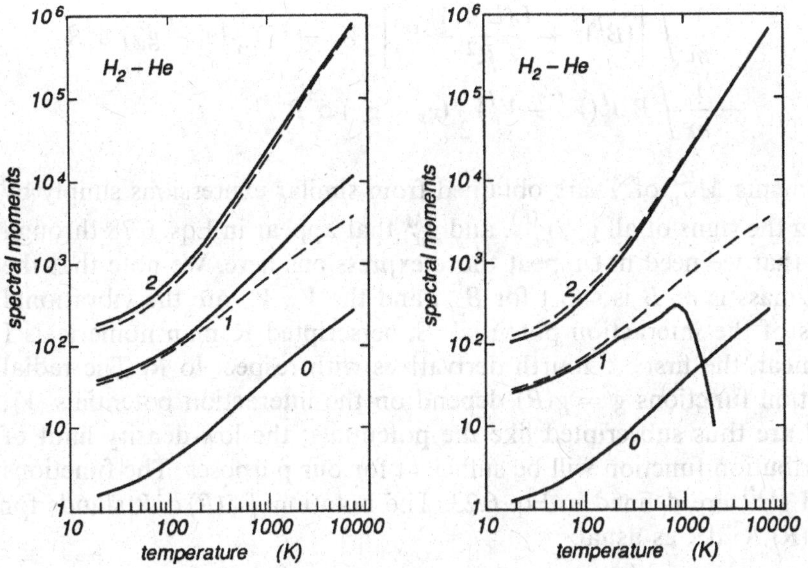

Fig. 6.21. To the left, the spectral moments $M^{01}_{vv';n}$ of the isotropic dipole component ($\lambda L = 01$) of the fundamental band of H_2–He, shown as function of temperature (solid lines). The zeroth moment ($n = 0$, bottom) is given in units of 10^{-64} erg, the first moment ($n = 1$, center) in 10^{-52} erg s^{-1}, and the second moment ($n = 2$, top) in 10^{-39} erg s^{-2}. Also shown are the moments computed without accounting for the vibrational dependence of the potential. To the right, the same data are shown for the inverse transition, $v = 1 \rightarrow 0$ [295].

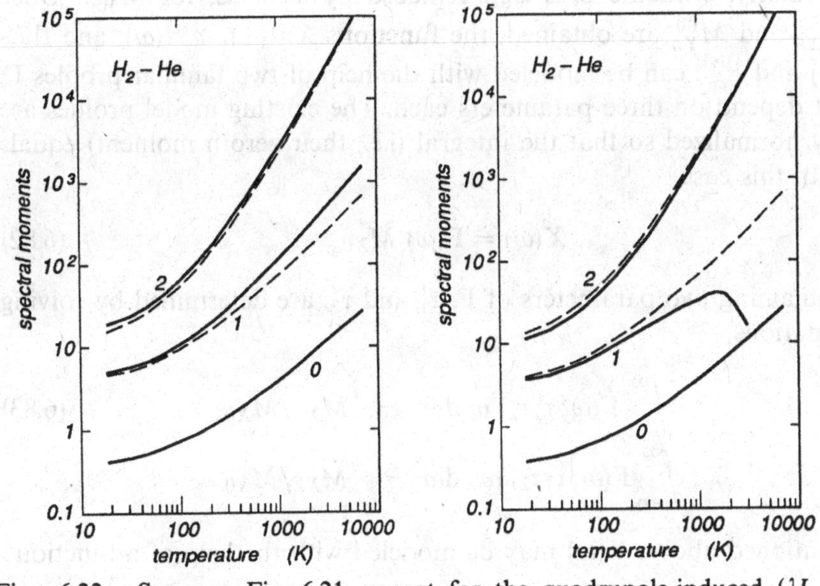

Fig. 6.22. Same as Fig. 6.21, except for the quadrupole-induced ($\lambda L = 23$) component [295].

$$+\frac{1}{m}\int\left[(B^I)^2+\frac{L(L+1)}{R^2}B^2\right](V_{v'}-V_v)\frac{1}{2}(g_v-g_{v'})\,\mathrm{d}^3R$$
$$+\frac{1}{m}\int BB^I(V^I_{v'}-V^I_v)\frac{1}{2}(g_v-g_{v'})\,\mathrm{d}^3R\,.$$

The moments $M^{(c)}_{Yn}$ of Y are obtained from similar expressions simply by changing the signs of all $g_{v'}$, $g^{(E)}_{v'}$, and $g^M_{v'}$ that appear in Eqs. 6.78 through 6.80, so that we need not repeat those expressions here. We note that the reduced mass is m, B is short for $B^{(c)}_{vv'}$, and the V_v, $V_{v'}$ are the vibrational averages of the interaction potential. Superscripted Roman numericals I ... IV mean the first ... fourth derivatives with respect to R. The radial distribution functions $g = g(R)$ depend on the interaction potentials, V_v, $V_{v'}$, and are thus subscripted like the potentials; the low-density limit of the distribution function will be sufficient for our purposes. The functions $g^{(E)}$ and $g^{(M)}$ are defined in Eq. 6.23. The notation $\int f(R)\,\mathrm{d}^3R$ stands for $4\pi\int_0^\infty f(R)\,R^2\,\mathrm{d}R$ as usual.

For each profile $K^{(c)}_{vv'}(\omega)$, the spectral moments $M^{(c)}_{vv';n}$, with $n = 0$, 1, 2, can be calculated using a quantum or semi-classical formalism as was described above [291, 292]. From Eqs. 6.77 it is obvious that these are related to the M_{Xn}, according to

$$M^{(c)}_{Xn} = 0.5\left(M^{(c)}_{vv';n} + M^{(c)}_{v'v;n}\right)\,, \tag{6.81}$$

and similarly, with the plus sign replaced by a *minus*, for $M^{(c)}_{Yn}$. Once the $M^{(c)}_{Xn}$ and $M^{(c)}_{Yn}$ are obtained, the functions $X^{(c)}(\omega)$, $Y^{(c)}(\omega)$, and thus $K^{(c)}_{vv'}(\omega)$ and $F^{(c)}_{vv'}$, can be modeled with the help of two familiar profiles Γ, Γ' that depend on three parameters each. The existing model profiles are usually normalized so that the integral (i.e., their zeroth moment) equals unity. In this case

$$X(\omega) = \Gamma(\omega)\,M_{X0}\,. \tag{6.82}$$

The remaining two parameters of Γ, τ_1 and τ_2, are determined by solving the equations

$$\int_{-\infty}^{\infty}\Gamma(\omega;\tau_1\tau_2)\,\omega\,\mathrm{d}\omega = M_{X1}/M_{X0} \tag{6.83}$$
$$\int_{-\infty}^{\infty}\Gamma(\omega;\tau_1\tau_2)\,\omega^2\,\mathrm{d}\omega = M_{X2}/M_{X0}\,.$$

As mentioned above, $Y(\omega)$ may be modeled with the help of a function

$$Y(\omega) = \frac{\omega}{\Delta}\,\Gamma'(\omega;\tau_1'\tau_2')\,M_{Y0}\,.$$

In this case, we get

$$\begin{aligned} \int_{-\infty}^{\infty} \Gamma'(\omega;\tau_1'\tau_2')\,\omega\,\mathrm{d}\omega &= \Delta \qquad (6.84) \\ \int_{-\infty}^{\infty} \Gamma'(\omega;\tau_1'\tau_2')\,\omega^2\,\mathrm{d}\omega &= \Delta M_{Y1}/M_{Y0} \\ \int_{-\infty}^{\infty} \Gamma'(\omega;\tau_1'\tau_2')\,\omega^3\,\mathrm{d}\omega &= \Delta M_{Y2}/M_{Y0}\,, \end{aligned}$$

which determine the parameters τ_1', τ_2', and Δ. The functions X and Y, and thus K (Eq. 6.75), are fixed in this way once the models Γ and Γ' are chosen. K satisfies Eq. 6.73 as rotovibrational profiles should. The only question remaining at this point is which of the known functions Γ, Γ' to select for an optimal representation of an exact profile, Eq. 6.71. Such studies are under way and may currently be summarized by stating that for the rotovibrational profiles the BC and K0 models seem to work well for multipolar induction and overlap induction, respectively. We note that the left-hand sides of Eqs. 6.83 through 6.84 are simple algebraic functions of the parameters, see Eqs. 5.106 and 5.109.

The profiles represented by the X and Y functions have been used most successfully for the modeling of the rotovibrational bands of H_2–He and H_2–H_2 [62, 61, 64].

Improved vibrational line profile. In some of the modeling attempts based on the X and Y functions introduced above, numerical problems have been encountered which required special attention. Although in all practical cases simple solutions to these 'problems' could be found, an alternative approach is now preferred because it is easy to use [48].

The approach does not aim to satisfy the condition Eq. 6.73 exactly. The model functions consist of a sum of three functions whose parameters are related to three v'-independent and three v'-dependent terms of the quantum spectral moments, Eqs. 6.31 through 6.34; v' is the vibrational quantum number of the final states which differs from v, the initial vibrational state. As a result, the line profile consists of a core which is the same as for rototranslational spectra, and a v'-dependent 'correction'. It converges to the standard solution for potentials that do not depend on the vibrational excitation. The models are six parameter functions which are defined by the lowest three spectral moments [48, 65].

Instead of representing the model line profile $\Gamma(\omega)$ by two functions $X(\omega)$ and $Y(\omega)$ having the same center frequency, an *ad hoc* correlation function model is constructed whose Fourier transform suggests a representation of the line profiles by three functions which are shifted relative to the center frequency,

$$\Gamma(\omega) = \Gamma^{v}(\omega) + \Gamma^{vv'}(\omega_-) - \Gamma^{vv'}(\omega_+) \qquad (6.85)$$

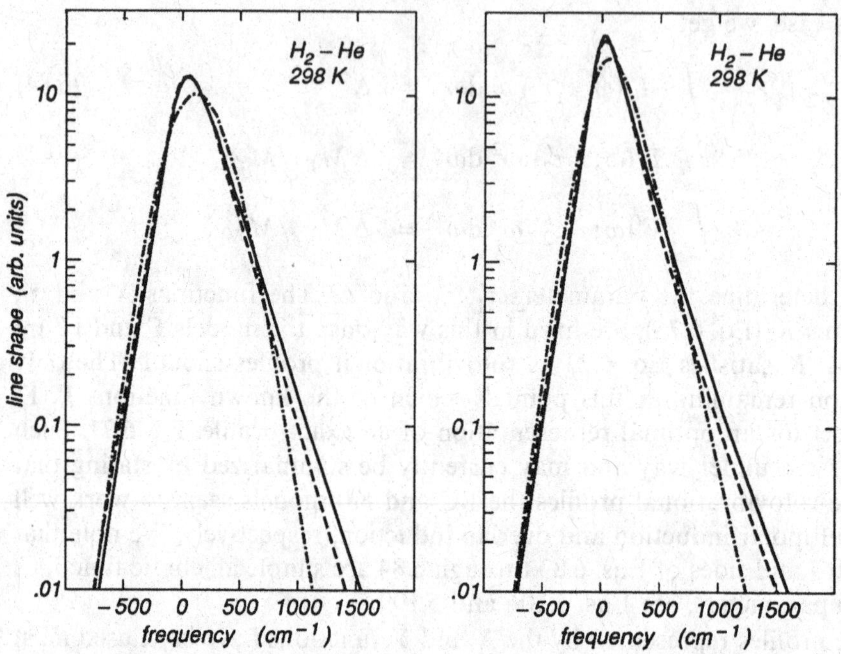

Fig. 6.23. Comparison of three rotovibrational line profiles; H_2–He 298 K, fundamental band of H_2. To the left, the isotropic overlap component ($\lambda L = 01$) is shown; to the right the quadrupole-induced ($\lambda L = 23$) is shown. Solid lines: from quantum line shape calculations (exact); dashed lines: K0 (left) and BC (right) profile satisfying Eq. 6.72 instead of Eq. 6.73; dash-dot line: the K0 and BC profiles obtained from vibration-independent potential [295].

with $\omega_\mp = \omega \mp (\omega_1 \pm \omega_2)$. The frequency shifts are related to the spectral moments M_n of $g(\omega)$, according to

$$\omega_1 = M_1^{vv'}/M_0^v \qquad \text{and} \qquad \omega_1\omega_2 = M_2^{vv'2}/2M_0 .$$

In these equations, $M_2^{vv'2}$ is the second term to the right of Eq. 6.34. The difference $\Gamma^{vv'}(\omega_-) - \Gamma^{vv'}(\omega_+)$ resembles the Y function and, remotely, the dispersion profile.

Further *ad hoc* assumptions are made, such as $\tau_1 = \tau_2$ in the BC shape used to represent the correction profile. The $\Gamma^{vv'}$ are thus given by

$$\Gamma^{vv'}(\omega_\pm) = \frac{M_0^v}{2\pi} \frac{\tau\sqrt{1.5}}{1 + (2/3)(\tau_0/\tau)^2} \exp\,(\hbar\omega_\pm/2kT)\ |s\omega_\pm|\, K_1\,(|s\omega_\pm|)\ ,$$

with $\tau_0 = \hbar/2kT$ and $s = (1.5\tau^2 + \tau_0^2)^{1/2}$. The function $\Gamma^v(\omega)$ is represented by the standard BC shape if multipole-induced profiles are to be modeled, and the K0 profile if overlap-induced components are to be modeled. While it is not clear whether the profiles, Eq. 6.85, satisfy the new detailed balance condition, Eq. 6.73, computed profiles have been consistent with exact quantum profiles obtained by numerical computations. The equations defining the frequency shifts ω_1 and ω_2 were inspired by similar relationships that are well known from the theory of pressure broadening of spectral lines.

6.4 Dipole autocorrelation function

The spectral density is given by Eq. 6.51. For a specific induction mechanism, $(c) = \lambda_1\lambda_2\Lambda L$, the correlation function may be written (Birnbaum, Guillot and Bratos 1982)

$$\langle \boldsymbol{\mu}(0) \cdot \boldsymbol{\mu}(t) \rangle_{(c)} = \sum_{ss'} P_s \; |\mu_{ss'}|^2 \; \exp(i\omega_{ss'}t) \; C_{(c)}(t) \,, \tag{6.86}$$

where $C_{(c)}(t)$ is the translational correlation function, Eq. 6.10 and the s, s' designate the set of the molecular quantum numbers of initial and final state. If one sets $C_{(c)}(t) \equiv 1$, Eq. 6.86 becomes the correlation function for free molecular rotations. For the rototranslational bands in the isotropic potential approximation, the correlation function, Eq. 6.86, is therefore a product of the correlation function of translation and free rotational motion. Alternatively, Eq. 6.86 may be regarded as the sum of correlation functions for each of the uncoupled lines of a band [36, 38].

6.5 Intercollisional dip

An unexplained 'component X' near the H_2 $Q_1(j)$ vibration frequency was discovered experimentally in 1950 in the collision-induced fundamental absorption band of hydrogen in hydrogen–rare gas mixtures [129]. Subsequent studies suggested that the feature should actually be considered a splitting of the Q branch due to a striking decrease of absorption near the line center; the main features (e.g., pressure and temperature dependence) of what is now generally called the intercollisional (absorption) dip have been investigated by Welsh and associates [120, 121, 175, 422]. Similar features were observed not only in hydrogen, but also in deuterium [335], deuterium–rare gas mixtures [305], and in nitrogen [336].

An explanation was offered by van Kranendonk many years after the experimental discovery. Van Kranendonk argued that anticorrelations exist between the dipoles induced in subsequent collisions [404], Fig. 3.4. If one assumed that the induced dipole function is proportional to the intermolecular force – an assumption that is certainly correct for the *directions* of the isotropic dipole component and the force, and it was then thought, perhaps even for the dipole *strength* – an interference is to be expected. The force pulses on individual molecules are correlated in such a way that the power spectrum of the net force goes to zero at zero frequency shift [328]. In other words, the autocorrelation function has a negative tail of an area comparable to that under the positive peak about zero lag. At low densities the characteristic relaxation time of the negative tail is of the order of the mean time between collisions so that the dip in the power spectrum (and thus of the observable spectrum) is broader at higher densities. Van Kranendonk expressed the line shape function in the region of the dip as

$$J(\omega) = \frac{1 - \gamma + \omega^2 \tau_c^2}{1 + \omega^2 \tau_c^2} \; G_{\text{intra}}(0) \,, \tag{6.87}$$

which may be considered an inverted Lorentzian (a 'dip' of absorption). In this expression, $1/\tau_c$ is the collision frequency and $G_{\text{intra}}(0)$ is the value the intracollisional profile would assume at zero frequency shift in the absence of intercollisional interference. The ratio of $G(0)$ to $G_{\text{intra}}(0)$ equals $1 - \gamma$. A dip that goes exactly to zero would be given by $\gamma = 1$. The theoretical profile was experimentally verified (Welsh 1972); γ values near unity were found in several cases.

Subsequently, a kinetic theory of intercollisional interference was developed (Lewis 1980 and 1985). The kinetic theory was based on the idea of pairwise additivity of intermolecular force and induced dipole moment. It traces the collisional history of an individual molecule of a highly diluted system. The traced molecule may be a vibrating molecule, surrounded by non-vibrating molecules, or else a dissimilar molecule of low concentration (gas mixtures).

Line shape. A theory of the intercollisional line shape was given in Chapter 5 for the case of gas mixtures A and B. The simplifying assumption was made that the masses $m_A \ll m_B$ be very different. For unmixed gases and for mixtures of gases of comparable mass, this assumption is not tenable. We will simply quote here the results derived from the kinetic theory without introducing a vanishing mass ratio m_A/m_B (Lewis 1985). If the dilute (or vibrating) species is labeled A, the other B, the principal result of the kinetic theory is given by [235]

$$J(\omega) = \overline{\nu_c} \int_0^\infty \mathrm{d}v' \, P(v') \left[K(v') - 2\Re \left\{ \frac{\nu_c(v')}{\nu_c(v') + i\omega} A(v') \, Q_c(\omega, v') \right\} \right] . \tag{6.88}$$

Here, $\overline{\nu_c}$ is the mean collision frequency of molecules of the type A with B; $\nu_c(v)$ is the collision frequency of molecule A with A's initial speed being v; $P(v)$ is the probability density for AB collisions with A's initial speed being v; $K(v)$ is the intracollisional spectrum at zero frequency, per collision, with initial speed v; $A(v)$ is the mean projection of the integrated dipole moment induced by collisions with initial speeds v on the velocity

v of A before the collision; and $Q_c(\omega, v)$ is the solution to the integral equation

$$Q_c(\omega, v) = A(v) + \int_0^\infty dv' \frac{\nu_c(v')}{\nu_c(v') + i\omega} \Delta(v'|v)\, Q_c(\omega, v') . \qquad (6.89)$$

The kernel $\Delta(v'|v)$ is a measure of the persistence of A's velocity in a collision with initial and final speeds, v and v', respectively. The expressions for K, A and Δ are given in [235]. Here, we mention that with a number of simplifying assumptions, such as velocity-independent collision frequencies, etc., the equivalent of Eq. 6.87, can be obtained from Eq. 6.88; computations based directly on Eq. 6.88 and realistic potential and dipole models have never been attempted.

The development of the kinetic theory of intercollisional interference has stimulated extensive experimental, computational and other theoretical work which we will briefly consider (Lewis 1985).

The dip in the Q branches of the H_2 fundamental band was studied in hydrogen–rare gas mixtures in the gas phase [245], and also in the liquid phase, slightly above the critical points. The results were compared with predictions of the kinetic theory, based on the hard sphere model for the calculation of the collision frequencies and the persistence of velocity. Good agreement of the theoretical predictions and the measurements was observed at small gas densities where the observed half-widths depended linearly on density; at higher densities and in the liquid poor agreement was observed – above roughly 700 amagats for hydrogen–helium mixtures, and above 250 amagat for hydrogen–xenon mixtures, with intermediate values for argon and krypton. (At high densities, the observed half-widths were greater than the prediction based on the kinetic theory.) Moreover, the persistence of velocity effect was found to be significant for the hydrogen–helium mixtures. The most important predictions of the kinetic theory were thus verified at the lower gas densities.

The intercollisional dip was seen most strikingly in the Q branch of various vibrational bands. To a large extent, it arises from the isotropic dipole component ($\lambda_1\lambda_2\Lambda L$ = 0001). The dip is also seen in the far infrared, in the purely translational band of helium–argon mixtures, see Fig. 3.5 (lower panel). However, a substantial dip was observed in the far infrared even in unmixed hydrogen, Fig. 3.5 (upper panel), where the $\lambda_1\lambda_2\Lambda L = 0001$ component of the induced dipole moment does not exist. The intercollisional effect seen in the translational band of pure hydrogen must be due to other dipole components, such as the 0223/2023 and 0221/2021 components whose main spectral manifestations are the rotational $S_0(j)$ lines. Apparently, for the orientational transitions at zero frequency, the $Q_0(j)$ lines, intercollisional interference is strong [130].

Intercollisional dips were also observed in the S lines of the vibrational bands. A weak dip in the $S_1(1)$ line was observed in mixtures of hydrogen with helium [188], hydrogen with neon [337], and HD in neon [327]. Systematic studies in hydrogen–helium with sufficient resolution to show the variation of the dip width with pressure were communicated by Poll, Hunt and Mactaggert [316]. These authors point out that the S_1 lines arise from the $\lambda L = 23$ and 21 dipole components, while the Q line arises from the $\lambda L = 01$ component. The $\lambda L = 23$ dipole component is strongly orientation dependent and this fact must reduce the anticorrelation of the dipoles induced in subsequent collisions; dipole components with $L > 1$ may be expected to show much weaker intercollisional interference than those with $L = 1$ [234]. The authors therefore have reasoned that the $\lambda L = 21$ dipole component, which is of similar functional form to the $\lambda L = 01$ component, should be mainly responsible for the observed intercollisional interference. Since the spectral intensities arising from the $\lambda L = 21$ component amount to just a few percent of the total S line intensity, the dips should be shallow but otherwise much like those seen in the Q branch, in agreement with the observations [316]. For the translational effect shown in the upper part of Fig. 3.5, however, the $\lambda L = 21$ cannot possibly be responsible for the strong observed dip at zero frequency; for purely orientational transitions the argument advanced above for the weakness of the $L > 1$ component may not be valid if the measurement is correct [130].

A basic assumption of the kinetic theory was originally that the binary mixture is dilute in one component. In more recent years, the theory has been extended to mixtures of arbitrary composition [117].

6.6 Interference of allowed and induced lines

HD–X pairs. In general, allowed lines are many orders of magnitude more intense than the induced lines. Any interference of allowed and induced lines is, therefore, not easy to detect. The case of HD is, however, exceptional in that the allowed dipole of HD is weak, roughly of the same strength as the induced dipole. As a consequence, striking Fano profiles have been seen in unmixed deuterium hydride, and in deuterium hydride mixed with a rare gas (as in Fig. 3.36), which are shaped by such an interference. A successful analysis of the observed profiles will not only give the desired information concerning the magnitude of the permanent dipole moment. It should also provide some significant information on the role of the anisotropy of the HD–X interactions. Unlike the collision-induced profile, which in the low-density limit is independent of density variation if normalized by density squared, the observed Fano profiles vary with density. As a consequence, pressure broadening and the complexity of anisotropic interactions enter into any analysis of such spectra (Tabisz *et al.*, 1990).

The induced dipole moment of the HD–X systems, with X = He, Ar, H_2, HD, is well known from the fundamental theory, for the purely rotational bands and also for most fundamental bands [59]. To the induced dipole, the permanent dipole moment of HD has to be added vectorially, accounting for the linear variation with density which differs from the density variation of the induced dipole components [391]. According to the theory of intracollisional interference (as the process was called, to be distinguished from the intercollisional interference considered elsewhere in this monograph), interference occurs for those induced components that are of the same symmetry as the allowed dipole, namely $\lambda_1\lambda_2\Lambda L = 0110$ and 1010 [178, 179, 321, 389]. These induced components are always parallel or antiparallel to the allowed dipole, causing constructive or destructive interference.

In the framework of the impact approximation of pressure broadening, the shape of an ordinary, allowed line is a Lorentzian. At low gas densities the profile would be sharp. With increasing pressure, the peak decreases linearly with density and the Lorentzian broadens in such a way that the area under the curve remains constant. This is more or less what we see in Fig. 3.36 at low enough density. Above a certain density, the $R_1(0)$ line shows an anomalous dispersion shape and finally turns upside down. The asymmetry of the profile increases with increasing density [258, 264, 345]. Besides the $R_1(j)$ lines, we see of course also a purely collision-induced background, which arises from the other induced dipole components which do not interfere with the allowed lines; its intensity varies as density squared in the low-density limit. In the $Q_1(j)$ lines, the intercollisional dip of absorption is clearly seen; at low densities, it may be thought to arise from three-body collisional processes. The spectral moments and the integrated absorption coefficient thus show terms of a linear, quadratic and cubic density dependence,

$$\int \alpha(\omega)(1/\omega)\,\mathrm{d}\omega = C_1\varrho + C_2\varrho^2 + C_3\varrho^3 + \ldots$$

In a recent review, Tabisz *et al.* (1990) summarize the state of our understanding of the deuterium hydride spectra obtainable with and without admixtures X.

i.) The measured value of the permanent dipole moment of HD shows a possible j dependence not predicted by the *ab initio* calculations.

ii.) Characterization of the density and temperature dependence of the absorption coefficient and spectral line parameters is reasonably complete. The integrated absorption coefficient has a more complicated behavior than predicted by present theory. The temperature dependence of the interference parameter C_2/C_1, particularly its change of sign for HD–HD, cannot be explained by a theory in which temperature enters only through the pair distribution function.

iii.) The temperature dependence of the line widths agrees with the predictions of the impact theory in the few quantitative assessments made to date. The frequency shifts are in qualitative agreement with theory.

iv.) In general, the HD–inert gas results conform better with present theory than do the pure deuterium hydride spectra.

More experimental and theoretical work is required, especially under conditions where the interference parameter C_2/C_1 shows strong temperature variations for a number of lines.

6.7 General references

See also the references given at the end of the previous Chapter.

L. C. Biedenharn and J. D. Louck, *Angular Momentum in Quantum Physics*, vol. 8 and *The Racah–Wigner Algebra in Quantum Theory*, vol. 9 of the *Encyclopedia of Mathematics and its Applications*, Addison-Wesley, Reading, Massachusetts (1981).

G. Birnbaum, S. I. Chu, A. Dalgarno, L. Frommhold, and E. L. Wright, Theory of collision-induced translation-rotation spectra: H_2–He, *Phys. Rev.* **A 29**, (1984) 595.

G. Birnbaum, B. Guillot and S. Bratos, Theory of Collision-induced Line Shapes – Absorption and Light Scattering at Low Densities, in *Advances in Chemical Physics*, vol. 51, I. Prigogine and S. A. Rice, eds., Wiley, New York, 1982.

P. R. Bunker, *Molecular Symmetry and Spectroscopy*, Academic Press, New York and London, 1979.

N. Halberstadt and K. C. Janda, eds., *Dynamics of Polyatomic van der Waals Complexes*, Plenum, New York, 1990.

J. van Kranendonk, Intermolecular Spectroscopy. *Physica* 73:156, 1974.

A. Weber, ed., *Structure and Dynamics of Weakly Bound Molecular Complexes*, NATO ASI Series C, vol. 212, D. Reidel, Dordrecht, 1987.

H. L. Welsh, Pressure-induced Absorption Spectra of Hydrogen, in *Spectroscopy*, MTP Int. Rev. Sci., Phys. Chem., Series one, vol. 3, A. D. Buckingham and D. A. Ramsay, eds., Butterworths, London, 1972.

R. N. Zare, *Understanding Spatial Aspects in Chemistry and Physics*, Wiley Interscience, New York (1988).

7
Related topics

In this Chapter, we will briefly look at a number of topics related to collision-induced absorption of infrared radiation in gases. Specifically, in Section 7.1, we consider collision-induced spectra involving *electronic* transitions in one or more of the interacting molecules. In Section 7.2, we focus on collision-induced light scattering, which is related to collision-induced absorption in the same way that Raman and infrared spectra of ordinary molecules are related. The collision-induced Raman process arises from the fact that the polarizability of interacting atoms/molecules differs from the sum of polarizabilities of the non-interacting species. Closely related to the collision-induced Raman and infrared spectroscopies are the second (and higher) virial coefficients of the dielectric properties of gases, which provide independent measurements of the collision-induced dipole moments, Section 7.3. Finally, we look at the astrophysical and other applications of collision-induced absorption in Sections 7.4 and 7.5.

7.1 Collision-induced electronic spectra

Collision-induced electronic spectra have many features in common with rovibrotranslational induced absorption. In this Section, we take a look at the electronic spectra. We start with a historical note on the famous forbidden oxygen absorption bands in the infrared, visible and ultraviolet. We proceed with a brief study of the common features, as well as of the differences, of electronic and rovibrotranslational induced absorption. Recent work is here considered much of which was stimulated by the advent of the laser – hence the name laser-assisted collisions. The enormous available laser powers stimulated new research on laser-controlled, reactive collisions and interactions of supermolecules with intense radiation fields. In conclusion, we attempt a simple classification of various types of electronic collision-induced spectra.

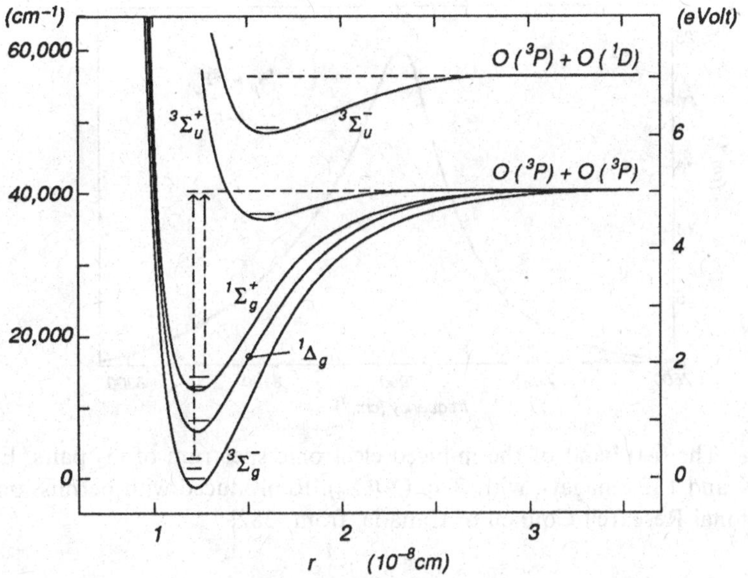

Fig. 7.1. The low-lying electronic states of the O_2 molecule. The lowest vibrational levels of each electronic state are also indicated; after [180].

1 Early observations

In 1885, Janssen [197, 198, 199, 200] found that in oxygen at pressures of tens or hundreds of atmospheres new absorption bands appear which are unknown from such studies at atmospheric pressures. The associated absorption coefficients increase as the *square* of density, in violation of Beer's law (pp. 6ff.). The observed quadratic dependence suggests an absorption by *pairs* of molecules; Beer's law, in contrast, attempts to describe absorption by individual molecules.

Interaction-induced absorption (as the new features were called early on [353]) has stimulated considerable interest. For a long time, explanations were attempted in terms of weakly bound 'polarization' molecules (that is, van der Waals molecules, $(O_2)_2$), but some of the early investigators argued that unbound *collisional pairs* might be responsible for the observed absorption. For example, Ångström suggested a free-pair origin of the absorption in 1908; Steiner, Salow and Finkelnburg early on suggested that the absorption was by free pairs [353, 147]. More recently, a study of the temperature dependence of the induced intensities has provided evidence for the significance of collisional complexes [382, 383, 260]. Nevertheless, it seems fair to say that the idea of absorption by unbound molecular pairs was not widely accepted for many decades.

It is well known that, besides the electronic ground state, $X\ ^3\Sigma_g^-$, of the O_2 molecule, there are two low lying excited electronic states labeled

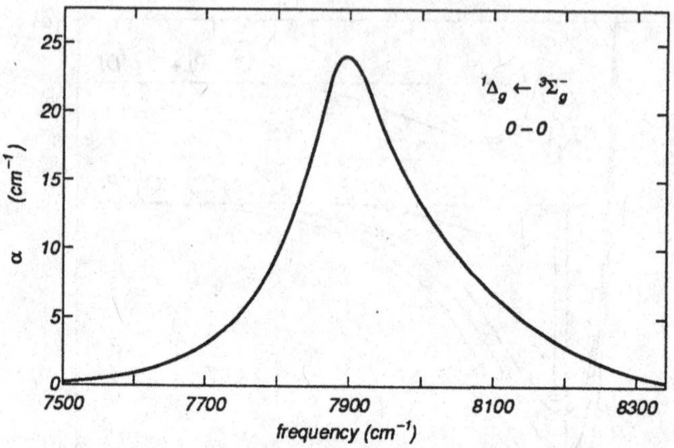

Fig. 7.2. The 0–0 band of the induced electronic spectrum of O_2 pairs, Eq. 7.1, at 297 K and 132 amagats, with $\mathscr{X} = O_2(^3\Sigma_u)$. Reproduced with permission from the National Research Council of Canada, from [382].

$a\,^1\Delta_g$ and $b\,^1\Sigma_g^+$. Each of these is associated with a number of vibrational levels; the ones with the lowest energies are sketched in Fig. 7.1. For the $a\,^1\Delta_g$ and $b\,^1\Sigma_g^+$ states, the vibrational ground states are about 1 and 1.6 eV above the vibrational ground state of $X\,^3\Sigma_g^-$, respectively. In the isolated molecule, transitions from the electronic groundstate to either excited electronic state are forbidden by both spin and electric dipole selection rules. The absence of atmospheric absorption in the wavelength regions near 1260 and 761 nm (which correspond to photon energies of 1 and 1.6 eV, respectively) is thus understandable.

At not too high temperatures, all O_2 molecules are in the electronic ground state. Under such conditions, absorption by *pairs* of O_2 molecules may be symbolically represented by schemes like

$$\begin{aligned}
\left(X\,^3\Sigma_g^-\,;\,a\,^1\Delta_g\right) &\leftarrow \left(X\,^3\Sigma_g^-\,;\,X\,^3\Sigma_g^-\right) \quad (1260\text{ nm}) \qquad (7.1)\\
\left(X\,^3\Sigma_g^-;\,b\,^1\Sigma_g^+\right) &\leftarrow \left(X\,^3\Sigma_g^-\,;\,X\,^3\Sigma_g^-\right) \quad (760.6\text{ nm})\\
\left(a\,^1\Delta_g\,;\,a\,^1\Delta_g\right) &\leftarrow \left(X\,^3\Sigma_g^-\,;\,X\,^3\Sigma_g^-\right) \quad (629.9\text{ nm})\\
\left(b\,^1\Sigma_g^+\,;\,b\,^1\Sigma_g^+\right) &\leftarrow \left(X\,^3\Sigma_g^-\,;\,X\,^3\Sigma_g^-\right) \quad (380.3\text{ nm})
\end{aligned}$$

etc. The approximate wavelengths λ given in parentheses to the right of each transition correspond to the 0–0 band which does not involve vibrational transitions of the molecules of the complex. These bands are in the infrared ($\lambda > 770$ nm), visible, and ultraviolet ($\lambda < 390$ nm) regions of the electromagnetic spectrum.

Figure 7.2 shows the measured absorption profile near the wavelength of 1260 nm (that is, near the frequency of $\approx 7900\ cm^{-1}$); it has the form of a very diffuse, asymmetric line. Other absorption bands are related to different transitions, Eqs. 7.1, and show similar contours [382]. Besides these, the so-called 1–0 (and higher) bands occur which involve vibrational transitions in one or both molecules; these are shifted to higher frequencies by the vibrational frequency, about 1556 cm^{-1}, or by multiples thereof.

The absorption coefficients of the 1260 and 761 nm bands were shown to vary accurately as the square of oxygen density if these are studied in pure oxygen. On the other hand, if large amounts of argon or nitrogen were added to a fixed, small amount of oxygen, several of the observed absorption coefficients of these bands vary linearly with foreign gas density. This behavior suggests that for the oxygen molecule that remains in the ground state, Eqs. 7.1, a foreign gas particle $\mathscr{X}$ may be substituted,

$$\left(\mathscr{X};\, a\,{}^1\Delta_g\right) \;\leftarrow\; \left(\mathscr{X};\, X\,{}^3\Sigma_g^-\right) \quad (1260\ \text{nm})$$

$$\left(\mathscr{X};\, b\,{}^1\Sigma_g^+\right) \;\leftarrow\; \left(\mathscr{X};\, X\,{}^3\Sigma_g^-\right) \quad (760.6\ \text{nm})$$

for example, $\mathscr{X} = N_2$, Ar, or almost anything else. For collisional induction of these bands a foreign molecule is more or less as expedient as an O_2 molecule. The specific properties of the collisional partner $\mathscr{X}$ hardly matter as long as it is not absent.

The 630 nm band, on the other hand, shows no significant intensity variation when the foreign-gas density is varied; an oxygen density squared behavior is observed regardless of the presence and concentration of the admixtures. This band arises from *simultaneous* transitions of both molecules of the O_2–O_2 complex. In this case, an O_2 molecule cannot be replaced by a foreign gas particle without losing the band, or perhaps shifting it to a different part of the spectrum.

The induced bands mentioned are known from spectral studies with liquid, solid and compressed gaseous oxygen. Condensed oxygen is blue, because of the red 0–0 absorption bands near 760 and 630 nm, and their 1–0 counterparts, that is, because of interaction-induced absorption.

We note that the induced absorption bands mentioned are associated with single or double *electronic* transitions. Such induced bands are not limited to oxygen; similar bands of a few other systems (benzene, for example) were known for some time. However, most of the common diatomic molecules have electronic states that are many eV above the ground state. As a consequence, for those molecules, electronic induced transitions commonly occur in the ultraviolet region of the spectrum where interference from allowed electronic spectra may be strong. Electronic induction is not nearly as common as rovibro-translational induced absorption.

2 Theory

A general, quantitative theory of the line shapes associated with electronic collision-induced absorption and emission has been advanced by Gallagher and Holstein [155]. Results were compared with the existing measurements.

Specifically, the collision-induced absorption and emission coefficients for electric-dipole forbidden atomic transitions were calculated for weak radiation fields and photon energies $\hbar\omega$ near the atomic transition frequencies, utilizing the concepts and methods of the traditional theory of line shapes for dipole-allowed transitions. The example of the S–D transition induced by a spherically symmetric perturber (e.g., a rare gas atom) is treated in detail and compared with measurements. The case of the radiative collision, i.e., a collision in which both colliding atoms change their state, was also considered.

The theoretical results were compared with the existing measurements. The features of the measured profiles were generally reproduced by the computations. Certain failures of the detailed calculations do exist which to a large extent are due to the existing uncertainties in the interatomic potentials used in the calculations. Only the leading long-range terms proportional to the inverse separation to the sixth power were utilized as potentials in most calculations; where better potentials existed, a much better agreement of theory and measurements was usually observed [155]. However, the neglect of rotational mixing had a strong influence, especially in the wings of the spectral profiles of systems involving molecules. It is clear that significant improvements of such calculations would be possible by the use of improved interaction potentials. Furthermore, the induced dipole transition moments are likely to differ from their long-range form employed in the calculations.

3 Laser-assisted collisions

The advent of the laser has stimulated new research in collisional physics. The term laser-assisted collision was coined to describe the various research activities that have evolved. These studies are concerned with electronic transitions in supermolecules and will be briefly considered here. Similarities between laser-assisted collisions (LAC) and collision-induced absorption (CIA) exist, both in the types of phenomena considered and in the calculations of the spectra [208].

A more general term for laser-assisted collision is photon-assisted collision (PAC), since it is not always necessary that *lasers* provide the light. Both photon-assisted collisions and collision-induced absorption may be symbolically written inthe form of a reaction equation like

$$A + B + \epsilon + \hbar\omega \rightarrow A' + B' + \epsilon' , \tag{7.2}$$

where ϵ represents translational energy. A and B are typically neutral atoms or molecules, but interesting cases where A represents an ion or an electron are known. The basic difference between photon-assisted collisions and collision-induced absorption is that the latter [406, 45, 47] has concentrated on transitions which involve translational, rotational ($J \rightarrow J'$), and vibrational ($v \rightarrow v'$) states of the electronic ground state of the interacting pair, but not of the electronic state. Such transitions normally occur in the microwave or infrared region of the spectrum, although certain induced spectra involving higher overtones or double transitions may be found in the visible region. On the other hand, photon-assisted collision processes involve transitions between electronic states in one or both of the interacting atoms [427, 285, 206], in addition to a change of translational energy, and possibly also rotovibrational states if molecules are involved. A change of electronic state normally requires a photon of optical frequency but we note that, for example, photon-assisted collisions between highly excited Rydberg states have been observed in the microwave region of the spectrum [156].

Like the collision-induced rovibrotranslational spectra, the photon-assisted collision spectra consist of diffuse wings near a transition frequency of A or B; the line widths $\Delta\omega$ reflect the short lifetime, $\Delta t \approx 10^{-12}$ s, of the transient supermolecule. We note that one or both of the species A and B in Eq. 7.2 can change its state. A photon-assisted collision can even be reactive, so that the nature of the species changes, as in a charge transfer process. The more common occurence in photon-assisted collisions is a collision in which both A and B undergo a transition (double transition). In that case new, diffuse lines are observed at sums or differences of transition frequencies of A and B. Examples are the LICET (laser-induced collisional energy transfer) collisions [427, 26], and pair transitions, in which two excited states are produced by absorption of a single photon from the electronic ground state. Pair transitions have been observed in metal vapors (see references listed in [206]) and these contribute to the opacity in the visible of dense gases. A double optical transition causes the blue color of liquid oxygen.

The right-hand side of Eq. 7.2 may alternatively contain a molecule, AB, instead of a free pair, $A + B$; such processes are sometimes referred to as half-collisions.

Collision-induced spectra are usually studied at high densities, indeed often at densities that are so high that many-body interactions (as opposed to just binary interactions, Eq. 7.2) dominate the spectra [45, 47]. For example, liquids have strong many-body components [364, 376]. On the

Table 7.1. Comparison of laser-assisted collisions (LAC) and collision-induced absorption (CIA) [208].

	laser-assisted collision (LAC)	collision-induced spectra (CIA)
process:	$A + B + \hbar\omega \rightarrow A^* + B + \Delta\epsilon$	$A_{vJ} + B + \hbar\omega \rightarrow A_{v'J'} + B + \Delta\epsilon$
	also reactive collisions;	also many-body collisions;
	also double excitations (of *A and B*)	
spectral range:	(mostly) visible	(mostly) infrared, microwave, also Raman
gas density:	$< 10^{19}$ cm^{-3}	$> 10^{19}$ cm^{-3}
transition element:	$\langle i \vert \boldsymbol{\mu}^I \vert f \rangle$	
induced dipole:	$\mu^I \sim R^{-3}, R^{-4} \ldots$	$\mu^I \sim R^{-4}, R^{-5} \ldots$
	also multipole induction modified by overlap	
	also optical collisions and redistribution	also interference of allowed and induced dipoles
theory:	classical path or quantum treatment	
	intense field effects	also molecular dynamics
applications:	laser physics	spectra of dense fluids
	collision dynamics probe and reaction control	stellar and planetary atmospheres

other hand, photon-assisted collisions are produced from, or result in excited electronic states, and are normally studied at much lower density, $n \ll 1$ amagat. As a consequence, collision-induced rovibrotranslational absoption is readily observed. For photon-assisted collisions, on the other hand, the spectral profile must often be observed by product fluorescence, e.g., by detecting a *product* of the reaction.

Both photon-assisted collisions and collision-induced absorption deal with transitions which occur because a dipole moment is induced in a collisional pair. The induction proceeds, for example, via the polarization of B in the electric multipole field of A. A variety of photon-assisted collisions exist: for example, the above mentioned LICET or pair absorption process, or the induction of a transition which is forbidden in the isolated atom [427]. All of these photon-assisted collision processes are characterized by long-range transition dipoles which vary with separation, R, as R^{-n} with $n = 3$ or 4, depending on the symmetry of the states involved. Collision-induced spectra, on the other hand, frequently arise from quadrupole ($n = 4$), octopole ($n = 5$) and hexadecapole ($n = 6$) induction, as we have seen. At near range, a modification of the inverse power law due to electron exchange is often quite noticeable. The importance of such overlap terms has been demonstrated for the forbidden oxygen $^1S \to {}^1D$ emission induced by collision with rare gases [206] and in collision-induced absorption studies (Chapters 4 through 6).

A laser-assisted collision in which one atom changes state in a transition which is dipole allowed for the separated atom is called an optical collision. This corresponds ordinary line broadening if the laser is not too intense, and gives rise to the usual spectral profile with a Lorentzian core due to multiple collisions and with quasi-static or anti-static wings due to single collisions. The methods for treating the far wings in terms of binary interactions are identical to those for treating collision-induced transitions; see, for example, Section 4 of [45].

The computations of the line shapes of photon-assisted collisions are quite similar as described above in Chapters 5 and 6. Time dependent, semi-classical theories may be used or full quantum mechanical treatments. The initial and final states $|i\rangle$, $|f\rangle$, of a supermolecule are usually represented as a product of a translational function, $\psi(\boldsymbol{R})$, and molecular functions, ψ_A and ψ_B. Semi-classical theories based on an atomic analysis of the long range potentials and induced dipoles give a good qualitative understanding of the spectral profiles, but generally do not suffice for fully quantitative studies. Fully quantum mechanical close coupling scattering treatments have been applied to photon-assisted collisions [206]. These require accurate knowledge of the intermolecular potentials, couplings, and induced dipoles, i.e., information that is often not available. The semi-classical long-range analysis is usually all one can do. The much longer range of the interatomic potential typically associated with excited state processes permits the qualitative features of spectral line shapes for photon-assisted collisions to be reproduced from the long-range parts of the potential and induced transition dipoles. To a molecular spectroscopist, the concept of a photon-assisted collision is really nothing new; the resulting molecular continua can be interpreted in terms of Franck–Condon factors and transition dipoles.

Experimental data of photon-assisted collisions are much sparser than those of collision-induced absorption. To some extent, this fact may be due to the greater difficulty in working with excited states [176]. There was an early hope that collision-induced emission or LICET processes could result in new laser sources. When these methods did not work well for various reasons, some of the interest in these phenomena waned.

One other area of interest in photon-assisted collisions concerns intense fields [9, 427], which are not well understood [26].

In collisions of ultra-cold atoms, intense field effects are expected to be important. Since experimentally temperatures of the order of 0.0001 K have been achieved, and since processes associated with resonance line broadening play an important role in optical trap dynamics, it will be necessary to examine the role of light interacting with very cold colliding atoms [207]. Ultra-cold collision dynamics will be modified by intense laser radiation, since $\boldsymbol{\mu}\cdot\boldsymbol{E}$ in an optical trap can be orders of magnitude larger than kT. Collision duration is so long that spontaneous emission during the collision will affect the collision dynamics [25].

4 *Classification of electronic transitions*

In a reaction like Eq. 7.2, one generally distinguishes several types of processes. For example, a transition may take place in just one of the interacting partners, or in two (or more).

Single transitions. If the photon energy, $\hbar\omega$ in Eq. 7.2, is *roughly* resonant with the energy difference between two eigenstates of one partner A, the excitation may occur if the energy difference can be made up by the energy of relative motion.

Forbidden transitions. If the transition is dipole forbidden in the isolated molecule, an induced electronic absorption line or spectrum may appear as was seen above in oxygen and in mixtures of oxygen with other gases, p. 357ff.

Allowed transitions. If the transition is dipole allowed, we have a line broadening collision. These are often studied in emission, i.e., a photon $\hbar\omega'$ appears on the right-hand side of Eq. 7.2, but otherwise the physics is the same.

If no collision interferes with the emission process, the line is very sharp. In the presence of collisions, on the other hand, the line will be broadened. The intensity at the line center may rise by several orders of magnitude, depending on various factors, such as the presence of unresolved hyperfine structure components, etc. [119]. The core of the line still remains sharp; the collisional broadening is most striking in the wings, especially the far wings where the intensities have fallen off by orders of magnitude.

At small frequency shifts, $|\Delta\nu| = |\nu - \nu_0| \ll \nu_c$ (shift $\Delta\nu$ relative to the line center ν_0; ν_c is the collision frequency), the shape of the line reflects the effects of long-range collisions that perturb the emitter just slightly. The effects discernible at the core of the line are subtle. In the extreme wing, $\Delta\nu \gg \nu_c$, on the other hand, the dynamics of near central collisions shape the profile. Strong temperature dependences are here observed and valuable information concerning the interaction potential can often be obtained from the analysis. The quasi-static theory is used to compute the wings of the profiles. Besides pressure broadening, certain 'satellite' structures are often seen in the wings [154].

Collisional redistribution of radiation. A system $A + B$ of two atoms /molecules may be excited by absorption of an off-resonant photon, in the far wing of the (collisionally) broadened resonance line of species A. One may then study the radiation that has been redistributed into the resonance line – a process that may be considered the inverse of pressure-broadened emission. Interesting polarization studies provide additional insights into the intermolecular interactions [118, 388].

Double transitions. Simultaneous transitions in two interacting atoms /molecules occur if the incident photon is *roughly* resonant with the energy difference between an eigenstate of species A and another one of species B; any energy defect is made up of the energy of relative motion. The process has been referred to by several names, including light-induced collisional energy transfer (LICET), radiative collision, and radiatively aided inelastic collision (RAID). They are the electronic analog of the simultaneous transitions discussed in Chapters 3–6 above [155].

Reactive collisions. In reactive collisions chemical bonds form while existing bonds may be broken; electronic transitions take place generally in more than two of the interacting atoms/radicals [82]. A transition occurs from an initial reagent state to a final product state. One may be tempted to talk about a 'transition state' (i.e., the state between reagent and product state – if such a well defined state exists), but in most cases it would be more realistic to talk about the 'transition region species'. Spectroscopic methods have been developed that probe the transition region species and are amenable to a more or less rigorous theoretical analysis. Detailed studies of such transition region species may be expected to reveal significant facts concerning the chemical dynamics. The time scales of reactive collisions will often be the same as in collision-induced absorption, that is the duration of a fly-by collision.

A variety of process types exist. One of the best studied processes is the photon-assisted association,

$$A + B + \hbar\omega \rightarrow AB^* ,$$

where the ground state species AB is unstable. Photon-assisted bimolecular reactions are also well known,

$$A + BC + \hbar\omega \rightarrow AB + C .$$

These and a few other processes were reviewed by Brooks [82].

7.2 Collision-induced light scattering

An induced Raman process exists which is related to collision-induced absorption like the Raman spectra of ordinary molecules are related to the infrared rotovibrational spectra.

Collision-induced light scattering is a supermolecular Raman process that arises from the interaction-induced variations of the polarizability of a binary, ternary, . . . , clusters of atoms or molecules [60]. At low density, the pair properties that give rise to the collision-induced Raman spectra are directly measurable. Non-reactive atoms or molecules in thermal (i.e., non-violent) collisions largely preserve their character throughout the encounter. The collisional complex may be viewed as a *supermolecule* which for the duration of the interaction possesses properties which differ from the sum of properties of the participating, well separated molecules. Specifically, the total polarizability of the complex differs slightly from the sum of the permanent polarizabilities of its parts, for various reasons. For example, in the presence of a polarized neighbor, local electric fields typically differ from the externally applied field of the light wave. Since local field strengths vary with the laser field and yet differ from these, one can describe the situation classically in terms of an interaction-induced polarizability tensor. Furthermore, at near range, exchange and dispersion forces cause a temporary rearrangement of electronic charge that affects the 'permanent' polarizabilities of the collisional partners. With increasing separation, the induced polarizability increment falls off to negligibly small values.

For example, in the case of two interacting atoms, an induced optical anisotropy of the supermolecule arises, given by

$$\beta(R_{12}) = \alpha_{\parallel}(R_{12}) - \alpha_{\perp}(R_{12}) .$$

Here, $\alpha_\parallel$ and $\alpha_\perp$ are the polarizabilities of the diatom parallel and perpendicular to the internuclear separation, $\boldsymbol{R}_{12}$. The electrostatic theory accounts for the distortions of the local field by the proximity of a point dipole (the polarized collisional partner) and suggests that the anisotropy is given by $\beta(R_{12}) \approx 6\alpha_0^2/R_{12}^3$; α_0 designates the permanent polarizability of an unperturbed atom (assumed to be unaffected by intermolecular interactions). This is the so-called dipole-induced dipole (DID) model, which approximates the induced anisotropy of such diatoms often fairly well. It gives rise to pressure-induced depolarization of scattered light, and to depolarized, collision-induced Raman spectra in general.

The depolarization of light by dense systems of spherical atoms or molecules has been known as an experimental fact for a long time. It is, however, discordant with Smoluchowski's and Einstein's celebrated theories of light scattering which were formulated in the early years of this century. These theories consider the effects of fluctuation of density and other thermodynamic variables [371, 144].

The concept of supermolecules allows a rather rigorous treatment of the aspects of collision-induced light scattering in the binary regime. The frequency dependence of the scattered light is described by the differential photon scattering cross section per unit angular frequency and per pair,

$$\left.\frac{\partial^2\sigma}{\partial\Omega\,\partial\omega}\right|_{\hat{\boldsymbol{n}}_0\hat{\boldsymbol{n}}_S} = k_0\,k_S^3\,g(\omega;T)\,,$$

where $k_0 = 2\pi/\lambda_0$ and $k_S = 2\pi/\lambda_S$ are the magnitudes of incident and scattered wave vectors. The spectral function is given by the 'golden rule',

$$g(\omega;T) = \sum_{i,f} P_i(T)\ |\langle i\,|\hat{\boldsymbol{n}}_0\cdot\boldsymbol{\alpha}\cdot\hat{\boldsymbol{n}}_S|\,f\rangle|^2\ \delta(\omega_{if}-\omega)\,,$$

an equation very similar to the spectral density of the absorption profile, Eqs. 5.67 and 6.51 above. Here, $|i\rangle$ and $|f\rangle$ are initial and final states of the supermolecular complex; $\boldsymbol{\alpha}$ is the induced incremental polarizability (a second rank tensor); $\hat{\boldsymbol{n}}_0$ and $\hat{\boldsymbol{n}}_S$ are unit vectors in the direction of the electric polarization of incident and scattered waves which are often specified in the form of subscripts V and H, for vertical and horizontal polarization; $\omega_{if} = (E_f - E_i)/\hbar$ is the energy difference of initial and final state in units of angular frequency; $P_i(T)$ is the population of the initial state (a function of temperature); $\delta(\omega)$ is Dirac's delta function; and the summation is over all initial and final states of the collisional complex.

Several types of collision-induced light scattering spectra are known. We have already mentioned the depolarized translational spectra of rare gas pairs and bigger complexes which arise from the *anisotropy* of the diatom polarizability. Contrary to the infrared inactivity of like pairs, e.g., Ar–Ar; like pairs are Raman active. Furthermore, polarized translational spectra have also been observed. The polarized spectra arise from the *trace* of the diatom polarizability; polarized spectra are relatively weak but have been recorded for a number of like pairs. Various rotational and rotovibrational spectra that are Raman inactive in non-interacting molecules have also been observed, much as this was seen in collision-induced absorption. Moreover, induced Raman components have been seen in the far wings (i.e., in the higher spectral moments) of certain allowed transitions and are widely thought to be common; they are however weak and are not readily identified. Induction also gives rise to simultaneous transitions in interacting molecules. These occur at sums and differences of the rotovibrational frequencies of the molecules involved.

Collision-induced 'lines', and the induced far wings of allowed lines, are usually very diffuse, typically of a width of $\Delta\omega \approx 1/\Delta t \approx 10^{12}$ Hz reflecting the short duration ($\Delta t \approx \rho/v_{\text{rms}} \approx 10^{-12}$ s) of intermolecular interactions, just as was seen in the case of collision-induced absorption above. In the low-density limit, induced intensities vary as the *square* of the gas density, which reflects the supermolecular origin. With increasing density, cubic and higher components have been identified.

In the low-density limit, collision-induced light scattering arises from binary complexes. In this case, a fairly rigorous theory of induced polarizabilities, spectral moments and spectral line shapes exists. In recent years, many measurements have been shown to be in agreement with the fundamental theory. With increasing gas density, ternary and higher spectral contributions (which may be associated with negative integrated intensities) are superposed with the binary spectra and modify these to a greater or lesser extent. Work at moderate densities showing the onset of ternary (and sometimes of higher-order) contributions is also known.

Induced polarizability. Induced polarizability increments differ from zero for a variety of reasons. A most important influence is the deviation of the local electric field from the externally applied laser field if another polarized atom or molecule is in close proximity. The classical dipole-induced dipole model considers the interacting pair as point dipoles and leads to separation-dependent invariants of the incremental polarizabilities. For the more highly polarizable particles, the DID anisotropy model is fairly accurate. Even for the less polarizable atoms such as He and Ne, at most separations of interest, the DID component of the anisotropy is known to be more important than all other induced contributions combined. However, a quantum treatment of the interacting pair is indispensable and has pointed out various corrections to the DID model, especially for the induced trace; the shape and intensity of the far wings of depolarized collision-induced light scattering spectra also indicate the presence of significant non-classical electron exchange contributions to the induced anisotropy. Deviations from the DID model arise at near range (where the electron shells of interacting molecules overlap); electron exchange lowers the polarizability of the pair [195]. At distant range, dispersion forces enhance the polarizabilities [194]. If molecules are involved, collisional frame distortion [346], multipole- polarizabilities, etc., [91, 92, 193] may also be important influences. Nonlinear polarization in the presence of strong electric fields from the permanent multipole moments of neighboring molecules is also significant [87, 194, 226] because of the strong fields of surrounding molecules. Non-uniform fields from neighboring induced dipoles interacting with higher multipole polarizabilities contribute [92]. Knowledge of the irreducible components of the incremental polarizability of higher than binary complexes is sparse [311].

The spectra. Collision-induced light scattering spectroscopy is concerned with the frequency-, density-, polarization- and temperature-dependences of the supermolecular Raman process. Pure and mixed gases are considered, i.e., complexes of like and of dissimilar molecules, are of interest. Collision-induced light scattering spectra are described by a photon scattering cross section, σ, the spectral function, $g(\omega; T)$, or the Fourier transformation of a correlation function, $C(t)$. For the most part, collision-induced light scattering spectra are diffuse and show little structure. An intercollisional process is predicted but was never clearly seen in any existing measurement.

In gases near the low-density limit, at all frequencies where significant collision-induced light scattering intensities are observed, the spectral intensity is proportional to the *square* of the density, $g(\omega; T) \propto n^2$, in harmony with the bimolecular origin. Under such circumstances, the normalized spectra, $g(\omega; T)/n^2$, i.e., the *spectral shapes*, do not depend on density. It had been noticed in the first experimental studies [272, 273] that at higher densities the shapes of most collision-induced light scattering spectra vary with density. This fact reveals the presence of three-body and possibly higher collision-induced light scattering components. The onset of discernible many-body spectral components is best dealt with in the form of a virial expansion of the spectral moments [17, 18, 318] that at least in principle permits the separation of the binary, ternary etc., spectral components; virial expansions of spectral line shapes have also been reported recently [297]. Most interesting attempts have been made at intermediate densities to separate the ternary from the binary (and sometimes, from the higher) contributions [16, 17, 19, 20, 429].

We note that the theory of collision-induced, nonlinear Raman processes exists [212, 211] but experimental investigations are virtually non-existent.

7.3 Dielectric properties of gases

The dielectric properties of gases are closely related to collision-induced absorption. It is well known that collisions modify molecular properties. Specifically, we are here interested in the dipole moments induced by collisions (Chapter 4) which manifest themselves not only in collision-induced absorption, but also in the dielectric virial properties of gases.

The dielectric constant ε of a gas sample depends on the total dipole moment induced in response to an applied electric field. The dipole moment is the vector sum of the partially oriented permanent dipoles $\boldsymbol{\mu}_i$ which individual molecules i may possess, *plus* the field-induced dipoles $\boldsymbol{\alpha}_i \cdot \boldsymbol{E}$ arising from the polarizability $\boldsymbol{\alpha}_i$ of the molecules i, *plus* all interaction-induced dipoles $\boldsymbol{\mu}_{ik}$, *plus* the field-induced dipoles which arise from the interaction-induced polarizability $\boldsymbol{\mu}_{ik}$ [93]. The dielectric constant ε depends, therefore, on the density ϱ of the gas, according to

$$\frac{\varepsilon - 1}{\varepsilon + 2} = A_\varepsilon \varrho + B_\varepsilon \varrho^2 + C_\varepsilon \varrho^3 + \cdots \tag{7.3}$$

(Clausius–Mossotti equation). The A_ε, B_ε, ..., are the first, second, ..., dielectric virial coefficients, given by

$$A_\varepsilon = \frac{4\pi}{3} N_A \frac{1}{4\pi\varepsilon_0} \left(\alpha_0 + \frac{\mu^2}{3kT} \right) \tag{7.4}$$

$$B_\varepsilon = \frac{4\pi}{3} N_A^2 \frac{1}{4\pi\varepsilon_0} \times \tag{7.5}$$

$$\int \left[\left(\frac{1}{2}\alpha_{12} - \alpha_0 \right) + \frac{1}{3kT} \left(\frac{1}{2}\mu_{12}^2 - \mu^2 \right) \right] \exp\left(-\frac{V_{12}(R)}{kT} \right) \, dv_2 ,$$

where N_A is Avogadro's number and ϱ is the density in units of moles per unit volume. A_ε gives the contributions of the non-interacting (i.e., well separated) molecules and B_ε summarizes the contributions due to the pairwise interactions. The integration is over the positions and orientations of molecule 2 in a large, spherical volume centered on molecule 1. The virial coefficients A_ε, B_ε, ... are functions of temperature.

The dielectric constant ε is measured with static fields, i.e., at zero frequency. At optical frequencies, the refractive index n is of interest, which can be represented by a similar expression,

$$\frac{n^2 - 1}{n^2 + 2} = A_n \varrho + B_n \varrho^2 + \cdots \tag{7.6}$$

(Lorentz–Lorenz equation). The A_n, B_n, ... differ from the A_ε, B_ε, ... by the absence of the permanent dipole terms. At sufficiently high frequencies, the molecules have not sufficient time to reorient themselves in response to the high-frequency electric field and the contributions of the permanent dipoles simply disappear. The polarizabilities α_0 are now the dynamic polarizabilities which differ slightly from the static polarizabilities. The Clausius–Mossotti and Lorentz–Lorenz equations are analogous (and of course related) to the virial expansions of the spectral moments in collision-induced absorption (Chapters 5 and 6) and light scattering [60].

The B_ε and B_n virial coefficients have been measured for a number of common gases [73, 90, 380]. Useful experimental tests of empirical or theoretical induced dipole models derived from other data are thus possible.

7.4 Astrophysical applications

For much of their work, astronomers rely on spectroscopy and spectroscopic methods. Hot objects like the stars may be studied by more or less conventional atomic and plasma spectroscopy. In any case, in the hot environments, electronic spectra are excited by energetic electrons which are abundant; intensities are high and even the cool, outer regions of hot objects can be studied by their absorption spectra, e.g., the dark Fraunhofer lines of the solar spectrum, etc. The cooler regions of space, on the other hand, have long been known for their molecular bands arising from polar molecules. The non-polar species as H_2, He, etc., do not generate band spectra in the infrared despite their abundance; non-polar species in cool environments were long thought not to exist.

It is, therefore, noteworthy that almost immediately upon Welsh and associates' discovery of collision-induced absorption in hydrogen [128, 129, 420], Herzberg found *the first direct evidence* of the H_2 molecule in the atmospheres of the outer planets [181, 182]. He was able to reproduce in the laboratory the unidentified diffuse feature at 827.0 nm observed by Kuiper in the spectra of Uranus and Neptune, using an 80 m path of hydrogen at 100 atmospheres pressure and a temperature of 78 K. The feature is the $S_3(0)$ line of the $3 \leftarrow 0$ collision-induced rotovibrational band of the H_2 molecule [182].

Herzberg was able to point out another line at 816.6 nm which he identified as a double transition, partially overlapped by an adjacent CH_4 band in the Uranus spectra. In the laboratory spectra recorded with unmixed hydrogen, this double transition was relatively strong, but in the photographic plates of Uranus the feature was much weaker relative to the $S_3(0)$ line. This observation led Herzberg to conclude that sizeable He concentrations exist in these atmospheres (albeit the estimates of [He]:[H_2] abundance ratio seem high), because the $S_3(0)$ feature is enhanced by the presence of He, but H_2–He pairs cannot undergo double transitions; these features thus appear weak in Kuiper's plates relative to the $S_3(0)$ feature.

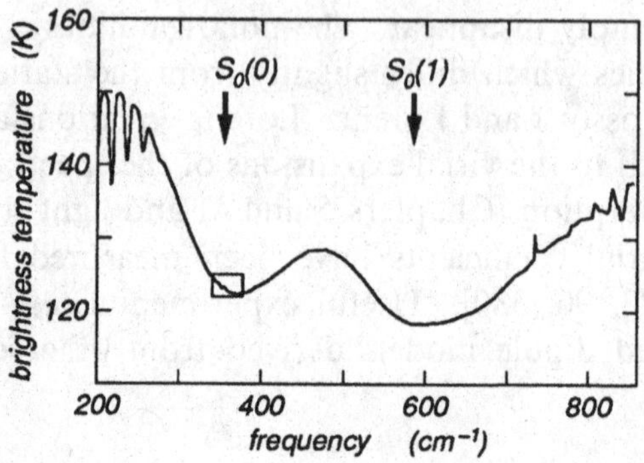

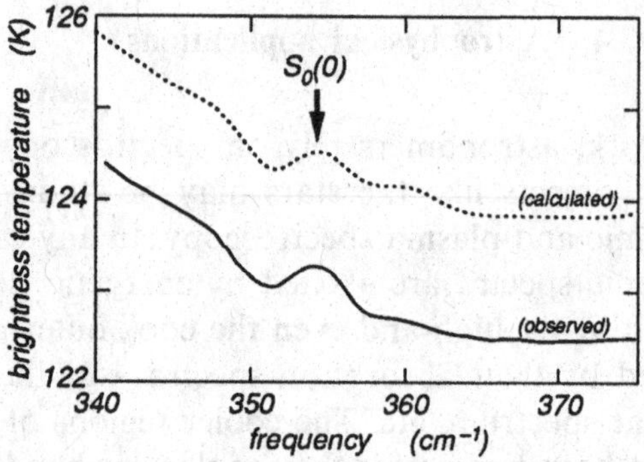

Fig. 7.3. Upper figure: Emission spectrum of Jupiter in the far infrared; two diffuse, dark fringes are seen at the H_2 $S_0(0)$ and $S_0(1)$ rotational transition frequencies, caused by collision-induced absorption in the upper, cool regions. The lower figure presents an enlarged portion which shows the dimer structures near the $S_0(0)$ transition frequency [150].

We know today that hydrogen and helium are overwhelmingly the most abundant species in the atmospheres of the outer planets, but direct evidence for their presence was virtually absent prior to the work mentioned [145]. Supermolecular spectroscopy had to be discovered before such evidence could be understood and it comes as no surprise that soon after Welsh's discovery many other uses of collision-induced absorption were pointed out in various astrophysical studies. Supermolecular absorption and emission have become the spectroscopy of the neutral, dense regions, especially where non-polar gases prevail.

Trafton has shown in 1964 that the opacity in the far infrared of the atmospheres of the outer planets is due to the rototranslational band of H_2–H_2 and H_2–He pairs [393]. It is now clear that collision-induced absorption plays a major role in the thermal balance and atmospheric structure of the major planets. The Voyager emission spectra of Jupiter and Saturn show dark fringes in the vicinity of the $S_0(0)$ and $S_0(1)$ lines of H_2, Fig. 7.3, which are due to collision-induced absorption in the upper, cool layers of the atmospheres (Trafton 1989). They are the planets' equivalent of the dark Fraunhofer lines of the solar spectrum. Curious features were discovered in these dark lines [157] and shown to be due to the $(H_2)_2$ van der Waals dimer [150, 263].

Collision-induced absorption spectra are also of considerable interest for studies of the atmospheres of late stars [241, 307, 112]; certain white dwarfs [366, 300, 239, 240]; low-mass stars, brown dwarfs, certain cool white dwarfs [240]; and the hypothetical 'population III' stars [306, 372].

Collision-induced absorption has been studied in the laboratory in various dense gases besides hydrogen and mixtures of hydrogen and helium, especially in oxygen, nitrogen, methane, etc., and mixtures of such gases which are of interest in the atmospheres of the inner planets [5, 131, 58].

The Earth's atmosphere is composed primarily of non-polar molecules like N_2 and O_2, especially at greater altitudes where the H_2O concentrations are small. One would therefore expect collision-induced contributions to the absorption of the Earth's atmosphere from N_2–N_2, N_2–O_2 and O_2–O_2 pairs. The induced rototranslational absorption of nitrogen has not been detected in the Earth's atmosphere, presumably because of strong interference by water absorption bands, but absorption in the various induced vibrational bands is well established (Tipping 1985). Titan (the large moon of Saturn) has a nitrogen atmosphere, somewhat like the Earth; methane is also present. Collision-induced absorption by N_2–N_2 and N_2–CH_4 is important in the far infrared.

Venus' atmosphere consists mainly of CO_2 of high density. It is perhaps the least well understood atmosphere, because the existing laboratory studies of collision-induced absorption in carbon dioxide and the theoretical analyses attempted have revealed some unexpected complexity. Some of the problems mentioned have to do with the strong ternary components observed; furthermore, the pair interaction is strongly anisotropic and the anisotropy has never been accounted for. More work is required for a better understanding (Tipping 1985).

Supermolecular absorption determines significant features of the atmospheres of the planets and their large moons, such as the vertical temperature profile and the high-altitude haze distribution, and offers opportunities for the determination of abundance ratios of helium and hydrogen, ortho- and para-H_2, etc. [390, 396]. In certain spectral bands the spectra may sometimes be obtained by Earth-based observations. More commonly, the spectra will be obtained in space missions, such as IRIS of Voyager I and II; future missions (Infrared Space Observatory) will doubtlessly enhance the available information significantly.

Data base. For the analyses of such spectra, absorption data need to be known accurately as functions of frequency, density, and temperature for all molecular pairs normally encountered in such atmospheres. Most significant information of that kind is obtained by laboratory measurements of the absorption of compressed gases, but the use of laboratory spectra for the modeling of planetary atmospheres is not straightforward. The problem is one of interpolating (and usually also extrapolating) the absorption coefficients measured in the laboratory at a number of fixed temperatures (say 77, 195, and 300 K) to the temperatures of the atmospheres, with little or no loss of the precision of the measurement. Temperatures of interest to the planetary scientist do not generally coincide with temperatures that are conveniently maintained in the laboratory, considering the bulk of long-path absorption cells. More or less empirical interpolation procedures have been attempted over the years by various authors for certain molecular systems and bands of interest. However, the results thus obtained are often inconsistent with predictions based on the fundamental theory, especially where extrapolation to low temperatures is involved. In short, purely empirical procedures do not in general provide the most reliable input for the purpose. The information has been collected in the form of a detailed data base of pressure-induced absorption spectra of the H_2–X systems of interest in the atmospheres of the outer planets [66].

7.5 Other applications

Liquids. For a long time, the study of infrared absorption by liquids and solutions has been a convenient way of determining rotovibrational spectra. The goal has often been to just determine peak frequencies, without paying much attention to the band shapes. In more recent years, attention has been devoted to a study of the shapes of vibrational bands and the dynamics of the molecules in the liquid. Only a crude understanding of the dynamics exists which is based on often highly simplified models of real liquids.

Liquids are difficult to model because, on the one hand, many-body interactions are complicated; on the other hand, liquids lack the symmetry of crystals which makes many-body systems tractable [364, 376, 94]. No rigorous solutions currently exist for the many-body problem of the liquid state. Yet the molecular properties of liquids are important; for example, most chemistry involves solutions of one kind or another. Significant advances have recently been made through the use of spectroscopy (i.e., infrared, Raman, neutron scattering, nuclear magnetic resonance, dielectric relaxation, etc.) and associated time correlation functions of molecular properties.

Because of intermolecular induced effects, characteristic low-frequency spectra are expected in liquids that are the many-body analog of the translational spectra of rarefied gases discussed above. It has been widely assumed that pairwise additive interactions describe the main effects in liquids. The situation is, however, much more complicated and it is clear that irreducible three-body and higher-order interactions become important in liquids. Nevertheless, it is probably fair to say that the studies of more or less rarefied gases described above provide some important, initial basis for the construction of model liquids, i.e., for the advancement of an understanding of liquids (Clarke 1978; Gray and Gubbins 1984; Guillot 1990; Steele 1989).

Solids. In general, solids are somewhat more tractable many-body systems than liquids because of the symmetry of crystals. Induced effects are certain to exist because of the density of solids.

A special case of considerable current interest is solid hydrogen and its isotopes (van Kranendonk 1983; Poll 1990). At low temperatures and pressures, the molecules H_2, D_2, HD, T_2, HT, and DT all form molecular crystals; these are true quantum solids. A rough model of the solid is obtained if one pictures the unperturbed molecules at rest at the lattice sites. The molecules are separated enough from their nearest neighbors that the properties of the free molecules are nearly unaffected; for example, the molecules in the lattice rotate nearly freely. The properties of the hydrogen and isotopic molecules are of course accurately known and a rigorous theoretical treatment from first principles is possible for the hydrogen solids. The same types of interaction-induced infrared spectra are observed as in the gases, supplemented in the cases of HD, HT and DT by the spectroscopic effects of the (weak) permanent dipole moment. Quadrupolar induction is the main induction mechanism, just as this was seen in the gas phase spectra.

7.6 Summary

Collision-induced absorption is a well developed science. It is also ubiquitous, a common spectroscopy of neutral, dense matter. It is of a supermolecular nature. Near the low-density limit, molecular pairs determine the processes that lead to the collision-induced interactions of electromagnetic radiation with matter. Collision-induced absorption by non-polar fluids is particularly striking, but induced absorption is to be expected universally, regardless of the nature of the interacting atoms or molecules. With increasing density, ternary absorption components exist which are important especially at the higher temperatures. Emission and stimulated emission by binary and higher complexes have also been observed. A collision-induced Raman process exists which is related to rovibrotranslational absorption like the Raman spectra of ordinary molecules are related to the rotovibrational monomolecular spectra in the infrared.

Collision-induced absorption is the principal spectroscopy of dense, cool (i.e., basically neutral) matter, especially of matter composed of non-polar atoms and molecules.

7.7 General references

A. Borysow and L. Frommhold, Collision-induced light scattering: a Bibliography, *Adv. Chem. Phys.* **75** (1989) 439–505.

J. H. R. Clarke, Band shapes and molecular dynamics in liquids, Ch. 4 of *Advances in Infrared and Raman Spectroscopy*, vol. 4, R. J. H. Clark and R. E. Hester, eds., Heyden, London, 1978.

R. M. Goody and Y. L. Yung, *Atmospheric Radiation – Theoretical Basis*, 2nd ed., Oxford University Press, 1989.

C. G. Gray and K. E. Gubbins, *Theory of Molecular Fluids*, vol. 1, Clarendon, Oxford, 1984. The second volume is of special interest here (to be published).

B. Guillot, Line shapes in dense fluids: the problem, some answers, future directions. In L. Frommhold and J. W. Keto, eds., *Spectral Line Shapes* **6**, p. 453, Am. Inst. Phys., New York, 1990.

J. D. Poll, Infrared and Raman spectroscopy of the solid hydrogens. In L. Frommhold and J. W. Keto, eds., *Spectral Line Shapes* **6**, p. 473, Am. Inst. Phys., New York, 1990.

W. A. Steele, Computer simulation studies of the induced far-infrared absorption of liquids. In J. Szudy, ed., *Spectral Line Shapes* **5**, p. 653, Ossolineum, Warsaw, 1989.

R. H. Tipping, Collision-induced effects in planetary atmospheres. In G. Birnbaum, *Phenomena Induced by Intermolecular Interactions*, p. 727, Plenum, New York, 1985.

L. Trafton, Observational studies of collision-induced absorption in the atmospheres of the major planets. In J. Szudy, ed., *Spectral Line Shapes* **5**, p. 755, Ossolineum, Warsaw, 1989.

J. van Kranendonk, *Solid Hydrogen: Theory of the Properties of Solid H_2, HD, and D_2*, Plenum, New York, 1983.

Appendix

In this Appendix we attempt to briefly review developments since the early 1990ies in the field of collision-induced absorption in gases. Many of the new contributions were announced, and numerous references to current literature were given in the proceedings of periodic conferences and special workshops. We mention especially the Proceedings of the biennial *International Conferences on Spectral Line Shapes* [1–7] and the annual *Symposia on Molecular Spectroscopy* [8]. New work in collision-induced absorption in gases has been reviewed in the Proceedings of a *NATO Advanced Study Institute* [9], a *NATO Advanced Research Workshops* [10], and in a recent monograph *Molecular Complexes in Earth's, Planetary, Cometary, and Interstellar Atmospheres* [11]. A multi-authored volume, a significantly augmented treatment of bremsstrahlung [12], is also of interest here, for example when electrically charged particles exist in dense, largely neutral and hot environments, e.g., in shock waves [13], in the atmospheres of "cool" white dwarf stars [14], in sonoluminescence studies [15, 16], etc.

Binary Interaction-Induced Dipoles. *Ab initio* quantum chemical calculations of interaction-induced dipole surfaces are known for some time (Section 4.4, pp. 159 *ff.*) Such calculations were recently extended for the H_2–He [17] and H_2–H_2 [18] systems, to account more closely for the dependencies of such data on the rotovibrational states of the H_2 molecules [19]. New calculations of the kind and quality are now available also for the H–He system [20], the H_2–H system [21], and the HD–He system [22]. New computational methods, called symmetry adapted perturbation theory (SAPT), were shown to be also successful for calculating interaction-induced dipole surfaces of such simple, binary van der Waals systems [23–26].

A careful theoretical description and evaluation of the long-range behavior of the spherical dipole tensor components was given for interacting pairs of homonuclear diatomic molecules [27, 28], and for tetrahedral molecules (CH_4, CF_4) interacting with linear molecules (H_2, N_2, CO_2, CS_2) [29]. New empirical dipole data of interacting pairs of dissimilar rare gas atoms were also recently reported [30–32]. An empirical value for the quadrupole transition moment in the ν_2 fundamental band of CH_4 has been obtained [33, 34] for modeling the important quadrupole-induced dipole component of collisional pairs containing the CH_4 molecule as a partner. The role of the induced dipole moment in the collisional interference in the pure rotational spectrum of HD–He and HD–Ar was studied [35].

The dipole-inducing mechanims of simple binary systems, such as atom-atom, atom-diatom, and diatom-diatom complexes with limited numbers of electrons, are well understood. For example, the exchange force-induced and dispersion-force-induced dipole components have been accurately computed for several simple systems, using quantum chemical methods (Chapter 4). The classical multipolar induction scheme is well known since the discovery of interaction-induced spectroscopy in 1949, and the mechanisms that involve electronic transitions are understood (Section 7.1), for example, when molecular pairs involving an O_2 molecule are considered [36–38]. However, there is a somewhat elusive, but apparently important dipole induction mechanism which seems to have defied even crude attempts of quantitative modeling, namely the dipole induction process by collisional frame distortion of molecules such as CO_2 or CH_4, that is of polyatomic molecules which have strong internal dipole moments along the bonding directions that add up to zero in the undistorted frame. A transient, collisional displacement of one O atom of CO_2, or one H atom of CH_4, will almost certainly result in a sizable, transient electrical dipole moment in the molecule which will more or less contribute to collision-induced absorption and emission, over and above what the other, better known induction mechanisms cause. A new study compares the measured rototranslational absorption spectra of interacting CH_4–X pairs, where X = He, H_2, N_2, or CH_4, with calculations based on the currently most advanced long-range dipole components [29] whenever possible. The study [39–41] finds various unresolved rotational bands of CH_4 (to be discussed below), that must be due to dipole induction mechanisms unaccounted for in the long-range induction model. Collisional frame distortion-induced dipoles of CH_4 molecules are likely to be important for an understanding of the observed excess absorption of methane-X gas mixtures.

It is well known that the tetrahedral frame of the CH_4 molecule is easily distorted. If the tetrahedral frame of CH_4 were robust, the purely rotational infrared spectra of CH_4 would not exist. However, even at temperatures as low as room temperature, the CH_4 molecule features hundreds of very weak, dipole-allowed rotovibrational lines at frequencies from 42 to 208 cm^{-1}, the so-called groundstate to groundstate (gs→gs) transitions. Moreover, more than 1500 weak, dipole-allowed transitions exist within the polyad system $\nu_2/\nu_4 \rightarrow \nu_2/\nu_4$, at frequencies from 14 to 500 cm^{-1} [42]. These allowed transitions arise from distortions of the tetrahedral frame by rotation and the internal dynamics of the CH_4 molecule, due to the coupling of normal modes of the flexible CH_4 frame. Collisional frame distortion should probably be associated with unresolved gs→gs and similar polyad bands. Some evidence of such collision-induced bands of CH_4 in CH_4–X complexes has been pointed out [39–41]. Besides these collision-induced bands that presumably are due to collisional frame distortion of CH_4, fairly significant, unexplained collision-induced bands also exist that are shaped by rotovibrational transitions of the collisional partner X = H_2, N_2, or CH_4, and by double transitions of the bimolecular CH_4–X complex [39–41].

In hot and dense environments, at temperatures of several thousand kelvin, e.g., in shockwave [13], astrophysical [14, 43], and sonoluminescence studies [15, 16], small concentrations of electrically charged particles (electrons and ions) exist which give rise to some strong emission (absorption) continua. In mostly neutral environments a substantial part of the absorption arising from the presence of charged particles would be electron-neutral bremsstrahlung and radiative attachment [13, 44–46], which customarily are not considered interaction-induced processes. The original theory of bremsstrahlung was developed in the early 1920ies, that is long before collision-induced absorption was discovered. However, these early treatments considered only the photon emission when an electrically charged particle (usually an electron) accelerates in the static Hartree-Fock field of the target atom. In more recent years, since about 1975, numerous theoretical and experimental studies have pointed out that actually there are other radiative processes, called polarization bremsstrahlung [12, 47] that are equally important. In other words, significant photon emission also occurs by virtual excitation (polarization) of the target electrons during an electron-molecule collision. In the simplest case, when a charged particle encounters an atom or molecule, the atom (molecule) will be temporarily polarized in the electric field of the charged particle. The field-induced dipole interacts with electromagnetic radiation like any other multipole-induced dipole, Eqs. (4.47) through (4.49), p. 191 *ff.* The resulting emission is surprisingly strong [48–50]. Emission and absorption by dipoles induced by polarization of the target atom or molecule is simply a monopole-induced dipole mechanism and may thus be considered a classical collision-induced dipole processes.

We note that the new science of polarization bremsstrahlung also considers the induced dipoles and the resulting interactions with electromagnetic radiation that are commonly found in energetic atoms impacting on a solid target, as well as in many other types of energetic collisions; collective radiative polarization effects in very dense neutral, or in more or less highly ionized environments have also been studied and were shown to be substantial [12].

Ternary and Other Interaction-induced Dipoles. It is well known that besides the pairwise additive ternary dipole components, several more or less significant irreducible ternary dipole components exist (p. 188). It is convenient to express the irreducible three-body dipole moment in terms of irreducible spherical components, to take advantage of the rotational symmetry properties of molecular complexes. Such a representation is particularly useful for systems consisting of molecules whose interactions may be modeled by an isotropic potential. As a basis set spherical harmonics of the orientations of the three molecules are used, together with rotation matrices of a set of Euler angles, specifying the orientation in space of the centers of mass of the three molecules. To this end we consider a reference frame whose origin coincides with the center of mass of molecule 1. The z axis is determined by the direction of the vector joining molecule 1 with molecule 2. The x, z plane contains the centers of mass of the three molecules and the direction of x is such that the vector joining molecules 1 and 3 has a zero azimuth. The Euler angles specifying the orientation of such a reference system, which we collectively indicate by $\mathbf{\Omega}$, are the arguments of the rotation matrices. We can thus construct linear combinations of products of such functions which transform under rotation like a dipole moment. These functions have the form [51]

$$\Psi^{1\nu}_{\lambda_1\lambda_2\lambda_3\Lambda J;LN}(\widehat{\mathbf{u}}_1, \widehat{\mathbf{u}}_2, \widehat{\mathbf{u}}_3, \mathbf{\Omega}) \qquad (1)$$
$$= \sqrt{2L+1} \sum_{MM'} Y^{JM}_{\lambda_1\lambda_2\lambda_3\Lambda}(\widehat{\mathbf{u}}_1, \widehat{\mathbf{u}}_2, \widehat{\mathbf{u}}_3)\, D^{L*}_{M'N}(\mathbf{\Omega})\, C(JL1; MM'\nu) \ ,$$

with

$$Y^{JM}_{\lambda_1\lambda_2\lambda_3\Lambda}(\hat{\mathbf{u}}_1, \hat{\mathbf{u}}_2, \hat{\mathbf{u}}_3) = \sum_{m_1m_2m_3m} Y_{\lambda_1m_1}(\hat{\mathbf{u}}_1)\, Y_{\lambda_2m_2}(\hat{\mathbf{u}}_2)\, Y_{\lambda_3m_3}(\hat{\mathbf{u}}_3) \times C(\lambda_1\lambda_2\Lambda; m_1m_2m)\, C(\Lambda\lambda_3J; mm_3M) \,. \quad (2)$$

The $\Psi^{1\nu}$ form a basis set of functions of the orientations of three molecules which transform as a vector. The ν-th spherical component of the dipole moment μ may be expanded in terms of this basis, according to

$$\mu_\nu(123) = \frac{(4\pi)^{3/2}}{\sqrt{3}} \sum_{\lambda_1\lambda_2\lambda_3\Lambda J;LN} \mathcal{B}^{LN}_{\lambda_1\lambda_2\lambda_3\Lambda J}(R, R', \cos\theta) \times \Psi^{1\nu}_{\lambda_1\lambda_2\lambda_3\Lambda J;LN}(\hat{\mathbf{u}}_1, \hat{\mathbf{u}}_2, \hat{\mathbf{u}}_3, \boldsymbol{\Omega}) \,. \quad (3)$$

The $\mathcal{B}^{LN}_{\lambda_1\lambda_2\lambda_3\Lambda J}$ are functions of the variables which describe the relative configuration of the centers of mass of the three molecules. For such variables we have chosen the lengths of the intermolecular vectors joining molecule 1 with molecules 2 and 3 ($\mathbf{R}$ and $\mathbf{R}'$, respectively) and the angle θ subtended by these vectors. We denote by R'' the distance between centers of mass of molecules 2 and 3 which can be expressed in terms of R, R' and θ, as $R'' = \sqrt{R^2 + (R')^2 - 2RR'\cos\theta}$. Equation (3) is the general expression for the dipole induced in a system of three rigid rotors and consequently it is appropriate for the treatment of the rototranslational spectra. If vibrational bands are considered, the functions $\mathcal{B}$ become the transition matrix element between the states of the vibrating molecule. We assume that molecule 1 is the vibrating molecule. We account for the possibility of other vibrating molecules by introducing appropriate statistical weights.

For the pairwise additive component of molecules 1 and 2 we have, from Eqs. (4.18) – (4.20),

$$\mu_\nu(12) = \frac{(4\pi)^{3/2}}{\sqrt{3}} \sum_{\lambda_1\lambda_2\Lambda L} B_{\lambda_1\lambda_2\Lambda L}(R) \times\sqrt{2L+1} \sum_{MM'} Y^{\Lambda M}_{\lambda_1\lambda_2 0\Lambda}(\hat{\mathbf{u}}_1, \hat{\mathbf{u}}_2, \hat{\mathbf{u}}_3,)\, D^{L*}_{M'0}(\boldsymbol{\Omega})\, C(\Lambda L1; MM'\nu) \,. \quad (4)$$

The comparison of Eqs. (1,2,3) shows that the ternary dipole functions $\mathcal{B}$ arising from the dipole moment induced in molecules 1,2 are related to the pair dipole functions B, according to

$$\mathcal{B}^{LN}_{\lambda_1\lambda_2\lambda_3\Lambda J} = B_{\lambda_1\lambda_2\Lambda L}(R)\, \delta_{N0}\, \delta_{\lambda_3 0}\, \delta_{\Lambda J} \,. \quad (5)$$

The pair dipole moment induced in molecules 1,3 may be similarly written,

$$\mu_\nu(13) = \frac{(4\pi)^{3/2}}{\sqrt{3}} \sum_{\lambda_1\lambda_3\Lambda L} B_{\lambda_1\lambda_3\Lambda L}(R') \times\sqrt{2L+1} \sum_{MM'} Y^{\Lambda M}_{\lambda_1 0\lambda_3\Lambda}(\hat{\mathbf{u}}_1, \hat{\mathbf{u}}_2, \hat{\mathbf{u}}_3)\, D^{L*}_{M'0}(\boldsymbol{\Omega}')\, C(\Lambda L1; MM'\nu) \,, \quad (6)$$

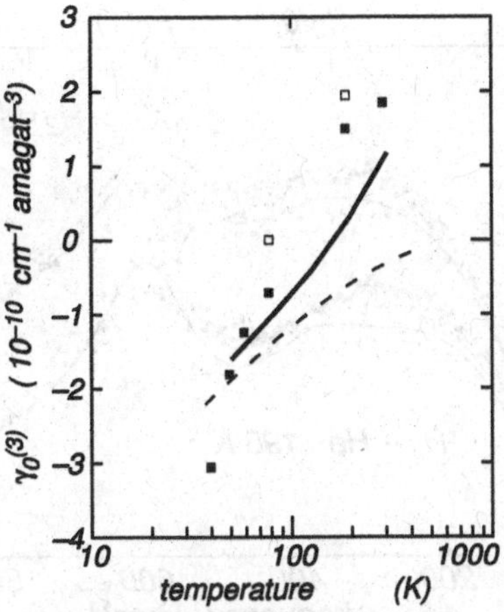

Figure 1: Comparison of observed (□,■) and computed ternary spectral moments of collision-induced absorption of compressed hydrogen in the H_2 fundamental band. The dashed curve represents the moments computed on the basis of the pairwise additive dipole components only. The solid line also accounts for the irreducible dipole components; from Ref. [53]. (This figure is an update of Fig. 3.46.)

where $\mathbf{\Omega}'$ represents the set of Euler angles of an intrinsic frame with its origin at the center of mass of molecule 1, its z axis parallel to $\mathbf{R}'$ and the x, z plane containing the centers of mass of the three molecules. Note that this frame can be obtained from the above frame defined by the Euler angles $\mathbf{\Omega}$ by a rotation about the y axis through the angle θ. Taking into account the transformation properties of the matrices under rotation it is seen at once that the $\mathcal{B}$ functions for the dipole moment induced in molecules 1,3 may be written

$$\mathcal{B}^{LN}_{\lambda_1\lambda_2\lambda_3\Lambda J} = B_{\lambda_1\lambda_3\Lambda L}(R')\, D^L_{N0}(0\theta 0)\, \delta_{\lambda_2 0}\, \delta_{\Lambda J}\,. \tag{7}$$

The B functions for the pair dipole $\mu(12)$ couple to only a small subset of the ternary dipole components $\mathcal{B}$ [51, 52].

The comparison of theoretical predictions with measurements of the third virial spectroscopic coefficients [51, 54], and a triple transition of the type $Q_1(j_1)+Q_1(j_2)+Q_1(j_3)$ at 12,466 cm^{-1} that was observed in compressed hydrogen gas [54–56], illustrates the presence of a significant irreducible ternary dipole contribution. The spectral contributions by the non-additive dipole components of hydrogen increase sharply with increasing temperatures, Fig. 1. For example, at temperatures exceeding $\approx$ 100 K, the third virial spectroscopic coefficient is already affected by the non-additive dipole mechanisms, and with increasing temperature those contributions become more and more important [53]. Such a strong temperature dependence suggests overlap and exchange force-induced ternary dipole components as the principal mechanisms. Treatment of these requires quantum mechanical methods. Since exact quantum chemical methods have not been applied to the case of ternary induced dipoles yet, a one-effective

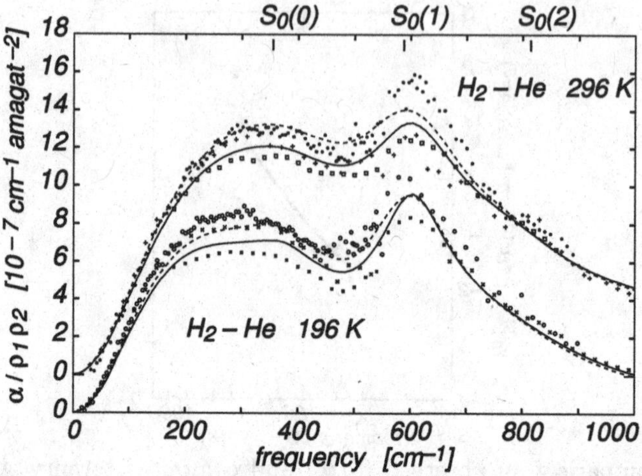

Figure 2: The absorption coefficient α, normalized by the helium and hydrogen gas densities, ρ_1 and ρ_2, respectively, as function of frequency in the H_2 roto-translational band, at the temperature of 296 K (upper trace) and 196 (lower trace, shifted downward one step for clarity). Solid and dashed curves represent calculations with and without accounting for the anisotropy of the intermolecular interactions, respectively. Also shown are measurements (as in Fig. 3.12, p. 85); from Ref. [17]

electron model is used to define the overlap-induced dipole (OVID) component, the exchange quadrupole-induced dipole (EQID) component, and the exchange dipole-induced dipole (EDID) component, p. 189; refined models were later introduced for the EQID and EDID mechanisms, which are called antiparallel dipoles-induced dipole (ADID) and opposite charges-induced dipole (OCID) mechanisms, respectively [53].

A convenient theoretical framework for understanding the variety of observational results related to ternary and higher-order dipoles, rather than providing a basis for accurate analyses, has been presented [57].

Non-additive ternary dipole calculations of non-overlapping molecular clusters of the type A-A-A have been reported, where A stands for H, H_2, He, Ne, or Ar. The induced dipoles of clusters of the type A-A-B and A-B-C, with A, B, C, designating any one of the species H, H_2, and He, were computed [58, 59].

Collision-induced Spectra of Binary Systems. A brief overview of new measurements of collision-induced spectra is given in Table 1. Theoretical calculations of such spectra are also important, e.g., for the comparison of measurement with the fundamental theory, for easy frequency and/or temperature interpolation (and, sometimes, extrapolation) for many applications, or to predict or estimate such spectra that have not been measured or cannot be measured readily.

In the past, most calculations of collision-induced absorption spectra were

Table 1: New measurements of collision-induced spectra

system	freq. band	remarks	Ref.
He–Ar	translational		[31]
He–Xe	translational		[30, 60]
Ar–Xe	translational		[32]
Ne–Xe, Ar–Xe, Kr–Xe	far IR	dimer	[61]
H_2–He	H_2 fundamental		[62, 63]
pH_2–Ar	H_2 fundamental	dimer	[64]
H_2–Kr	first H_2 overtone		[65]
H_2–Xe	first H_2 overtone		[65]
H_2–H_2	pure translational	T = 21–38 K	[66–68]
H_2–H_2	rototranslational	dimer	[69, 70]
H_2–H_2	H_2 fundamental		[71]
H_2–N_2	N_2 fundamental		[72]
N_2–N_2	rototranslational	$T \approx 80$ K	[73]
N_2–rare gas	rototranslational	$T \approx 80$ K	[73]
CO–Ar	CO fundamental	dimer	[64, 74]
CO–H_2O	CO fundamental	dimer	[64]
O_2–O_2	O_2 fundamental		[75–77]
O_2–CO_2	O_2 fundamental		[78]
CO_2–CO_2	CO_2 $3\nu_3$ wing		[79]
CO_2–CO_2	CO_2 $\nu_1, 2\nu_2$		[80, 81]
CO_2–CO_2	CO_2 Fermi triad		[82]
CO_2–He	rototranslational		[83]
CO_2–Ar	rototranslational		[83]
CO_2–Xe	rototranslational		[83]
N_2–N_2	N_2 fundamental		[84]
N_2–O_2	N_2 fundamental		[85]
N_2–O_2	O_2 fundamental		[85]
N_2–CO_2	N_2 fundamental		[86, 87]
N_2–H_2O	N_2 fundamental		[87, 88]
N_2–CH_4	rototranslational		[89]
CH_4–Ar	CH_4 ν_2		[33]
CH_4–H_2	CH_4 ν_2		[34]
CH_4–N_2	CH_4 ν_2		[34]
CH_4–CO_2	CH_4 ν_2		[34]
CF_4–CF_4	CF_4 ν_2		[90]
CF_4–X	CF_4 ν_2		[90]
O_2–O_2	${}^3\Sigma_g^-, {}^3\Sigma_g^- \to {}^3\Sigma_g^-, {}^3\Delta_u$		[37, 91]
O_2–O_2	visible+near UV		[36, 38, 92, 93]
O_2–CO_2	${}^3\Sigma_g^-, {}^3\Sigma_g^- \to {}^3\Sigma_g^-, {}^3\Delta_u$		[94]

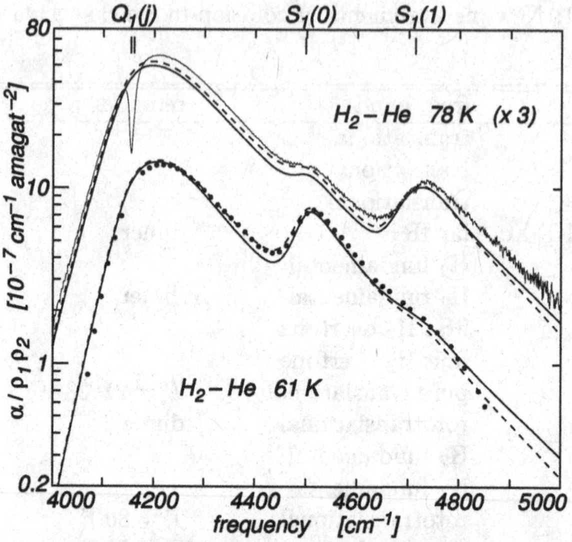

Figure 3: Rotovibrational collision-induced absorption spectra of H_2–He pairs at 78 K (upper trace, multiplied by 3 for better readability of the graph) and 61 K (lower trace); solid and dashed curves represent calculations with and without accounting for the anisotropy of the intermolecular interaction, respectively. The dots (at 61 K) and the thin line (at 78 K) are measurements [62]; from [17].

based on the isotropic interaction approximation (IIA). In order to estimate the effects of the anisotropy of the intermolecular interactions on the spectra, sum formulas ("spectral moments") were developed which account for the anisotropy of the potential energy surface [95–99]. The far wings of induced spectra often need to be known reliably, but spectral moments are not very sensitive to the far wing intensities, which may be much affected by the anisotropy ot the intermoleculat potential. For simple systems, i.e., molecular pairs with small numbers of electrons, new *ab initio* line shape calculations appeared that account for the anisotropies of the intermolecular interactions. Of course, the rototranslational collision-induced band shapes of H_2–H_2 and H_2–He complexes, which account for the anisotropy of intermolecular forces, were known for some time (pp. 329*ff.*) Such quantum line shape calculations solve the Schrödinger equation of the molecular scattering process in the so-called close-coupled scheme. Recently, new close-coupled quantum line shape calculations for H_2–H_2 and H_2–He complexes have been reported in the rototranslational band, and for the first time also in the H_2 fundamental band [17, 100], which permit detailed studies of the effects of the anisotropy of the potential energy surfaces. We note that close-coupled solutions of the Schrödinger equation, and the calculations of the spectra (radial transition matrix elements), are computation intensive. For systems and conditions involving large numbers of partial waves and/or rotational excitations, e.g., massive systems at elevated temperatures, close-coupled solutions are presently not known.

For the H_2–He system, the accounting for the anisotropy of the intermolecular forces introduces small, but significant modifications of the calculated spectral shapes, especially in the range of the H_2 S lines and their high-frequency wings. Accounting for the anisotropy of the interaction generally improves the agreement of measurement and fundamental theory, see Figs. 2 and 3. The theoretical profiles appear to be bracketed by existing measurements, Fig. 2.

Similar close-coupled calculations of the H_2–H_2 rototranslational spectrum show almost no differences at frequencies near the S line centers, when compared with results of the isotropic interaction approximation [100]. At the frequencies of the H_2 S lines the quadrupolar induction is much stronger than the hexadecapolar induction. However, in the far wings of the S lines, where absorption by the hexadecapolar induction process becomes more and more important relative to the quadrupole induction, the close-coupled scheme gives stronger absorption with increasing frequencies, by factors up to 2.5 at 2,400 cm^{-1} [100]. (A similar effect may be expected for the H_2–He absorption profiles at very high frequencies, in the far wings.) We note that, to some extent, this difference could be an artifact, due to the presently less well known induction components of hexadecapolar symmetry: efforts to obtain the $\lambda_1\lambda_2\Lambda L = 0443$, 0445, 4043, 4045 induced dipole components of hydrogen pairs from first principles with better accuracy are underway [101]. Unfortunately, the laboratory measurements in compressed hydrogen at such high frequencies are difficult and also somewhat uncertain because of the weak absorption. Measurements of the extremely weak absorption of hydrogen at ≈ 2000 cm^{-1} and above requires relatively high pressures so that ternary contributions may affect the measurements in ways that are not readily corrected with the necessary precision [100]. Accurate knowledge of the absorption by hydrogen in the 2000 cm^{-1} band is of great importance in planetary science.

It is interesting to note that in high-resolution studies of the spectra of solid hydrogen transitions were seen with a change of the rotational quantum numbers ΔJ of 6 and 8 [102]. The suggestion was made that these could be caused by H_2 multipole moments of higher order than the hexadecapole (or 2^4) moment, e.g., by the H_2 2^6 and 2^8 multipole moments. Such transitions are weak and have hitherto not been included in any treatments of collision-induced absorption in gases.

The rototranslational and fundamental absorption spectra of the H_2–H complex have been obtained from first principles, for temperatures from 200 to 2500 K [21, 103]. Close-coupled and isotropic interaction approximation calculations give nearly identical values at frequencies from 0 to 6000 cm^{-1}. No laboratory measurements exist for comparison with the calculations. The H_2–H system is of considerable interest in stellar environments at such temperatures.

Calculations of the rototranslational absorption spectra of nitrogen, oxygen, and nitrogen-oxygen mixtures were communicated [104]. Similar work for the fundamental band of oxygen was reported [77]. Collision-induced absorption spectra of O_2–N_2, O_2–CO_2, and O_2–H_2O pairs involving $a \leftarrow X$ and $b \leftarrow X$ electronic transitions of O_2 have been studied theoretically [105, 106].

New quantum line shape calculations of the rototranslational enhancement

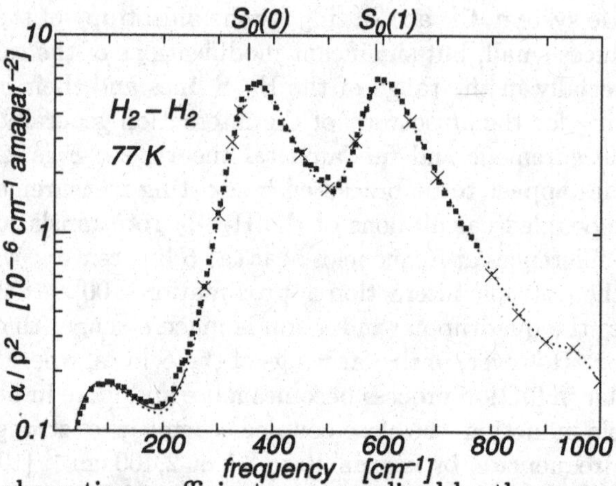

Figure 4: The absorption coefficient α, normalized by the square of the hydrogen gas density ρ, as function of frequency in the H_2 rototranslational band, at 77 K. The solid line at low frequencies (< 240 cm^{-1}) and the crosses ($\times$) represent calculations that account for the anisotropy of the intermolecular interaction potential. The dotted curve is calculated in the isotropic interaction approximation; from Ref. [100].

spectra of CH_4–X systems, with X = He, H_2, N_2, or CH_4, which are based on advanced, long-range dipole surfaces [29] and the isotropic interaction approximation, have been reported [39–41]. The work was undertaken in the hope of resolving a long-standing problem, the intensity defect of calculated spectra relative to measured absorption spectra of methane-X gas mixtures. However, these defects persist even if these improved (but limited to long range) dipole surfaces are input. Apparently, it will be necessary to include additional dipole induction mechanisms that are of a short-range nature and are at present not well known. Even advanced long-range dipole component surfaces, e.g., Ref. [29], are not sufficient when a CH_4 molecule (and probably many other polyatomic "nonpolar" molecules with strong internal dipole moments) are involved. For diatom-atom and diatom-diatom systems, on the other hand, simple *at hoc* empirical exchange force-induced dipole components were virtually always found to satisfactorily account for the measured intensities and spectral profiles. A different induction mechanism seems to matter when CH_4–X pairs interact. We think that besides some unknown contributions due to exchange force-induced dipoles, significant collisional frame distortion of the CH_4 molecule may be responsible for the excess absorption seen in these measurements. More work is required for more definite conclusions.

Interesting "ripples" were first seen in the interaction-induced absorption spectra of compressed nitrogen [107] and later also of oxygen and in various gas mixtures in the infrared [73, 78, 84]. It has been suggested that these weak, roughly sinusoidal structures that are superimposed with the quasi-continuous, induced background arise possibly from line mixing, due to the anisotropy of the intermolecular interaction of the monomers ("hindered rotation") [76, 108]. Other researchers have pointed out that the accounting for transitions involving predissociating ("metastable") dimers reproduces the observed structures reasonably well [109]. Such dimers exist in these gases at the few percent level at the temperatures and gas densities of interest.

Crude estimates of the absorption spectra of hydrogen-helium mixtures at temperatures from 2,000 to 20,000 K were attempted [110, 111]. The motivation for that work was to better understand the effects of highly rotovibrationlly excited molecules on the collision-induced absorption spectra. Since small basis sets were used for the quantum chemical calculations of the induced dipole components, the results should not be used for comparisons with measurements. Theoretical estimates of the emission of colliding high-speed ($\approx$ 10 km/s) neutral atoms, based on classical trajectories and certain quantum corrections, were reported [112].

Interference of Permanent and Induced Dipoles. The HD–X systems, where X may be a rare gas atom or another HD molecule, has long been of a special interest, because the HD molecule possesses a weak non-adiabatic, permanent electrical dipole moment and thus various allowed rototranslational and rotovibrational bands. At high enough gas densities, the wings of the allowed rotational lines show more or less striking Fano-type patterns which arise from interferences of the permanent dipoles with certain collision-induced dipole components. These Fano profiles have been studied closely for some time, pp. 115 *ff*, and experimental studies and analyses continue to appear [113–120]. New spectral line shape calculations have been done for the rototranslational and rotovibrational bands of HD–He pairs that account for the anisotropy of the intermolecular potential energy surface [22] which for HD–He is much stronger than for the H_2–He system [22]. The anisotropy causes rotational level mixing and gives rise to pressure broadening of the allowed lines. It also couples the various induced dipole components with the permanent HD dipole, which in turn gives rise to Fano profiles. *Ab initio* line shape calculations without adjustable parameters reproduce the observed wings of the HD R lines quite well [121].

A simulation of the far infrared spectrum of liquid water and steam at temperatures from 273 to 473 K was undertaken, based on molecular dynamics techniques [122].

Ternary and Other Induced Spectra. Three-particle induced dipoles and the associated ternary collision-induced absorption spectra and dipole autocorrelation functions have been studied for fluids composed of mixtures of rare gases, and for neat fluids of nonpolar molecules — that is for systems that are widely thought to interact with radiation only by virtue of interaction-induced properties. A convenient framework is thus obtained for understanding the variety of experimental observations. The computer simulation studies permit an insight into the involved basic processes, but were not intended for direct comparison with measurements [57]. Methods have been developed for computer simulations of contributions to the dipole autocorrelation function of systems of three interacting particles. Ternary contributions are studied in rare gas mixtures, in mixtures of quadrupolar molecules with rare gases, and in pure rare gases [123, 124]. The anisotropy of the interaction potential is accounted for in some of these studies. For the molecular gases simultaneous triple transitions during the absorption of a single photon have been predicted for some time, p. 129. In compressed hydrogen, a triple transition of the form $3Q_1$ was detected [55], and in solid hydrogen a triple $3S_0$ was reported [125, 126]. Such simultaneous triple transitions are thought to be manifestations of the existence of irreducible ternary dipole components.

Ternary spectral moments of collision-induced absorption in hydrogen gas are analyzed in the H_2 fundamental band in terms of pairwise additive and irreducible contributions to the interaction-induced dipole moment, Eqs. (1 – 7) [51]. Numerical results show that irreducible dipole components, especially of the exchange quadrupole-induced ternary dipole component, are significant for agreement with spectroscopic measurements, such as ternary spectral moments (Fig. 1) [53], an observed diffuse triple transition $3Q_1$ centered at 12,466 cm^{-1} [52, 54, 55], and the intercollisional dip in compressed hydrogen gas, pp. 188 – 190.

Interaction-induced absorption by the vibrational or rotational motion of an atom, ion, or molecule trapped within a C_{60} cage, so-called endohedral buckminsterfullerene, has excited considerable interest, especially in astrophysics. The induced bands of such species are unusual in the sense that they are discrete, not continuous; they may also be quite intense [127]. Other carbon structures, such as endohedral carbon nanotubes, giant fullerenes, etc., should have similar induced band spectra [128], but current theoretical and computational research is very much in flux while little seems to be presently known from actual spectroscopic measurements of such induced bands.

Applications. For detailed spectroscopic modeling of the atmospheres of planets and stars simple analytical model profiles exist (see Section 5.10, p. 270 *ff.*) which often permit surprisingly accurate reproductions of measured and calculated translational profiles [129]. An accurate description of the absorption by interacting molecular pairs is essential in virtually all attempts of the kind. To that end, especially at the low temperatures of planetary atmospheres, the separation of pair states in the phase space is necessary for a correct assessment of the various spectral contributions, see Chapters 5 and 6, e.g., Eq. (5.67) on page 240 above, and Refs. [108, 130, 131]. There are the *free pairs*, that is molecules in collisional interaction, when the duration of the collision is of the order of the fly-by time. But especially at the lower temperatures and higher gas densities, a significant fraction of the individual molecules will be *bound*, as van der Waals dimers, i.e., in a rotovibrating supramolecular system that is stable unless it is destroyed in collisions with a third body. Besides these free and bound states, there are the *predissociating states*, sometimes also called *metastable states* (which must not be confused with electronic metastable states of atoms and molecules), and other *scattering resonances*, the so-called

Feshbach or compound resonances. These latter systems are bound rotovibrational supramolecular states that are coupled to the dissociation continuum in some way so that they have a finite lifetime: these states will dissociate on their own, even in the absence of third-body collisions, unless they undergo a radiative transition first into some other pair state. The free-to-free state transitions are associated with broad profiles, which may often be approximated quite closely by certain model line profiles, Section 5.2, p. 270*ff*. If bound states are involved, the resulting spectra show more or less striking structures: pressure broadened rotovibrational bands of bound-to-bound transitions, e.g., the "sharp" lines shown in Fig. 3.41 on p. 120, and more or less diffuse structures arising from bound-to-free and free-to-bound transitions which are also noticeable in that figure and in Figs. 6.5 and 6.19. At low spectroscopic resolution or at high pressures, these structures flatten, often to the point of disappearance. Spectral contributions of bound dimer states show absorption dips at the various monomer Raman lines, as in Fig. 6.5.

In recent years the role of interaction-induced absorption processes in the earth's atmosphere have finally received the appropriate attention [10, 11, 132]. Especially the various spectroscopic effects of water vapor are the subject of intense continuing studies [133, 134], with a focus on the far wings of the allowed bands, the interaction-induced contributions by bound and free pairs, and their self and foreign absorption continua [88, 132, 135–160]. The self and foreign absorption contiunua of bound and free complexes involving the oxygen molecule have likewise received much attention [38, 134, 161, 162]. The O_2 molecule possesses low-lying electronic states that participate in various spectral bands to the collision-induced absorption, see Section 7.1, p. 356*ff*. and Refs. [36–38, 91–94, 105]. Several other molecular pairs that contribute to interaction-induced absorption in the earth's atmosphere were investigated in great detail, especially systems containing nitrogen (N_2) and/or carbon dioxide (CO_2) [11, 72, 77, 84, 98, 99, 109, 163–168].

The significance of collision-induced absorption for the planetary sciences is well established (Chapter 7); reviews and updates appeared in recent years [115, 165, 166, 169–173]. Numerous efforts are known to model experimental and theoretical spectra of the various hydrogen bands for the astrophysical applications [170, 174–181]. More recently, important applications of collisional absorption in astrophysics were discovered in the "cool" and extremely dense stellar atmospheres of white dwarf stars [14, 43, 182–184], at temperatures from roughly 3000 to 6000 K. Under such conditions, large populations of vibrationally excited H_2 molecules exist and collision-induced absorption extends well into the visible region of the spectrum and beyond. Numerous "hot bands," high H_2 overtone bands, and H_2 rotovibrational sum and difference spectral bands due to simultaneous transitions that were never measured in the laboratory must be expected. *Ab initio* calculations of the collisional absorption processes in the dense atmospheres of such stars have yet to be provided so that the actual stellar emission spectra may be obtained more accurately than presently known.

There is a resurgence of interest in very cold white dwarf stars as they are at the heart of the method of cosmochronology. This method provides an independent means to determine the ages of old stellar populations [185, 186]. White dwarfs evolve by simple cooling and their present temperatures (or blackbody spectra) can be used as "clocks" telling their age [187, 188]. The goal is to construct accurate white dwarf luminosity functions by liberating the observable spectra from the influence of the dense atmospheres surrounding the stars. In such environments, several of the absorption mechanisms are well known, e.g., bremsstrahlung and radiative recombination and attachment processes. However, collision-induced absorption at temperatures of 3000 to 6000 K at frequencies up to the visible and near ultraviolet is virtually unknown, but it is clearly of great importance for the proper determination of the age of these stars and probably of the universe.

Acknowledgment. Valuable communications with Drs. G. Birnbaum, A. Borysow, J.-M. Hartmann, A. Kouzov, M. Moraldi, I. Ozier, G. C. Tabisz, R. H. Tipping, M. V. Tonkov, and A. A. Vigasin are acknowledged.

Corrections.
pp. 44–47: The differentials $\partial\nu$ and $d\nu$ appearing to the left of the equal sign should read $\partial\omega$ and $d\omega$, respectively, which differ by the factor $2\pi c$: $\omega = 2\pi c\nu$. The equation at the bottom of p. 45 should be divided by $2\pi^2$.
Eq. (4.39), p. 168, should be modified so that the $B_{\lambda L}^{(0)}$ coefficient is not a function of R.
Eq. (5.79), p. 254, should read

$$C_{\rm cl}(t) \to C_{\rm cl}\left(\left[t^2 + \hbar^2/(2kT)^2\right]^{1/2}\right) = \widetilde{C}(t) \to \widetilde{C}\left(t - \imath\hbar/(2kT)\right)$$

Eq. (6.47), p. 308: the third line of this equation should read

$$= N\left[Y^{JM}\left(\widetilde{\mathbf{r}}_1, \widetilde{\mathbf{r}}_2, \widetilde{\mathbf{R}}; j_1 j_2 j \ell\right) \pm Y^{JM}\left(-\widetilde{\mathbf{r}}_1, -\widetilde{\mathbf{r}}_2, -\widetilde{\mathbf{R}}; j_2 j_1 j \ell\right)\right]$$

References

[1] R. Stamm and B. Talin, editors. *Spectral Line Shapes*, 1993. Nova Science Pub. vol. **7**.
[2] A. D. May, J. R. Drummond, and E. Oks, editors. *Spectral Line Shapes*, 1995. Am. Inst. Physics. vol. **8**.
[3] M. Zoppi and L. Ulivi, editors. *Spectral Line Shapes*, 1997. Am. Inst. Physics. vol. **9**.
[4] R. M. Herman, editor. *Spectral Line Shapes*, 1999. Am. Inst. Physics. vol. **10**.
[5] J. Seidel, editor. *Spectral Line Shapes*, 2001. Am. Inst. Physics. vol. **11**.
[6] Ch. A. Back, editor. *Spectral Line Shapes*, 2002. Am. Inst. Physics. vol. **12**.
[7] E. Dalimier, editor. *Spectral Line Shapes*, 2004. Frontier Group.
[8] http://molspect.chemistry.ohio-state.edu.

[9] G. C. Tabisz and M. N. Neuman, editors. *Collision- and Interaction-Induced Spectroscopy.* NATO ASI series C, vol. 452. Kluwer, Dordrecht, 1995.

[10] C. Camy-Peyret and A. A. Vigasin, editors. *Weakly Interacting Molecular Pairs: Unconventional Absorbers of Radiation in the Atmosphere,* Dordrecht, 2003. Kluwer. Proceedings of the ARW held in Fontevraud, France, from 4/29 – 5/2/2002, NATO Science Series, Earth and Environmental Sciences, vol.27.

[11] A. A. Vigasin and Z. Slanina, editors. *Molecular Complexes in Earth's, Planetary, Cometary, and Interstellar Atmospheres.* World Sci., Singapore, 1998.

[12] V. N. Tsytovich and I. M. Ojringel, editors. *Polarization Bremsstrahlung.* Plenum, New York, 1992.

[13] Ya. B. Zeldovich and Yu. P. Raizer. *Physics of Shock Waves and High-Temperature Hydrodynamic Phenomena.* Dover, New York, 2002.

[14] U. G. Jørgensen, D. Hammer, A. Borysow, and J. Falkesgaard. The atmospheres of cool, helium-rich white dwarfs. *Astronomy and Astrophysics,* 361:283, 2000.

[15] D. Hammer and L. Frommhold. Sonoluminescence: how bubbles glow. *J. Modern Optics,* 48:239, 2001.

[16] M. P. Brenner, S. Hilgenfeldt, and D. Lohse. Single bubble sonoluminescence. *Rev. Mod. Phys.,* 74:425, 2002.

[17] M. Gustafsson, L. Frommhold, and W. Meyer. Infrared absorption spectra by H_2–He complexes: The effect of the anisotropy of the interaction potential. *J. Chem. Phys.,* 113:3641, 2000.

[18] W. Meyer, A. Borysow, and L. Frommhold. Collision-induced first overtone band of gaseous hydrogen from first principles. *Phys. Review A,* 47:4065, 1993.

[19] W. Meyer and L. Frommhold. Ab initio interaction-induced dipoles and absorption spectra. p. 441, 1995. In Ref. [9].

[20] W. Meyer and L. Frommhold. Ab initio potential and dipole surfaces and the absorption of H–He pairs in the infrared. *Theoretica Chimica Acta,* 88:201, 1994.

[21] M. Gustafsson, L. Frommhold, and W. Meyer. The H_2–H complex: Interaction-induced dipole surface and infrared absorption spectra. *J. Chem. Phys.,* 118:1667, 2003.

[22] M. Gustafsson and L. Frommhold. The HD–He complex: Interaction-induced dipole surface and infrared absorption spectra. *J. Chem. Phys.,* 115:5427, 2001.

[23] R. Moszynski, T.G.A. Heijmen, and A. v.d. Avoird. *Chem. Phys. Letters,* 247:440, 1995.

[24] T.G.A. Heijmen, R. Moszynski, P.E.S. Wormer, and A. v.d. Avoird. Symmetry-adapted perturbation theory applied to interaction-induced properties of collisional complexes. *Molec. Phys.,* 89:81, 1996.

[25] R. Moszynski, T.G.A. Heijmen, P.E.S. Wormer, and A. v.d. Avoird. Theoretical modeling of spectra and collisional processes of weakly interacting complexes. *Adv. Quantum Chem.*, 28:119, 1997.

[26] T. Korona, R. Moszynski, and B. Jeziorski. Convergence of symmetry-adapted perturbation theory for the interaction between helium atoms and between a hydrogen molecule and a helium atom. *Adv. Quantum Chem.*, 28:171, 1997.

[27] X. Li and K. L. C. Hunt. Transient collision-induced dipoles in pairs of centrosymmetric, linear molecules at long range: Results from spherical-tensor analysis. *J. Chem. Phys.*, 100:9276, 1994.

[28] K. L. C. Hunt and X. Li. Dielectric properties of dense fluids. 1995. in Ref. [9].

[29] X. Li, M. H. Champagne, and K. L. C. Hunt. Long-range, collision-induced dipoles of T_d–$D_{\infty h}$ molecule pairs: theory and numerical results for CH_4 or CF_4 interacting with H_2, N_2, CO_2, or CS_2. *J. Chem. Phys.*, 109:8416, 1998.

[30] L. Frommhold and P. Dore. Analysis of the He–Xe absorption spectrum in the far infrared. *Phys. Review A*, 54:1717, 1996.

[31] L. Frommhold and M. V. Tonkov. Interaction-induced dipole and absorption spectra of collisional He–Ar pairs. *Phys. Review A*, 54:5438, 1996.

[32] I. M. Grigoriev, M. V. Tonkov, and L. Frommhold. Interaction-induced dipole and absorption spectra of collisional Ar–Xe pairs. *Phys. Review A*, 58:4978, 1998.

[33] R. H. Tipping, A. Brown, Q. Ma, J.M. Hartmann, C. Boulet, and J. Liévin. Collision-induced absorption in the ν_2 fundamental band of CH_4. Determination of the quadrupole transition moment. *J. Chem. Phys.*, 115:8852, 2001.

[34] J. M. Hartmann, C. Brodbeck, P.-M. Flaud, R. H. Tipping, A. Brown, Q. Ma, and J. Liévin. Collision-induced absorption in the ν_2 fundamental band of CH_4. Dependence on the perturber gas. *J. Chem. Phys.*, 116:123, 2002.

[35] B. McQuarrie, G. C. Tabisz, B. Gao, and J. Cooper. Role of the induced dipole moment in the collisional interference in the pure rotational spectrum of HD–He and HD–Ar. *Phys. Review A*, 52:1976, 1995.

[36] G. D. Greenblatt, J. J. Orlando, J. B. Burkholder, and A. R. Ravishankara. Absorption measurements of oxygen between 330 and 1140 nm. *J. Geophys. Res.*, 95:18577, 1990.

[37] M. B. Kiseleva, G. Ya. Zelikina, M. V. Buturlimova, and A. B. Burtsev. Collision-induced absorption of gaseous oxygen in the Herzberg continuum. p. 183, 2003. In Ref. [10].

[38] K. Pfeilsticker, H. Bösch, R. Fitzenberger, and C. Camy-Peyret. Spectroscopic and thermochemical information on the O_2–O_2 collisional complex inferred from atmospheric UV/visible O_4 absorption band profile measurements. p. 273, 2003. in Ref. [10].

[39] M. Buser, L. Frommhold, M. Gustafsson, M. Moraldi, M. H. Champagne, and K. L. C. Hunt. Far-infrared absorption by collisionally interacting nitrogen and methane molecules. *J. Chem. Phys.*, 121:2617, 2004.

[40] M. Buser and L. Frommhold. Infrared absorption by collisional CH_4–X pairs, with X=He, H_2, or N_2. *J. Chem. Phys.*, 122:024301-1, 2005.

[41] M. Buser and L. Frommhold. Work in preparation.

[42] L. S. Rothman *et al.* The HITRAN 2004 molecular spectroscopic data base. *J. Quant. Spectroscopy and Rad. Transfer*, to appear, 2005. HITRAN.

[43] A. Borysow and U. G. Jørgensen. Collision-induced absorption in dense atmospheres of cool stars. In R. Herman, editor, *Spectral Line Shapes, vol. 10*, p. 207, Woodbury, N.Y., 1999. Am. Inst. Physics.

[44] L. Frommhold. Electron-atom bremsstrahlung and the sonoluminescence of rare gas bubbles. *Phys. Review E*, 58:1899, 1998.

[45] D. Hammer and L. Frommhold. Electron-ion bremsstrahlung spectra calculations for sonoluminescence. *Phys. Review E*, 66:056303-1, 2002.

[46] L. A. Bureyeva and V. S. Lisitsa. Polarization recombination, a new recombination channel. *Phys. Scripta*, T80A:163, 1999.

[47] A. V. Korol and A. V. Solov'yov. Polarizational bremsstrahlung of electrons in collisions with atoms and clusters. *J. Phys. B*, 30:1105, 1997.

[48] V. A. Astapenko, L. A. Bureyeva, and V. S. Lisitsa. Incoherent polarization bremsstrahlung of a fast charged particle scattered by an atom. *Laser Phys.*, 10:960, 2000.

[49] V. A. Astapenko, L. A. Bureyeva, and V. S. Lisitsa. Classical and quantum theories of the polarization bremsstrahlung in the local density model. *J. Exp. Theor. Phys.*, 90:434, 2000.

[50] D. Hammer and L. Frommhold. Polarization bremsstrahlung spectra of electron-rare-gas atom collisions at temperatures from 5 to 40 kK. *Phys. Review A*, 64:024705-1, 2001. Erratum: ibid. **64**, 059901-1 (2001).

[51] M. Moraldi and L. Frommhold. Three-body induced dipole moments and infrared absorption: The H_2 fundamental band. *Phys. Review A*, 49: 4508, 1994.

[52] M. Moraldi and L. Frommhold. Irreducible three-body dipole moments in hydrogen. p. 41, 1995. in Ref. [9].

[53] M. Moraldi and L. Frommhold. Dipole moments induced in three interacting molecules. *J. Molec. Liquids*, 70:143, 1996.

[54] M. Moraldi and L. Frommhold. Triple transition $Q_1(j_1)+Q_1(j_2)+Q_1(j_3)$ near 12,466 cm^{-1} in compressed hydrogen. *Phys. Rev. Letters*, 74:363, 1995.

[55] S. P. Reddy, Fan Xiang, and G. Varghese. Observation of the triple transition $Q_1(J_1) + Q_1(J_2) + Q_1(J_3)$ in molecular hydrogen in its second overtone region. *Phys. Rev. Letters*, 74:367, 1995.

[56] M. Moraldi and L. Frommhold. Irreducible dipole components of three interacting H_2 molecules and the triple Q_1 transition near 12,466 cm^{-1}. *J. Chem. Phys.*, 103:2377, 1995.

[57] G. Birnbaum and B. Guillot. Cancellation effects in collision-induced phenomena. p. 1, 1995. in Ref. [9].

[58] X. Li and K. L. C. Hunt. Nonadditive, three-body dipoles and forces on nuclei: New interrelations and an electrostatic interpretation. *J. Chem. Phys.*, 105:4076, 1996.

[59] X. Li and K. L. C. Hunt. Nonadditive three-body dipoles of inert gas trimers and $H_2 \cdots H_2 \cdots H_2$: Long-range effects in far infrared absorption and triple vibrational transitions. *J. Chem. Phys.*, 107:4133, 1997.

[60] P Dore, L. Finzi, A. Nucara, P. Postorino, and M. Rovere. Translational absorption band in low-density mixtures of noble gases: the He–Xe case. *Molec. Phys.*, 84:1065, 1995.

[61] W. Jäger, Y. Xu, and M. C. L. Gerry. Pure rotational spectra of the mixed rare gas van der Waals complexes Ne–Xe, Ar–Xe, and Kr–Xe. *J. Chem. Phys.*, 99:919, 1993.

[62] C. Brodbeck, Nguyen-van-Thanh, J. P. Bouanich, and L. Frommhold. Collision-induced absorption by H_2–He pairs in the H_2 fundamantal band at 78 and 298 K. *Phys. Review A*, 51:1209, 1995.

[63] J. P. Bouanich, C. Brodbeck, and Nguyen-van-Thanh. Collision-induced absorption by H_2–He pairs in the H_2 fundamental band at 78 and 298 K. 1993.

[64] A. R. W. McKellar. Infrared spectra of weakly bound complexes and collision-induced effects involving atmospheric molecules. p. 223, 2003. In Ref. [10].

[65] R. D. G. Prasad, P. G. Gillard, and S. P. Reddy. Collision-induced first overtone band of H_2 in H_2–Kr and H_2–Xe mixtures. *J. Chem. Phys.*, 107:4906, 1997.

[66] E. H. Wishnow. *The far-infrared absorption spectrum of low temperature hydrogen gas.* PhD thesis, U. British Columbia, Vancouver, Canada, 1993.

[67] E. H. Wishnow, I. Ozier, and H. P. Gush. The pure translational spectrum of low temperature hydrogen gas. p. 495, 1995. In Ref. [9].

[68] E. H. Wishnow, I. Ozier, H. P. Gush, and J. Schaefer. Translational band of gaseous hydrogen at low temperature. *Astrophys. J.*, 492:843, 1998.

[69] A. R. W. McKellar. Infrared studies of van der Waals complexes. p. 467, 1995. In Ref. [9].

[70] J. Schäfer. Dimer features of H_2–H_2 and isotopomers at low temperatures. p. 485, 1995. In Ref. [9].

[71] C. Brodbeck, Nguyen-van-Thanh, A. Jean-Louis, J. P. Bouanich, and L. Frommhold. Collision-induced absorption by H_2 pairs in the fundamental band at 78 and 298 K. *Phys. Review A*, 50:484, 1994.

[72] J. Boissoles, A. Domanskaya, C. Boulet, R. H. Tipping, and Q. Ma. New experimental measurements and theoretical analysis of the collision-induced absorption of N_2–H_2 pairs. *J. Quant. Spectroscopy and Rad. Transfer*, 95:489, 2005.

[73] E. H. Wishnow, H. P. Gush, and I. Ozier. Far-infrared spectrum of N_2 and N_2-noble gas mixtures near 80 K. *J. Chem. Phys.*, 104:3511, 1996.

[74] Y. Xu and A. R. W. McKellar. The infrared spectrum of the Ar–CO complex: comprehensive analysis including van der Waals stretching and bending states. *Molec. Phys.*, 88:859, 1996.

[75] E. J. Mlawer, S. A. Clough, P. D. Brown, T. M. Stephens, J. C. Landry, A. Goldman, and F. J. Murcray. Observed atmospheric collision-induced absorption in near infrared oxygen bands. *J. Geophys. Res.*, 103:3859, 1998.

[76] A. A. Vigasin. Collision-induced absorption in the region of the O_2 fundamental band: bandshapes and dimeric features. *J. Molec. Spectrosc.*, 202:59, 2000.

[77] G. Moreau, J. Boissoles, C. Boulet, R. H. Tipping, and Q. Ma. Theoretical study of the collision-induced fundamental absorption spectra of O_2–O_2 pairs for temperatures between 193 and 273 K. *J. Quant. Spectroscopy and Rad. Transfer*, 64:87, 2000.

[78] Y. I. Baranov, W. J. Lafferty, and G. T. Fraser. Infrared spectrum of the continuum and dimer absorption in the vicinity of the O_2 vibrational fundamental in O_2/CO_2 mixtures. *J. Molec. Spectrosc.*, 228:432, 2004.

[79] N. N. Filippov, J.-P. Bouanich, C. Boulet, M. V. Tonkov, R. Le Doucen, and F. Thibault. Collision-induced double transition effects in the $3\nu_3$ CO_2 band wing region. *J. Chem. Phys.*, 106:2067, 1997.

[80] Y. I. Baranov and A. A. Vigasin. Collision-induced absorption by CO_2 in the region of ν_1, $2\nu_2$. *J. Molec. Spectrosc.*, 193:319, 1999.

[81] A. A. Vigasin, Y. I. Baranov, and G. V. Chlenova. Temperature variations of the interaction induced absorption of CO_2 in the ν_1,$2\nu_2$ region: FTIR measurements and dimer contribution. *J. Molec. Spectrosc.*, 213:51, 2002.

[82] Y. I. Baranov, G. T. Fraser, W. J. Lafferty, and A. A. Vigasin. Collision-induced absorption in the CO_2 Fermi triad for temperatures from 211 to 296 K. p. 149, 2003. In Ref. [10].

[83] M. V. Tonkov. Far infrared absorption spectrum of CO_2 with He, Ar, and Xe: Experiment and calculations. p. 457, 1995. In Ref. [9].

[84] W. J. Lafferty, A. M. Solodov, A. Weber, W. B. Olson, and J.-M. Hartmann. Infrared collision-induced absorption by N_2 near 4.3 μm for atmospheric applications: measurements and empirical modeling. *Appl. Opt.*, 35:5911, 1996.

[85] G. Moreau, J. Boissoles, R. Le Doucen, C. Boulet, R. H. Tipping, and Q. Ma. Experimental and theoretical study of the collision-induced fundamental absorption spectra of N_2–O_2 and O_2–N_2 pairs. *J. Quant. Spectroscopy and Rad. Transfer*, 69:245, 2001.

[86] B. Maté, G. T. Fraser, and W. J. Lafferty. Intensity of the simultaneous vibrational absorption $CO_2(\nu_3 = 1) + N_2(\nu = 1) \leftarrow CO_2(\nu_3 = 0) + N_2(\nu = 0)$ at 4680 cm^{-1}. *J. Molec. Spectrosc.*, 201:175, 2000.

[87] A. Brown and R. H. Tipping. Collision-induced absorption in dipolar molecule – homonuclear diatom pairs. p. 93, 2003. In Ref. [10].

[88] D. C. Tobin, L. L. Strow, W. J. Lafferty, and B. Olson. Experimental investigation of the self- and N_2-broadened continuum within the ν_2 band of water vapor. *Appl. Optics*, 35:4724, 1998.

[89] G. Birnbaum, A. Borysow, and A. Buechele. Collision-induced absorption in mixtures of symmetrical linear and thetrahedral molecules: methane-nitrogen. *J. Chem. Phys.*, 99:3234, 1993.

[90] I. M. Grigoriev, A. V. Domanskaya, and M. V. Tonkov. Intra- and intermolecular components of the ν_2 forbidden band of CF_4 in pure gas and in He, Ar, Xe, and N_2 mixtures. *Molec. Phys.*, 2:1851, 2004.

[91] B. Maté, C. Lugez, G. T. Fraser, and W. J. Lafferty. Absolute intensities for the O_2 1.27 μm continuum absorption. *J. Geophys. Res.*, 104: 30,585, 1999.

[92] A. Campargue, L. Biennier, A. Kachonov, R. Jost, B. Bussery-Honvault, V. Veyret, S. Churassy, and R. Bacis. Rotationally resolved absorption spectrum of the O_2 dimer in the visible range. *Chem. Phys. Letters*, 288:734, 1998.

[93] D. A. Newnham and J. Ballard. Visible absorption cross section and integrated absorption intensities of molecular oxygen O_2 and O_4. *J. Geophys. Res.*, 103:28801, 1998.

[94] G. T. Frazer and W. J. Lafferty. The 1.27-μm O_2 continuum absorption in O_2/CO_2 mixtures. *J. Geophys. Res.*, 106:31,749, 2001.

[95] A. Borysow and M. Moraldi. On the role of anisotropic interactions on collision induced absorption of systems containing linear molecules: the CO_2–Ar case. *J. Chem. Phys.*, 99:8424, 1993.

[96] M. Moraldi and L. Frommhold. Collision-induced infrared absorption by H_2–He complexes: Accounting for the anisotropy of the interaction. *Phys. Review A*, 52:274, 1995.

[97] M. Gruszka and A. Borysow. New analysis of the spectral moments of collision induced absorption in gaseous N_2 and CO_2. *Molec. Phys.*, 88:1173, 1996.

[98] M. Gruszka and A. Borysow. Rototranslational collision-induced absorption of CO_2 for the atmosphere of Venus at frequencies from 0 to 250 cm^{-1} at temperatures from 200 to 800 K. *Icarus*, 129:172, 1997.

[99] M. Gruszka and A. Borysow. Computer simulation of the far infrared collision-induced absorption spectra of gaseous CO_2. *Molec. Phys.*, 93:1007, 1998.

[100] M. Gustafsson, L. Frommhold, D. Bailly, J.-P. Bouanich, and C. Brodbeck. Collision-induced absorption in the rototranslational band of dense hydrogen gas. *J. Chem. Phys.*, 119:12264, 2003.

[101] K. L. C. Hunt and associates, 2005. In preparation.

[102] A. P. Mishra, T. K. Balasubramanian, R. H. Tipping, and Q. Ma. Absorption spectroscopy in solid hydrogen: challenges to experimentalists and theorists. *J. Molec. Structure*, 695-696:103, 2004.

[103] M. Gustafsson and L. Frommhold. The H_2–H infrared absorption bands at temperatures from 1000 to 2500 K. *A&A*, 400:1161, 2003.

[104] J. Boissoles, C. Boulet, R. H. Tipping, A. Brown, and Q. Ma. Theoretical calculation of the translation-rotation collision-induced absorption in N_2–N_2, O_2–O_2, and N_2–O_2 pairs. *J. Quant. Spectroscopy and Rad. Transfer*, 82:505, 2003.

[105] R. H. Tipping, A. Brown, Q. Ma, and C. Boulet. Collision-induced absorption in the $a^1\Delta_g(v' = 0) \leftarrow X^3\Sigma_g^-(v = 0)$ transition in O_2–CO_2, O_2–N_2, and O_2–H_2O mixtures. *J. Molec. Spectrosc.*, 209:88, 2001.

[106] R. H. Tipping, Q. Ma, C. Boulet, and J.-M. Hartmann. Theoretical analysis of the collision induced electronic absorption in O_2–N_2 and O_2–CO_2 pairs. *J. Molec. Structure*, 742:83, 2005.

[107] A. R. W. McKellar. Low temperature infrared absorption of gaseous N_2 and N_2+H_2 in the 2.0–2.5 μm region: Application to the atmospheres of Titan and Triton. *Icarus*, 80:361, 1989.

[108] A. A. Vigasin. On the nature of collision-induced absorption in gaseous homonuclear diatomics. *J. Quant. Spectroscopy and Rad. Transfer*, 56:409, 1996.

[109] G. Moreau, J. Boissoles, R. Le Doucen, C. Boulet, R. H. Tipping, and Q. Ma. Metastable dimer contributions to the collision-induced fundamental absorption spectra of N_2 and O_2 pairs. *J. Quant. Spectroscopy and Rad. Transfer*, 70:99, 2001.

[110] D. Hammer, L. Frommhold, and W. Meyer. Emission by collisional H_2–He pairs for temperatures from 2 to 20 kK. *J. Chem. Phys.*, 111: 6283, 1999.

[111] D. Hammer and L. Frommhold. Emission by collisional D_2–He pairs at temperatures from 2 to 20 kK. *J. Chem. Phys.*, 112:654, 2000.

[112] A. Gross and R. D. Levine. Spectroscopic characterization of collision-induced electronic deformation energy using sum rules. *J. Chem. Phys.*, 119:4283, 2003.

[113] L. Ulivi, Z. Lu, and G. C. Tabisz. Interference of allowed and collision-induced transitions in hd: experiment. p. 407, 1995. In Ref. [9].

[114] G. C. Tabisz, B. Gao, and J. Cooper. Interference of allowed and interaction-induced transitions in HD. p. 417, 1995. In Ref. [9].

[115] G. C. Tabisz. Interference effects in the infrared spectrum of HD: Atmospheric implications. p. 83, 2003. In Ref. [10].

[116] B. Gao, G. C. Tabisz, M. Trippenbach, and J. Cooper. Spectral line shape arising from collisional interference between electric-dipole-allowed and collision-induced transitions. *Phys. Review A*, 44:7379, 1991.

[117] B. Gao, J. Cooper, and G. C. Tabisz. Rotational spectrum of HD perturbed by He or Ar gases. *Phys. Review A*, 46:5781, 1992.

[118] Z. Lu, G. C. Tabisz, and L. Ulivi. Temperature dependence of the pure rotational band of HD: Interference, widths and shifts. *Phys. Review A*, 47:1159, 1993.

[119] B. McQuarrie and G. C. Tabisz. Collisional interference in the infrared spectrum of HD: calculation of the line shape of vibrorotational transitions for HD–He. *J. Molec. Liquids*, 70:159, 1996.

[120] W. Herrebout, B. Van der Veken, A. P. Kouzov, and M. O. Bulanin. Collision-induced absorption of hydrogen deuteride dissolved in liquid neon. *Phys. Rev. Letters*, 92:023002, 2004.

[121] M. Gustafsson and L. Frommhold. Intracollisional interference of *R* lines of HD in mixtures of deuterium hydride and helium gas. *Phys. Review A*, 63:052514-1, 2001.

[122] B. Guillot and Y. Guissani. Simulation of the far infrared spectrum of liquid water and steam. p. 129, 1995. In Ref. [9].

[123] S. Weiss. Ternary effects in far IR absorption in the gas phase. p. 51, 1995. In Ref. [9].

[124] S. Weiss. Simulation of gas phase collision-induced spectra: quadrupole-induced spectra including ternary effects. *Molec. Phys.*, 72:987, 1991.

[125] M. Mengel, B. P. Winnewisser, and M. Winnewisser. *J. Molec. Spectrosc.*, 188:221, 1998.

[126] M. Mengel, B. P. Winnewisser, and M. Winnewisser. *J. Low Temp. Phys.*, 111:757, 1998.

[127] C. G. Joslin, C. G. Gray, J. D. Poll, S. Goldman, and A. D. Buckingham. Interaction-induced spectra of endohedral complexes of buckminsterfullerene. p. 261, 1995. In Ref. [9].

[128] Z. Slanina, L. Adamowicz, J.-P. Francois, and E. Osawa. chapter Fullerenes and othe carbon aggregates: and the diffuse interstellar bands, p. 133. 1998. in Ref. [11].

[129] W. Glaz and G. C. Tabisz. Modelling the far wings of collision-induced translational spectral profiles. *Can. J. Phys.*, 79:801, 2001.

[130] A. A. Vigasin. Bound, metastable and free states of bimolecular complexes. *Infrared Phys.*, 32:461, 1991.

[131] S. Yu. Epifanov and A. A. Vigasin. Contributions of bound, metastable and free states of bimolecular complexes to collision-induced intensity of absorption. *Chem. Phys. Letters*, 225:537, 1994.

[132] A. A. Vigasin. Bimolecular absorption in atmospheric gases. p. 23, 2003. In Ref. [10].

[133] P. F. Bernath. The spectroscopy of water vapor: Experiment, theory and applications. *Phys. Chem. Chem. Phys*, 4:1501, 2002.

[134] Water dimers and weakly interacting species in atmospheric modeling, 2005. CECAM Workshops, www.tech.ing.unipg.it/WISPA.

[135] Q. Ma and R. H. Tipping. A far wing line shape theory and its application to the water vibrational bands II. *J. Chem. Phys.*, 96:8655, 1992.

[136] A. Bauer, M. Godon, J. Carlier, Q. Ma, and R. H. Tipping. Absorption by H_2O and H_2O–N_2 mixtures at 153 GHz. *J. Quant. Spectroscopy and Rad. Transfer*, 50:463, 1993.

[137] Q. Ma and R. H. Tipping. A near-wing correction to the quasistatic far-wing line shape theory. *J. Chem. Phys.*, 100:2537, 1994.

[138] R. H. Tipping and Q. Ma. Calculation of far wings of allowed spectra: The water continuum. p. 369, 1995. In Ref. [9].

[139] R. H. Tipping and Q. Ma. Theory of the water vapor continuum and validations. *Atmosph. Res.*, 36:69, 1995.

[140] A. Bauer, M. Godon, J. Carleer, and Q. Ma. Water vapor absorption in the atmospheric window at 239 GHz. *J. Quant. Spectroscopy and Rad. Transfer*, 53:411, 1995.

[141] A. A. Vigasin. chapter Dimeric absorption in the atmosphere, p. 60. 1998. In Ref. [11].

[142] F. Huisken. chapter Infrared spectroscopy of size-selected free and adsorbed water complexes, p. 238. 1998. In Ref. [11].

[143] A. A. Vigasin. Water vapor continuous absorption in various mixtures: possible role of weakly bound complexes. *J. Quant. Spectroscopy and Rad. Transfer*, 64:25, 2000.

[144] Q. Ma and R. H. Tipping. The frequency detuning corrections and the asymmetry of line shapes: the far wings of H_2O–H_2O. *J. Chem. Phys.*, 116:4102, 2002.

[145] J. G. Cormier, R. Ciurylo, and J. R. Drummond. Cavity ringdown spectroscopy measurements of the infrared water vapor continuum. *J. Chem. Phys.*, 116:1030, 2002.

[146] T. Kuhn, A. Bauer, M. Godon, S. Bühler, and K. Künzi. Water vapor continuum: absorption measurement at 350 GHz and model calculations. *J. Quant. Spectroscopy and Rad. Transfer*, 74:545 – 562, 2002.

[147] M. V. Tonkov and N. N. Filippov. Collision-induced far wings of CO_2 and H_2O bands in IR spectra. p. 125, 2003. In Ref. [10].

[148] R. H. Tipping and Q. Ma. Far wing line shapes: application to the water continuum. p. 137, 2003. In Ref. [10].

[149] A. N. Maurellis, R. Lang, J. E. Williams, W. J. van der Zande, K. Smith, D. A. Newnham, and R. N. Tolchenov. The impact of new water vapor spectroscopy on satellite retrievals. p. 259, 2003. In Ref. [10].

[150] J. M. Hartmann, M. Y. Perrin, Q. Ma, and R. H. Tipping. The infrared continuum of pure water vapor: Calculations and high-temperature measurements. *J. Quant. Spectroscopy and Rad. Transfer*, 49:675, 2003.

[151] X. Wang, A. Senchuk, and G. C. Tabisz. The far-infrared continuum in the spectrum of water vapor. p. 233, 2003. In Ref. [10].

[152] K. Pfeilsticker, A. Lotter, C. Peters, and H. Bösch. Atmospheric detection of water dimers via near-infrared absorption. *Science*, 300:2078, 2003.

[153] D. P. Schofield and H. G. Kjaergaard. Calculated OH-stretching and HOH-bending vibrational transitions in the water dimer. *Phys. Chem. Chem. Phys*, 5:3100, 2003.

[154] M. Carleer, M. Kiseleva, S. Fally, P.-F. Coheur, C. Clerbaux, R. Colin, L. Daumont, A. Jenouvrier, M.-F. Merienne, C. Hermans, and A. C. Vandaele. Laboratory Fourier transform spectroscopy of the water absorption continuum from 2500 to 22500 cm^{-1}. p. 213, 2003. In Ref. [10].

[155] I. V. Ptashnik, K. M. Smith, K. P. Shine, and D. A. Newham. Laboratory measurements of water vapour continuum absorption in spectral region 5000 – 5600 cm^{-1}: Evidence for water dimers. *Quart. J. Roy. Meteor. Soc.*, 130:2392, 2004.

[156] J. G. Cormier, J. T. Hodges, and J. R. Drummond. Infrared water vapor continuum absorption at atmospheric temperatures. *J. Chem. Phys.*, 122:114309, 2005.

[157] A. A. Vigasin, T. G. Adiks, E. G. Tarakanova, and G. V. Yukhnevich. Simultaneous infrared absorption in a mixture of CO_2 and H_2O: the role of hydrogen-bonded aggregates. *J. Quant. Spectroscopy and Rad. Transfer*, 52:295, 1994.

[158] Q. Ma and R. H. Tipping. The density matrix of H_2O–N_2 in the coordinate representation: a Monte Carlo calculation of the far-wing line shape. *J. Chem. Phys.*, 112:574, 2000.

[159] Q. Ma and R. H. Tipping. Water vapor millimeter wave foreign continuum: a Laczos calculation in the coordinate representation. *J. Chem. Phys.*, 117:10581, 2002.

[160] Q. Ma and R. H. Tipping. A simple analytical parametrization for the water vapor millimeter wave foreign continuum. *J. Quant. Spectroscopy and Rad. Transfer*, 82:517, 2003.

[161] Y. I. Baranov, G. T. Fraser, W. J. Lafferty, B. Maté, and A. A. Vigasin. Laboratory studies of oxygen continuum absorption. p. 159, 2003. In Ref. [10].

[162] C. Hermans, A. C. Vandaele, S. Fally, M. Carleer, R. Colin, B. Coquart, A. Jenouvrier, and M. F. Merienne. Absorption cross section of the collision-induced bands of oxygen in the from the UV to the NIR. p. 193, 2003. In Ref. [10].

[163] G. V. Tchlenova, A. A. Vigasin, J.-P. Bouanich, and C. Boulet. The nature of the absorption bandshape density evolution for the first overtone of CO compressed by N_2. *Infrared Phys.*, 34:289, 1993.

[164] A. Borysow and C. Tang. Far infrared spectra of N_2–CH_4 pairs for modeling of Titan's atmosphere. *Icarus*, 105:175, 1993.

[165] Z. Slanina, S. J. Kim, K. Fox, F. Uhlik, and A. Hinchliffe. chapter Dimers in earth's and planetary atmospheres, p. 100. 1998. In Ref. [11].

[166] L. Schriver-Mazzuoli. chapter Chemical and optical properties of molecular complexes using matrix isolation spectroscopy, p. 194. 1998. In Ref. [11].

[167] Y. I. Baranov, W. J. Lafferty, G. T. Fraser, and A. A. Vigasin. On the origin of the band structure observed in the collision-induced absorption bands of CO_2. *J. Molec. Spectrosc.*, 218:260, 2003.

[168] A. A. Vigasin. Equilibrium constants for the formation of weakly bound dimers. p. 111, 2003. in Ref. [10].

[169] L. M. Trafton. Induced spectra in planetary atmospheres. p. 517, 1995. In Ref. [9].

[170] G. Birnbaum, A. Borysow, and G. S. Orton. Collision-induced absorption of H_2–H_2 and H_2–He in the rotational and fundamental bands for planetary applications. *Icarus*, 123:4, 1996.

[171] L. M. Trafton. chapter Planetary Atmospheres: The role of collision-induced absorption, p. 177. 1998. In Ref. [11].

[172] M. Sneep. *The atmosphere in the laboratory: Cavity ring-down measurements on scattering and absorption.* PhD thesis, Free University of Amsterdam, 2004.

[173] B. E. Carlson, Q. Ma, and A. A. Lacis. On the inclusion of the hydrogen dimer in the analysis of the Voyager IRIS spectra. *Astrophys. J.*, 394: L29, 1992.

[174] Y. Fu, C. Zheng, and A. Borysow. First quantum mechanical computations of collision induced absorption in the second overtone band of H_2. *J. Quant. Spectroscopy and Rad. Transfer*, 67:303, 2000.

[175] C. Brodbeck, J.-P. Bouanich, Nguyen-Van-Thanh, Y. Fu, and A. Borysow. Collision induced absorption by H_2 pairs in the second overtone band at 298 and 77.5 K. *J. Chem. Phys.*, 10:4750, 1999.

[176] C. Brodbeck, J.-P. Bouanich, Nguyen-Van-Thanh, and A. Borysow. The binary collision induced second overtone band of gaseous hydrogen: modeling and laboratory measurements. *Planetary and Space Sci.*, 47: 1285, 1999.

[177] C. Zheng and A. Borysow. Rototranslational CIA spectra of H_2–H_2 at temperatures from 600 to 7000 K. *Astrophys. J.*, 441:960, 1995.

[178] C. Zheng and A. Borysow. Modeling of collision-induced infrared absorption spectra of H_2–H_2 in the first overtone band at temperatures from 20 to 500 K. *Icarus*, 113:84, 1995.

[179] A. Borysow, J. Borysow, and Y. Fu. Semiempirical model of cllision-induced infrared absorption spectra of H_2–H_2 complexes in the second overtone band of H_2 at temperatures from 50 to 500 K. *Icarus*, 145: 601, 2000.

[180] A. Borysow, U. G. Jørgensen, and Y. Fu. High temperature (1000 – 7000 K) collison-induced absorption spectra of H_2 pairs from first principles with applications to dense stellar atmospheres. *J. Quant. Spectroscopy and Rad. Transfer*, 68:235, 2001.

[181] A. Borysow, J. P. Champion, U. G. Jørgensen, and C. Wenger. Preliminary CH_4 line list data for stellar atmospheres. *ASP Conf. Series*, 288:352, 2003.

[182] A. Borysow. Pressure-induced molecular absorption in stellar atmospheres. In U. G. Jørgensen, editor, *Molecules in the Stellar Environment*, volume 428 of *Lecture Notes in Physics*, p. 209, Springer, 1 edition, 1994.

[183] A. Borysow. Collision-induced molecular absorption in stellar atmospheres. p. 529, 1995. In Ref. [9].

[184] A. Borysow, U. G. Jørgensen, and Ch. Zheng. Model atmospheres of cool, low-metallicity stars: The importance of collision-induced absorption. *Astronomy and Astrophysics*, 324:185, 1997.

[185] D.E. Winget, C.J. Hansen, J. Liebert, H.M. Van Horn, G. Fontaine, R.E. Nather, S.O. Kepler, and D.Q. Lamb. *ApJ*, 315:L77, 1987.

[186] G. Fontaine, P. Brassard, and P. Bergeron. *PASP*, 113:409, 2001.

[187] P. Bergeron, A. Ruiz, and B. Leggett. The chemical evolution of white dwarfs and the age of the local galactic disk. *Astrophys. J.*, 108:339, 1997.

[188] B. M. S. Hansen. Cooling models for old white dwarfs. *Astrophys. J.*, 520:680, 1999.

References

[1] G. Adam and J. Katriel. Dipole moment of ternary homonuclear systems. *Chem. Phys. Lett.*, 21:451, 1973.

[2] A. D. Afanasev, M. O. Bulanin, and M. V. Tonkov. Collision anisotropy in the CH_4–He system. *Sov. Tech. Phys. Lett.*, 6:1444, 1980.

[3] A. D. Afanasev, M. O. Bulanin, and M. V. Tonkov. The collision-induced translational spectrum of CF_4–He gas mixture. *Can. J. Phys.*, 58:836, 1980.

[4] E. J. Allin, W. F. J. Hare, and R. E. MacDonald. Infrared absorption of liquid and solid hydrogen. *Phys. Rev.*, 98:554, 1955.

[5] E. J. Allin, A. D. May, B. P. Stoicheff, J. C. Stryland, and H. L. Welsh. Spectroscopy research at the McLennan Physical Laboratories of the University of Toronto. *Appl. Opt.*, 6:1597, 1967.

[6] R. D. Amos. An accurate *ab initio* study of the multipole moments and polarizability of methane. *Molec. Phys.*, 38:33, 1979.

[7] G. V. Andreeva, A. A. Kudryavtsev, M. V. Tonkov and N. N. Filippov. Investigation of the integral characteristics of far-IR absorption of mixtures of CO_2 with inert gases. *Opt. Spectrosc. (USSR)* 68:623, 1990.

[8] R. L. Armstrong and H. L. Welsh. The infrared spectrum of CO–He mixtures at high pressures. *Can. J. Phys.*, 43:547, 1965.

[9] H. F. Arnoldus, T. F. George, K. S. Lam, F. F. Scipione, P. L. DeVries, and J. M. Yuan. Recent progress in the theory of laser-assisted collisions. In D. K. Evans, ed., *Laser Applications in Physical Chemistry*, Marcel Dekker, N.Y., 1986.

[10] A. M. Arthurs and A. Dalgarno. The theory of scattering by a rigid rotator. *Proc. Roy. Soc. (London)*, Ser. A, 256:540, 1960.

[11] R. A. Aziz and J. Chen. *J. Chem. Phys.*, 67:5719, 1977.

[12] R. A. Aziz, P. W. Riley, U. Buck, G. Maneke, J. Schleussner, G. Scoles, and U. Valbusa. *J. Chem. Phys.*, 71:2637, 1979.

[13] R. A. Aziz. Interatomic potentials for rare gases: pure and mixed interactions. In M. L. Klein, ed., *Inert Gases – Potentials, Dynamics, and Energy Transfer in Doped Crystals*, p. 5, Springer, Berlin, 1984.

[14] G. Bachet. Experimental observation of the intercollisional interference effect on the S(1) pure rotational line of the collision induced spectrum of the H_2–He mixture. *J. Physique Lett.*, 44:183, 1983.

[15] G. Bachet, E. R. Cohen, P. Dore, and G. Birnbaum. The translational rotational absorption spectrum of hydrogen. *Can. J. Phys.*, 61:591, 1983.

[16] U. Bafile, L. Ulivi, M. Zoppi, M. Moraldi, and L. Frommhold. The third virial coefficients of collision-induced, depolarized light scattering of hydrogen. *Phys. Rev. A*, 44:4450, 1991.

[17] F. Barocchi, M. Neri, and M. Zoppi. Derivation of three-body collision induced light scattering spectral moments for argon, krypton, xenon. *Molec. Phys.*, 34:1391, 1977.

[18] F. Barocchi, M. Neri, and M. Zoppi. Moments of the three-body collision induced light scattering spectrum. *J. Chem. Phys.*, 66:3308, 1977.

[19] F. Barocchi and M. Zoppi. Collision induced light scattering: Three- and four-body spectra of gaseous argon. *Phys. Lett.*, A 69:187, 1978.

[20] F. Barocchi, M. Celli and M. Zoppi. Interaction-induced translational Raman scattering in dense krypton gas: Evidence of irreducible many-body effects. *Phys. Rev.* A 38:3984, 1988.

[21] E. Bar-Ziv and S. Weiss. Translational spectra due to collision-induced overlap moments in mixtures of He with CO_2, N_2, CH_4, and C_2H_6. *J. Chem. Phys.*, 57:34, 1972.

[22] E. Bar-Ziv and S. Weiss. Collision-induced spectra of rare gas mixtures: Experimental and empirical relations. *J. Chem. Phys.*, 64:2412, 1976.

[23] E. Bar-Ziv and S. Weiss. Uniform reduced interaction dipole for all rare gas pairs. *J. Chem. Phys.*, 64:2417, 1976.

[24] W. E. Baylis. *J. Chem. Phys.*, 51:2665, 1969.

[25] A. Ben-Reuven. Radiatively damped collisions of ultracold atoms. In L. Frommhold and J. W. Keto, eds., *Spectral Line Shapes* **6**, p. 206, Am. Inst. Phys., N.Y., 1990.

[26] P. R. Berman. Three-state model for laser-assisted collisions. In J. Szudy, ed., *Spectral Line Shapes* **5**, Ossolineum, Warsaw, 1989.

[27] N. Bernardes and H. Primakoff. Molecule formation in the inert gases. *J. Chem. Phys.*, 30:691, 1959.

[28] B. J. Berne, J. Jortner, and R. G. Gordon. Vibrational relaxation of diatomic molecules in gases and liquids. *J. Chem. Phys.*, 47:1600, 1967.

[29] H. J. Bernstein and G. Herzberg. *J. Chem. Phys.*, 16:30, 1948.

[30] R. B. Bernstein. Quantum mechanical analysis of differential elastic scattering of molecular beams. *J. Chem. Phys.*, 33:795, 1960.

[31] R. B. Bernstein, ed. *Atom–Molecule Collision Theory*. Plenum, New York, 1979.

[32] M. V. Berry and K. E. Mount. *Rep. Prog. Phys.*, 35:315, 1972.

[33] G. Birnbaum, A. A. Maryott, and P. F. Wacker. Microwave absorption by the nonpolar gas CO_2. *J. Chem. Phys.*, 22:1782, 1954.

[34] G. Birnbaum, W. Ho, and A. Rosenberg. Far-infrared collision-induced absorption in CO_2. II. Pressure dependence in the gas phase and absorption in the liquid. *J. Chem. Phys.*, 55:1039, 1971.

[35] G. Birnbaum and E. R. Cohen. Determination of molecular multipole moments and potential function parameters of non-polar molecules from far infrared spectra. *Molec. Phys.*, 32:161, 1976.

[36] G. Birnbaum and E. R. Cohen. Theory of line shapes in pressure induced absorption. *Can. J. Phys.*, 54:593, 1976.

[37] G. Birnbaum. Far-infrared absorption in H_2 and H_2–He mixtures. *J.Q.S.R.T.* 19:51, 1978.

[38] G. Birnbaum. Determination of molecular constants from collision-induced far-infra-red spectra and related methods. In [406], p. 111.

[39] G. Birnbaum, M. S. Brown, and L. Frommhold. Lineshapes and dipole moments in collision-induced absorption. *Can. J. Phys.*, 59:1544, 1981.

[40] G. Birnbaum and H. Sutter. Collision induced absorption in a highly symmetric molecule: SF_6. *Molec. Phys.*, 42:21, 1981.

[41] G. Birnbaum. A study of atomic and molecular interactions from collision-induced spectra. *Proc. 8 Symp. Therm. Phys.*, 1:8, 1982.

[42] G. Birnbaum, L. Frommhold, L. Nencini, and H. Sutter. The collision-induced far-infrared absorption band of gaseous methane in the region 30–900 cm^{-1}. *Chem. Phys. Lett.*, 100:292, 1983.

[43] G. Birnbaum, Shih-I Chu, A. Dalgarno, L. Frommhold, and E. L. Wright. Theory of collision-induced translation–rotation spectra: H_2–He. *Phys. Rev.*, A 29:595, 1984.

[44] G. Birnbaum, M. Krauss, and L. Frommhold. Collision-induced dipoles of rare gas mixtures. *J. Chem. Phys.*, 80:2669, 1984.

[45] G. Birnbaum, ed., *Phenomena Induced by Intermolecular Interactions*. Plenum Press, New York, 1985.

[46] G. Birnbaum, A. Borysow, and H. G. Sutter. Measurement and analysis of the far infrared absorption spectrum of the gaseous mixture H_2–CH_4. *J.Q.S.R.T.* 38:189, 1987.

[47] G. Birnbaum, L. Frommhold, and G.C. Tabisz. Collision induced spectroscopy: absorption and light scattering. In J. Szudy, ed., *Spectral Line Shapes* **5**, p. 623, Ossolineum, Warsaw, 1989.

[48] G. Birnbaum and A. Borysow. On the problem of detailed balance and model line shapes in collision-induced rotovibrational bands: H_2–H_2 and H_2–He. *Molec. Phys.*, 73:57, 1991.

[49] B. L. Blaney and G. E. Ewing. Van der Waals molecules. In B. S. Rabinovitch, H. S. Johnston, and J. M. Schurr, eds., *Ann. Rev. Phys. Chem.*, p. 553, Annual Reviews, Palo Alto, 1976.

[50] J. E. Bohr and K. L. C. Hunt. Dipoles induced by van der Waals interactions during collisions of atoms with heteroatoms or centrosymmetric linear moleculs. *J. Chem. Phys.*, 86:5441, 1987.

[51] J. E. Bohr and K. L. C. Hunt. Dipoles induced by long-range interactions between centrosymmetric linear molecules: H_2–H_2, H_2–N_2, and N_2–N_2. *J. Chem. Phys.*, 87:3821, 1987.

[52] A. Borysow, M. Moraldi, and L. Frommhold. Modeling of collision-induced absorption spectra. *J.Q.S.R.T.* 31:235, 1984.

[53] A. Borysow and L. Frommhold. Collision induced rototranslational absorption spectra of N_2–N_2 pairs for temperatures from 50 to 300 K. *Astrophys. J.*, 311:1043, 1986.

[54] A. Borysow and L. Frommhold. Theoretical collision induced rototranslational absorption spectra for modeling Titan's atmosphere: H_2–N_2 pairs. *Astrophys. J.*, 303:495, 1986.

[55] A. Borysow and L. Frommhold. Theoretical collision induced rototranslational absorption spectra for the outer planets: H_2–CH_4 pairs. *Astrophys. J.*, 304:849, 1986.

[56] A. Borysow and L. Frommhold. Collision induced rototranslational absorption spectra of binary methane complexes (CH_4–CH_4). *J. Molec. Spectrosc.*, 123:293, 1987.

[57] A. Borysow and L. Frommhold. Collision induced rototranslational absorption spectra of CH_4–CH_4 pairs at temperatures from 50 to 300 K. *Astrophys. J.*, 318:940, 1987.

[58] A. Borysow, L. Frommhold, and P. Dore. Far infrared absorption by pairs of nonpolar molecules. *Intern. J. Infrared and Millimeter Waves*, 8:381, 1987.

[59] A. Borysow, L. Frommhold, and W. Meyer. Dipoles induced by the interactions of HD with He, Ar, H_2, or HD. *J. Chem. Phys.*, 88:4855, 1988.

[60] A. Borysow and L. Frommhold. Collision-induced light scattering: a bibliography. In I. Prigogine and S. A. Rice, eds., *Adv. Chem. Phys.* LXXV:439, 1989.

[61] A. Borysow and L. Frommhold. Collision-induced infrared spectra of H_2–He pairs at temperatures from 18 to 7,000 K: Overtone and 'hot' bands. *Astrophys. J.*, 341:549, 1989.

[62] A. Borysow, L. Frommhold, and M. Moraldi. Collision induced infrared spectra of H_2–He pairs involving $0 \leftrightarrow 1$ vibrational transitions and temperatures from 18 to 7,000 K. *Astrophys. J.*, 336:495, 1989.

[63] A. Borysow, L. Frommhold, and W. Meyer. Collision induced spectra of H_2–He: dependence of the radial dipole transition elements on the rotational states. *Phys. Rev.* A 41:264, 1990.

[64] A. Borysow and L. Frommhold. A new computation of the infrared absorption by H_2 pairs in the fundamental band at temperatures from 600 to 5000 K. *Astrophys. J.*, 348:L41, 1990.

[65] A. Borysow. Modeling of collision-induced infrared absorption spectra of H_2–H_2 pairs in the fundamental band at temperatures from 20 to 300 K. *Icarus*, 92:273, 1991.

[66] A. Borysow and L. Frommhold. Pressure-induced absorption in the infrared: a data base for the modeling of planetary atmospheres. *J. geophys. Res.*, 96:17501, 1991.

[67] A. Borysow and M. Moraldi. Effects of anisotropic interaction on collision-induced absorption by pairs of linear molecules. *Phys. Rev. Lett.* 68:3686, 1992.

[68] J. Borysow and L. Frommhold. The infrared and Raman line shapes of pairs of interacting molecules. In [45], p. 67.

[69] J. Borysow, L. Trafton, L. Frommhold, and G. Birnbaum. Modelling of pressure-induced far infrared absorption spectra: Molecular hydrogen pairs. *Astrophys. J.*, 296:644, 1985.

[70] J. Borysow, M. Moraldi, and L. Frommhold. The collision-induced spectroscopies. Concerning the desymmetrization of classical line shape. *Molec. Phys.*, 56:913, 1985.

[71] J. Borysow, M. Moraldi, L. Frommhold, and J.D. Poll. Spectral line shape in collision induced absorption: An improved constant acceleration approximation. *J. Chem. Phys.*, 84:4277, 1986.

[72] J. Borysow, L. Frommhold, and G. Birnbaum. Collision induced rototranslational absorption spectra of H_2–He pairs at temperatures from 40 to 3000 K. *Astrophys. J.*, 326:509, 1988.

[73] T. K. Bose. A comparative study of the dielectric, refractive and Kerr virial coefficients. In [45], p. 49.

[74] D. R. Bosomworth and H. P. Gush. Collision-induced absorption of compressed gases in the far infrared, Part I. *Can. J. Phys.*, 43:729, 1965.

[75] D. R. Bosomworth and H. P. Gush. Collision-induced absorption of compressed gases in the far infrared, II. *Can. J. Phys.*, 43:751, 1965.

[76] C. Bottcher, A. Dalgarno, and E. L. Wright. Collision-induced absorption in alkali-metal-atom–inert-gas mixtures. *Phys. Rev.*, A 7:1606, 1973.

[77] J. P. Bouanich, C. Brodbeck, P. Drossart, and E. Lellouch. Collision-induced absorption for H_2–H_2 and H_2–He interactions at 5 μm. *J.Q.S.R.T.* 42:141, 1989.

[78] J. P. Bouanich, C. Brodbeck, Nguyen-van-Thanh and P. Drossart. Collision-induced absorption by H_2–H_2 and H_2–He pairs in the fundamental band – an experimental study. *J.Q.S.R.T.* 44:393, 1990.

[79] S. L. Brenner and D. A. McQuarrie. On the theory of collision-induced absorption in rare gas mixtures. *Can. J. Phys.*, 49:837, 1971.

[80] D. M. Brink and G. R. Satchler. *Angular Momentum.* Oxford, 1962.

[81] F. R. Britton and M. F. Crawford. Theory of collision-induced absorption in hydrogen and deuterium. *Can. J. Phys.*, 36:761, 1958.

[82] P. R. Brooks. Spectroscopy of transition region species. *Chem. Rev.*, 88:407, 1988.

[83] M. S. Brown, L. Frommhold, and G. Birnbaum. About an information theoretical spectral shape proposed for the collision induced spectroscopies. *Molec. Phys.*, 62:907, 1987. See also *Molec. Phys.*, 64:1001, 1988.

[84] L. W. Bruch, C. T. Corcoran, and F. Weinhold. On the dipole moment of three identical spherical atoms. *Molec. Phys.*, 35:1205, 1978.

[85] L. W. Bruch and T. Osawa. Perturbation theory of interacting dipoles. *Molec. Phys.*, 40:491, 1980.

[86] A. D. Buckingham and J. A. Pople. Electromagnetic properties of compressed gases. *Discussions Faraday Soc.*, 22:17, 1956.

[87] A. D. Buckingham. The molecular refraction of an imperfect gas. *Trans. Faraday Soc.*, 52:747, 1956.

[88] A. D. Buckingham. L'absorption des ondes micrométriques induite par la pression dans des gaz non polaires. In *Propriétés Optiques et Acoustiques des Fluides Comprimés et Actions Intermoléculaires*, p. 57, Centre National de la Recherche Scientifique, Paris, 1959. Colloques Internationaux du C.N.R.S., vol. 77.

[89] A. D. Buckingham. Permanent and induced molecular moments and long-range interactions. In I. Prigogine and S. A. Rice, eds., *Adv. Chem. Phys.* **12**, p. 107, Wiley, New York, 1967.

[90] A. D. Buckingham. Theory of dielectric polarization. In A. D. Buckingham and G. Allen, eds., *MTP Int. Rev. of Science*, vol. 2, p. 241, Butterworths, London, 1972.

[91] A. D. Buckingham and G. C. Tabisz. Collision induced rotational Raman scattering. *Opt. Lett.*, 1:220, 1977.

[92] A. D. Buckingham and G. C. Tabisz. Collision-induced Raman scattering by tetrahedral and octahedral molecules. *Molec. Phys.*, 36:583, 1978.

[93] A. D. Buckingham. The effects of collisions on molecular properties. *Pure and Appl. Chem.*, 52:2253, 1980.

[94] A. D. Buckingham. General Introduction – Faraday Symposium 22 on Interaction Induced Spectra in Dense Fluids and Disordered Solids. *J. Chem. Soc. Faraday Trans. 2*, 83(10):1743ff., 1987.

[95] V. I. Bukhtoyarova and M. V. Tonkov. Binary and ternary absorbances in the translational spectrum of a He–Ar mixture at 165 K. *Opt. Spectrosc.*, 42:14, 1977.

[96] V. I. Bukhtoyarova and M. V. Tonkov. Intermolecular interactions in compressed gases from translational absorption spectra I: Spectral moments of translation bands. *Opt. Spectrosc.*, 43:27, 1977.

[97] V. I. Bukhtoyarova and M. V. Tonkov. Intermolecular interactions in compressed gases from translational absorption spectra II: Temperature dependence of intensity. *Opt. Spectrosc.*, 43:133, 1977.

[98] M. O. Bulanin. Electrostatic induction in van der Waals complexes. In J. Szudy, ed., *Spectral Line Shapes* **5**, p. 597, Ossolineum, Warsaw, 1989.

[99] U. Buontempo, S. Cunsolo, and G. Jacucci. Far infrared absorption in N_2–Ar mixtures. *Molec. Phys.*, 21:381, 1971.

[100] U. Buontempo, S. Cunsolo, and G. Jacucci. Memory effects in the collision-induced spectra of the noble gases in liquid argon. *Can. J. Phys.*, 49:2870, 1971.

[101] U. Buontempo, S. Cunsolo, and P. Dore. Intercollisional memory effects and short-time behavior of the velocity-autocorrelation function from translational spectra of liquid mixtures. *Phys. Rev.*, A 10:913, 1974.

[102] U. Buontempo, S. Cunsolo, G. Jacucci, and J. J. Weis. The far infrared absorption spectrum of N_2 in the gas and liquid phases. *J. Chem. Phys.*, 63:2570, 1975.

[103] U. Buontempo, S. Cunsolo, and P. Dore. Far infrared spectra of H_2 in liquid Ar and memory effects in the translational bands. *J. Chem. Phys.*, 62:4062, 1975.

[104] U. Buontempo, S. Cunsolo, P. Dore, and P. Maselli. Analysis of translational band observed in gaseous and liquid Kr–Ar mixtures. *J. Chem. Phys.*, 66:1278, 1977.

[105] U. Buontempo, S. Cunsolo, P. Dore, and P. Maselli. Density narrowing of a rotational line of H_2 solved in Ar. *Molec. Phys.*, 37:779, 1979.

[106] U. Buontempo, S. Cunsolo, P. Dore, and P. Maselli. Molecular motions in liquids from infrared spectra. In [406], p. 211.

[107] U. Buontempo, S. Cunsolo, P. Dore, and P. Maselli. Analysis of the translational band measured in Ne–Ar mixtures with a new line-shape profile. *Can. J. Phys.*, 59:1499, 1981.

[108] U. Buontempo, P. Codastefano, S. Cunsolo, P. Dore, and P. Maselli. Density effects on the rotational lines of D_2 in D_2–Ar mixtures. *Can. J. Phys.*, 59:1495, 1981.

[109] U. Buontempo, P. Codastefano, S. Cunsolo, P. Dore, and P. Maselli. New analysis of the density effects observed on the rotational line profile of induced spectra of H_2 and D_2 dissolved in argon. *Can. J. Phys.*, 61:156, 1983.

[110] U. Buontempo, P. Maselli, and L. Nencini. Density effects on the translational motion from infrared induced rotational spectra of N_2. *Can. J. Phys.*, 61:1498, 1983.

[111] U. Buontempo, A. Filabozzi, and P. Maselli. Collision-induced fundamental band of N_2 and N_2–Ar mixtures. *Molec. Phys.*, 67:517, 1989.

[112] A. Burrows, W. B. Hubbard, and J. I. Lunine. Brown dwarfs. *Astrophys. J.*, 345:439, 1989.

[113] W. Byers Brown and D. M. Whisnant. Long range interatomic forces. *Chem. Phys. Lett.*, 7:239, 1970.

[114] W. Byers Brown and D. M. Whisnant. Interatomic dispersion dipole. *Molec. Phys.*, 25:1385, 1973.

[115] G. E. Caledonia, R. H. Krech, and T. Wilkerson. Measurements of the CO first overtone bandstrength at elevated temperatures. *J.Q.S.R.T.* 34:183, 1985.

[116] G. E. Caledonia, R. H. Krech, T. D. Wilkerson, R. L. Taylor and G. Birnbaum. Collision-induced emission in the fundamental roto-vibration band of H_2. *Phys. Rev.*, A 43:6010, 1991.

[117] R. R. A. Campbell and J. Lewis. Kinetic theory of intercollisional effects in arbitrary gas mixtures. Dissertation, Memorial University of Newfoundland, St. John's, 1982.

[118] J. L. Carlsten, A. Szöke, and M. G. Raymer. Collisional redistribution of near resonance scattered light. *Phys. Rev. A* 15:1029, 1977.

[119] R. H. Chatham, A. Gallagher, and E. L. Lewis. Broadening of the sodium D lines by rare gases. *J. Phys.*, B 13:17, 1980.

[120] D. A. Chisholm, J. C. F. McDonald, M. F. Crawford, and H. L. Welsh. Induced infrared absorption in hydrogen and hydrogen–foreign gas mixtures at pressures up to 1500 atmospheres. *Phys. Rev.*, 88:957, 1952.

[121] D. A. Chisholm and H. L. Welsh. Induced infrared absorption in hydrogen and hydrogen–foreign gas mixtures at pressures up to 1500 atmospheres. *Can. J. Phys.*, 32:291, 1954.

[122] E. R. Cohen. Infrared spectral moment and molecular multipoles in collision-induced absorption. *Can. J. Phys.*, 54:475, 1976.

[123] E. R. Cohen, L. Frommhold, and G. Birnbaum. Analysis of the far-infrared H_2–He spectrum. *J. Chem. Phys.*, 77:4933, 1982. Erratum: *ibid.*, 78:5283, 1983..

[124] E. R. Cohen and P. Giacomo. Symbols, units, nomenclature and fundamental constants in physics–1987 revision. *Physica*, A 146:1, 1987.

[125] J. P. Colpa and J. A. A. Ketelaar. The pressure-induced rotational absorption spectrum of hydrogen: I. *Molec. Phys.*, 1:14, 1958.

[126] J. P. Colpa. Induced absorption in the infrared. In A. van Itterbeck, ed., *Physics of High Pressure and the Condensed Phase*, Ch. 12, p. 490, North-Holland, Amsterdam, 1965.

[127] E. U. Condon. Production of infrared spectra with electric fields. *Phys. Rev.*, 41:759, 1932.

[128] M. F. Crawford, H. L. Welsh, and J. L. Locke. Infrared absorption of oxygen and nitrogen induced by intermolecular forces. *Phys. Rev.*, 75:1607, 1949.

[129] M. F. Crawford, H. L. Welsh, J. C. F. MacDonald, and J. L. Locke. Infrared absorption of hydrogen induced by foreign gases. *Phys. Rev.*, 80:469, 1950.

[130] S. Cunsolo and H. P. Gush. Collision-induced absorption of H_2 and He–A mixtures near $\lambda = 1$ mm. *Can. J. Phys.*, 50:2058, 1972.

[131] I. R. Dagg. Collision-induced absorption in the microwave region. In [45], p. 95.

[132] I.R. Dagg, G.E. Reesor, and M. Wong. Collision-induced microwave absorption in Ne–Xe and Ar–Xe gaseous mixtures. *Can. J. Phys.*, 56:1046, 1978.

[133] G. Danby and D. R. Flower. Theoretical studies of van der Waals molecules: the $(H_2)_2$ dimer. *J. Phys.*, B 16:3411, 1983.

[134] G. Danby and D. R. Flower. On the $S_0(0)$ and $S_0(1)$ spectra of the H_2–H_2 dimer. *J. Phys.* B 17:L867, 1984.

[135] J. De Boer. Molecular distribution and equation of state of gases. *Rep. on Prog. in Phys.*, 12:305, 1948.

[136] J. De Boer. Development of probability densities in power series of density. *Physica*, 15:680, 1949.

[137] J. De Remigis, J. W. Mactaggart, and H. L. Welsh. Pressure narrowing of the rotational lines of the fundamental infrared band of H_2 in collision-induced absorption. *Can. J. Phys.*, 49:381, 1971.

[138] P. Dore, L. Nencini, and G. Birnbaum. Far infrared absorption in normal H_2 from 77 to 298 K. *J.Q.S.R.T.*, 30:245, 1983.

[139] P. Dore, A. Filabozzi, and G. Birnbaum. Measurements and analysis of rototranslational absorption spectra of low density H_2–Ar mixtures. *Can. J. Phys.*, 66:803, 1988.

[140] P. Dore, A. Filabozzi, and G. Birnbaum. Rototranslational absorption in gaseous H_2–Ar mixtures at intermediate densities. *Can. J. Phys.*, 67:599, 1989.

[141] P. Dore, M. Moraldi, J. D. Poll, and G. Birnbaum. Analysis of rototranslational absorption spectra induced in low-density gases of non-polar molecules: The methane case. *Molec. Phys.*, 66:335, 1989.

[142] S. Dossou, D. Clermontel, and H. Vu. Recent results on induced spectra of hydrogen at high pressures. *Physica*, 139&140B:541, 1986.

[143] P.A. Egelstaff. Neutron scattering studies of liquid diffusion. *Adv. Phys.*, 11:203, 1962.

[144] A. Einstein. *Annalen der Physik*, 33:1275, 1910.

[145] G. B. Field, W. B. Somerville, and K. Dressler. Hydrogen molecules in astronomy. *Annual Review of Astronomy and Astrophysics*, 4:207, 1966.

[146] V. N. Filimonov. Induced absorption of infrared radiation by molecules. *Sov. Phys. USPEKHI*, 2:894, 1960.

[147] W. Finkelnburg. *Kontinuierliche Spektren.* Springer, Berlin, 1938.

[148] B. C. Freasier and N. D. Hamer. Density effects in collision-induced light absorption in helium–argon mixtures. *Chem. Phys*, 58:347, 1981.

[149] L. Frommhold, K. H. Hong, and M. H. Proffitt. Collision-induced scattering of light by Ar pairs. *Molec. Phys.*, 35:665, 1978.

[150] L. Frommhold, R. Samuelson, and G. Birnbaum. Hydrogen dimer structures in the far-infrared spectra of Jupiter and Saturn. *Astrophys. J.*, 283:L79, 1984.

[151] L. Frommhold and W. Meyer. Collision induced rotovibrational spectra of H_2-He pairs from first principles. *Phys. Rev.*, A 35:632, 1987. Erratum: *Phys. Rev.*, A 41:534, 1990.

[152] L. Frommhold. Unpublished work.

[153] H. S. Gabelnick and H. L. Strauss. Dielectric loss in liquid carbon tetrachloride at far-infrared frequencies. *J. Chem. Phys.*, 46:396, 1966.

[154] A. Gallagher. The spectra of colliding atoms. In G. zu Putlitz *et al.*, eds., *Atomic Physics*, Plenum, N.Y., 1975.

[155] A. Gallagher and T. Holstein. Collision-induced absorption in atomic transitions. *Phys. Rev.*, A 16:2413, 1977.

[156] T. F. Gallagher, P. Pillet, R. Kachru, N. H. Tran, and W. W. Smith. In J. Eichler, I. V. Hertel, and N. Stolterfoht, eds., *Electronic and Atomic Collisions*, Elsevier, New York, 1984.

[157] D. Gautier, A. Marten, J. P. Baluteau, and G. Bachet. About unexplained features in the Voyager far infrared spectra of Jupiter and Saturn. *Can. J. Phys.*, 61:1455, 1983.

[158] Z. Gburski, C. G. Gray, and D. E. Sullivan. Higher-order spectral moments in collision-induced absorption: Inert gas mixtures. *Chem. Phys. Lett.*, 95:430, 1983.

[159] Z. Gburski, C. G. Gray, and D. E. Sullivan. Information theory of line shape in collision-induced absorption. *Chem. Phys. Lett.*, 100:383, 1983.

[160] D. Goorvitch, P. M. Silvaggio, and R. W. Boese. Investigation of the 1-0 pressure-induced vibrational absorption spectrum of hydrogen ot temperatures below ambient. *J.Q.S.R.T.*, 25:237, 1981.

[161] R. G. Gordon and J. K. Cashion. Intermolecular potentials and the infrared spectrum of the molecular complex $[H_2]_2$. *J. Chem. Phys.*, 44:1190, 1966.

[162] R. G. Gordon. New method for constructing wavefunctions for bound states and scattering. *J. Chem. Phys.*, 51:14, 1969.

[163] C. G. Gray. Theory of collision-induced absorption for spherical top molecules. *J. Phys. B*, 4:1661, 1971.

[164] C. G. Gray and B. W. N. Lo. Long-range induced dipole moment of three interacting atoms. *Chem. Phys. Lett.*, 25:55, 1974.

[165] C. G. Gray, K. E. Gubbins, B. W. N. Lo, and J. D. Poll. Theory of collision-induced absorption in liquids I: Rare gas liquids. *Molec. Phys.*, 32:989, 1976.

[166] C. G. Gray and K. E. Gubbins. *Theory of Molecular Fluids*. Vol. 1: Fundamentals, vol. 2 to appear. Clarendon Press, Oxford, 1984.

[167] C. G. Gray, B. G. Nickel, J. D. Poll, S. Singh, and S. Weiss. Line shape in collision induced absorption: comparison of theory and computer simulation. *Molec. Phys.*, 58:253, 1986.

[168] C. G. Gray, B. G. Nickel, J. D. Poll, Y. S. Sainger, S. Singh, and S. Weiss. Line shape in collision induced absorption spectra: theory and computer simulation for atomic and simple molecular species. *Molec. Phys.*, 60:951, 1987.

[169] H. S. Green. *The Molecular Theory of Fluids*. Interscience, New York, 1952.

[170] B. Guillot and G. Birnbaum. Theoretical interpretation of the far infrared absorption spectrum in molecular liquids: nitrogen. In [45], p. 437.

[171] B. Guillot, R. D. Mountain, and G. Birnbaum. Theoretical study of the 3-body absorption spectrum in pure rare-gas fluids. *Molec. Phys.*, 64:747, 1988.

[172] B. Guillot. Triplet dipoles in the absorption spectra of dense rare gas fluids: II. Long range interactions. *J. Chem. Phys.*, 91:3456, 1989.

[173] B. Guillot, R. D. Mountain, and G. Birnbaum. Triplet dipoles in the absorption spectra of dense rare gas mixtures: I Short range interactions. *J. Chem. Phys.*, 90:650, 1989.

[174] R. Hanel, B. Conrath, M. Flasar, V. Kunde, P. Lowman, W. McGuire, J. Pearl, J. Pirraglia, R. Samuelson, D. Gautier, P. Gierasch, S. Kumar, and C. Ponnamperuma. *Science*, 204:972, 1979.

[175] W. J. F. Hare and H. L. Welsh. Pressure-induced infrared absorption of hydrogen and hydrogen–foreign gas mixtures in the range 1500–5000 atmospheres. *Can. J. Phys.*, 36:88, 1958.

[176] H. Harima. Some aspects of collision-induced transitions in spectral line shape experiments. In L. Frommhold and J. W. Keto, eds., *Spectral Line Shapes* **6**, Am. Inst. Phys., N.Y., p. 269, 1990.

[177] R. W. Hartye, C. G. Gray, J. D. Poll, and M. S. Miller. Moment analysis and quantum effects in collision-induced absorption. *Molec. Phys.*, 29:825, 1975.

[178] R. M. Herman, R. H. Tipping, and J. D. Poll. Shape of the R and P lines in the fundamental band of gaseous HD. *Phys. Rev.*, A 20:2006, 1979.

[179] R. M. Herman. Scalar and vector collisional interference in the vibration–rotation absorption spectra of H_2 and HD. In R. J. Exton, ed., *Spectral Line Shapes* **4**, p. 351, Deepak, Hampton, VA, 1987.

[180] G. Herzberg. *Molecular Spectra and Molecular Structure. I. Spectra of Diatomic Molecules.* van Nostrand, Princeton, 1950.

[181] G. Herzberg. *J. Roy: Astronom. Soc. Canada* 45:100, 1951.

[182] G. Herzberg. Spectroscopic evidence of molecular hydrogen in the atmospheres of uranus and neptune. *Astrophys. J.*, 115:337, 1952.

[183] T. L. Hill. Molecular clusters in imperfect gases. *J. Chem. Phys.*, 23:617, 1955.

[184] T. L. Hill. *Statistical Mechanics.* McGraw-Hill, New York, 1956.

[185] J. O. Hirschfelder, C. F. Curtiss, and R. B. Bird. *Molecular Theory of Gases and Liquids.* John Wiley & Sons, Inc., New York, 1964.

[186] W. Ho, G. Birnbaum, and A. Rosenberg. Far infrared collision-induced absorption in CO_2. I. Temperature dependence. *J. Chem. Phys.*, 55:1028, 1971.

[187] J. L. Hunt. *The pressure-induced vibrational spectrum of hydrogen in the temperature range 300 to 40 K.* Dissertation, University of Toronto, 1959.

[188] J. L. Hunt and H. L. Welsh. Analysis of the profile of the fundamental infrared band of hydrogen in pressure-induced absorption. *Can. J. Phys.*, 42:873, 1964.

[189] J. L. Hunt and J. D. Poll. Lineshape analysis of collision-induced spectra of gases. *Can. J. Phys.*, 56:950, 1978.

[190] J. L. Hunt, J. D. Poll, and L. Wolniewicz. Ab initio calculation of properties of the neutral diatomic hydrogen molecules H_2, HD, HT, DT and T_2. *Can. J. Phys.*, 62:1719, 1984.

[191] J. L. Hunt and J. D. Poll. A second bibliography on collision induced absorption. *Molec. Phys.*, 59:163, 1986. Publication 1/86, Department of Physics, University of Guelph.

[192] J. L. Hunt and J. D. Poll. *First update for A Second Bibliography on Collision-induced Absorption.* Guelph-Waterloo Program for Graduate Work in Physics, University of Guelph, Canada, 1990.

[193] K. L. C. Hunt. Classical multipole models: Comparison with *ab initio* and experimental results. In [45], p. 1.

[194] K. L. C. Hunt. Long-range dipoles, quadrupoles, and hyperpolarizabilities of interacting inert-gas atoms. *Chem. Phys. Lett.*, 70:336, 1980.

[195] K. L. C. Hunt. Nonlocal polarizability densities and the effects of short-range interactions on molecular dipoles, quadrupoles and polarizabilities. *J. Chem. Phys.*, 80:393, 1984.

[196] K. Imre, F. Ozizmir, M. Rosenbaum, and P. F. Zweifel. *J. Math. Phys.*, 8:1097, 1967.

[197] J. Janssen. Analyse spectrale des éléments de l'atmosphere terrestre. *Compt. Rend.*, 101:649, 1885.

[198] J. Janssen. Sur les spectres d'absorption de l'oxygène. *Compt. Rend.*, 102:1352, 1886.

[199] J. Janssen. Sur les spectres de l'oxygène. *Compt. Rend.*, 106:1118, 1888.

[200] J. Janssen. Sur l'origine tellurique des raies de l'oxygène dans le spectre solaire. *Compt. Rend.*, 108:1035, 1889.

[201] M. C. Jones. Far infrared absorption in liquefied gases. *NBS Tech.*, 390, 1970.

[202] M. C. Jones. Far infrared absorption in liquid hydrogen. *J. Chem. Phys.*, 51:3833, 1969.

[203] C. G. Joslin and C. G. Gray. Information theory of the lineshape for collision induced absorption: Nitrogen gas. *Chem. Phys. Lett.*, 107:249, 1984.

[204] C. G. Joslin, C. G. Gray, and Z. Gbursky. Far-infrared absorption in nitrogen gas: A theoretical study. *Molec. Phys.*, 53:203, 1984.

[205] C. G. Joslin, C. G. Gray, and S. Singh. Far-infrared absorption in gaseous CH_4 and CF_4: A theoretical study. *Molec. Phys.*, 54:1469, 1985.

[206] P. S. Julienne. Collision-induced radiative transitions at optical frequencies. In [45], p. 749.

[207] P. S. Julienne, R. Heather, and J. Vigué. Laser assisted collisions at ultracold temperatures. In L. Frommhold and J. W. Keto, eds., *Spectral Line Shapes* **6**, p. 191, Am. Inst. Phys., N.Y., 1990.

[208] P. S. Julienne and L. Frommhold. Notes and comments on roundtable discussion on laser-assisted collisions and collision induced spectra. In J. Szudy, ed., *Spectral Line Shapes* **5**, University of Torun Press, 1989.

[209] G. Karl, J. D. Poll, and L. Wolniewicz. Multipole moments of the hydrogen molecule. *Can. J. Phys.*, 53:1781, 1975.

[210] J. A. A. Ketelaar. Infrared spectra of compressed gases. *Rec. Chem. Progress*, 20:1, 1959.

[211] S. Kielich. Multi-photon scattering molecular spectroscopy. *Prog. Optics*, 20:155, 1983.

[212] S. Kielich. The determination of molecular electric multipoles and their polarizabilities by methods of nonlinear intermolecular spectroscopy of scattered laser light. In [406], p. 146.

[213] Z. J. Kiss, H. P. Gush, and H. L. Welsh. The pressure-induced rotational absorption spectrum of hydrogen. I. A study of the absorption intensities. *Can. J. Phys.*, 37:362, 1959.

[214] Z. J. Kiss and H. L. Welsh. Pressure-induced absorption of mixtures of rare gases. *Phys. Rev. Lett.*, 2:166, 1959.

[215] Z. J. Kiss and H. L. Welsh. The pressure-induced rotational absorption spectrum of hydrogen. II. Analysis of the absorption profiles. *Can. J. Phys.*, 37:1249, 1959.

[216] W. Kolos and L. Wolniewicz. *J. Chem. Phys.*, 41:3663, 1964.

[217] W. Kolos and L. Wolniewicz. Potential energy curves for the $X^1\Sigma_g^+$, $b^3\Sigma_u^+$ and $C^1\Pi_u$ states of the hydrogen molecule. *J. Chem. Phys.*, 43:2429, 1965.

[218] M. Krauss, P. Maldonado, and A. C. Wahl. *J. Chem. Phys.*, 54:4944, 1971.

[219] R. Krech, G. Caledonia, S. Schertzer, K. Ritter, T. W. Wilkerson, L. Cotnoir, R. Taylor, and G. Birnbaum. Laboratory observation of collision induced emission in the fundamental vibration rotation band of H_2. *Phys. Rev. Lett.*, 49:1913, 1982.

[220] R. H. Krech and G. E. Caledonia. Configuring systems for high-speed data acquisition and analysis. *J. of Engin. Computing Appl.*, 1:25, 1987.

[221] A. Kudian, H. L. Welsh, and A. Watanabe. Direct spectroscopic evidence of bound states of H_2–Ar complexes at 100 K. *J. Chem. Phys.*, 43:3397, 1965.

[222] A. A. Kudryavtsev and M. V. Tonkov. Causes of the long-wavelength IR absorption spectrum of the gaseous mixture CO_2+He. *Opt. Spectrosc. (USSR)* 61:614, 1986.

[223] A. A. Kudryavtsev and M. V. Tonkov. Invariance of translational band conturs in induced spectra of noble gas mixtures. *Opt. Spectrosc. (USSR)* 65:158, 1988.

[224] A. A. Kudryavtsev and M. V. Tonkov. Mechanisms of formation of the contours of induced CO_2 bands in mixtures with inert gases. *Opt. Spectrosc. (USSR)* 68:616, 1990.

[225] A. J. Lacey and W. Byers Brown. Long range overlap dipoles for inert gas diatoms. *Molec. Phys.*, 27:1013, 1974.

[226] B. M. Ladanyi and T. Keyes. Theory of the static Kerr effect in dense fluids. *Molec. Phys.*, 34:1643, 1977.

[227] R. J. Le Roy and J. S. Carley. *Adv. Chem. Phys.*, 42:353, 1980.

[228] R. J. Le Roy and J. M. Hutson. Improved potential energy surfaces for the interaction of H_2 with Ar, Kr, and Xe. *J. Chem. Phys.*, 86:837, 1986.

[229] R. J. Le Roy and J. S. Carley. Spectroscopy and potential energy surfaces of van der Waals molecules. In K. P. Lawley, ed., *Potential Energy Surfaces*, p. 353, Wiley, 1980.

[230] D. Levesque, J. J. Weis, Ph. Marteau, J. Obriot, and F. Fondere. Collision induced far infrared spectrum of liquid N_2: Computer simulations and experiment. *Molec. Phys.*, 54:1161, 1985.

[231] H. B. Levine. Quantum theory of collision-induced absorption in rare-gas mixtures. *Phys. Rev.*, 160:159, 1967.

[232] H. B. Levine and G. Birnbaum. Classical theory of collision-induced absorption in rare gas mixtures. *Phys. Rev.*, 154:86, 1967.

[233] H. B. Levine. Role of dispersion in collision-induced absorption. *Phys. Rev. Lett.*, 21:1512, 1968.

[234] J. C. Lewis and J. van Kranendonk. Intercollisional interference effects in collision-induced light scattering. *Phys. Rev. Lett.*, 24:802, 1970.

[235] J. C. Lewis. Theory of intercollisional interference effects. II. Induced absorption in a real gas. *Can. J. Phys.*, 50:2881, 1972.

[236] J. C. Lewis and J. van Kranendonk. Theory of intercollisional interference effects. I. Induced absorption in a Lorentz gas. *Can. J. Phys.*, 50:352, 1972.

[237] J. C. Lewis. Intercollisional interference effects. In [406], p. 91.

[238] J. C. Lewis. Intercollisional interference – Theory and experiment. In [45], p. 215.

[239] J. Liebert. White dwarf stars. *Ann. Rev. Astron. Astrophys*, 18:363, 1980.

[240] J. Liebert, M. J. Lebofsky, and G. H. Rieke. Infrared photometry and the atmospheric composition of cool white dwarfs: the lowest luminosity candidates. *Astrophys. J.*, 246:L73, 1981.

[241] J. L. Linsky. On the pressure-induced opacity of molecular hydrogen in late-type stars. *Astrophys. J.*, 156:989, 1969.

[242] R. P. Lowndes and A. Rastogi. Collision induced far infrared absorption in helium–argon mixtures and nitrogen at high pressures. *J. O. S. A.*, 67:905, 1977.

[243] Q. Ma, R. H. Tipping and J. D. Poll. Mixing of rotational levels and intracollisional interference in the pure rotational $R_0(J)$ transitions of gaseous HD. *Phys. Rev.* A 38:6185, 1988.

[244] J. W. Mactaggart and J. L. Hunt. An analysis of the profile of the pressure-induced pure rotational spectrum of hydrogen gas. *Can. J. Phys.*, 47:65, 1969.

[245] J. W. Mactaggart and H. L. Welsh. Studies in molecular dynamics by collision-induced infrared absorption in H_2–rare gas mixtures. I Profile analysis and the intercollisional interference effect. *Can. J. Phys.*, 51:158, 1973.

[246] P. A. Madden. The interference of molecular and interaction-induced effects in liquids. In [45], p. 643.

[247] P. A. Madden. Workshop report: The interference of induced and allowed molecular moments in liquids. In [45], p. 695.

[248] G. C. Maitland, M. Rigby, E. B. Smith, and W. A. Wakeham. *Intermolecular Forces.* Clarendon Press, Oxford, 1981.

[249] Ph. Marteau, H. Vu, and B. Vodar. Probabilité de transition des bandes de transition pure des couples de gaz rares aux basses frequences. *Compt. Rend.*, 266:1068, 1968.

[250] Ph. Marteau, H. Vu, and B. Vodar. Etude experimental des effets de masse et de temperature sur le profil et la position du maximum des bandes de translation induitessous pression au sein couples de gaz rares. *J.Q.S.R.T.* 10:283, 1970.

[251] Ph. Marteau and F. J. Schuller. Traitment théorique du spectre d'absorption translationelle induit dans les mélanges de gaz rares. *J. Physique*, 33:645, 1972.

[252] Ph. Marteau. Far infrared induced absorption in highly compressed atomic and molecular systems. In [45], p. 415.

[253] Ph. Marteau, J. Obriot, and F. Fondere. First experimental observation of far-infrared translational absorption in He–Ne mixtures. *Can. J. Phys.*, 64:822, 1986.

[254] P. H. Martin. The long-range dipole moment of three identical atoms. *Molec. Phys.*, 27:129, 1974.

[255] A. A. Maryott and G. Birnbaum. Collision induced microwave absorption in compressed gases. I Dependence on density, temperature, and frequencies. *J. Chem. Phys.*, 36:2026, 1962.

[256] R. L. Matcha and R. K. Nesbet. Electric dipole moments of rare gas diatomic molecules. *Phys. Rev.*, 160:72, 1967.

[257] G. T. McConville. A consistent spherical potential function for para-hydrogen. *J. Chem. Phys.*, 74:2201, 1981.

[258] A. R. W. McKellar. Intensities and the Fano line shape in the infrared spectrum of HD. *Can. J. Phys.* 51:389, 1973.

[259] A. R. W. McKellar and H. L. Welsh. Spectra of $[H_2]_2$, $[D_2]_2$, and H_2–D_2 van der Waals complexes. *Can. J. Phys.*, 52:1082, 1974.

[260] A. R. W. McKellar, N. H. Rich, and H. L. Welsh. Collsion-induced vibrational and electronic spectra of gaseous oxygen at low temperatures. *Can. J. Phys.*, 50:1, 1972.

[261] A. R. W. McKellar, J. W. Mactaggart, and H. L. Welsh. Studies in molecular dynamics by collision induced infrared absorption in H_2–rare gas mixtures III. H_2–He. *Can. J. Phys.*, 53:2060, 1975.

[262] A. R. W. McKellar. Infrared spectra of hydrogen–rare gas van der Waals molecules. *Faraday Discuss. Chem. Soc.*, 73:89, 1982.

[263] A. R. W. McKellar. Possible identification of sharp features in the Voyager far-infrared spectra of Jupiter and Saturn. *Can. J. Phys.*, 62:760, 1984.

[264] A. R. W. McKellar and N. H. Rich. Interference effects in the spectrum of HD: II. The fundamental band for HD–rare gas mixtures. *Can. J. Phys.* 62:1665, 1984.

[265] A.R.W. McKellar. The collision-induced first overtone band of hydrogen at low temperature. *Can. J. Phys.*, 66:155, 1987.

[266] A. R. W. McKellar. Experimental verification of hydrogen dimers in the atmospheres of Jupiter and Saturn from Voyager IRIS far-infrared spectra. *Astrophys. J.*, 326:L75, 1988.

[267] A. R. W. McKellar. Infrared spectra of the H_2–N_2 and H_2–CO van der Waals complexes. *J. Chem. Phys.*, 93:18, 1990.

[268] A.R.W. McKellar. Infrared spectra of van der Waals molecules using long paths at low temperatures. In L. Frommhold and J. W. Keto, eds., *Spectral Line Shapes* **6**, p. 369. Am. Inst. Phys., New York, 1990.

[269] A. R. W. McKellar and J. Schäfer. Far-infrared spectra of hydrogen dimers: comparisons of experiment and theory for $(H_2)_2$ and $(D_2)_2$. *J. Chem. Phys.*, 95:3081, 1991.

[270] D. A. McQuarrie. *Statistical Mechanics.* Harper and Row, New York, 1976.

[271] D. A. McQuarrie and R. B. Bernstein. Calculated collision-induced absorption spectrum for He–Ar. *J. Chem. Phys.*, 49:1958, 1968.

[272] J. P. McTague and G. Birnbaum. Collision induced light scattering in gaseous Ar and Kr. *Phys. Rev. Lett.*, 21:661, 1968.

[273] J. P. McTague and G. Birnbaum. Collision induced light scattering in gases: I. Rare gases Ar, Kr and Xe. *Phys. Rev.*, A 3:1376, 1971.

[274] N. Meinander and G. C. Tabisz. Information theory and line shape engineering approaches to spectral profiles of collision induced depolarized light scattering. *Chem. Phys. Lett.*, 110:388, 1984.

[275] W. Meyer. *Chem. Phys.*, 17:27, 1976.

[276] W. Meyer, P. C. Hariharan, and W. Kutzelnigg. Refined *ab initio* calculation of the potential energy surface of the H_2–He interaction with special emphasis to the region of the van der Waals minimum. *J. Chem. Phys.*, 73:1880, 1980.

[277] W. Meyer. *Ab initio* calculations of collision-induced dipole moments. In [45], p. 29.

[278] W. Meyer and L. Frommhold. *Ab initio* calculations of the dipole moment of He–Ar and the collision induced absorption spectra. *Phys. Rev.*, A 33:3807, 1986.

[279] W. Meyer and L. Frommhold. Collision induced rototranslational spectra of H_2–He from an accurate *ab initio* induced dipole surface. *Phys. Rev.* A 34:2771, 1986.

[280] W. Meyer and L. Frommhold. Collision induced rototranslational spectra of H_2–Ar from an accurate *ab initio* potential surface. *Phys. Rev.*, A 34:2936, 1986.

[281] W. Meyer, A. Borysow, and L. Frommhold. Absorption spectra of H_2–H_2 pairs in the fundamental band. *Phys. Rev.*, A 40:6931, 1989.

[282] W. Meyer, L. Frommhold, and G. Birnbaum. Rototranslational absorption spectra of H_2–H_2 pairs in the far infrared. *Phys. Rev.*, A 39:2434, 1989.

[283] W. Meyer and L. Frommhold. *Ab initio* potential and dipole surfaces and the absorption of H–He pairs in the infrared. *Theoret. Chimica Acta*, to appear.

[284] W. Meyer, A. Borysow, and L. Frommhold. The binary collision-induced first overtone band of hydrogen. To be published.

[285] F. H. Mies. *Theoretical Chemistry: Advances and Perspectives*, p. 127, Vol. 6B, Academic Press, New York, 1981.

[286] M. S. Miller, D. A. McQuarrie, G. Birnbaum, and J. D. Poll. Constant acceleration approximation in collision induced absorption. *J. Chem. Phys.*, 57:618, 1972.

[287] M. S. Miller and J. D. Poll. On the pair correlation function for hard spheres at low density and temperature. *Can. J. Phys.*, 46:879, 1968.

[288] M. Mizushima. A theory of pressure absorption. *Phys. Rev.*, 76:1268, 1949. Erratum: *ibid.* 77:149, 1950.

[289] M. Mizushima. On the infrared absorption of the hydrogen molecule. *Phys. Rev.*, 77:150, 1950.

[290] M. Moon and D.W. Oxtoby. Collision-induced absorption in gaseous N_2. *J. Chem. Phys.*, 84:3830, 1986.

[291] M. Moraldi. Quantum mechanical spectral moments in collision induced light scattering and collision induced absorption in rare gases at low densities. *Chem. Phys.*, 78:243, 1983.

[292] M. Moraldi, A. Borysow, and L. Frommhold. Quantum sum formulae for the collision induced spectroscopies: Molecular systems as H_2–H_2. *Chem. Phys.*, 86:339, 1984.

[293] M. Moraldi, A. Borysow, and L. Frommhold. Effects of the anisotropic interaction on collision induced rototranslational spectra of H_2–He pairs. *Phys. Rev.*, A 35:3679, 1987.

[294] M. Moraldi, J. Borysow, and L. Frommhold. Spectral moments for the collision induced rotovibrational absorption bands of nonpolar gases and mixtures. *Phys. Rev.*, A 36:4700, 1987.

[295] M. Moraldi, A. Borysow, and L. Frommhold. Rotovibrational collision induced absorption by nonpolar gases and mixtures (H_2–He pairs): About the symmetry of line profiles. *Phys. Rev.* A 38:1839, 1988.

[296] M. Moraldi and L. Frommhold. Three-body components of collision induced absorption. *Phys. Rev.*, A 40:6260, 1989.

[297] M. Moraldi, M. Celli, and F. Barocchi. Theory of virial expansion of correlation functions and spectra: Application to interaction induced spectroscopy. *Phys. Rev.*, A 40:1116, 1989.

[298] M. Moraldi. Virial expansion of correlation functions for collision induced spectroscopies. In L. Frommhold and J. W. Keto, eds., *Spectral Line Shapes* **6**, p. 438, *AIP Conference Proceedings 216*, Am. Inst. Phys., New York, 1990.

[299] M. Moraldi and L. Frommhold. Second and third virial coefficients of collision-induced absorption and light scattering. J. J. C. Teixeira-Dias, ed., *Molecular Liquids: New Perspectives in Physics and Chemistry*, p. 423. *C: Math. and Phys. Sci.*, Kluwer, 1992.

[300] J. Mould and J. Liebert. Infrared photometry and the atmospheric composition of cool white dwarfs. *Astrophys. J.*, 266:L29, 1978.

[301] R. D. Mountain and G. Birnbaum. Molecular dynamics study of intercollisional interference in collision induced absorption in compressed fluids. *J. Chem. Soc. Faraday Trans. 2*, 83:1791, 1987.

[302] G. Nienhuis. Theory of quantum corrections to the equation of state and the particle distribution function. *J. Math. Phys.*, 11:239, 1970.

[303] K. Okada, T. Kajikawa, and T. Yamamoto. Theory of collision induced infrared absorption in gases. I. *Prog. Theoret. Phys.*, 39:863, 1968.

[304] I. Oppenheim and M. Bloom. *Can. J. Phys.*, 39:845, 1961.

[305] S. T. Pai, S. P. Reddy, and C. W. Cho. Induced infrared absorption in deuterium–foreign gas mixtures. *Can. J. Phys.*, 44:2893, 1966.

[306] F. Palla. Low-temperature Rosseland mean opacities for zero-metal gas mixtures. In G. H. F. Diercksen, W. F. Huebner, and P. W. Langhoff, eds., *Molecular Astrophysics – State of the Art and Future Directions*, p. 687, D. Reidel, Dordrecht, 1985.

[307] R. W. Patch. Theory of pressure induced vibrational and rotational absorption of diatomic molecules at high temperatures. *J.Q.S.R.T.* 11:1311, 1971.

[308] R. W. Patch. Absorption coefficients for hydrogen. II. Calculated pressure induced H_2–H_2 vibrational absorption in the fundamental region. *J.Q.S.R.T.* 11:1331, 1971.

[309] E. W. Pearson, M. Waldman, and R. G. Gordon. Dipole moments and polarizabilities of rare gas atom pairs. *J. Chem. Phys.*, 80:1543, 1984.

[310] R. J. Penney, R. D. G. Prasad, and S. P. Reddy. Collision induced absorption spectra of the fundamental band of gaseous deuterium: overlap parameters of D_2–D_2. *J. Chem. Phys.*, 77:131, 1982.

[311] J. J. Perez, J. H. R. Clarke, and A. Hinchliffe. Three-body contributions to the dipole polarizability of He_3. *Chem. Phys. Lett.*, 104:583, 1984.

[312] M. Perrot and J. Lascombe. Collision-induced effects in allowed infrared and Raman spectra of molecular fluids. In [45], p. 613.

[313] J. D. Poll. *Theory of translational effects in induced infrared spectra.* Dissertation, University of Toronto, 1960.

[314] J. D. Poll and J. van Kranendonk. Theory of translational absorption in gases. *Can. J. Phys.*, 39:189, 1961.

[315] J. D. Poll and M. S. Miller. Pair correlation function for gaseous H_2 at low density and temperature. *J. Chem. Phys.*, 54:2673, 1971.

[316] J. D. Poll, J. L. Hunt, and J. W. Mactaggart. Intercollisional interference in the s lines of H_2–He mixtures. *Can. J. Phys.*, 53:954, 1975.

[317] J. D. Poll and J. L. Hunt. On the moments of pressure induced spectra of gases. *Can. J. Phys.*, 54:461, 1976.

[318] J. D. Poll. Intermolecular spectroscopy of gases. In [406], p. 45.

[319] J. D. Poll. Spectral moments. presentation at the *Conference on Collision-induced Phenomena, Absorption, Light Scattering, and Static Properties*, held at Florence, Italy, September 2–5, 1980.

[320] J. D. Poll and J. L. Hunt. Analysis of the far infrared spectrum of gaseous N_2. *Can. J. Phys.*, 59:1448, 1981.

[321] J. D. Poll. The infrared spectrum of HD. In [45], p. 677.

[322] J. D. Poll, M. Attia and R. H. Tipping. Induced dipole moment function of HD. *Phys. Rev.*, B 39:11378, 1989.

[323] J. D. Poll and R. H. Tipping. Intensities of collision-induced infrared and Raman spectra of diatomic gases. To be submitted.

[324] J. D. Poll and A. Weyland. unpublished work.

[325] S. R. Polo and M. K. Wilson. *J. Chem. Phys.*, 23:2376, 1955.

[326] J. G. Powles and B. Carazza. *J. Phys.*, A 3:335, 1970.

[327] R. D. G. Prasad and S. P. Reddy. Infrared absorption spectra of gaseous HD II: Collision induced fundamental band of HD in HD–Ne and HD–Ar mixtures at room temperature. *J. Chem. Phys.*, 65:83, 1976.

[328] E. M. Purcell. *Phys. Rev.*, 117:828, 1960.

[329] J. Quazza, Ph. Marteau, H. Vu, and B. Vodar. Induced far infrared absorption in rare gas mixtures. *J.Q.S.R.T.* 16:491, 1976.

[330] A. Raczynski. Collision induced absorption in a He–Ar mixture. *Chem. Phys.*, 72:321, 1982.

[331] A. Raczynski. Density expansion of correlation functions. *Acta Phys. Polonia*, 62 A:303, 1982.

[332] A. Raczynski. Collision induced absorption in He–Xe: Quantum mechanical and classical approach. *Chem. Phys.*, 88:129, 1984.

[333] A. Raczynski and G. Staszewska. Density dependence of moments and lineshape in collision-induced absorption in He + Ar. *Molec. Phys.*, 58:919, 1986.

[334] D. H. Rank, B. S. Rao, P. Sitaram, A. F. Slomba, and T. A. Wiggins. Quadrupole and induced dipole spectrum of molecular hydrogen. *J. O. S. A.*, 52:1004, 1962.

[335] S. P. Reddy and C. W. Cho. Infrared absorption of deuterium induced by intermolecular forces. *Can. J. Phys.*, 43:793, 1965.

[336] S. P. Reddy and C. W. Cho. Induced infrared absorption of nitrogen and nitrogen-foreign gas mixtures. *Can. J. Phys.*, 43:2331, 1965.

[337] S. P. Reddy and W. F. Lee. Pressure induced infrared absorption of the fundamental band of hydrogen in H_2–Ne and H_2–Kr mixtures at room temperature. *Can. J. Phys.*, 46:1373, 1968.

[338] S. P. Reddy and C. Z. Kuo. Collision induced 1'st overtone infrared absorption band of deuterium. *J. Molec. Spectrosc.*, 37:327, 1971.

[339] S. P. Reddy and K. S. Chang. Collision induced fundamental bands of H_2–He and H_2–Ne mixtures at different temperatures. *J. Molec. Spectrosc.*, 47:22, 1973.

[340] S. P. Reddy and R. D. G. Prasad. Infrared absorption spectra of gaseous HD IV: Analysis of the collision induced fundamental band of the pure gas. *J. Chem. Phys.*, 66:5259, 1977.

[341] S. P. Reddy, G. Varghese, and R. D. G. Prasad. Overlap parameters of H_2–H_2 molecular pairs from the absorption spectra of the collision induced fundamental band. *Phys. Rev.*, A 15:975, 1977.

[342] S. P. Reddy. Induced vibrational absorption in the hydrogens. In [45], p. 129.

[343] N. H. Rich and A. R. W. McKellar. *A Bibliography on Collision-induced Absorption.* Nat. Research Council Canada, No. 15145, Herzberg Institute of Astrophysics, Ottawa, 1975.

[344] N. H. Rich and A. R. W. McKellar. A bibliography on collision induced absorption. *Can. J. Phys.*, 54:486, 1976.

[345] N. H. Rich and A. R. W. McKellar. Interference effects in the spectrum of HD I: The fundamental band of HD. *Can. J. Phys.*, 61:1648, 1983.

[346] D. Robert and L. Galatry. Influence of molecular non-rigidity on the infrared absorption and Raman scattering line shape in dense media. *J. Chem. Phys.*, 64:2721, 1976.

[347] A. Rosenberg, I. Ozier, and A. Kudian. Pure rotational spectrum of CH_4. *J. Chem. Phys.*, 57:568, 1972.

[348] A. Rosenberg and I. Ozier. The forbidden $J \rightarrow J+1$ spectrum of CH_4 in the ground vibronic state. *J. Molec. Spectrosc.*, 56:124, 1975.

[349] A. Rosenberg and K. M. Chen. Rotation-induced rotational lines of CH_4. *J. Chem. Phys.*, 64:5304, 1976.

[350] M. Rotenberg, N. Metropolis, R. Birins, and J.K. Wooten, jr. *The 3-j and 6-j Symbols.* Technology Press–MIT, Cambridge, Mass., 1959.

[351] W. E. Russell, S. P. Reddy, and C. W. Cho. Collision induced fundamental band of D_2 in D_2–He and D_2–Ne mixtures at different temperatures. *J. Molec. Spectrosc.*, 52:72, 1974.

[352] V. A. Ryzhov and M. V. Tonkov. Translational absorption in the spectra of noble gases. *Opt. Spectrosc.*, 37:606, 1974.

[353] H. Salow and W. Steiner. Die durch Wechselwirkungskräfte bedingten Absorptionsspektren des Sauerstoffes. I. Die Absorptionsbanden des O_2–O_2-Moleküls. *Z. Physik.*, 99:137, 1936.

[354] J. Schäfer and W. Meyer. Theoretical studies of H_2–H_2 collisions: I. Elastic scattering of ground state para- and ortho-H_2 in the rigid rotor approximation. *J. Chem. Phys.*, 70:344, 1979.

[355] J. Schäfer and W. Meyer. Collision induced dipole radiation of normal hydrogen gas in frequency range of the cosmic background. In J. Eichler, I. V. Hertel, and N. Stolterfoht, eds., *Electronic and Atomic Collisions*, p. 524, North-Holland, Amsterdam, 1984.

[356] J. Schäfer. *Astron. Astrophys.*, 182:L40, 1987.

[357] J. Schäfer. Faint features of the collision induced absorption spectra of gaseous hydrogen. In J. Szudy, ed., *Spectral Line Shapes* **5**, p. 807, Ossolineum, Warsaw, 1989.

[358] J. Schäfer and A. R. W. McKellar. Faint features of the rotational $S_0(0)$ and $S_0(1)$ transitions of H_2: a comparison of calculations and measurements at 77 K. *Z. Physik D*, 15:51, 1990. Erratum: *Z. Physik D*, 17:231, 1990.

[359] S. B. Schneideman and H. H. Michel. *J. Chem. Phys.*, 43:3706, 1965.

[360] P. Schofield. *Phys. Lett.*, 4:39, 1960.

[361] F. Schuller and Ph. Marteau. Theory of field induced translational absorption. *Phys. Lett.*, 49A:229, 1974.

[362] V. F. Sears. On the form of the intracollisional dipole correlation function. *Can. J. Phys.*, 46:1501, 1968.

[363] V. F. Sears. Theory of collision induced translational absorption. *Can. J. Phys.*, 46:1163, 1968.

[364] G. Seeley and T. Keyes. Normal mode analysis of the second order light scattering spectrum in liquids. In J. Szudy, ed., *Spectral Line Shapes* **5**, p. 649, Ossolineum, Warsaw, 1985.

[365] R. D. Sharma and R. R. Hart. Collision-induced absorption in He–Ar mixture. *Phys. Rev.*, A 12:85, 1975.

[366] H. L. Shipman. Masses, radii and model atmospheres for cool white-dwarf stars. *Astrophys. J.*, 213:138, 1977.

[367] G. V. Shlyapnikov and I. P. Shmatov. Translational radiative transitions in collisions between atoms. *J. E. T. P.*(USSR), 79:2978, 1980.

[368] I. F. Silvera and G. Birnbaum. *Appl. Opt.*, 9:617, 1970.

[369] A. L. Smith, W. E. Keller, and H. L. Johnston. Infrared spectra of condensed oxygen and nitrogen. *Phys. Rev.*, 79:728, 1950.

[370] F. T. Smith. Atomic distortion and the combining rule for repulsive potentials. *Phys. Rev.*, A 5:1708, 1972.

[371] M. Smoluchowski. *Annalen der Physik*, 25:205, 1908.

[372] S. W. Stahler, F. Palla, and E. E. Salpeter. Primordial stellar evolution: the protostar phase. *Astrophys. J.*, 302:590, 1986.

[373] J. M. Steed, T. A. Dixon, and W. Klemperer. *J. Chem. Phys.*, 70:4095, 1979.

[374] W. A. Steele. *Molec. Phys.*, 39:1411, 1980.

[375] W. A. Steele and G. Birnbaum. Molecular calculations of moments of the induced spectra for N_2, O_2, and CO_2. *J. Chem. Phys.*, 72:2250, 1980.

[376] W. A. Steele. Computer simulation studies of the induced far infrared absorption in liquids. In J. Szudy, ed., *Spectral Line Shapes* **5**, p. 653, Ossolineum, Warsaw, 1989.

[377] D. E. Stogryn and J. O. Hirschfelder. Contribution of bound, metastable and free molecules to the second virial coefficient. *J. Chem. Phys.*, 31:1531 (1959).

[378] D. E. Stogryn and A. P. Stogryn. *Molec. Phys.*, 11:371, 1966.

[379] H. L. Strauss and S. Weiss. Collision induced line shapes calculated with a hard-sphere model. *J. Chem. Phys.*, 70:5788, 1979.

[380] H. Sutter. Dielectric polarization of gases. In M. Davies, ed., *A Specialist Periodic Report – Dielectric and Related Molecular Processes* 1, Ch. 3, p. 65, Chem. Soc., London, 1972.

[381] K. Szalewicz and B. Jeziorski. Symmetry adapted, double perturbation analysis of intramolecular correlation effects in weak intermolecular interactions. *Molec. Phys.*, 38:191, 1979.

[382] G. C. Tabisz, E. J. Allin, and H. L. Welsh. Interpretation of the visible and near infrared absorption spectra of compressed oxygen as collision induced electronic transitions. *Can. J. Phys.*, 47:2859, 1969.

[383] G. C. Tabisz. Pressure dependence of the electronic transition $^1\Sigma_g^+ \leftarrow {}^3\Sigma_g^-$ in the absoption spectrum of compressed oxygen. *Chem. Phys. Lett.*, 9:581, 1971.

[384] G. C. Tabisz, L. Ulivi, P. Drakopoulos, and Z. Lu. Collisional interference in the infrared absorption spectrum of HD. In L. Frommhold and J. W. Keto, eds., *Spectral Line Shapes* **6**, p. 421, Am. Inst. Phys., 1990.

[385] K. T. Tang and J. P. Toennies. *J. Chem. Phys.*, 80:3726, 1984.

[386] O. Tanimoto. Band shape of the collision induced infrared absorption by rare gas mixtures. *Prog. Theoret. Phys.*, 33:585, 1965.

[387] R. H. Taylor, A. Borysow, and L. Frommhold. Concerning the rototranslational absorption spectra of He–CH_4 pairs. *J. Molec. Spectrosc.*, 129:45, 1988.

[388] P. Thoman, K. Burnett, and J. Cooper. Observation of dynamic correlations in collisional redistribution and depolarization of light. *Phys. Rev. Lett.*, 45:1325, 1980.

[389] R. H. Tipping, J. D. Poll, and A. R. W. McKellar. The influence of intracollisional interference on the dipole spectrum of HD. *Can. J. Phys.* 56:75, 1978.

[390] R. H. Tipping. Collision induced effects in planetary atmospheres. In [45], p. 727.

[391] R. H. Tipping and J. D. Poll. Multipole moments of hydrogen and its isotopes. In K. N. Rao, ed., *Molecular Spectroscopy: Modern Research* **3**, p. 421, Academic Press, 1985.

[392] R. H. Tipping, Q. Ma and J. D. Poll. Predicted intensity of the $S_0(0) + S_0(0) + S_0(0)$ triple transition in the infrared spectrum of solid orthodeuterium. *Phys. Rev.* B 44:12314, 1991.

[393] L. M. Trafton. The thermal opacity in the major planets. *Astrophys. J.*, 140:1340, 1964.

[394] L. M. Trafton. The pressure induced monochromatic translational absorption coefficients for homopolar and nonpolar gases and gas mixtures with particular application to H_2. *Astrophys. J.*, 146:558, 1966.

[395] L. M. Trafton. On the He–H_2 thermal opacity in planetary atmospheres. *Astrophys. J.*, 179:971, 1973.

[396] L. Trafton. Observational studies of collision induced absorption in the atmospheres of the outer planets. In J. Szudy, ed., *Spectral Line Shapes* **5**, p. 755, Ossolineum, Warsaw, 1989.

[397] G. E. Uhlenbeck and E. Beth. Pair distribution function. *Physica*, 3:729, 1936.

[398] L. Ulivi, Z. Lu, and G. C. Tabisz. Temperature dependence of the collisional interference in the pure rotational far-infrared spectrum of HD. *Phys. Rev.*, A 40:641, 1989.

[399] B. T. Ulrich, L. Ford, and J. C. Browne. Translational absorption of HeH. *J. Chem. Phys.*, 57:2906, 1972.

[400] J. van Kranendonk. Theory of induced infrared absorption. *Physica*, 23:825, 1957.

[401] J. van Kranendonk. Induced infrared absorption in gases. Calculation of the binary absorption coefficients of symmetrical diatomic molecules. *Physica*, 24:347, 1958.

[402] J. van Kranendonk. Induced infrared absorption in gases. Calculation of the ternary absorption coefficients of symmetrical diatomic molecules. *Physica*, 25:337, 1959.

[403] J. van Kranendonk and Z. J. Kiss. Theory of the pressure induced rotational spectrum of hydrogen. *Can. J. Phys.*, 37:1187, 1959.

[404] J. van Kranendonk. Intercollisional interference effects in pressure induced spectra. *Can. J. Phys.*, 46:1173, 1968.

[405] J. van Kranendonk. Intermolecular spectroscopy. *Physica*, 73:156, 1974.

[406] J. van Kranendonk, ed., *Intermolecular Spectroscopy and Dynamical Properties of Dense Systems – Proceedings of the Int. School of Physics "Enrico Fermi", Course LXXV*. North-Holland, Amsterdam, 1980.

[407] J. van Kranendonk. Intercollisional interference effects. In [406], p. 77.

[408] G. Varghese and S. P. Reddy. Further studies on the collision induced absorption of the fundamental band of hydrogen at room temperature. *Can. J. Phys.*, 47:2745, 1969.

[409] B. Vodar and H. Vu. Intensités absolues des transitions induites par la pression. *J.Q.S.R.T.*, 3:397, 1963.

[410] A. Watanabe. *Pressure-induced infrared absorption of gaseous hydrogen at low temperatures*. Dissertation, University of Toronto, 1964.

[411] A. Watanabe, J. L. Hunt, and H. L. Welsh. Structure of the pressure induced infrared spectrum of hydrogen in the first overtone region. *Can. J. Phys.*, 49:860, 1971.

[412] A. Watanabe and H. L. Welsh. Direct spectroscopic evidence of bound states of $[H_2]_2$ complexes. *Phys. Rev. Lett.*, 13:810, 1964.

[413] A. Watanabe and H. L. Welsh. Pressure induced infrared absorption of gaseous hydrogen and deuterium at low temperatures. I. The integrated intensities. *Can. J. Phys.*, 43:818, 1965.

[414] A. Watanabe and H. L. Welsh. Pressure induced infrared absorption of gaseous hydrogen and deuterium at low temperatures. II. Analysis of the band profiles for hydrogen. *Can. J. Phys.*, 45:2859, 1967.

[415] S. Weiss. Translational spectrum due to ternary collisions in pure rare gases. *Chem. Phys. Lett.*, 19:41, 1973.

[416] S. Weiss. The interaction dipole. *Can. J. Phys.*, 54:584, 1976.

[417] S. Weiss. Infrared absorption in rare gas mixtures. In [406], p. 202.

[418] S. Weiss. Frame distortion as a possible mechanism for collision induced infrared absorption. *J. Phys. Chem.*, 86:429, 1982.

[419] S. Weiss. Frame distortion in some linear and tetrahedral molecules. *Chem. Phys. Lett.*, 114:536, 1985.

[420] H. L. Welsh, M. F. Crawford, J. C. F. McDonald, and D. A. Chisholm. Induced infrared absorption of H_2, N_2, and O_2 in the first overtone regions. *Phys. Rev.* 83:1264, 1951.

[421] H.L. Welsh and J.L. Hunt. Line shapes in pressure induced absorption. *J.Q.S.R.T.* 3:385, 1963.

[422] H.L. Welsh. Pressure induced absorption spectra of hydrogen. Ch. 3, p. 33, Vol. III, Spectroscopy, *MTP Int. Review of Science – Physical Chemistry, Series one*, Butterworths, London, 1972.

[423] J. U. White. *J. O. S. A.*, 32:285, 1942.

[424] E. H. Wishnow, I. Ozier and H. P. Gush. Submillimeter spectrum of low-temperature hydrogen: The pure translational band of H_2 and the $R(0)$ line of HD. *Astrophys. J.*, 392:L43, 1992.

[425] L. Wolniewicz. Vibrational–rotational study of the electronic ground state of the hydrogen molecule. *J. Chem. Phys.*, 45:515, 1966.

[426] P. E. S. Wormer and G. van Dijk. *Ab initio* calculations of the collision induced dipole in H_2–He II: S.C.F. results and comparison with experiment. *J. Chem. Phys.*, 70:5695, 1979.

[427] S. I. Yakovlenko. Optical collisional phenomena. In J. Szudy, ed., *Spectral Line Shapes* **5**, p. 695, Ossolineum, Warsaw, 1989.

[428] H. R. Zaidi and J. van Kranendonk. Diffusional pressure narrowing in induced infrared spectra. *Can. J. Phys.*, 49:385, 1971.

[429] M. Zoppi, F. Barocchi, D. Varshneya, M. Neumann, and T. A. Litovitz. Density dependence of the collision induced light scattering spectral moments of argon. *Can. J. Phys.*, 59:1475, 1981.

Index